(*Continued*)

RAO • Linear Statistical Inference and Its Applications, *Second Edition*
ROBERTSON, WRIGHT, and DYKSTRA • Order Restricted Statistical Inference
ROGERS and WILLIAMS • Diffusions, Markov Processes, and Martingales, Volume II: Îto Calculus
ROHATGI • Statistical Inference
ROSS • Stochastic Processes
RUBINSTEIN • Simulation and the Monte Carlo Me
RUZSA and SZEKELY • Algebraic Probability Theo
SCHEFFE • The Analysis of Variance
SEBER • Linear Regression Analysis
SEBER • Multivariate Observations
SEBER and WILD • Nonlinear Regression
SEN • Sequential Nonparametrics: Invariance Principles and Statistical Inference
SERFLING • Approximation Theorems of Mathematical Statistics
SHORACK and WELLNER • Empirical Processes with Applications to Statistics
STOYANOV • Counterexamples in Probability
WHITTAKER • Graphical Models in Applied Multivariate Statistics

Applied Probability and Statistics

ABRAHAM and LEDOLTER • Statistical Methods for Forecasting
AGRESTI • Analysis of Ordinal Categorical Data
AGRESTI • Categorical Data Analysis
AICKIN • Linear Statistical Analysis of Discrete Data
ANDERSON and LOYNES • The Teaching of Practical Statistics
ANDERSON, AUQUIER, HAUCK, OAKES, VANDAELE, and WEISBERG • Statistical Methods for Comparative Studies
ARTHANARI and DODGE • Mathematical Programming in Statistics
ASMUSSEN • Applied Probability and Queues
*BAILEY • The Elements of Stochastic Processes with Applications to the Natural Sciences
BARNETT • Interpreting Multivariate Data
BARNETT and LEWIS • Outliers in Statistical Data, *Second Edition*
BARTHOLOMEW • Stochastic Models for Social Processes, *Third Edition*
BARTHOLOMEW and FORBES • Statistical Techniques for Manpower Planning
BATES and WATTS • Nonlinear Regression Analysis and Its Applications
BHATTACHARYA and WAYMIRE • Stochastic Processes with Applications
BECK and ARNOLD • Parameter Estimation in Engineering and Science
BELSLEY • Conditioning Diagnostics: Collinearity and Weak Data in Regression
BELSLEY, KUH, and WELSCH • Regression Diagnostics: Identifying Influential Data and Sources of Collinearity
BHAT • Elements of Applied Stochastic Processes, *Second Edition*
BHATTACHARYA and WAYMIRE • Stochastic Processes with Applications
BLOOMFIELD • Fourier Analysis of Time Series: An Introduction
BOLLEN • Structural Equations with Latent Variables
BOX • R. A. Fisher, the Life of a Scientist
BOX and DRAPER • Empirical Model-Building and Response Surfaces
BOX and DRAPER • Evolutionary Operation: A Statistical Method for Process Improvement
BOX, HUNTER, and HUNTER • Statistics for Experimenters: An Introduction to Design, Data Analysis, and Model Building
BROWN and HOLLANDER • Statistics: A Biomedical Introduction
BUCKLEW • Large Deviation Techniques in Decision, Simulation, and Estimation
BUNKE and BUNKE • Nonlinear Regression, Functional Relations and Robust Methods: Statistical Methods of Model Building
BUNKE and BUNKE • Statistical Inference in Linear Models, Volume I
CHAMBERS • Computational Methods for Data Analysis
CHATTERJEE and HADI • Sensitivity Analysis in Linear Regression
CHATTERJEE and PRICE • Regression Analysis by Example
CHOW • Econometric Analysis by Control Methods
CLARKE and DISNEY • Probability and Random Processes: A First Course with Applications, *Second Edition*
COCHRAN • Sampling Techniques, *Third Edition*
COCHRAN and COX • Experimental Designs, *Second Edition*

*Now available in a lower priced paperback edition in the Wiley Classics Library.

Applied Probability and Statistics (Continued)

CONOVER · Practical Nonparametric Statistics, *Second Edition*
CONOVER and IMAN · Introduction to Modern Business Statistics
CORNELL · Experiments with Mixtures, Designs, Models, and the Analysis of Mixture Data, *Second Edition*
COX · A Handbook of Introductory Statistical Methods
COX · Planning of Experiments
DANIEL · Applications of Statistics to Industrial Experimentation
DANIEL · Biostatistics: A Foundation for Analysis in the Health Sciences, *Fourth Edition*
DANIEL and WOOD · Fitting Equations to Data: Computer Analysis of Multifactor Data, *Second Edition*
DAVID · Order Statistics, *Second Edition*
DAVISON · Multidimensional Scaling
DEGROOT, FIENBERG, and KADANE · Statistics and the Law
*DEMING · Sample Design in Business Research
DILLON and GOLDSTEIN · Multivariate Analysis: Methods and Applications
DODGE · Analysis of Experiments with Missing Data
DODGE and ROMIG · Sampling Inspection Tables, *Second Edition*
DOWDY and WEARDEN · Statistics for Research, *Second Edition*
DRAPER and SMITH · Applied Regression Analysis, *Second Edition*
DUNN · Basic Statistics: A Primer for the Biomedical Sciences, *Second Edition*
DUNN and CLARK · Applied Statistics: Analysis of Variance and Regression, *Second Edition*
ELANDT-JOHNSON and JOHNSON · Survival Models and Data Analysis
FLEMING · Counting Processes and Survival Analysis
FLEISS · The Design and Analysis of Clinical Experiments
FLEISS · Statistical Methods for Rates and Proportions, *Second Edition*
FLURY · Common Principal Components and Related Multivariate Models
FRANKEN, KÖNIG, ARNDT, and SCHMIDT · Queues and Point Processes
GALLANT · Nonlinear Statistical Models
GIBBONS, OLKIN, and SOBEL · Selecting and Ordering Populations: A New Statistical Methodology
GNANADESIKAN · Methods for Statistical Data Analysis of Multivariate Observations
GREENBERG and WEBSTER · Advanced Econometrics: A Bridge to the Literature
GROSS and HARRIS · Fundamentals of Queueing Theory, *Second Edition*
GROVES · Survey Errors and Survey Costs
GROVES, BIEMER, LYBERG, MASSEY, NICHOLLS, and WAKSBERG · Telephone Survey Methodology
GUPTA and PANCHAPAKESAN · Multiple Decision Procedures: Theory and Methodology of Selecting and Ranking Populations
GUTTMAN, WILKS, and HUNTER · Introductory Engineering Statistics, *Third Edition*
HAHN and SHAPIRO · Statistical Models in Engineering
HALD · Statistical Tables and Formulas
HALD · Statistical Theory with Engineering Applications
HAND · Discrimination and Classification
HEIBERGER · Computation for the Analysis of Designed Experiments
HOAGLIN, MOSTELLER, and TUKEY · Exploring Data Tables, Trends and Shapes
HOAGLIN, MOSTELLER, and TUKEY · Understanding Robust and Exploratory Data Analysis
HOCHBERG and TAMHANE · Multiple Comparison Procedures
HOEL · Elementary Statistics, *Fourth Edition*
HOEL and JESSEN · Basic Statistics for Business and Economics, *Third Edition*
HOGG and KLUGMAN · Loss Distributions
HOLLANDER and WOLFE · Nonparametric Statistical Methods
HOSMER and LEMESHOW · Applied Logistic Regression
IMAN and CONOVER · Modern Business Statistics
JESSEN · Statistical Survey Techniques
JOHN · Statistical Methods in Engineering and Quality Assurance
JOHNSON · Multivariate Statistical Simulation

Continued on back end papers

*Now available in a lower priced paperback edition in the Wiley Classics Library.

Conditioning Diagnostics

Conditioning Diagnostics

Collinearity and Weak Data in Regression

DAVID A. BELSLEY

Professor of Economics
Department of Economics
Boston College, and

Senior Research Associate
Center for Computational Research
in Economics and Management Science
Massachusetts Institute of Technology

A Wiley-Interscience Publication
JOHN WILEY & SONS
New York • Chichester • Brisbane • Toronto • Singapore

In recognition of the importance of preserving what has been written, it is a policy of John Wiley & Sons, Inc. to have books of enduring value published in the United States printed on acid-free paper, and we exert our best efforts to that end.

Library of Congress Cataloging-in-Publication Data:

Belsley, David A.
Conditioning diagnostics: collinearity and weak data in regression/David A. Belsley.
p. cm. -- (Probability and mathematical statistics. Applied probability and statistics section)
"A Wiley-Interscience publication."
Includes bibliographical references (p.) and indexes.
1. Regression analysis. I. Title. II. Title: Collinearity and weak data in regression. III. Series.
QA278.2.B437 1990
519.5'36--dc20 90-12574
CIP

ISBN 0-471-52889-7

Printed in the United States of America

10 9 8 7 6 5 4 3 2 1

To

Judith

Contents

Preface

In the decade that has passed since the publication of the collinearity diagnostics in *Regression Diagnostics: Identifying Influential Data and Sources of Collinearity* by David A. Belsley, Edwin Kuh, and Roy E. Welsch, (BKW), much additional research has taken place that has substantively extended and filled out that body of knowledge. This monograph integrates this research with a complete revision of the original material. The result is a greatly expanded, more nearly self-contained treatment of the problems of ill conditioning and data weaknesses as they affect the least-squares estimation of the linear model along with extensions to nonlinear models and simultaneous-equations estimators.

While this monograph encompasses the material relating to the conditioning diagnostics found in BKW, this material has been substantively revised in response to the many constructive comments I have received and in light of changes in points of view that have occurred over the intervening years.[1] Moreover, a substantial amount of new material has been added, including background material and data sets, and many related elements have been brought together that were previously available only in disparate sources.

Specifically, Chapter 1 is almost all new. Of interest is the development of the various geometric techniques that, while used extensively in BKW, were assumed known to the reader—an assumption I have discovered to be tenuous. Their inclusion here makes this book more suitable for academic course work as well as more generally useful as a research monograph.

Chapter 2 is based on Section 3.1 of BKW. The intuitive discussion of collinearity has been greatly expanded, making this work accessible to a wider audience. An "n-dimensional" geometric view of collinearity has been added, which is drawn on repeatedly in subsequent chapters. The discussion of alternative approaches to collinearity diagnostics has been substantially

[1]Those parts of this work that have been adapted from *Regression Diagnostics: Identifying Influential Observations and Sources of Collinearity*, by David A. Belsley, Edwin Kuh, and R. E. Welsch, copyright © 1980 by John Wiley & Sons, Inc., are reprinted by permission of John Wiley & Sons, Inc.

expanded and unified and includes the material from Belsley (1984a). Efforts are made to answer questions that have arisen over the years by users of BKW.

Chapter 3 is based on Section 3.2 of BKW. The discussion has been revised and expanded, and geometric views have been added, allowing an intuitive understanding of what was completely analytical in BKW. The material dealing with the conditioning of inexact linear systems has been newly added. The development has been altered to presage the test for harmful collinearity of Chapter 7.

Chapter 4 is based on Section 3.3 of BKW, but new experiments have been added that deal with truly simultaneous near dependencies—the absence of which was of concern to some readers of BKW. The results are gratifyingly consistent.

Chapter 5 is based on Section 3.4 of BKW. The section "Some Hints on Usage" is completely new and hopefully will please those readers and users who have asked for such. This section details how to display, digest, and interpret the diagnostic information and should help to make these diagnostics more accessible to statistical and forecasting practitioners. Also, data sets have been added for the examples, which will increase the usefulness of this monograph for academic course work and for those who like to experiment.

Chapter 6 unifies material from various sources. The first two sections are from Appendix 3B of BKW, after being revised, tightened, and better integrated. The material on alternate scalings is new. Sections 6.3–6.9 are adapted from Belsley (1984b, 1986a), which benefit from a fresh look, and provide a substantially deeper interpretative basis for the diagnostics than was possible in BKW. These interpretations are exploited in the remaining chapters. Their placement here rather than earlier is a natural heuristic that attempts to keep the initial development as simple as possible, adding complications only as the structure becomes sufficiently strong to support them. The final section is taken from Appendix 3A of BKW.

Chapter 7 is the proper development of the material whose seeds are to be found in Appendix 3D of BKW. It is adapted from Belsley (1982) and carries the collinearity diagnostics of BKW from the realm of data analysis into that of statistics. The collinearity diagnostics can determine the presence of collinear relations but cannot determine when they are causing statistical harm. This material provides the needed notions and tests to define and analyze harmful collinearity and the very closely related concept of short data.

Since the publication of BKW, much work has been done on the effects of individual rows (or small sets of rows) on the conditioning of a given data matrix. These observations have been dubbed collinearity-influential observations, and they and various diagnostics for them are discussed in Chapter 8.

In Chapter 9, the material on conditioning in models with logs is adapted from Belsley (1988c). The niceties of this case make a good bridge between the standard diagnostics of the earlier chapters and the more general and technically more demanding diagnostics in Chapter 11. The material on assessing the conditioning of first-differenced data is adapted from Belsley (1986a).

The first section of Chapter 10 encompasses Section 4.1 of BKW, although the section "A Nonsolution: Occam's Hatchet" is a new and important addition adapted from an example in Belsley (1988b) that shows the sterility (indeed harm) of one of the most popularly advocated solutions to the collinearity problem: discarding collinear variates. Otherwise, however, the thrust of this chapter is really quite new and different. The seeds for "remedial" action are to be found in Chapter 4 of BKW but are not fully formed. Here, however, a consistent philosophy is advanced in which collinearity is noted to be akin to the identification problem. Thus, legitimate solutions are those that add the needed identifying information. Such means are discussed in the first section and then three complete case studies are given in three subsequent sections showing how remedial activities can be integrated into a statistical study. The first of these, using prior information to improve conditioning in the estimation of the consumption function, comes from the example in Section 4.1 of BKW. The second, using collinearity diagnostics and prior information to improve forecasting accuracy, is adapted from Belsley (1984c), and the third, improving the specification of a model of energy consumption, is from Belsley and Welsch (1988). I feel this to be an important and interesting new chapter and one that will make this material more accessible and useful to the everyday statistical practitioner.

Chapter 11 extends the notion of a conditioning analysis into far broader statistical reaches than the collinearity diagnostics of BKW. Seeds for some of this material are to be found in Chapter 5 of BKW. The first section, adapted from Belsley and Oldford (1986), defines a general conditioning analysis applying both to nonlinear models and to simultaneous systems. The second and third sections are new, giving, respectively, a more general measure of conditioning based on an analysis of the variance–covariance matrix and a measure (related to that of Chapter 7) of the degree of effective identification of a system of simultaneous equations. The fourth section, based on Belsley (1988a), applies the conditioning measures as a means for determining when to adopt three-stage least squares instead of two-stage least squares.

This book has been written with the theorist, the practitioner, and the experimentalist in mind. Included are the theoretical bases for the diagnostics as well as straightforward means for implementing and interpreting them. The practitioner who chooses to ignore some of the theoretical complexities can do so without jeopardizing the usefulness of the diagnostics. Data sets have been included wherever reasonable to allow for replication and experimentation.

At the risk of being a bore, I would like to attempt to mention those to whom, over the years, I have been grateful for comments and suggestions that are reflected in this work. At a minimum this list includes Ernst Berndt, Gregory Chow, R. D. Cook, Paolo Corsi, John Dennis, Jean-Marie Dufour, James Durbin, Harry Eisenpress, Ray Fair, Robert Fieldes, David Gay, Gene Golub, Clive Granger, Richard Gunst, David Hendry, Ullrich Heilemann, David Hoaglin, Paul Holland, Philip Howrey, Saul Hymans, Peter Kempthorne, Robert Kennard, Virginia Klema, Jan Kmenta, Edward Leamer, G. S. Maddala,

Spyros Makridakis, Donald Marquardt, John Meyer, Grayham Mizon, Forest Nelson, Jean-François Richard, Alexander Sarris, Ronald Snee, Pete Stewart, John Tukey, Walter Vandaele, Achilles Venetoulias, Paul Velleman, Kent Wall, Kenneth Wallis, Thomas Wonnacott, Fred Wood, and (last as usual) Arnold Zellner.

My special thanks go to Ali Hadi for the invaluable feedback he provided on an earlier draft of this manuscript that substantially strengthened the final product. No amount of thanks can express my appreciation to my colleagues Roy Welsch, Wayne Oldford, and the late Edwin Kuh; their many conversations have contributed greatly to the development of this work over the years, and elements of my joint research with Roy Welsch and Wayne Oldford are reflected directly in Sections 10.4 and 11.1. The same also is true regarding the painstaking editorial efforts of my wife Judith, which cost me, or some other member of our family, the opportunity of at least one hand-made sweater.

In addition, I am grateful to the following for aid and technical support: Ana Aizcorbe, Josh Charap, Sean Doyle, Mark Gelfand, David Jones, Karen Martel, Lisa Newton, Michael Ozog, Stephen Peters, Aik Quek, Stephen Swartz, Fatma Taskin, Nham Vu, Rick Wilk, and Yan Yu.

This work was funded in part through National Science Foundation grants IST-8420614 and SES-8420614; the Center for Computational Research in Economics and Management Science, Massachusetts Institute of Technology; a Mellon Grant to Boston College; and a guest scholarship at the Rheinisch-Westfälisches Institut für Wirtschaftsforschung, Essen, West Germany. I would also like to express my continuing appreciation to the Center for Advanced Study in the Behavioral Sciences at Stanford, where my serious inquiry into collinearity began during my tenure there as a Fellow in 1970–1971.

David A. Belsley

CHAPTER 1

Introduction and Overview

There are few statistical practitioners who have escaped the "collinearity problem" in their work, this certainly being true for those who have attempted to use linear regression to estimate models using nonexperimental data. Its symptoms are tell-tale: high standard errors, low t-statistics, nonsensical or overly sensitive parameter estimates. Its presence is highly frustrating, often preventing precise statistical knowledge about the most important parameters under study. Its exact diagnosis, care, and cure are problematic; many textbooks describe the collinearity problem, but few provide appropriate diagnostics, and fewer still offer proper counsel in understanding, assessing, and handling its presence. Actually, collinearity is a special case of a more general problem: conditioning. And in this monograph we examine these two related notions with an eye toward diagnosing, understanding, assessing, and handling the problems they can cause estimation of statistical models.

This monograph is entitled *Conditioning Diagnostics*. The statistical use of the term *diagnostic* is relatively recent and is, so far as I am aware, without formal definition in this context. However, it is quite clear that it is intended to encompass all statistical measures on sample outcomes, both graphical and numerical, that have obvious use in signaling or apprising the practitioner of the presence of specific conditions or circumstances that are relevant to interpreting or assessing a statistical analysis. Diagnostics are distinguished from other types of statistics in that no pretense is made regarding distributional attributes that would give them inferential content—even though many diagnostics have their roots or motivation in some underlying statistical theory. Their value, however, stems from their practical ability to notify the practitioner that certain interesting or important conditions prevail. Conditioning diagnostics warn of data weaknesses—an intuitively meaningful term that awaits Chapter 7 for a formal definition—and the deleterious effects they can have on many aspects of a statistical analysis.

1.1 BASIC CONVENTIONS

Terminology

This work draws on topics coming from a variety of disciplines that includes statistics, econometrics, data analysis, and numerical analysis. Each of these disciplines has its own terminological and notational preferences, which are unfortunately not all the same. Therefore, some compromises must be made, which is again unfortunate since such compromises are likely to please nobody. The reader's indulgence is therefore requested as we adopt the following.

Econometricians traditionally have T observations on K variates, whereas statisticians have n observations (or cases) on p variates. Since data analysts and numerical analysts also adopt the $n \times p$ convention, I shall use that here on the basis of a majority rule. Also, econometricians tend to speak, at least in the single-equation case, of dependent and independent variates, whereas data analysts and some statisticians tend to employ the somewhat more neutral terminology of response and explanatory variates. This latter choice seems a reasonable choice for this work, although different contexts seem to ask for different usage, and nobody should be confused by whatever terminology is adopted.

Differences also arise over the notation for matrix transposition. Econometricians and some mathematicians denote the transpose by a prime, as in $\mathbf{A}'$, while statisticians and data and numerical analysts use a superscript T, as in $\mathbf{A}^{\mathrm{T}}$. The feeling, as I understand it, is that the prime can get lost on the page. I have never had this trouble, and were I a typist, I would surely argue in the prime's favor. But if there is anyone out there who loses primes, that is one too many, so I will adopt the superscript T notation here. This also avoids any ambiguity with the well-established use of the prime to denote derivatives.

One of the most confusing differences in notation is that adopted for a linear equation. Numerical analysts typically write $\mathbf{Ax} = \mathbf{b}$, while statisticians and econometricians employ $\mathbf{y} = \mathbf{Xb}$. Thus, one person's b is another's x, whose b is y, while the first's X is the second's A. Conversations between practitioners of these disciplines that use shorthand phrases assuming commonality of notation can become humorous indeed. Since this work is directed principally at applications in statistical models, the statistical/econometric notation of $\mathbf{y} = \mathbf{Xb}$ will be used.

The terms *variable* and *variate* are used somewhat interchangeably in the literature, depending upon the discipline. The term variable seems preferred in statistics, while variate is often used in econometrics. There is, however, a distinction between these two terms that is rarely stated—primarily, I would suppose, because it is not widely suspected even by those that implicitly or unconsciously make it. The term variable refers to the argument of a mathematical function. Because mathematical functions are abstract concepts, they and their arguments can and do exist without context and without real-life meaning or interpretation. When, however, a function is used as a statistical model, a representation of a specific real-life situation, it and its arguments inherit a context and an interpretation. The function's variable now stands for something in the real world that has meaning and units and perhaps even some distributional properties. It is no longer possible to speak meaningfully about the value of the function at, say, 3; it must be at 3 so-and-so units. And the single term variate is used to denote both this variable and its context.[1] Since this work inevitably deals with applied concepts, there is always some underlying context, and the data are thus measurements on the variates, not the variables, of these contexts.

Least Squares and the Regression Model

For the most part, the statistical model of interest will be the standard linear regression model

$$\mathbf{y} = \beta_1 \mathbf{X}_1 + \cdots + \beta_p \mathbf{X}_p + \boldsymbol{\varepsilon} \tag{1.1a}$$

or in matrix terms,

$$\mathbf{y} = \mathbf{X}\boldsymbol{\beta} + \boldsymbol{\varepsilon}, \tag{1.1b}$$

estimated by least squares,

$$\mathbf{b} = (\mathbf{X}^{\mathrm{T}}\mathbf{X})^{-1}\mathbf{X}^{\mathrm{T}}\mathbf{y}. \tag{1.2}$$

But extensions to simultaneous-equations models, nonlinear models, and to other estimators and more general statistical analyses will also be made. The garden-variety, least-squares estimator (1.2) is often called the ordinary least-squares estimator (OLS) to distinguish it from such variants as generalized least squares (GLS), two-stage least squares (2SLS), or three-stage least squares (3SLS). In this work, however, the unadorned term *least squares* will denote this basic estimator.

The previous notation is common usage. $\mathbf{y}$ is an n-vector of observations on a response (dependent) variate, the $\mathbf{X}_i$ ($i = 1, \ldots, p$) are n-vectors of observations on the p explanatory (independent) variates, $\mathbf{X} \equiv [\mathbf{X}_1 \cdots \mathbf{X}_p]$ is an $n \times p$ data matrix whose columns are the $\mathbf{X}_i$, $\boldsymbol{\beta}$ is a p-vector of model parameters to be estimated, and $\boldsymbol{\varepsilon}$ is an n-vector of stochastic terms (typically assumed to have mean $\mathbf{0}$ and scalar variance–covariance matrix $\boldsymbol{\sigma}^2\mathbf{I}$). If the model is assumed to have an intercept (or constant) term, one of the $\mathbf{X}_i$'s, usually the first, will be assumed to be the column vector $\boldsymbol{\iota}$ of n ones.

In general it is not assumed that the $\mathbf{X}_i$'s have been centered or scaled. When a

[1]This usage directly parallels the notion of a "fluent," as seen, e.g., in Menger (1959).

result is stated that depends upon such transformations, it will be so noted. We shall have more to say on this issue later.

The p-vector $\mathbf{b}$ denotes the least-squares estimator (1.2) of $\boldsymbol{\beta}$.

Reporting of Regression Results

In the literature, estimated regression equations are sometimes reported in tables and sometimes written out as equations such as

$$\underset{(0.221)}{\mathbf{y} = 1.454\boldsymbol{\iota}} - \underset{(1.224)}{0.331\mathbf{X}_2} + \underset{(1.753)}{2.674\mathbf{X}_3}.$$

In this latter case, the figures in the parentheses below the estimates are sometimes the estimated standard errors and sometimes the t-statistics. And, unless the author tells you, it is often difficult to know which is being reported, as is the case here. So we shall adopt the convention (which I wish were universal) that, when the least-squares estimates are written out in equation form, if the figures beneath are in parentheses, they are standard errors, and if they are in square brackets, like

$$\underset{[0.221]}{\mathbf{y} = 1.454\boldsymbol{\iota}} - \underset{[1.224]}{0.331\mathbf{X}_2} + \underset{[1.753]}{2.674\mathbf{X}_3},$$

they are t-statistics.

We shall also adopt the standard (which I, again, wish were universal) that t-statistics are reported only when they are the central and primary focus of the analysis. This will often be the case in this work where regression is frequently used descriptively, with t-statistics exploited for their diagnostic value and not for any formal inference. When, however, valid regression results of general interest are being reported and the very specific (and often irrelevant) test of significance that is measured by the t-statistic is not of prime interest, the standard errors will be reported.

1.2 AN OPENING EXAMPLE

Before going on, it seems appropriate to provide a simple illustration of collinearity to demonstrate its effects as well as some of the difficulties that arise in detecting its presence. I fear there will be few practitioners so fortunate not to recognize in this example similarities with some of their own problems, but I hope that this recognition will whet the appetite for all that follows.

Consider, then, the data given in Exhibit 1.1. These data have been constructed to simulate those that might come from a physical process whose operation is to be estimated from readings taken under differing, nonexperimental conditions. This might be a chemical process in which a weight yield $\mathbf{y}$ is

Exhibit 1.1 Data for Opening Example

		$\mathbf{X}_2$		$\mathbf{X}_3$		$\mathbf{X}_4$	
Case	y	Mr. A	Ms. B	Mr. A	Ms. B	Mr. A	Ms. B
1	3.3979	−3.138	−3.136	1.286	1.288	0.169	0.170
2	1.6094	−0.297	−0.296	0.250	0.251	0.044	0.043
3	3.7131	−4.582	−4.581	1.247	1.246	0.109	0.108
4	1.6767	0.301	0.300	0.498	0.498	0.117	0.118
5	0.0419	2.729	2.730	−0.280	−0.281	0.035	0.036
6	3.3768	−4.836	−4.834	0.350	0.349	−0.094	−0.093
7	1.1661	0.065	0.064	0.208	0.206	0.047	0.048
8	0.4701	4.102	4.103	1.069	1.069	0.375	0.376

determined given readings on, say, temperature $\mathbf{X}_2$, pressure differential $\mathbf{X}_3$, and excitation voltage bias $\mathbf{X}_4$. Or it could be an electronic process in which field strength is determined given readings on load impedance (centered on 72 Ω), bias voltage, and coupling distance. In any event, we will note for our own information, but for our information only, that the **y** data have in fact been generated from the true **X**'s according to the model

$$\mathbf{y} = 1.2\boldsymbol{\iota} - 0.4\mathbf{X}_2 + 0.6\mathbf{X}_3 + 0.9\mathbf{X}_4 + \boldsymbol{\varepsilon}, \tag{1.3}$$

with $\boldsymbol{\varepsilon}$ coming from an independently and identically distributed (iid) normal with mean zero and variance 0.01.

Now, in this exercise, we assume that two investigators, Mr. A. and Ms. B, are separately interested in estimating this process and, to this end, take their own eight readings—which they do together but without looking at each other's data. It will be noted that they both collect exactly the same **y** data because these come from a digital devise but get slightly different values for the **X** data, which they read from various analog scales.

Mr. A's regression analysis produces

$$\begin{aligned} \mathbf{y} = 1.255&\boldsymbol{\iota} + 0.974\mathbf{X}_2 + 9.022\mathbf{X}_3 - 38.440\mathbf{X}_4, \\ (0.091&) \quad (3.818) \quad (23.602) \quad (108.97) \\ R^2 = .992, &\qquad \text{SER} = 0.162, \qquad \text{DW} = 2.55, \end{aligned} \tag{1.4}$$

where the figures in parentheses, we recall, are standard errors, SER is the standard error of the regression $\sqrt{\mathbf{e}^T\mathbf{e}/(n-p)}$, and DW is the Durbin–Watson statistic, a measure of first-order serial correlation defined in any econometrics textbook. These results are disappointing for a number of reasons. First, even though Mr. A does not have our knowledge of the true model (1.3), if he had even basic prior information about the true parameters, such as their signs and the likelihood that they were smaller than 1.0 in absolute value, it would be clear that these estimates are far off the mark—some wrong signs and absurd magnitudes. Second, even though the R^2 is quite high, all of the so-called slope coefficients are hopelessly insignificant. Only the intercept term appears to have explanatory power. These conditions are classically symptomatic of collinearity. And so, to investigate this possibility, Mr. A examines the pairwise correlations among the **X** data, which are given in Exhibit 1.2.

Here we see the third of Mr. A's disappointments, for none of the variate pairs has a correlation that greatly exceeds .6 in absolute value. While these correlations are not small, neither are they large. If collinearity is a problem, then this commonly employed diagnostic, using pairwise correlations (or, equivalently, examining pairwise scatter plots of the explanatory variates), is not able to see it in this case.

But Mr. A is in for yet one more disappointment, for he is now shown the regression results obtained by Ms. B, which are

$$\begin{aligned} \mathbf{y} = \underset{(0.093)}{1.275\boldsymbol{\iota}} + \underset{(2.307)}{0.247\mathbf{X}_2} + \underset{(14.207)}{4.511\mathbf{X}_3} - \underset{(65.709)}{17.644\mathbf{X}_4}, \\ R^2 = .992, \qquad \text{SER} = 0.163, \qquad \text{DW} = 2.73. \end{aligned} \tag{1.5}$$

These results too are poor, but of far greater interest and concern is the fact that they are radically different from Mr. A's, even though it is clear that the data upon which they are based differ in what would appear to be the most inconsequential way.

In this example, then, we can see some of the effects of collinearity—wrong signs, low significance, absurd and overly sensitive parameter estimates—, but we cannot readily see the collinearity itself. Its presence is certainly not betrayed by any high pairwise correlations. How, then, do we know these data are collinear? and how do we know which parameter estimates are being adversely affected by collinearity's presence? The tools to answer these and other questions are, of course, forthcoming. In the meantime we note that the data series in this

Exhibit 1.2 Pairwise Correlations for Mr. A's Data

	$\mathbf{X}_2$	$\mathbf{X}_3$	$\mathbf{X}_4$
$\mathbf{X}_2$	1.000		
$\mathbf{X}_3$	−.346	1.000	
$\mathbf{X}_4$	.533	.610	1.000

example have been kept intentionally short to make this an inviting context for those who would eventually care to return to try their skills.

1.3 COLLINEARITY, CONDITIONING, AND WEAK DATA: A PRELIMINARY DISCUSSION

In much recent literature (my own included), the two terms *collinearity* and *ill conditioning* have been used virtually synonymously to denote a particular form of data weakness (near linear dependencies) that severely reduces the precision with which some or all parameter estimates can be known and causes the estimated parameters **b** to possess undue sensitivity to small perturbations in the data. This usage, while not fundamentally incorrect, is nevertheless misleading, for the concept of conditioning as used in statistical contexts possesses far greater generality than does that of collinearity and actually encompasses this latter term. In the linear, least-squares context that occupies our attention for most of this monograph, however, the two concepts effectively coincide, and we can safely await the final Chapter 11 before truly rigorous definitions of the two concepts are given. It is nonetheless useful to keep the distinction in mind throughout, so we anticipate the difference by noting loosely that collinearity deals with the existence of a nearly linear relationship among a set of variates whereas conditioning is concerned with the sensitivity (or insensitivity) of a given relation to shifts (perturbations) in the underlying data. If the relation whose sensitivity is being examined is, for example, the least-squares estimator (1.2), it is well known that collinearity among the $\mathbf{X}_i$'s can result in ill conditioning (sensitivity) in the least-squares estimates to changes in the data, but the converse need not be true. Thus, in Chapter 11, we shall encounter data conditioning, estimator conditioning, and criterion conditioning. And collinearity will be seen to be directly associated only with the first of these, data conditioning.

A related concept is weak data. Data weaknesses, which are dealt with formally in Chapter 7, are characteristics of the data that rob them of the information needed for statistical analysis to proceed in some dimensions with adequate precision. Clearly, collinearity is a data weakness, but we shall discover that it is only one form of weak data, the other being "short data." And while collinear data are ill-conditioned data, short data need not be. Thus, data weaknesses and data ill conditioning are similar, but not identical, concepts.

In passing, I note that the term *multicollinearity* is also often used interchangeably with *collinearity*, particularly in econometric contexts. I was weaned on this term, and it slips out occasionally. But I tend to avoid its use because it makes an unnecessary distinction. Its original intent was to distinguish between the case of collinearity involving only two variates and that involving more than two. However, there is no conceptual difference between these cases, and both of them are adequately encompassed by the single, shorter term collinearity, much as the term *sphere* encompasses a ball of any dimension, usage rarely requiring the more awkward term *hypersphere*.

1.4 WHY STUDY COLLINEARITY AND ILL CONDITIONING?

To listen to some scholars, a study of this sort would appear to be unnecessary. The arguments, as I understand them, are two. The first is that collinearity is simply not a problem: one need only conduct one's experiments with data that are not collinear. This is clearly the argument of those who have the luxury of selecting their data by experimental design, as is indeed the case in many sciences. But, in nonexperimental sciences, such as economics, oceanography, astrophysics, education, social psychology, and even some elements of biology, physics, and chemistry, collinearity is a natural flaw in the data set resulting from the uncontrollable operations of the data-generating mechanism and is simply a painful and unavoidable fact of life.

The second argument admits that collinearity or conditioning problems exist but considers them of little statistical interest because, once discovered, there is little or nothing one can do with or about them—so why bother? There is no doubt that this nihilistic way of thinking provides a simple, albeit a somewhat ineffective, solution to the collinearity problem, but fortunately, it is not one with which we must remain content.

Consider, for example, an investigator who runs a regression and, to his dismay, discovers an important coefficient estimate to be insignificant. This could be, and often is, passed off with a mild wave of the hand as the result of collinearity. But collinearity is not the only reason this estimate could be insignificant. There could be data errors or entry errors or the model could be misspecified. All these must be checked. And if collinearity is finally to be blamed, it is clear that collinearity must not only be shown to be present but also shown to be adversely affecting the estimate of the given coefficient. Thus, the appropriate assessment of the model and the proper reporting of its results require means both for demonstrating the existence of collinearity and for pinpointing the estimates that are affected by it.

A similar situation arises when considering whether a "marginal" variate should be kept in a given model. In a commonly employed testing sequence, a regression is run and the t-statistic for this variate is examined. If it proves to be insignificant, the variate is dropped. But the presence of collinearity can severely reduce the power of this standard test of significance, producing high odds that this variate will be deemed insignificant and tossed from the analysis when in fact it should not be. Again, appropriate assessment of tests of hypotheses can only be made in the light of the further information obtained through collinearity diagnostics.

Indeed, quite to the contrary of the preceding disheartening objection to the value of collinearity diagnostics, we shall see in later examples that such diagnostics are invaluable for directing a solution to the collinearity problem and thus are of great importance to many users of least-squares regression and, indeed, to users of many other statistical procedures as well. This is because the diagnostics not only determine whether a collinearity or conditioning problem exists but can also determine the number of relations involved and often pinpoint the variates involved in each.

This information can be used fruitfully in many ways. It can be used to help determine when a data problem rather than, say, model misspecification is the source of poor-quality estimates. It can be used not only to determine which estimates are degraded by the presence of collinearity, but also, just as importantly, in many cases it can be used to determine estimates that remain undegraded despite the presence of collinearity. It can be used to help pinpoint what sort of better-conditioned data, if possible to obtain, are most needed. It can be used to determine when collinearity is not likely to affect forecasts adversely or when collinearity could cause forecasting techniques to go astray. And it can be used to help an investigator determine whether the introduction of prior information, the best source of corrective action, will be worthwhile and, if so, to help point to those places where it can most appropriately be placed.

1.5 SOME USEFUL GEOMETRY

In what follows, as would seem fitting for a study dealing with collinearity, we shall often have recourse to geometric arguments or illustrations that draw on the reader's geometric intuition. The value of being able to keep relevant geometric images in mind is immeasurable in a study of this sort. Many important statistical results are readily visualized and mentally manipulated geometrically without the need for algebra or calculation. Similarly, the geometry is useful in motivating statistical concepts and calculations. For this reason, a brief review of both the p- and n-dimensional geometries associated with linear regression and least-squares estimation is given here. Readers already familiar with these notions are invited to skip directly to the next section.

Distance, Length, and Angle

Any collection of q observed values $x_1, \ldots, x_q$ can be represented mathematically as a q-vector $\mathbf{x} \equiv (x_1, \ldots, x_q)$, that is, an element of a q-dimensional vector space. Without confusion, the vector $\mathbf{x}$ can also be viewed as a $q \times 1$ matrix $\mathbf{x} \equiv (x_1, \ldots, x_q)^{\mathrm{T}}$, also called a column vector, or a $1 \times q$ matrix, also called a row vector. In this work, as a matter of convention, all vectors explicitly lacking a transpose operator are assumed to be column vectors.

This vector can, in turn, be represented geometrically as a point, or, more properly, as a directed line between the origin $\mathbf{O}$ and this point in a q-dimensional geometric space. Two such vectors $\mathbf{x}$ and $\mathbf{y}$ are shown in Exhibit 1.3. In almost all applied statistical work, as here, the relevant space is $\mathscr{R}^q$, the q-dimensional space of reals.

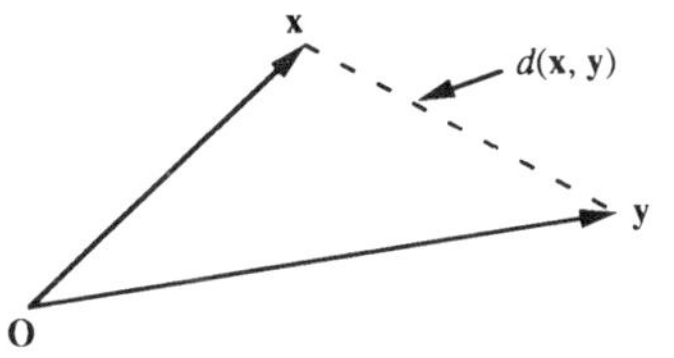

Exhibit 1.3 Distance between two vectors **x** and **y**.

Distance

A distance measure between two vectors $\mathbf{x}, \mathbf{y} \in \mathscr{R}^q$ is any real-valued functional ($\mathscr{R}^{q \otimes q}$ into $\mathscr{R}^1$) $d(\mathbf{x}, \mathbf{y})$ obeying the intuitively clear restrictions

(i) $d(\mathbf{x}, \mathbf{y}) \geqslant 0$ for all $\mathbf{x}, \mathbf{y} \in \mathscr{R}^q$,
(ii) $d(\mathbf{x}, \mathbf{y}) = 0$ if and only if $\mathbf{x} = \mathbf{y}$,
(iii) $d(\mathbf{x}, \mathbf{y}) = d(\mathbf{y}, \mathbf{x})$ for all $\mathbf{x}, \mathbf{y} \in \mathscr{R}^q$, and
(iv) $d(\mathbf{x}, \mathbf{y}) \leqslant d(\mathbf{x}, \mathbf{z}) + d(\mathbf{z}, \mathbf{y})$ for all $\mathbf{x}, \mathbf{y}, \mathbf{z} \in \mathscr{R}^q$.

The last of these will be recognized to be the triangle inequality. While there are many such measures that satisfy these conditions, the familiar Euclidean distance

$$d(\mathbf{x}, \mathbf{y}) = \left[\sum_{i=1}^{q} (x_i - y_i)^2 \right]^{1/2} \tag{1.6}$$

is the most well-known and is quite adequate for many of our needs.

The Euclidean distance between **x** and **y** is depicted geometrically in Exhibit 1.3. The "plane" of Exhibit 1.3 can be viewed as a two-dimensional subspace of the q-space, that is, the subspace generated by (or spanned by) **x** and **y**. One could also include a pair of orthogonal "basis" vectors going horizontally and vertically through the origin **O**, but they are rarely of use and tend only to clutter the picture. We therefore omit them, but the reader is invited to place them in his mind's eye if desired. Actually, this geometry is not limited to the two dimensions of the printed page, or even the three dimensions of space, because **x** and **y** could themselves represent, say, k_1- and k_2-dimensional subspaces of $\mathscr{R}^q$.

The sum of squared differences within the square brackets in (1.6) is denoted in several different ways. Thus, one can write

$$\begin{aligned} \sum_{i=1}^{q} (x_i - y_i)^2 &\equiv (\mathbf{x} - \mathbf{y}) \cdot (\mathbf{x} - \mathbf{y}) \\ &\equiv \langle (\mathbf{x} - \mathbf{y}), (\mathbf{x} - \mathbf{y}) \rangle \\ &\equiv (\mathbf{x} - \mathbf{y})^{\mathrm{T}} (\mathbf{x} - \mathbf{y}). \end{aligned} \tag{1.7}$$

The first expression in (1.7) is called the dot or scalar product and is used by many natural scientists and engineers. The second is called an inner or scalar product and is found mostly among mathematicians. The last is called the matrix or inner product, and it is widely employed by statisticians and econometricians and will be the notation typically used here. But they all denote the same sum of squares.

Length

The Euclidean length $l(\mathbf{x})$ of a vector $\mathbf{x}$ can be defined now simply as the Euclidean distance between $\mathbf{x}$ and the origin $\mathbf{O}$, that is,

$$l(\mathbf{x}) \equiv d(\mathbf{x}, \mathbf{O}). \tag{1.8}$$

A commonly employed notation for $l(\mathbf{x})$ is $\|\mathbf{x}\|$, also called the *Euclidean norm* or simply the *norm* of $\mathbf{x}$. Actually, care must be taken here, for there are many norms of $\mathbf{x}$; the Euclidean norm, for example, is merely an important special case of the family of p-norms of $\mathbf{x}$ defined as

$$\|\mathbf{x}\|_p \equiv \left[\sum_{i=1}^{q} |x_i|^p\right]^{1/p}. \tag{1.9}$$

When this distinction is important, the Euclidean norm is written $\|\mathbf{x}\|_2$ and is called the 2-norm. We shall encounter other norms later on, but for the moment, the Euclidean norm is wholly adequate.

Clearly, for Euclidean length we have

$$l(\mathbf{x}) = \|\mathbf{x}\| = (\mathbf{x}^T\mathbf{x})^{-1/2} = \left[\sum_{i=1}^{q} x_i^2\right]^{1/2}. \tag{1.10}$$

Angle

The angle ϕ between two nonzero vectors $\mathbf{x}$ and $\mathbf{y}$ can be defined in terms of the distance measure given above. We do this by first defining a "right angle" in terms of a minimum-distance or orthogonal projection of a vector $\mathbf{x}$ onto a vector $\mathbf{y}$. This is illustrated in Exhibit 1.4, where we seek to find the point along $\mathbf{y}$

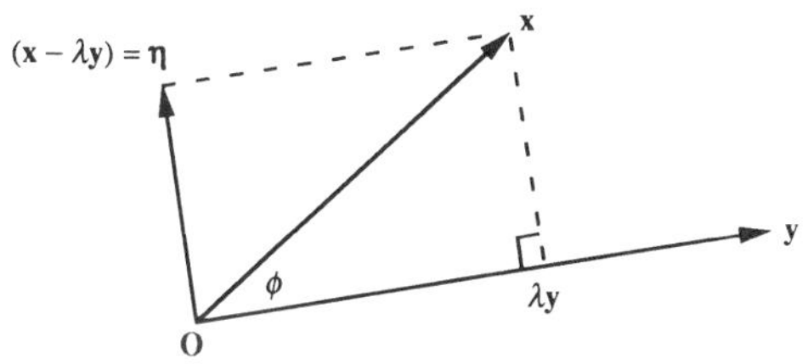

Exhibit 1.4 Angle ϕ between two vectors $\mathbf{x}$ and $\mathbf{y}$.

that is closest to $\mathbf{x}$. This occurs at the point $\lambda\mathbf{y}$ that minimizes $l(\mathbf{x} - \lambda\mathbf{y}) = d(\mathbf{x}, \lambda\mathbf{y})$ and is obtained from solving

$$\min_{\lambda} (\mathbf{x} - \lambda\mathbf{y})^{\mathrm{T}}(\mathbf{x} - \lambda\mathbf{y}). \tag{1.11}$$

Setting the derivative of (1.11) with respect to λ to zero and solving for λ gives

$$\lambda = \frac{\mathbf{x}^{\mathrm{T}}\mathbf{y}}{\mathbf{y}^{\mathrm{T}}\mathbf{y}}, \tag{1.12}$$

which is clearly a minimum since the second derivative, $2\mathbf{y}^{\mathrm{T}}\mathbf{y}$, is necessarily positive. The dotted line from $\mathbf{x}$ to $\lambda\mathbf{y}$ that results from this process of projecting $\mathbf{x}$ orthogonally onto $\mathbf{y}$ is called the *orthogonal projector* (more simply, the projector), and the stretch along $\mathbf{y}$ from $\mathbf{O}$ to $\lambda\mathbf{y}$ is called the *orthogonal projection* (more simply, the projection).

Now, recalling basic trigonometry, we define the cosine of the angle ϕ between $\mathbf{x}$ and $\mathbf{y}$ as the ratio of the length of the adjacent side to that of the hypotenuse, that is,

$$\cos\phi = \lambda\frac{\|\mathbf{y}\|}{\|\mathbf{x}\|} = \frac{\mathbf{x}^{\mathrm{T}}\mathbf{y}\|\mathbf{y}\|}{\|\mathbf{y}\|^2\|\mathbf{x}\|} = \frac{\mathbf{x}^{\mathrm{T}}\mathbf{y}}{\|\mathbf{x}\|\,\|\mathbf{y}\|}. \tag{1.13}$$

It is clear from this expression that if $\mathbf{x}$ and $\mathbf{y}$ have already been normalized to have unit length, that is, $\|\mathbf{x}\| = \|\mathbf{y}\| = 1$, then the angle between $\mathbf{x}$ and $\mathbf{y}$ is simply related to the inner product $\mathbf{x}^{\mathrm{T}}\mathbf{y}$ of these two vectors, which provides a geometric interpretation of the inner product.

Statistical Applications

The Two Geometries: p- and n-Dimensional

In statistics and econometrics, two complementary geometries arise naturally from the $n \times p$ data matrix $\mathbf{X}$. Let us denote the ith row of $\mathbf{X}$ by $\mathbf{x}_i^{\mathrm{T}}$ and the jth column by $\mathbf{X}_j$. Then, $\mathbf{X}$ may be written either in terms of its rows or its columns:

$$\mathbf{X} = \begin{bmatrix} \mathbf{x}_1^{\mathrm{T}} \\ \vdots \\ \mathbf{x}_n^{\mathrm{T}} \end{bmatrix} \equiv [\mathbf{X}_1 \cdots \mathbf{X}_p]. \tag{1.14}$$

In the first expression, we see that each row of $\mathbf{X}$ is a p-vector ($\mathbf{x}_i^{\mathrm{T}} \in \mathscr{R}^p$), so $\mathbf{X}$ may be interpreted geometrically as a collection of n points in p-space. Since each row of $\mathbf{X}$ is an observation in a statistical analysis, this space is often called observation space. In the second expression, we see that each column of $\mathbf{X}$ is an n-vector ($\mathbf{X}_j \in \mathscr{R}^n$), so $\mathbf{X}$ is also interpretable geometrically as a collection of p points in n-space. Each column of $\mathbf{X}$ is, of course, a variate in a statistical context, so this space is often called variate space.

Notice in the preceding that there can be a natural tendency to confuse the statistical terms *observation* or *variate* with the mathematical term *vector* or the geometric term *point*, but it is always good to keep in mind that these are not really the same things. Statistical variates have properties, such as units and interpretations derived from the underlying reality that the variates are measuring, properties that extend beyond the abstract mathematical concept of a vector, and conversely, the mathematical properties of vectors and computational algorithms that manipulate them can extend beyond the statistical meaning of the variates or observations these vectors represent. The vector of ones, $\boldsymbol{\iota} \equiv (1, \ldots, 1)^T$, for example, is often used as a column of **X** to represent the constant effects in a linear model, that is, the intercept term. There is, therefore, a sense in which one could argue that $\boldsymbol{\iota}$ is not a variate since it does not vary. We shall see later that this rather misses the point, but there is nevertheless no sense in which one can argue that $\boldsymbol{\iota}$ is not a vector. Devoid of its statistical meaning, $\boldsymbol{\iota}$ is a vector just like any other vector and hence must be treated exactly the same as any vector in assessing the computational properties of an algorithm that acts on an arbitrary set of vectors.

Each of the two geometries, the observation-space geometry and the variate-space geometry, has its uses, often complementary. We shall use them both shortly to picture least squares. But first, we use the n-dimensional geometry to provide a graphical interpretation of centering and of correlation.

The Geometry of Mean Centering (Deviations from the Mean)

In this exercise, we are in the n-dimensional variate space: each point represents a set of n observations on some measurable quantity. Consider, then, the variate $\mathbf{x} \equiv (x_1, \ldots, x_n)^T$, which could be **y** or any column of **X**. It is often useful to transform **x** into its mean-centered (deviation-from-mean) form

$$\tilde{\mathbf{x}} \equiv \mathbf{x} - \bar{x}\boldsymbol{\iota}, \tag{1.15}$$

where $\boldsymbol{\iota} \equiv (1, \ldots, 1)^T$, the vector of n ones introduced above, and $\bar{\mathbf{x}} = n^{-1} \Sigma_j x_j$ is the arithmetic mean of the elements of **x**. Geometrically, this transformation may be pictured as in Exhibit 1.5, which is just Exhibit 1.4 with $\mathbf{y} = \boldsymbol{\iota}$. In this case, we note from (1.12) that the value for λ is just $\bar{x}$, that is, $\lambda = \mathbf{x}^T\boldsymbol{\iota}/\boldsymbol{\iota}^T\boldsymbol{\iota} = n^{-1}\Sigma_j x_j$, and therefore that $\boldsymbol{\eta}$ in Exhibit 1.4 becomes $\tilde{\mathbf{x}} \equiv \mathbf{x} - \bar{x}\boldsymbol{\iota}$, just exactly the transformed vector we seek.

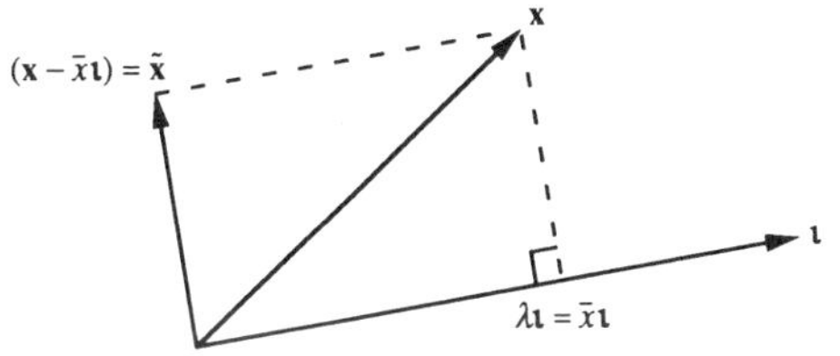

Exhibit 1.5 Geometry of mean centering.

Thus, the geometric interpretation of mean centering the variate $\mathbf{x}$ is simply to take the orthogonal projector of $\mathbf{x}$ onto $\boldsymbol{\iota}$ (the dotted line perpendicular to $\boldsymbol{\iota}$ in Exhibit 1.5) and "slide" it parallel along $\boldsymbol{\iota}$ to the origin; the result is $\tilde{\mathbf{x}}$. Equivalently, one can say that $\tilde{\mathbf{x}}$ is the orthogonal projection of $\mathbf{x}$ onto the orthogonal complement of $\boldsymbol{\iota}$.

Algebraically, the matrix $\mathbf{M}$ of the linear transformation that achieves this projection is derived as

$$\tilde{\mathbf{x}} \equiv \mathbf{x} - \bar{x}\boldsymbol{\iota} = \mathbf{x} - \boldsymbol{\iota}\bar{x} = \mathbf{x} - \boldsymbol{\iota}\left(\frac{\boldsymbol{\iota}^{\mathrm{T}}\mathbf{x}}{n}\right) = \left(\mathbf{I} - \frac{\boldsymbol{\iota}\boldsymbol{\iota}^{\mathrm{T}}}{n}\right)\mathbf{x} \equiv \mathbf{M}\mathbf{x}. \qquad (1.16)$$

That is, the matrix $\mathbf{M}$ that maps any variate $\mathbf{x}$ into its mean-centered form $\tilde{\mathbf{x}}$ is simply

$$\mathbf{M} \equiv \mathbf{I} - \frac{\boldsymbol{\iota}\boldsymbol{\iota}^{\mathrm{T}}}{n}, \qquad (1.17)$$

which is therefore appropriately called the *centering matrix*.

The Geometry of Correlation

The preceding may be used to provide geometric measures of the correlation between two variates $\mathbf{x}$ and $\mathbf{y}$. We again are using the n-dimensional, variate-space geometry. First, consider the angle between the mean-centered variates $\tilde{\mathbf{x}}$ and $\tilde{\mathbf{y}}$. This is pictured in Exhibit 1.6.

From (1.13), we have

$$\cos\phi = \frac{\tilde{\mathbf{x}}^{\mathrm{T}}\tilde{\mathbf{y}}}{\|\tilde{\mathbf{x}}\|\,\|\tilde{\mathbf{y}}\|} = \frac{\Sigma_i(x_i - \bar{x})(y_i - \bar{y})}{\sqrt{\Sigma_i(x_i - \bar{x})^2}\sqrt{\Sigma_i(y_i - \bar{y})^2}} \equiv r_{\mathbf{xy}}, \qquad (1.18)$$

where $r_{\mathbf{xy}}$ is seen to be the sample correlation between the elements of $\mathbf{x}$ and those of $\mathbf{y}$. Thus, the geometric interpretation of the statistical concept of the correlation between two vectors $\mathbf{x}$ and $\mathbf{y}$ is simply the cosine of the angle between the two mean-centered variates $\tilde{\mathbf{x}}$ and $\tilde{\mathbf{y}}$. If the angle between them is either zero or π, the correlation is $\cos 0$ or $\cos\pi$, that is, ± 1. Similarly, if the angle is $\pi/2$ (a right angle, so that $\tilde{\mathbf{x}}$ and $\tilde{\mathbf{y}}$ are orthogonal), then the correlation is $\cos\pi/2 = 0$.

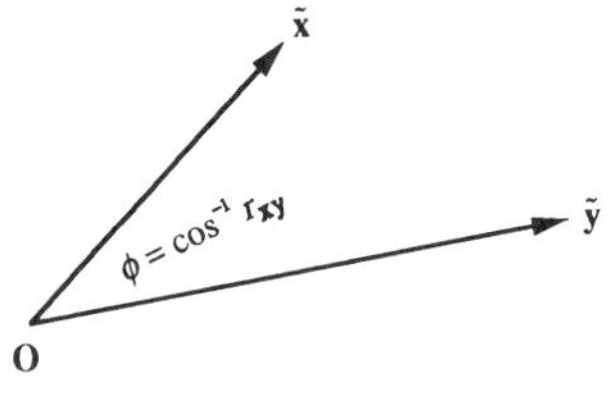

Exhibit 1.6 Correlation between two centered variates $\tilde{\mathbf{x}}$ and $\tilde{\mathbf{y}}$.

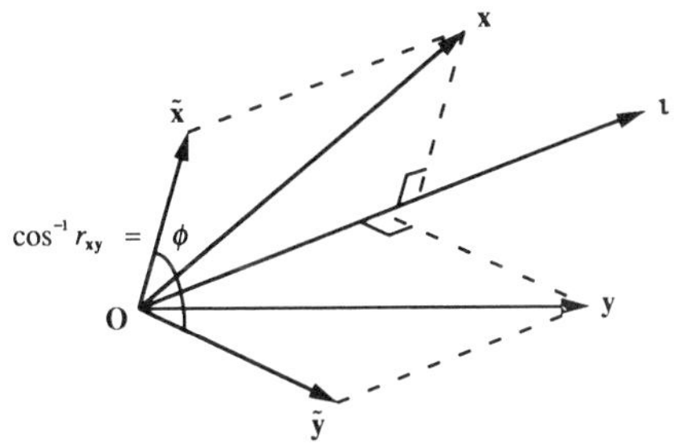

Exhibit 1.7 Correlation between two variates **x** and **y**.

We can also examine this correlation directly, letting the geometry do the mean centering. Thus, consider Exhibit 1.7. Here we assume the graph depicts the three-dimensional subspace of n-space that is determined by the three variates (vectors) $\mathbf{x}$, $\mathbf{y}$, and $\boldsymbol{\iota}$. We know from above that we can mean-center $\mathbf{x}$ and $\mathbf{y}$ simply by projecting them orthogonally onto the orthogonal complement of $\boldsymbol{\iota}$ and then we need only look to the angle between these projections, $\tilde{\mathbf{x}}$ and $\tilde{\mathbf{y}}$, to depict the correlation geometrically. All this is shown in Exhibit 1.7.

The Geometry of Least Squares

Both the p- and n-dimensional geometries can be used to depict least squares. A form of the p-dimensional geometry is found in virtually every textbook that introduces least-squares estimation, at least for the $p = 2$ case. Here one considers the model $\mathbf{y} = \beta_1 + \beta_2\mathbf{x} + \boldsymbol{\varepsilon}$, and the rows of the data matrix $\mathbf{Z} \equiv [\mathbf{y}\ \mathbf{x}]$ are plotted as in Exhibit 1.8. The least-squares line is that which minimizes the sum of squared vertical distances between the points and the line. The y-intercept is the least-squares estimate b_1 of the constant term β_1, and the slope of the line is the least-squares estimate b_2 of the parameter β_2. Such plots are so common that it is unnecessary to elaborate further upon them here; we mention them principally for completeness and to point out that this geometry has some (limited) value for elementary illustrations with $p = 2$, becomes somewhat unwieldy for $p = 3$, and is all but impossible for $p > 3$. We shall,

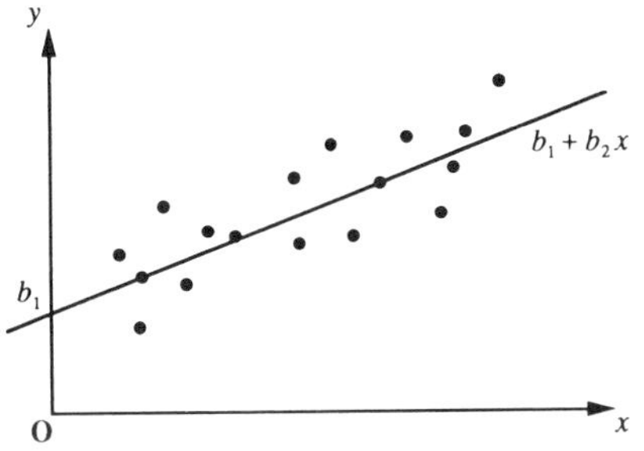

Exhibit 1.8 Conventional p-dimensional geometry of least squares.

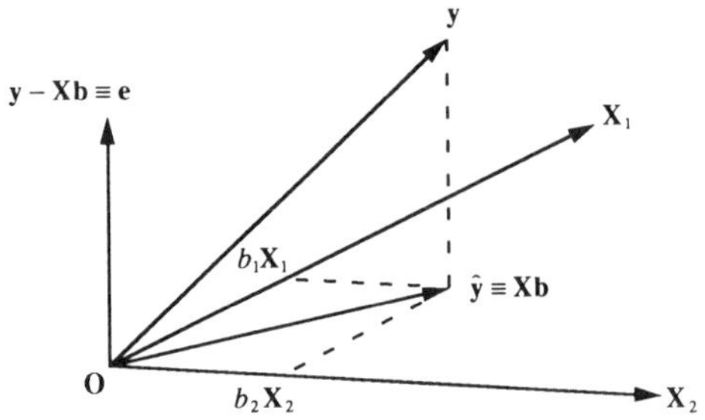

Exhibit 1.9 The n-dimensional geometry of least squares.

however, be able to put a $p = 3$ version of this geometry to good use in the next chapter in illustrating the problems collinearity can cause least-squares estimation.

The n-dimensional geometry of least squares is, by contrast, a far more powerful tool, both for illustrative and analytic purposes. This geometry is developed in some textbooks, such as Theil (1971), Wonnacott and Wonnacott (1979), or Johnston (1984), and is used with great power in such works as Malinvaud (1970). We shall have much opportunity to draw on the intuition it affords. Suppose, then, we wish to estimate the ($p = 2$) model $\mathbf{y} = \beta_1\mathbf{X}_1 + \beta_2\mathbf{X}_2 + \boldsymbol{\varepsilon} \equiv \mathbf{X}\boldsymbol{\beta} + \boldsymbol{\varepsilon}$ by least squares. By definition, we seek that linear combination $b_1\mathbf{X}_1 + b_2\mathbf{X}_2 \equiv \mathbf{Xb}$ that minimizes the sum of squared residuals $\mathbf{e} \equiv \mathbf{y} - \mathbf{Xb}$. But we know from above that this sum of squared residuals, $\mathbf{e}^T\mathbf{e}$, is simply $l^2(\mathbf{e}) = \|\mathbf{e}\|^2$. Thus, geometrically the least-squares estimator is that which minimizes the length of $\mathbf{e}$, that is, that linear combination of the $\mathbf{X}$'s that is closest to $\mathbf{y}$. This is pictured in Exhibit 1.9, where we show the minimum-distance (orthogonal) projection of $\mathbf{y}$ onto the space (plane) spanned by the $\mathbf{X}$'s. This is to be viewed with the illusion of three dimensions: the $\mathbf{y}$ vector in one dimension is projected orthogonally (at right angles) to the two-dimensional space spanned by the two $\mathbf{X}$'s (viewed as "going into" the plane of the page). The result of this is the projection $\hat{\mathbf{y}} \equiv \mathbf{Xb}$ of "fitted values," which in turn determines the least-squares combinations $b_1\mathbf{X}_1$ and $b_2\mathbf{X}_2$ by projecting $\hat{\mathbf{y}}$ along the $\mathbf{X}$'s. If this regression has a constant term, then one of the $\mathbf{X}$'s is $\boldsymbol{\iota}$. A good exercise uses the techniques developed so far to provide a geometric interpretation of R^2, which is simply the squared correlation between $\mathbf{y}$ and $\hat{\mathbf{y}}$.

It would seem that this n-dimensional geometry would fall prey to the same limitation as the p-dimensional geometry, namely, that it is awkward to apply for $p > 2$ or 3. This, however, is not true; with a little ingenuity and imagination this geometry can be used in many higher-dimensional problems. In tests of hypotheses on subsets of coefficients, for example, $\mathbf{X}_1$ can stand for the p_1 variates whose coefficients are not subject to test and $\mathbf{X}_2$ can stand for the p_2 variates whose coefficients are subject to test.[2]

[2]See Malinvaud (1970).

1.6 OVERVIEW

The plan of this book is as follows. In Chapter 2, the notion of collinearity is introduced and geometric views, both p-dimensional and n-dimensional, of its nature are presented. A review is then given of numerous techniques that have been offered for diagnosing collinearity, and their strengths and weaknesses are discussed. This leaves us with a well-defined suggestion for a collinearity diagnostic based on the information in the eigensystem of the data matrix $\mathbf{X}$. This suggestion is picked up in Chapter 3, which first provides the technical background and tools needed to develop the collinearity diagnostic: condition indexes and variance–decomposition proportions. Then a procedure based on these tools is specified that allows diagnosis of the presence and composition of collinear relations. This is accompanied by a discussion of the empirical elements needed to make the test practical.

Chapter 4 is devoted entirely to a series of controlled experiments that allow the empirical content of the collinearity diagnostic to be discovered. The essential results of these tests are summarized in the beginning of Chapter 5, which then continues with a detailed description of how to carry out the tests, how best to report the results, and most importantly, how best to go about digesting and interpreting the diagnostic information once it is obtained. Several examples are then given that show the application of the collinearity diagnostics to various sets of real-life data.

In Chapter 6, several important refinements to interpreting the collinearity diagnostics are developed. The information obtained from any diagnostic procedure is necessarily context dependent, so this chapter discusses the ways for configuring both data and model so that the resulting collinearity diagnostics provide information that is most meaningful and appropriate to assessing the desired context.

Diagnosing the presence and composition of collinear relations is one thing; determining whether those collinear relations are causing statistical harm to an ensuing regression analysis is another. So in Chapter 7 a test based on measuring signal-to-noise is first developed to determine the presence of *weak data*, a term that encompasses both collinearity and short data. This test is then combined with the previously developed collinearity diagnostics to define *harmful collinearity* and to devise a test for its presence. An illustration of this diagnostic procedure is given.

While collinear relations pertain to the columns of a data matrix, much interesting work has been done lately regarding the effects of observations, or rows of the data matrix, on those collinear relations. Such collinearity-influential observations and diagnostics for them are examined, exemplified, and discussed in Chapter 8.

Chapter 9 examines the conditioning of two important special cases, models with logged variates and models with first differences. Simple solutions are offered for both cases, but the first presages some of the difficulties that accompany nonlinear transformations and the second allows us to see how easily one can be misled by diagnostics applied to the wrong data.

In Chapter 10, general considerations for corrective, or remedial, action are discussed, that is, "the fix." These notions are exemplified by three lengthier case studies in which the conditioning diagnostics and remedial activities are integrated into a more complete statistical analysis. Finally, in Chapter 11, the general notion of conditioning in statistical analysis is defined, and its relation to collinearity is made clear. Extensions of the conditioning diagnostics to nonlinear and simultaneous-equations models are discussed.

CHAPTER 2

Collinearity

In this and the next several chapters, we focus on collinearity in the context of estimating the linear model $\mathbf{y} = \mathbf{X}\boldsymbol{\beta} + \boldsymbol{\varepsilon}$ with the least-squares estimator $\mathbf{b} = (\mathbf{X}^T\mathbf{X})^{-1}\mathbf{X}^T\mathbf{y}$. Ultimately, we want to develop (1) means for detecting the presence of one or more collinear relations among the columns of the data matrix $\mathbf{X}$ used in a linear regression, (2) means for identifying the subsets of explanatory variates involved in each such relation, and (3) means for assessing the extent to which each of the least-squares estimates $\mathbf{b}$ is harmed (potentially and actually) by the presence of the collinear relations. We begin here with preliminary notions of collinearity and illustrate them through simple graphical views. We then discuss different approaches that have been suggested for detecting collinearity, which helps to motivate the direction finally taken.

2.1 A SIMPLE INTRODUCTION TO COLLINEARITY

Literally, two variates are collinear if the data vectors representing them lie on the same line, that is, in a subspace of dimension 1. More generally, k variates are collinear, or linearly dependent, if one of the vectors that represents them is an exact linear combination of the others, that is, if the k vectors lie in a subspace of dimension less than k. In practice, such "exact collinearity" rarely occurs, and it is certainly not necessary for collinearity to be exact in order for problems to exist. A broader notion of collinearity is therefore needed to deal with the problem as it affects statistical estimation.

More loosely, then, two variates are collinear if they lie almost on the same line, that is, if the angle between them is small. We will ignore for the moment just what "small" means here. This notion is readily generalized to more than two variates by saying that k variates are collinear, or nearly dependent, if one of them lies almost in the space spanned by the remaining $k - 1$, that is, if the angle between the one and its orthogonal projection on the others is small. While this intuitive view of collinearity is adequate for the immediate discussion, we eventually define collinearity in terms of the conditioning of the data matrix $\mathbf{X}$.

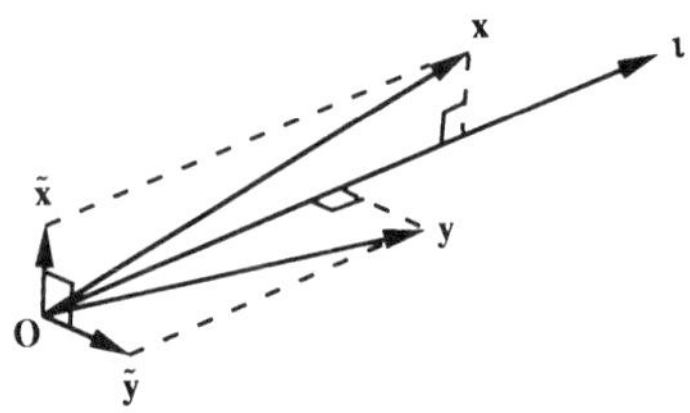

Exhibit 2.1 Two uncorrelated collinear variates.

In any event, the terms *collinearity* and *near dependency* will be used synonymously in this monograph to denote these inexact collinear relations. When "exact collinearity" is intended, the qualifying adjective will be employed.

Interestingly, we can see immediately from the geometry of Section 1.5 that a small angle between two vectors (or a vector and its orthogonal projection) is not equivalent to a high correlation (or multiple correlation) between them; a high correlation surely implies a low angle, but the converse need not be true. In Exhibit 2.1, for example, the angle between $\mathbf{x}$ and $\mathbf{y}$ is small, thereby indicating collinearity, while that between their centered variates $\tilde{\mathbf{x}}$ and $\tilde{\mathbf{y}}$ is a right angle, thereby indicating zero correlation. It is also clear that we may "slide" the tips of $\mathbf{x}$ and $\mathbf{y}$ along their projectors toward $\boldsymbol{\iota}$ so as to make the angle between them as small as desired without affecting the zero correlation.[1]

Thus, collinearity and correlation are not the same thing. This is an important realization because it conflicts with many popular points of view. In later examples, we shall see that the two terms are frequently confused, often to the detriment of any subsequent analysis and discussion.

The preceding discussion also allows us to see rather more generally that while collinearity is concerned with numerical or geometric characteristics of the given data matrix $\mathbf{X}$, it is not concerned with any of the statistical aspects that may generate that data matrix $\mathbf{X}$ or may otherwise be relevant to the linear regression model $\mathbf{y} = \mathbf{X}\boldsymbol{\beta} + \boldsymbol{\varepsilon}$—which has, as yet, not really even entered the discussion. That is, collinearity is a data problem, not a statistical problem. Collinearity can clearly exist among the columns of a matrix $\mathbf{X}$ whether or not those columns represent anything statistical. But when they do represent something statistical, the presence of collinearity can create a situation in which statistical estimation by least squares (and other estimators as well) is severely hampered. Collinearity, then, is a nonstatistical problem that can nevertheless have serious implications for the efficacy of some statistical procedures. This, loosely, is the reason one must diagnose, rather than test, for the presence of collinearity. Whether a given degree of collinearity can cause statistical

[1]For those who prefer mathematical illustrations, consider the following situation constructed in $\mathscr{R}^n$ for $n > 3$. Let $\aleph_\iota$ be the orthogonal complement of $\boldsymbol{\iota}$ in $\mathscr{R}^n$, and let $\mathbf{u}$ and $\mathbf{v}$ be two orthonormal vectors in $\aleph_\iota$, so $\mathbf{u}^T\mathbf{v} = \mathbf{u}^T\boldsymbol{\iota} = \mathbf{v}^T\boldsymbol{\iota} = 0$ and $\mathbf{u}^T\mathbf{u} = \mathbf{v}^T\mathbf{v} = 1$. Define $\mathbf{x}(\alpha) \equiv \boldsymbol{\iota} + \alpha\mathbf{u}$ and $\mathbf{y}(\alpha) \equiv \boldsymbol{\iota} + \alpha\mathbf{v}$. Clearly the angle ϕ between $\mathbf{x}(\alpha)$ and $\mathbf{y}(\alpha)$ goes to zero with α, i.e., $\cos\phi \to 1$ in (1.13) as $\alpha \to 0$, but the correlation between $\mathbf{x}(\alpha)$ and $\mathbf{y}(\alpha)$ is simply $\mathbf{u}^T\mathbf{v} = 0$ for all nonzero α. Here, then, are two vectors that can become arbitrarily collinear while remaining absolutely uncorrelated.

problems for least-squares estimation depends, as we shall see in Chapter 7, upon the statistical context.

2.2 SOME VIEWS OF COLLINEARITY

Intuitively we can understand the potential for harm that results from collinear data by realizing that a collinear variate, being nearly a linear combination of other variates, does not provide information that is very different from that already inherent in these others. It can become difficult, therefore, to infer the separate influence of such explanatory variates on the response variate. The phrase *potential for harm* is used here because the mere presence of collinearity need not cause problems; it depends on the circumstances. The distinction between the potential harm and the actual harm due to collinearity will be dealt with at various points as we proceed.

We can view the nature of collinearity geometrically using either the p- or n-dimensional geometry. We will do both, beginning with the p-dimensional geometry in Exhibits 2.2, where we have pictured several situations relevant to the model

$$y_i = \beta_1 + \beta_2 x_{i2} + \beta_3 x_{i3} + \varepsilon_i, \qquad i = 1, \ldots, n, \tag{2.1}$$

that is, the $p = 3$ case. In these exhibits, we show scatters of the n data points. In the x_2, x_3 "floor" are the (x_2, x_3) scatters of the explanatory variates (points denoted by solid circles), while above we show the data "cloud" that results when the y dimension is included (points denoted by open circles). Exhibit 2.2*a* depicts the well-behaved case where there is no collinearity between x_2 and x_3. The data scatter on the floor provides a broad and stable base in both x

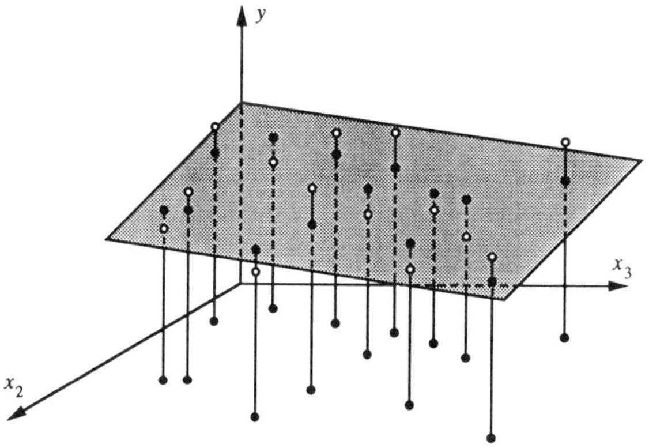

Exhibit 2.2*a* No collinearity; all regression coefficients well determined. A small change in any parameter of the regression plane will cause a relatively large change in the residual sum of squares.

dimensions, and the data cloud above provides a well-defined, least-squares plane, that is, the plane that minimizes the sum of squared errors in the y direction between the actual y_i and the plane. The y-intercept of this plane is b_1, the least-squares estimate of the intercept term β_1, and the partial slopes in the x_2 and x_3 directions are b_2 and b_3, the least-squares estimates of β_2 and β_3, respectively. It is clear that an attempt to tilt this plane in any direction would substantially alter (increase) the sum of squared errors, producing a situation that is clearly distinguishable from the least-squares solution, and hence the various regression parameters are all estimated with precision.

By contrast, Exhibit 2.2*b* depicts a case of perfect collinearity between x_2 and x_3. The scatter on the floor has length but no width, as is also the case for the data cloud above. The resulting least-squares plane is therefore not defined; tilting the plane along the "axis" of the cloud leaves the minimum sum of squared errors unchanged, and hence any of these planes is a least-squares solution. This illustrates the well-known fact that perfect collinearity destroys the uniqueness of the least-squares estimator. Exhibit 2.2*c* depicts strong (but not perfect) collinearity. Here, the data scatter on the floor has little width relative to its length, and the resulting least-squares plane is ill defined in the sense that tilting it along its axis will produce little change in the sum of squared errors. The fact that the plane is ill defined in this manner translates statistically into the fact that the least-squares estimates (the y-intercept and the partial slopes) are imprecise; that is, they have high variance.

These simple illustrations serve also to show that collinearity need not harm all parameter estimates. Exhibit 2.2*d*, for example, shows a case where the partial slopes are ill defined—and hence one has imprecise estimates of β_2 and β_3—but the y-intercept remains well defined as the plane is tilted along the cloud axis. Here, then, the intercept β_1 remains estimated with precision. Likewise, in Exhibit 2.2*e* we have a case where the partial slope in the x_3 direction remains well defined. Here, the estimates of β_1 and β_2 will lack precision while that of β_3 will not. It is easy to see in this case that the collinear relation is no longer between x_2 and x_3 but rather is now between x_2 and the constant term (even though these two are necessarily uncorrelated).

Exhibit 2.2*f* shows a possible interaction between collinear data and outliers. Except for the single highly influential observation to the far right, this data set would be highly collinear, and if for some reason this observation were shown to be erroneous, these data would be quite weak for estimation purposes. If, however, this observation can be shown to be a good data point, then its value to any subsequent statistical analysis is obvious. This example also illustrates the potential for interaction between diagnostics for collinearity and those for influential-data points. We examine this issue in greater detail in Chapter 8.

The effects of collinearity on the least-squares estimates may also be viewed quite dramatically and usefully with the n-dimensional geometry of least squares, as we can see from Exhibit 2.3. Here we have two variates, $\mathbf{X}_1$ and $\mathbf{X}_2$, relevant to the model $\mathbf{y} = \beta_1\mathbf{X}_1 + \beta_2\mathbf{X}_2 + \boldsymbol{\varepsilon}$. In Exhibit 2.3*a*, these two variates are well conditioned, the angle between them being nearly a right angle. In

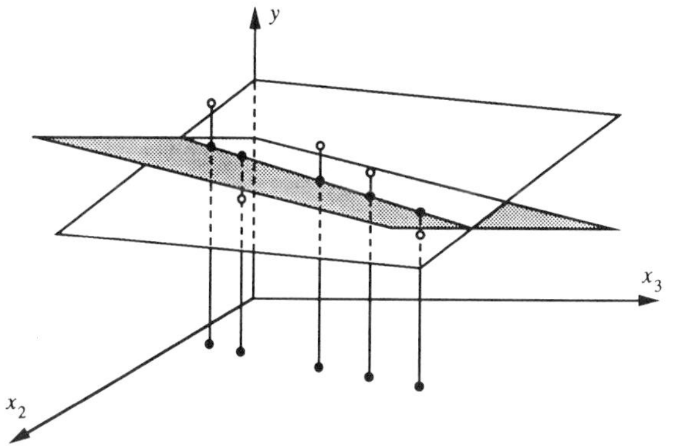

Exhibit 2.2*b* Exact collinearity; all regression coefficients undetermined. A simultaneous change in all the parameters of the regression plane can leave residual sum of squares unchanged.

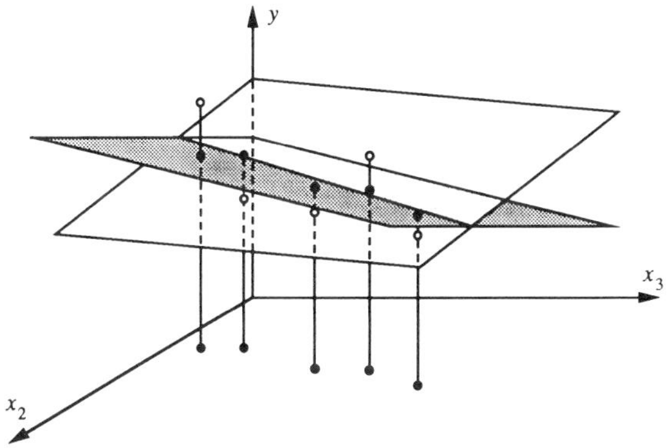

Exhibit 2.2*c* Strong collinearity; all regression coefficients ill determined. A simultaneous change in all the parameters of the regression plane can cause little change in the residual sum of squares.

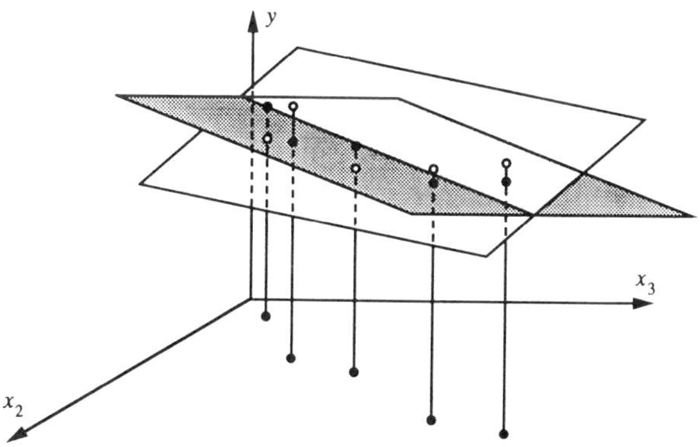

Exhibit 2.2*d* Strong collinearity but constant term well determined. Only changes in the slope parameters of the regression plane can leave the residual sum of squares little affected.

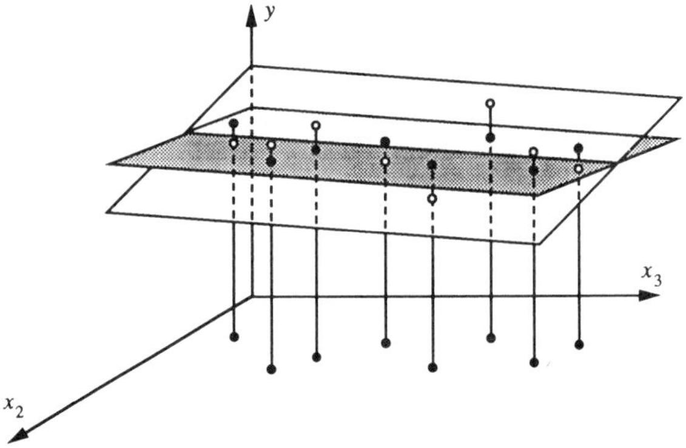

Exhibit 2.2e Strong collinearity but b_3 well determined. Only changes in the intercept and b_2 parameters of the regression plane can leave the residual sum of squares little affected.

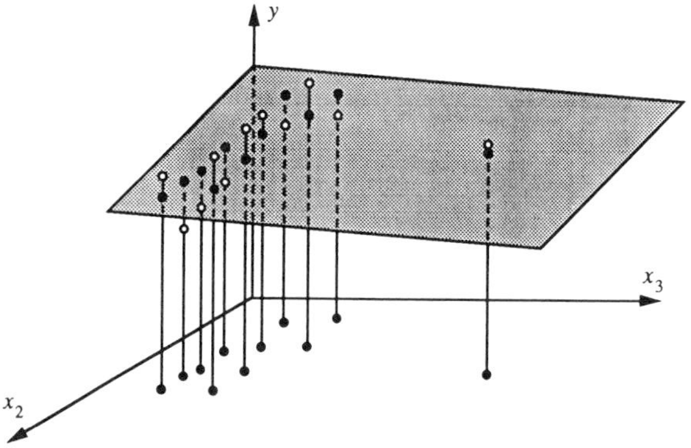

Exhibit 2.2f Strong collinearity, except for an outlier.

Exhibit 2.3*b* the conditioning is worse, and finally, in Exhibit 2.3*c* the two variates are quite ill conditioned.

We have not shown a **y** dimension here since it is really unnecessary and would only clutter the picture. Rather, we recognize that **y** is some linear combination of the **X**'s plus an error term $\boldsymbol{\varepsilon}$ assumed to be iid normal with scalar variance–covariance matrix $\sigma^2\mathbf{I}$. Geometrically, this means $\boldsymbol{\varepsilon}$ is spherically distributed: the isodensity contours of its distribution are the family of n-dimensional hyperspheres $\boldsymbol{\varepsilon}^T\mathbf{I}\boldsymbol{\varepsilon} = r^2$. Now, if we were to choose r^2 to be an appropriate multiple of σ^2 so that the sphere would contain $\boldsymbol{\varepsilon}$ with, say, probability .95, then the corresponding set of **y**'s that could occur from such a model would lie in that sphere centered on $E\mathbf{y} = \beta_1\mathbf{X}_1 + \beta_2\mathbf{X}_2 = \mathbf{X}\boldsymbol{\beta}$, and the orthogonal projections of that set of **y**'s onto the space spanned by the **X**'s would

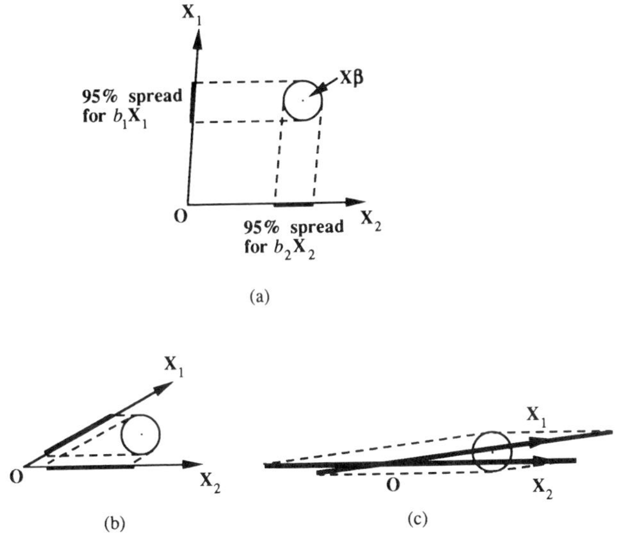

Exhibit 2.3 Depiction of the effects of collinearity on ordinary least-squares regression with the *n*-dimensional geometry: (*a*) well-conditioned data; (*b*) poorly conditioned data; (*c*) very ill-conditioned data.

necessarily be a circle centered on $\mathbf{X\beta}$, as is shown in Exhibit 2.3.[2] This circle, then, provides us with a geometric image of the likely least-squares outcomes for this model, which we suppose here is chosen to represent a probability level of .95.

Projecting these circles along the **X**'s allows us to see the likely range of least-squares estimates $\mathbf{b}_1$ and $\mathbf{b}_2$.[3] In the well-conditioned situation depicted in Exhibit 2.3*a*, this 95% spread is quite tolerable. In the less well-conditioned situation of Exhibit 2.3*b*, even though the **X**'s have the same lengths and the error term has the same variance, the range of probable least-squares estimates has been substantially spread. And in the ill-conditioned case depicted in Exhibit 2.3*c*, the spread is very broad, even encompassing the origin **O**. Further, it is clear that, as the angle between the two **X**'s get limitingly small, the spread extends indefinitely in both directions, and when the two **X**'s fully overlap, there is exact collinearity, and the least-squares estimator is no longer uniquely defined.

[2]The radius needed for this circle actually is somewhat less than that for the sphere containing 95% of the density of ε. This is because the **y**'s that lie outside this sphere (above it and below it relative to the space spanned by the **X**'s) can also be projected orthogonally into the same circle. We are thus interested in the circle determined as the intersection of the space spanned by the **X**'s with the cylinder whose central axis goes orthogonally through $\mathbf{X\beta}$ and whose volume encompasses 95% of the density of ε. The circles in Exhibit 2.3 represent this intersection.

[3]A similar argument could be made for confidence intervals by considering the point $\mathbf{X\beta}$ to be the least-squares estimate $\mathbf{Xb}$ and the radius of the circle to represent some appropriate multiple of s^2.

It is clear from these simple geometric views that the ability to diagnose collinearity is important to many users of least squares. To be most useful, such diagnostics should be able to determine when coefficient estimates are degraded—a term to be defined in the next chapter—by the presence of collinearity, when they are not, and when they are, where and how corrective help would best be applied. Such a set of collinearity diagnostics is the subject of the next chapter. First we examine other important efforts at devising collinearity diagnostics.

2.3 SETTING THE SCENE FOR A MEANINGFUL COLLINEARITY DIAGNOSTIC

Throughout the years, many procedures have been employed to detect collinearity. There is no need for us to survey this literature intensively; we need here only discuss the main classes of these procedures and indicate certain of their strengths and weaknesses.[4] This helps to set the scene for the collinearity diagnostic finally adopted.

1. *Hypothesized Signs for Some Regression Estimates Are Incorrect, "Important" Explanatory Variates Have Low* t*-statistics, or Various Regression Results are Sensitive to Deletion of a Row or a Column of* **X**. Any of these conditions is frequently cited as evidence of collinearity, and even worse, collinearity is often cited as an explanation for these conditions. Unfortunately for those looking for simple answers, while these conditions are often associated with collinearity, none of them is either necessary or sufficient for the existence of collinearity. We have already seen that collinearity need not cause harm, and a combination of model misspecification and influential data are quite capable of explaining any of these problems without the presence of collinearity. Thus, more refined techniques are required both to detect the presence of collinearity and to assess its potentially harmful effects.

2. *Examine the Correlation Matrix* **R** *of the Explanatory Variates.* Examining the correlation matrix of the explanatory variates is a commonly employed procedure, since a high correlation would seem intuitively to imply collinearity and since this matrix is a standard output of most regression packages. If we assume the columns of the data matrix **X** have been centered and scaled to have unit length, the correlation matrix **R** is simply $\mathbf{X}^T\mathbf{X}$.

While a high correlation coefficient between two explanatory variates can indeed point to a possible collinearity problem, the absence of high correlations cannot be viewed as evidence of the absence of collinearity problems. First, as we have already seen, collinearity and correlation are not the same; a high correlation implies collinearity, but the converse is not true. Second, it is possible for three or more variates to be collinear while no two of the variates taken together are highly correlated. Indeed, this is the case for the data

[4] Readers interested in a more extensive survey are referred to Kumar (1975a).

illustrated in Exhibit 1.1. Rather more generally, it is shown in Section 11.4 that for $p \geqslant 3$, p variates can be perfectly collinear (linearly dependent) and still have no absolute pairwise correlation between any two of them that exceeds $1/(p-1)$. Thus, 11 variates that together are perfectly collinear could have a largest absolute pairwise correlation among them of only .10. The correlation matrix is clearly incapable of diagnosing either of the preceding situations.

The correlation matrix is also unable to reveal the presence or number of several coexisting collinear relations. High pairwise correlations, for example, between $\mathbf{X}_1$ and $\mathbf{X}_2$ on the one hand and $\mathbf{X}_3$ and $\mathbf{X}_4$ on the other could occur along with a third near dependency involving all four variates or could result from just two separate near dependencies, one between $\mathbf{X}_1$ and $\mathbf{X}_2$ and one between $\mathbf{X}_3$ and $\mathbf{X}_4$. The information contained in the correlation matrix need not be able to distinguish between these two situations. By contrast, the diagnostic procedure that we shall develop shortly does allow the number of separate near dependencies to be discovered as well as provide information on the involvement of each variate in each such near dependency.

Correlation measures also rather generally have another weakness as a collinearity diagnostic; namely, there is no obvious cutoff for how large a correlation must be to indicate collinearity. Would you, for example, suggest a value of .5 or .9 or .99? Our experience and intuition surely leave us cold here. In many statistical studies, correlations of the magnitude .3 are considered large. These are studies that are interested in testing the hypothesis that no correlation exists. In other studies, those that presume the existence of a relation and are interested in knowing if it is strong, correlations like .3 are considered small. Thus, econometricians are often upset (sometimes quite incorrectly) with R^2s or Rs this low. Eventually we shall find that high values of correlations are important for collinearity analyses, but this is not something we can know from any prior understanding of correlations. Rather it is something we can learn inferentially about correlations from more appropriate collinearity diagnostics.

3. *Examine the Inverse,* $\mathbf{R}^{-1}$, *of the Correlation Matrix of the Explanatory Variates: Variance Inflation Factors and Multiple Correlations,* R_i^2. Clearly, some of the shortcomings just mentioned with regard to the use of the correlation matrix $\mathbf{R}$ as a diagnostic measure for collinearity would seem also to limit the usefulness of its inverse, $\mathbf{R}^{-1}$, and this is greatly the case. However, one diagnostic based on $\mathbf{R}^{-1}$, the variance inflation factor (VIF), is frequently advanced as a collinearity diagnostic, and it is not without merit.

Recall that we are assuming for the moment—but only for the moment—that the columns of $\mathbf{X}$ have been centered and scaled for unit length, so $\mathbf{R}^{-1} = (\mathbf{X}^{\mathrm{T}}\mathbf{X})^{-1}$. The diagonal elements of $\mathbf{R}^{-1}$, the r^{ii}, are sometimes called the variance inflation factors VIF_i, and these are easily shown to be

$$\mathrm{VIF}_i = \frac{1}{1 - R_i^2}, \tag{2.2}$$

where R_i^2 is the multiple correlation coefficient of $\mathbf{X}_i$ regressed on the remaining explanatory variates.[5] The term *variance inflation factor* was coined by D. W. Marquardt in the 1960s and derives from the fact[6] that the variance of the ith regression coefficient $\sigma_{b_i}^2$ from a regression on the full set of **X**'s obeys the relation

$$\sigma_{b_i}^2 = \frac{\sigma^2}{\mathbf{X}_i^T\mathbf{X}_i}\,\mathrm{VIF}_i, \tag{2.3}$$

where σ^2 is the variance of the regression disturbance $\boldsymbol{\varepsilon}$ in (1.1).

Clearly, a high VIF_i indicates an R_i^2 near unity and hence points to collinearity (as well as inflated variances—one of collinearity's best known side effects in the least-squares context). Furthermore, it is shown in Berk (1977) and Stewart (1987) that the largest VIF_i bounds the condition number of **X** from below. We shall see that high values for this latter measure serve quite well as a collinearity diagnostic, and hence the VIFs can also provide useful information. The VIFs are also very easy to compute.

There are, however, several weaknesses of VIFs that make them less desirable than some other measures as a collinearity diagnostic. First, like any correlation-based measure, high VIFs are sufficient to collinearity but not necessary to it. Second, VIFs are not able to diagnose the number of near dependencies that are present in **X**. This is readily seen, for if there were two strong near dependencies among the columns of **X**, the first involving, say, $\mathbf{X}_1$, $\mathbf{X}_2$, $\mathbf{X}_3$, and $\mathbf{X}_4$, and the second involving $\mathbf{X}_3$ and $\mathbf{X}_4$, all four VIFs would be large. But this would also be true if the first of these near dependencies existed alone, so there is no way to glean from this information the existence of the two separate near dependencies. Third, no means have yet been found for determining a meaningful boundary between values for the VIF that can be considered high and those that can be considered low. This issue is clearly related to that of determining when a correlation is close enough to unity to denote collinearity. Cutoffs of 7 or 10 have been suggested for the VIF, but these suggestions have a completely ad hoc basis.

The first weakness mentioned above can be mitigated somewhat by using an uncentered rather than a centered VIF. As we have seen, one principal situation in which correlations fail to discover collinearity is that involving collinearity

[5]Consider the result relative to r^{11} by partitioning $\mathbf{X} \equiv [\mathbf{x}\ \ \mathbf{Z}]$, where **x** is the first column of **X** and **Z** is the remaining $p - 1$ columns. Then

$$\mathbf{R} = \mathbf{X}^T\mathbf{X} = \begin{bmatrix} \mathbf{x}^T\mathbf{x} & \mathbf{x}^T\mathbf{Z} \\ \mathbf{Z}^T\mathbf{x} & \mathbf{Z}^T\mathbf{Z} \end{bmatrix},$$

and from the standard result for partitioned inverses (Theil, 1971), $r^{11} = [\mathbf{x}^T\mathbf{x} - \mathbf{x}^T\mathbf{Z}(\mathbf{Z}^T\mathbf{Z})^{-1}\mathbf{Z}^T\mathbf{x}]^{-1}$. The expression in this last inverse will be recognized to be $\mathbf{e}^T\mathbf{e}$, the sum of squared residuals of **x** regressed on **Z**. Now, since $\mathbf{x}^T\mathbf{x} = 1$, $R_1^2 = 1 - \mathbf{e}^T\mathbf{e}$, and hence $r^{11} = 1/(1 - R_1^2)$. The expression (2.2) holds since the analogous result arises regardless of which column comes first.

[6]See Theil (1971, p. 166).

with the constant term (the constant column of ones). This would suggest using a VIF that is based on an uncentered R^2 rather than the usual centered one.

The uncentered R^2 is calculated without first taking deviations from the mean. Thus, consider a least-squares regression of any n-vector $\mathbf{z}$ on any set of p n-vectors $\mathbf{W}_i$ comprising the columns of the $n \times p$ matrix $\mathbf{W}$. The residuals from this regression are $\mathbf{e} \equiv \mathbf{z} - \hat{\mathbf{z}} \equiv \mathbf{z} - \mathbf{Wc}$, where $\mathbf{c} = (\mathbf{W}^T\mathbf{W})^{-1}\mathbf{W}^T\mathbf{z}$. The uncentered R^2 is defined to be

$$\hat{R}^2 \equiv 1 - \frac{\mathbf{e}^T\mathbf{e}}{\mathbf{z}^T\mathbf{z}}, \tag{2.4a}$$

while the centered R^2 is defined to be

$$\tilde{R}^2 \equiv 1 - \frac{\mathbf{e}^T\mathbf{e}}{\mathbf{z}^T\mathbf{Mz}}, \tag{2.4b}$$

where $\mathbf{M}$ is the centering matrix (1.17). Both of these measures are of use in different regression contexts,[7] but (2.4b) lacks generality since it is based on the least-squares decomposition $\mathbf{z}^T\mathbf{z} = \hat{\mathbf{z}}^T\mathbf{M}\hat{\mathbf{z}} + \mathbf{e}^T\mathbf{e}$ that holds only when there is an intercept term in the regression equation, whereas $\mathbf{z}^T\mathbf{z} = \hat{\mathbf{z}}^T\hat{\mathbf{z}} + \mathbf{e}^T\mathbf{e}$ holds always.

Thus VIFs can be centered or uncentered depending, respectively, upon whether the centered or uncentered R^2 is used in (2.2). Clearly, a VIF based on the centered $\tilde{R}^2$ has lost all its ability to discover collinearity involving the intercept or constant term, whereas this would not be true for the uncentered VIF.[8]

Much interesting work has been done with VIF-like measures, although the concept masquerades under various names. Stewart (1987), for example, proposes a set of collinearity diagnostics called *collinearity indices* that are simply the square roots of the VIFs. This excellent paper (despite the fact that it uses centered VIFs rather than uncentered ones [see Belsley (1987) for a critique]) provides a masterly introduction to and summary of much of the numerical analysis relevant to the collinearity problem. Simon and Lesage (1988b) provides one of several interesting studies of conditioning diagnostics in which the inverse of the VIF arises and is given the name *tolerance*. And finally, we reemphasize that the VIFs are a direct transformation of the multiple correlations R_i^2 of each of the $\mathbf{X}_i$'s regressed on the remaining columns of $\mathbf{X}$.

Thus, these two popular means for diagnosing collinearity, VIFs and R_i^2s, are seen to be one and the same, and they necessarily possess exactly the same strengths and weaknesses. The problem, for example, of adjudging when a VIF, a Stewart collinearity index, or a tolerance has appropriate magnitude to denote collinearity is precisely the same as that of adjudging when an R_i^2 is close enough

[7]See Theil (1971, p. 164).

[8]Expression (2.3) continues to hold for the uncentered VIF as long as the $\mathbf{X}_i$ in the denominator are also uncentered. In this instance the index i can also refer to that for the variance of the intercept term, i.e., for $i = 1$.

to unity, and the weakness of the R_i^2 for determining the number of near dependencies is similarly a weakness for VIFs.

A substantive contribution to the development of VIF-like measures has been made by Schall and Dunne (1988) in which they develop a factorization $\text{VIF}_i = \text{VIF}_i(\mathbf{X}_1) \cdot \text{VIF}_i(\mathbf{X}_2 | \mathbf{X}_1)$ of the VIF into traditional marginal and conditional components, where $\mathbf{X} = [\mathbf{X}_1 \; \mathbf{X}_2]$. This factorization has yet to be fully exploited, but it has the potential for making VIF-like measures much more useful. It does not, however, seem to overcome all of the VIF's shortcomings.

4. *Examine* $\det(\mathbf{X}^T\mathbf{X})$ *or* $\det(\mathbf{R})$. It is well known that a collinear relation among the columns of $\mathbf{X}$ will cause the determinant of $\mathbf{X}^T\mathbf{X}$ to become small, suggesting $\det(\mathbf{X}^T\mathbf{X})$ as a collinearity diagnostic. Unfortunately, the determinant has many lethal shortcomings in this role. First, how small must the determinant become before collinearity is shown? Second, even if it worked well otherwise, $\det(\mathbf{X}^T\mathbf{X})$ could not determine the number of linear dependencies among the columns of $\mathbf{X}$. The presence of any one linear dependency would make it small, thereby masking any others. Third, a small determinant may be necessary to collinearity, but it is not sufficient to it. This is seen through the simple example of a data matrix $\mathbf{X}$ with orthogonal columns of Euclidean length $\alpha^{1/2}$. In this case $\mathbf{X}^T\mathbf{X} \equiv \mathbf{A} = \alpha\mathbf{I}$, a scalar matrix whose determinant is $\det(\mathbf{A}) = \alpha^n$, where n is the size of $\mathbf{A}$. Clearly, α may be chosen to make $\det(\mathbf{A})$ as close to zero as is desired even though there is clearly no collinearity among the columns of $\mathbf{X}$.

This last problem is mitigated somewhat, but not eliminated, by using $\det(\mathbf{R})$. It is readily shown, for example, that $\det(\mathbf{R})$ lies between 0 and 1 (see Section 11.4) and is equal to 1 for all $\alpha > 0$ above. However, Stewart (1987) provides an excellent example showing that $\det(\mathbf{R})$ cannot be relied upon to be so well behaved in general. Furthermore, $\det(\mathbf{R})$ retains the first two problems given above for $\det(\mathbf{X}^T\mathbf{X})$. The determinant of $\mathbf{R}$ will also figure below in the procedure of Farrar and Glauber.

5. *Do All-Subsets Regression on the Columns of* $\mathbf{X}$. In all-subsets regression, each column of $\mathbf{X}$ is regressed on all possible combinations of the other columns. Clearly this is a very direct and brute-force means for examining the collinear relations that may exist among the columns of a data matrix $\mathbf{X}$ and their content. Such information would substantially augment that from VIFs, which only provide information on the strength of the top level of such regressions, namely, the R_i^2 that result from regressing each column of $\mathbf{X}$ on all the remaining columns.

Much can be learned from all-subsets regressions; the various R^2's key tight relations, and the various t-tests key variate involvement. But the procedure has obvious disadvantages. First, the computational costs are relatively high. While there are efficient algorithms for conducting all-subsets regression, the computational burden is still considerably larger than alternative techniques (such as that developed in the next chapter) that, in my experience, provide better information. Second, interpretive costs are very large. Even for modest sized p, say, 8–10, large amounts of output must be digested and compared. Third, the collinearity that is being diagnosed is itself capable of causing this procedure to misfire diagnostically. This occurs when there are several coexisting collinear relations among the columns of **X**, making it possible for some of these regressions to be on subsets of the data that are themselves collinear. This, of course, inflates variances and allows the t-statistics to misdiagnose variate involvement. The only way around this is first to determine which subsets of the data are free from collinear relations; that is, in order to work well as a collinearity diagnostic, all-subsets regression would itself benefit from a collinearity diagnostic.

6. *The Technique of Farrar and Glauber.* Farrar and Glauber (1967) propose a diagnostic technique that employs information both from **X**'s correlation matrix $\mathbf{R} \equiv (r_{ij})$ and its inverse $\mathbf{R}^{-1} \equiv (r^{ij})$. This procedure is based on an assumption that the rows of the $n \times p$ data matrix **X** are a sample of size n from a p-variate Gaussian (normal) distribution. Under the further assumption that **X** has orthogonal columns, they show that a transformation of det(**R**) is approximately χ^2 distributed and, hence, provides a measure of the deviation from orthogonality, or the presence of collinearity. In addition, they make use of the VIFs, though not by that name, as indicators of the variates involved. They further propose the use of the measure $r_{ij.} \equiv -r^{ij}/(\sqrt{r^{ii}}\sqrt{r^{jj}})$, that is, the partial correlation between $\mathbf{X}_i$ and $\mathbf{X}_j$, adjusted for all other **X** variates, to investigate the patterns of interdependence in greater detail. Initially the Farrar and Glauber diagnostic procedure had a large and growing impact in the statistical literature but has since shown itself prey to many interesting criticisms.

First, the use of **R**, det(**R**), and VIFs lends to this diagnostic the significant weaknesses of those measures as already discussed. Second, Haitovsky (1969) rather generally criticizes a diagnostic measure based on a deviation from orthogonality of the columns of **X** and suggests a widely accepted change in emphasis to a measure of deviation from perfect singularity. This suggestion reflects the fact that the Farrar and Glauber measure frequently indicates collinearity when, as a practical matter, no problem exists; that is, the measure seems to lack discriminatory power. While the Haitovsky modification seemingly strengthens the Farrar and Glauber process, other criticisms have proved more troublesome.

Third, Kumar (1975b) highlights the obvious fact that the Farrar and Glauber technique, in assuming the data matrix **X** to be stochastic, need not be relevant to the regression model in which **X** is assumed fixed and the less obvious fact that, even when **X** is stochastic, a truly rigorous derivation of the statistic employed would depend on the typically untenable assumption that the rows of the **X** matrix, that is, the different observations in a regression analysis, are independent.

Fourth, in quite another vein, O'Hagan and McCabe (1975) question the validity of Farrar and Glauber's "statistical" interpretation of a measure of collinearity, concluding that their procedure misinterprets the use of a t-statistic as providing a cardinal measure of the severity of collinearity. This criticism is correctly placed, for it is not proper, as some are wont, to interpret the Farrar and Glauber procedure as a statistical test for collinearity. Indeed, Farrar and

Glauber make no such claim for their procedure. A statistical test, of course, must be based on a *testable hypothesis*; that is, the probability for the outcome of a relevant test statistic, calculated at the actual sample data, is assessed in light of the distribution implied for it *by the model* under a specific null hypothesis. The Farrar and Glauber technique differs critically from this procedure exactly in the fact that the linear regression model makes no testable assumptions on the data matrix **X**. The data matrix **X** is, of course, assumed to have full rank, but this is not testable, for its absence, the null hypothesis, renders the regression model inestimable. Thus, there are no distributional implications of the linear regression model for specific null hypotheses on the nature of the data matrix **X**, such as orthogonality, within which tests can be made.

All this is not to say that one could not make additional assumptions about the generation of the **X** data and make tests about any "inherent singularity" that might characterize this process. But this would not only be extending the regression analysis far beyond the usual single-equation setting considered here, it would also be making tests about the singularity of the *process that generates* the **X**'s (the degeneracy of the variance–covariance matrix of the error process involved) and not about the singularity of the *specific* **X** matrix obtained. It is this latter that is the subject of a collinearity analysis.

The preceding highlights the point made in Section 2.1 (and that arises again in subsequent chapters) that collinearity can cause computational problems and reduce the precision of statistical estimates, but in the context of the linear regression model, collinearity is not itself a statistical phenomenon subject to statistical test. The solution to the problem of diagnosing collinearity, then, must be sought elsewhere, in methods that deal directly with the numerical properties of the data matrix that can cause calculations based on it to be unstable or sensitive (in ways to be discussed later). Once this collinearity diagnostic has been developed, then, in Chapter 7, we shall see a way for it to be adjoined to a test for signal-to-noise to produce a statistical test for harmful collinearity, that is, a test to determine whether any collinearity that is shown to be present is adversely affecting specific aspects of the statistical estimation procedure.

7. *Partial Correlations of* **X** *and the Correlation Matrix*, **Correl(b)**, *of the Least-Squares Estimator* **b**. Two notions that are often suggested as collinearity diagnostics are the partial correlations of the **X** data and **Correl(b)**, the correlation matrix of the least-squares estimator **b**. They are grouped together here because they are really the same thing.

The partial correlations between any two columns of **X** are simply the correlations after their linear dependence on the other columns of **X** has been accounted for.[9] The correlation matrix of the least-squares estimator **b** is

$$\mathbf{Correl(b)} \equiv \mathbf{D}^{-1}\mathbf{V}(\mathbf{b})\mathbf{D}^{-1}, \tag{2.5}$$

where $\mathbf{V}(\mathbf{b})$ is the variance–covariance matrix of **b**,

$$\mathbf{V}(\mathbf{b}) \equiv \sigma^2(\mathbf{X}^{\mathrm{T}}\mathbf{X})^{-1}, \tag{2.6}$$

[9]See Theil (1971, pp. 171–175).

and where $\mathbf{D}^2 \equiv \text{diag}(\mathbf{V}(\mathbf{b}))$, that is, a diagonal matrix whose diagonal elements are those of $\mathbf{V}(\mathbf{b})$. We are no longer assuming here that $\mathbf{X}$ has been centered or scaled.

It is a straightforward exercise to show that the partial correlation between $\mathbf{X}_i$ and $\mathbf{X}_j$ is the negative of the correlation between $\mathbf{b}_i$ and $\mathbf{b}_j$.[10] Thus, intuitively, we might say $\mathbf{X}_i$ and $\mathbf{X}_j$ are collinear if they have a strong positive (negative) partial correlation or, equivalently, if the correlation between the corresponding least-squares estimates $\mathbf{b}_i$ and $\mathbf{b}_j$ is close to -1 $(+1)$. The drawbacks for both of these measures are just like those for $\mathbf{R}$ and $\mathbf{R}^{-1}$: these measures are sufficient to collinearity but not necessary to it; they cannot diagnose the presence of several near dependencies that may coexist among the columns of $\mathbf{X}$; and there is no natural cutoff to indicate when either correlation is sufficiently close to unity (in absolute value) to denote collinearity.

The seriousness of the first of these drawbacks is illustrated by the following example based on the consumption function data. These data are given in Exhibit 5.10 and figure in several examples elsewhere. It is unnecessary for current needs to know much about them except that they are used in a regression of consumption $\mathbf{C}(T)$, on an intercept $\boldsymbol{\iota}$, lagged consumption $\mathbf{C}(T-1)$, disposable personal income $\mathbf{DPI}(T)$, the interest rate $\mathbf{r}(T)$, and the change in disposable personal income $\Delta\mathbf{DPI}(T)$. Exhibit 2.4 gives $\mathbf{Correl(b)}$ for this regression.

[10] This result, so well known in back-parlor gossip, seems strangely absent from statistics texts, so perhaps a formal demonstration is in order. In doing so, I apologize to de Combrugghe (1983), whose efforts in this regard were, I now see, incorrectly discouraged.

Scale $\mathbf{X}$ as $\tilde{\mathbf{X}} \equiv \sigma^{-1}\mathbf{XD}$ so $\mathbf{Correl(b)} = (\tilde{\mathbf{X}}^{\mathrm{T}}\tilde{\mathbf{X}})^{-1}$. Now consider $\text{Cor}(b_1, b_2)$, the 1, 2-element of $\mathbf{Correl(b)}$, by first partitioning $\tilde{\mathbf{X}} \equiv [\mathbf{x}\ \mathbf{Z}]$, where $\mathbf{x} \equiv [\mathbf{x}_1\ \mathbf{x}_2]$ is the first two columns of $\tilde{\mathbf{X}}$, and $\mathbf{Z}$ is the remaining columns. Then

$$(\tilde{\mathbf{X}}^{\mathrm{T}}\tilde{\mathbf{X}})^{-1} = \begin{bmatrix} \mathbf{x}^{\mathrm{T}}\mathbf{x} & \mathbf{x}^{\mathrm{T}}\mathbf{Z} \\ \mathbf{Z}^{\mathrm{T}}\mathbf{x} & \mathbf{Z}^{\mathrm{T}}\mathbf{Z} \end{bmatrix}^{-1},$$

whose top left-hand block, by applying the matrix-inverse formula, is

$$(\mathbf{x}^{\mathrm{T}}\mathbf{M}_z\mathbf{x})^{-1} = \begin{bmatrix} \mathbf{x}_1^{\mathrm{T}}\mathbf{M}_z\mathbf{x}_1 & \mathbf{x}_1^{\mathrm{T}}\mathbf{M}_z\mathbf{x}_2 \\ \mathbf{x}_2^{\mathrm{T}}\mathbf{M}_z\mathbf{x}_1 & \mathbf{x}_2^{\mathrm{T}}\mathbf{M}_z\mathbf{x}_2 \end{bmatrix}^{-1},$$

where $\mathbf{M}_z \equiv \mathbf{I} - \mathbf{Z}(\mathbf{Z}^{\mathrm{T}}\mathbf{Z})^{-1}\mathbf{Z}^{\mathrm{T}}$. By the preceding scaling, both these diagonal elements must be 1, so using the adjoint method for the inverse, we must have $\det(\mathbf{x}^{\mathrm{T}}\mathbf{M}_z\mathbf{x}) = (\mathbf{x}_1^{\mathrm{T}}\mathbf{M}_z\mathbf{x}_1)(\mathbf{x}_2^{\mathrm{T}}\mathbf{M}_z\mathbf{x}_2) - (\mathbf{x}_1^{\mathrm{T}}\mathbf{M}_z\mathbf{x}_2)^2 = \mathbf{x}_1^{\mathrm{T}}\mathbf{M}_z\mathbf{x}_1 = \mathbf{x}_2^{\mathrm{T}}\mathbf{M}_z\mathbf{x}_2$, and the off-diagonal element, which is $\text{Cor}(b_1, b_2)$, reduces to $-\mathbf{x}_1^{\mathrm{T}}\mathbf{M}_z\mathbf{x}_2/\mathbf{x}_2^{\mathrm{T}}\mathbf{M}_z\mathbf{x}_2$. Now consider $\text{PartCor}(\mathbf{x}_1, \mathbf{x}_2)$, the partial correlation between $\mathbf{x}_1$ and $\mathbf{x}_2$, which is unaffected by the above scaling. This is simply $\text{Cor}(\mathbf{e}_1, \mathbf{e}_2)$, where $\mathbf{e}_i$ is the residual of $\mathbf{x}_i$ regressed on $\mathbf{Z}$, i.e., $\mathbf{e}_i = \mathbf{M}_z\mathbf{x}_i$, $i = 1, 2$. Thus,

$$\text{PartCor}(\mathbf{x}_1, \mathbf{x}_2) = \text{Cor}(\mathbf{e}_1, \mathbf{e}_2) \equiv \frac{\mathbf{x}_1^{\mathrm{T}}\mathbf{M}_z\mathbf{x}_2}{(\mathbf{x}_1^{\mathrm{T}}\mathbf{M}_z\mathbf{x}_1)^{1/2}(\mathbf{x}_2^{\mathrm{T}}\mathbf{M}_z\mathbf{x}_2)^{1/2}} = \frac{\mathbf{x}_1^{\mathrm{T}}\mathbf{M}_z\mathbf{x}_2}{\mathbf{x}_2^{\mathrm{T}}\mathbf{M}_z\mathbf{x}_2} = -\text{Cor}(b_1, b_2),$$

recalling that $\mathbf{x}_1^{\mathrm{T}}\mathbf{M}_z\mathbf{x}_1 = \mathbf{x}_2^{\mathrm{T}}\mathbf{M}_z\mathbf{x}_2$. Since this result does not depend on the order of the columns in $\mathbf{X}$, we have $\mathbf{Correl(b)} = -\mathbf{PartCorrel(X)}$.

Exhibit 2.4 Correl(b): consumption function data

	ι	**C**$(T-1)$	**DPI**(T)	**r**(T)	Δ**DPI**(T)
ι	1.000	−.685	.608	.469	−.508
C$(T-1)$	−.685	1.000	−.990	−.098	.884
DPI(T)	.608	−.990	1.000	−.037	−.916
r(T)	.469	−.098	−.037	1.000	.179
Δ**DPI**(T)	−.508	.884	−.916	.179	1.000

From this we can see that the partial correlation between **r**(T) and **C**$(T-1)$ is quite small, −.098, and certainly would not give any indication that these two variates might be involved jointly in a collinear relation. If, however, we regress **r**(T) on ι, **C**$(T-1)$, and Δ**DPI**(T), we obtain (recalling that t-statistics are in square brackets)

$$\begin{aligned} \mathbf{r}(T) = &-1.024\iota + 0.017\mathbf{C}(T-1) - 0.014\,\Delta\mathbf{DPI}(T), \\ &[-3.9] \quad\; [22.3] \qquad\qquad\quad [-1.9] \\ &R^2 = .9945, \end{aligned} \tag{2.7}$$

demonstrating that **C**$(T-1)$ is in fact quite significantly involved in a linear dependency with **r**(T).

Thus, collinearity diagnostics based on partial correlations and **Correl(b)** are quite capable of overlooking variate involvement in more complicated collinear relations, and these measures must therefore be adjudged less than adequate for the purpose. By contrast, we shall see that this case is correctly handled by the diagnostics we develop later.

8. *Examine Bunch Maps.* Historical completeness certainly requires inclusion of Frisch's (1934) bunch-map analysis in this review. Frisch's technique of graphically investigating the possible relations among a set of data series is among the first major attempts to uncover the sources of near linear dependencies in economic data series. Frisch's work addresses itself to one of collinearity's important diagnostic problems, the location of near dependencies, but makes no attempt to deal with other important diagnostic problems, such as determining the degree to which regression results are degraded by collinearity's presence. Bunch-map analysis has not become a major diagnostic tool in regression analysis because its extension to dependencies among more than two variates is both time consuming and very subjective.

9. *Determine the Essential Flats of the Matroid of Dependencies in* **X**. From Frisch's historically early suggestion for diagnosing collinearity, we turn briefly to the most recent suggestion, the theory of matroids, proposed by Greene (1985, 1986a, b, 1987, 1988). While this technique is as untried as it is new, it is worth mentioning because it points to an entirely new direction for research that

would seem to be a highly promising means for examining the possible linear structures underlying the columns of a data matrix $\mathbf{X}$. The method is based on determining the matroid of (near) linear dependencies among the columns of $\mathbf{X}$ defined by applying cluster analysis (some best-fit criterion) to the myriad of combinations of columns taken from $\mathbf{X}$. From these, *essential flats* can be identified as the nonoverlapping subsets of columns that can be considered the underlying sources of near dependencies. At first blush, this smacks of being an elaborate means for conducting and analyzing something akin to all-subsets regression, but this remains to be seen, and this line of research is definitely worth pursuing and watching.

10. *Examine the Eigenvalues and Eigenvectors (or Principal Components) of* $\mathbf{X}^T\mathbf{X}$ *or* $\mathbf{R}$. For many years, the eigensystem (eigenvalues and eigenvectors) of the cross-product matrix $\mathbf{X}^T\mathbf{X}$ (or the closely related correlation matrix $\mathbf{R}$) has been suggested as a means for diagnosing and dealing with collinearity. For those not comfortable with these important concepts, we note that they will be developed later. For the moment, we need only know that the eigenvectors of $\mathbf{X}^T\mathbf{X}$ are the set of p nonzero vectors $\boldsymbol{\zeta}$ that obey $\mathbf{X}^T\mathbf{X}\boldsymbol{\zeta} = \lambda\boldsymbol{\zeta}$; that is, they are the vectors that get transformed into some scalar multiple λ of themselves (stretched if $|\lambda| > 1$ or shrunk if $|\lambda| < 1$) when multiplied by $\mathbf{X}^T\mathbf{X}$. The "stretch factor" λ is the eigenvalue that corresponds to the given eigenvector. The relation of this to collinearity is clear, for if an eigenvector $\boldsymbol{\zeta}$ has an eigenvalue equal to zero, then $\mathbf{X}^T\mathbf{X}\boldsymbol{\zeta} = \mathbf{0}$, or equivalently, $\mathbf{X}\boldsymbol{\zeta} = \mathbf{0}$, and we have determined an exact linear dependency among the columns of $\mathbf{X}$.

A concept closely related to the eigensystem of $\mathbf{X}^T\mathbf{X}$ is that of the principal components of $\mathbf{X}$. These are simply the set of linear combinations of the columns of $\mathbf{X}$ formed by the eigenvectors, that is, the $\mathbf{X}\boldsymbol{\zeta}$.[11] Kloeck and Mennes (1960) depict several ways to use the principal components of $\mathbf{X}$ or related matrices to reduce some ill effects of collinearity.

In a direction more useful for diagnostic purposes, Kendall (1957) and Silvey (1969) suggest using the presence of "small" eigenvalues of $\mathbf{X}^T\mathbf{X}$ to signal the presence of collinearity among the columns of $\mathbf{X}$. This suggestion simply extends the relation noted previously between a zero eigenvalue and exact collinearity to the case of a near dependency. Now for the small eigenvalue and its corresponding eigenvector, we have $\mathbf{X}\boldsymbol{\zeta} \approx \mathbf{0}$. This suggestion has an advantage over all the preceding ones in that it is capable of dealing with several coexisting collinear relations; there will be one small eigenvalue for each such near dependency. Unfortunately, as it stands, this idea lacks practical value because we are never informed how to determine what small is.

Part of this problem arises from the natural tendency to define small relative to the wrong standard, namely, zero. By contrast, Chatterjee and Price (1977) indicate, but without justification, that collinearity may exist if "one eigenvalue is small in relation to the others." Here, small is interpreted in relation to larger eigenvalues rather than to zero. This fundamental distinction is good and lies at

[11]Most definitions of principal components assume that the columns of $\mathbf{X}$ have been centered, but this restriction is often too limiting.

the heart of the method we discuss later, but its justification requires much additional theory and discussion. This additional material comes from the extremely rich literature in numerical analysis showing the relevance of the *condition number* of a matrix (a measure related to the ratio of the maximal to minimal eigenvalues) to problems akin to collinearity.

Some use of the eigenvalues, then, would seem to have value for diagnosing the presence of collinearity. For similar reasons, it would seem that the elements of the eigenvectors would have value in diagnosing which variates are involved in collinear relations. This is in fact the case, but not in the obvious way that first comes to mind. Thus, suppose there is a small eigenvalue so we have the relation $\mathbf{X}\zeta \approx \mathbf{0}$. The first obvious assumption is that it is the nonzero elements of ζ that indicate which columns of $\mathbf{X}$ are in the collinear relation and the zero elements that indicate which are not. Of course, it will almost never be the case for any given $\mathbf{X}$ matrix that any of the elements of ζ will be calculated to be exactly zero. So the next obvious assumption (Hocking, 1983) is that it is the small elements of ζ that indicate which columns of $\mathbf{X}$ do not belong to the collinear relation—as long as we are able, once again, to define what small is.

However, this issue never really becomes important here, for the assumption turns out to be faulty. It is possible for an element of ζ to be arbitrarily small even though the corresponding column of $\mathbf{X}$ still belongs to the collinear relation. In other words, it is not possible to diagnose the involvement of a column of $\mathbf{X}$ in a collinear relation by the smallness of its corresponding element in an eigenvector. The following example from Belsley (1984a) demonstrates this point.

Suppose we were to use eigenvalues and eigenvectors to investigate the possibility of collinear relations among the variates given in Exhibit 2.5. The usual practice is to scale each variate so that it has unit Euclidean length and then to find the eigenvalues and eigenvectors of the scaled $\mathbf{X}^T\mathbf{X}$. The scaling merely reflects the fact that, a priori, we have no reason to believe that all variates should not have equal weight in the analysis. Of the four eigenvalues for these data, the smallest is 1.777×10^{-7}, a value that most would (not necessarily correctly) deem small.[12] And the eigenvector that corresponds to this smallest eigenvalue is $\zeta = (0.7033, -0.7109, -0.0089, -0.0001)$. The first two elements here are large and the second two are small, indicating a collinear relation involving $\mathbf{X}_1$ and $\mathbf{X}_2$ but not $\mathbf{X}_3$ and $\mathbf{X}_4$. However, a simple regression of, say, $\mathbf{X}_1$ on the other three belies this conclusion. Here we find (again, t-statistics are in the square brackets)

$$\begin{aligned} \mathbf{X}_1 = 1.011&\mathbf{X}_2 + 0.013\mathbf{X}_3 + 0.0002\mathbf{X}_4, \\ [391.4]& \qquad [4.8] \qquad\quad [0.1] \\ R^2 &= .9990, \end{aligned} \tag{2.8}$$

[12]I say not necessarily correctly because we have no idea what the units may be here. If, for example, they were in trillions of U.S. dollars, this small figure would represent $1,777,000! This, perhaps, gives some indication of the difficulty of determining when small is small.

Exhibit 2.5 Data demonstrating the inefficacy of eigenvector elements for diagnosing variate involvement in collinear relations

Index	X_1	X_2	X_3	X_4
1	−0.654023	−0.732817	0.553252	0.429669
2	0.119577	0.138641	−0.477180	0.501280
3	0.583241	0.653593	−0.497927	−0.119353
4	−0.101979	−0.118835	0.456433	−0.716115
5	0.026982	0.029709	0.055325	−0.167094
6	0.035132	0.039612	−0.082988	0.095482

where it is clear that $\mathbf{X}_3$ does indeed significantly enter the collinear relation. Thus, for an eigenvector corresponding to a small eigenvalue, relatively large eigenvector elements may indicate the involvement of a variate in a collinear relation, but relatively small eigenvector elements cannot be relied upon to show the absence of involvement.[13] Happily, the diagnostics soon to be developed are able to handle this case correctly.

2.4 A BASIS FOR A DIAGNOSTIC

We have seen that none of the preceding approaches is fully successful in diagnosing the presence of collinearity and variate involvement or in assessing collinearity's potential harm. The basis for a successful diagnostic is, however, close at hand, for various concepts developed in the field of numerical analysis are capable of putting real meaning to the last of the measures discussed above, that based on the eigensystem.

While the attention of numerical analysts has rarely been focused specifically on the problem of collinearity,[14] it has been directed at closely related topics. Numerical analysts, for example, are interested in the properties (conditioning) of a matrix $\mathbf{A}$ of a linear system of equations $\mathbf{Az} = \mathbf{c}$ that allow a solution for $\mathbf{z}$ to be obtained with numerical stability, or equivalently, the conditions under which the (generalized) inverse of $\mathbf{A}$ can be so obtained. The relevance of this to collinearity in the least-squares context is readily apparent, for the least-squares

[13]This example illustrates the fact that the eigenvector elements are discontinuous indicators of variate involvement. Clearly, a variate is not involved if its corresponding eigenvector element equals zero, but a variate may be involved even if its corresponding eigenvector element becomes arbitrarily close to zero. This discontinuity is fully demonstrated from the $n \times 3$ rank-deficient matrix $\mathbf{X} = [\mathbf{X}_1\ \mathbf{X}_2\ \mathbf{X}_3]$ (with unit-length columns) having the relation $\mathbf{X}_2 = -m^{-1}(\mathbf{X}_1 + \varepsilon\mathbf{X}_3)$, $m = \|\mathbf{X}_1 + \varepsilon\mathbf{X}_3\|$. Clearly, for all $\varepsilon \neq 0$, (a) $\mathbf{X}_3$ belongs to the relation and (b) $\zeta = (1, m, \varepsilon)^T$ is an eigenvector (not normalized) of $\mathbf{X}^T\mathbf{X}$ corresponding to the zero eigenvalue. Thus, by letting ε become arbitrarily close to zero, we can make the element of the eigenvector ζ that corresponds to $\mathbf{X}_3$ arbitrarily small even though $\mathbf{X}_3$ remains an integral part of the linear dependency among the three columns.

[14]See, however, Stewart (1987).

estimator is a solution to the linear system of (normal) equations $(\mathbf{X}^T\mathbf{X})\mathbf{b} = \mathbf{X}^T\mathbf{y}$. Furthermore, the variance–covariance matrix of this estimator is $\sigma^2(\mathbf{X}^T\mathbf{X})^{-1}$. To the extent, then, that collinearity among the columns of the data matrix $\mathbf{X}$ results in a matrix $\mathbf{A} \equiv \mathbf{X}^T\mathbf{X}$ whose ill conditioning causes both the solution for $\mathbf{b}$ and its variance–covariance matrix to be numerically unstable, the techniques of the numerical analyst have direct bearing on understanding the econometrician's and statistician's problems with collinearity. The important efforts in numerical analysis relevant to this study are contained in Businger and Golub (1965), Golub and Reinsch (1970), Golub and Van Loan (1983), Golub et al. (1976, 1980), Hanson and Lawson (1969), Lawson and Hanson (1974), Stewart (1973, 1987), and Wilkinson (1965).

Few of the techniques of the numerical analyst have been directly absorbed in applied statistical fields such as econometrics. This is almost as true today as when I first wrote it in Belsley, Kuh, and Welsch (1980). And it remains a phenomenon that is difficult to explain since one of the principal tools of numerical analysis, the singular-value decomposition (SVD) of the data matrix $\mathbf{X}$, has an intimate connection to the eigensystem of $\mathbf{X}^T\mathbf{X}$, a tool of great importance in statistical theory. This connection will be discussed in detail in the next chapter.

The lack of communication among these various disciplines is explained in part by the awkward differences in notation that were examined in Chapter 1 and in part by seemingly different interests. Numerical analysts tend, for example, to work with nonstochastic equation systems. Furthermore, they have placed much of their emphasis on determining which columns of a data matrix can be discarded with least sacrifice to subsequent analysis,[15] a solution that is rarely open to the econometrician or applied statistician whose theory has already determined those variates that must be present—or those that may be deleted, but not on grounds of numerical stability. Nevertheless, as we shall see, the numerical analysts' techniques, fused with the previously described Kendall–Silvey line of research employing eigenvalues, have much to offer the user of least squares in diagnosing collinearity.

We recall from above that Silvey concludes that collinearity is discernable from the presence of a small eigenvalue of $\mathbf{X}^T\mathbf{X}$, a fact first noted by Kendall. This conclusion is correct, but it falls short of the mark since Silvey fails to inform us how to determine when an eigenvalue is small. In the next chapter, we show that the numerical-analytic notion of the condition number can be applied to correct this shortcoming. Furthermore, Silvey provides the basis for diagnosing the involvement of individual explanatory variates; namely, he examines a decomposition of the estimated variance of each regression coefficient in a manner that can illuminate the degradation of each coefficient caused by collinear relationships. Silvey, however, fails to recognize and exploit this use of the decomposition.

In the next chapter, then, we (i) apply the relevant techniques of numerical

[15] See, e.g., Hawkins (1973), Golub et al. (1976), or Webster et al. (1974, 1976).

analysis to Silvey's suggestion in order to provide a set of indexes (condition indexes)[16] that signal the presence of one or more near dependencies among the columns of **X** and (ii) adapt the Silvey variance decomposition in a manner that can be combined with the preceding indexes to produce a set of measures (the variance–decomposition proportions) for discovering those variates that are involved in particular near dependencies and for assessing the degree to which the estimated coefficients are being degraded by the presence of the near dependencies.

[16]The term *condition indexes* is not to be confused with Stewart's (1987) *collinearity indices*, which, as we have seen, are a transform of the VIFs or, equivalently, of the R_i^2.

CHAPTER 3

A Collinearity Diagnostic

The various suggestions for collinearity diagnostics described in the preceding chapter are all seen to fall short in one or more important respects of being able to provide information on either the strength or number of collinear relations among the columns of the data matrix **X** or on the variates involved in each.[1] However, the last of these suggestions, that based on the eigensystem, is able to be extended into a collinearity diagnostic that does both: the eigenvalues can be used to form a set of condition indexes that allow us to determine the strength and number of near dependencies, and the eigenvalues and eigenvectors together can be used to form a set of variance–decomposition proportions that allow us to determine variate involvement. This information can also be used to determine which least-squares estimates have been degraded by the presence of collinearity. The material needed for this extension and the resulting collinearity diagnostic are the subject of this chapter.

3.1 THE CONDITION INDEXES: DETERMINING THE NUMBER OF COLLINEAR RELATIONS

To determine whether there are collinear relations among the columns $\mathbf{X}_i$ of an $n \times p$ data matrix $\mathbf{X} \equiv [\mathbf{X}_1 \cdots \mathbf{X}_p]$, we seek to put meaning to the question of whether there is a linear combination of these columns that is in some sense "tight." That is, we ask if there are values $\mathbf{v} \equiv (v_1, \ldots, v_p)$, not all zero, such that

$$v_1\mathbf{X}_1 + \cdots + v_p\mathbf{X}_p = \mathbf{a} \quad \text{or} \quad \mathbf{X}\mathbf{v} = \mathbf{a} \tag{3.1}$$

for an **a** that is somehow deemed small. Clearly no problem arises in the event that $\mathbf{a} = \mathbf{0}$, for then we have exact collinearity and the issue of small is well resolved. But if $\mathbf{a} \neq \mathbf{0}$, as will almost always be the case in practice, we must give

[1]A possible exception to this is that based on the matroid of dependencies currently being developed by Greene (1988).

ourselves a meaningful notion of small to gain a meaningful definition of inexact collinearity.

As a first attempt, we might say that a near dependency exists if the vector **a** in (3.1) has little length, that is, if $\|\mathbf{a}\|$ is small. This suggestion clearly begs the issue, but let us nevertheless follow its lead for the moment, for it can eventually put us in the right direction.

Eigenvalues and Eigenvectors of $\mathbf{X}^{\mathrm{T}}\mathbf{X}$

Finding a set of v's that makes the length of **a** small in (3.1) is not a well-defined question in itself, for the v's may be scaled together toward zero to make the length of **a** arbitrarily small. We therefore must normalize the problem, which we do by considering only **v** with unit length, that is, with

$$\|\mathbf{v}\| = 1. \tag{3.2}$$

We then seek the **v** that makes $\|\mathbf{a}\|$ smallest subject to (3.2), or equivalently,

$$\min_{\mathbf{v}} \mathbf{v}^{\mathrm{T}}\mathbf{X}^{\mathrm{T}}\mathbf{X}\mathbf{v} \quad \text{subject to } \mathbf{v}^{\mathrm{T}}\mathbf{v} = 1. \tag{3.3}$$

The solution to this problem is related to the well-known eigenvalue problem, which we briefly review here. Those wishing greater detail are referred to such texts as Theil (1971), Johnston (1984), or Strang (1980).

Thus, forming the Lagrangian

$$L(\mathbf{v}, \lambda) \equiv \mathbf{v}^{\mathrm{T}}\mathbf{X}^{\mathrm{T}}\mathbf{X}\mathbf{v} + \lambda(1 - \mathbf{v}^{\mathrm{T}}\mathbf{v}) \tag{3.4}$$

of the constrained minimum (3.3), taking the partial derivatives with respect to **v**, and setting them equal to zero produces as necessary conditions

$$\mathbf{X}^{\mathrm{T}}\mathbf{X}\mathbf{v} = \lambda\mathbf{v}. \tag{3.5}$$

From this we see that the vector **v** that we seek, when multiplied by $\mathbf{X}^{\mathrm{T}}\mathbf{X}$, must become some multiple λ of itself. Such a vector (other than **0**) is called an eigenvector (characteristic vector) of $\mathbf{X}^{\mathrm{T}}\mathbf{X}$, and the multiple λ is called its corresponding eigenvalue (characteristic value). The set of eigenvalues of $\mathbf{X}^{\mathrm{T}}\mathbf{X}$ are those values of λ that allow (3.5) to be solved, that is, those for which the homogeneous set of equations $(\mathbf{X}^{\mathrm{T}}\mathbf{X} - \lambda\mathbf{I})\mathbf{v} = \mathbf{0}$ has a nontrivial solution. But this can only happen if $\mathbf{X}^{\mathrm{T}}\mathbf{X} - \lambda\mathbf{I}$ is singular, for otherwise $\mathbf{X}^{\mathrm{T}}\mathbf{X} - \lambda\mathbf{I}$ is invertible, and the only solution would be the trivial one, $\mathbf{v} = \mathbf{0}$. Thus, we must have $\det(\mathbf{X}^{\mathrm{T}}\mathbf{X} - \lambda\mathbf{I}) = 0$. This determinantal equation (also called the characteristic equation) can be shown to be a pth-order polynomial in λ, having p roots in the complex field, so there are p (not necessarily distinct) eigenvalues of $\mathbf{X}^{\mathrm{T}}\mathbf{X}$ and p corresponding eigenvectors. But because $\mathbf{X}^{\mathrm{T}}\mathbf{X}$ is a nonnegative definite, real symmetric matrix, these eigenvalues can be shown to be real and nonnegative, and the eigenvectors can be shown to be real.[2]

Of course, we do not want just any eigenvector and eigenvalue; we want the one that corresponds to $\|\mathbf{a}\|$ with minimum length. But by premultiplying (3.5) by $\mathbf{v}^T$ and using (3.2), we see that $\|\mathbf{a}\|^2 = \lambda$. That is, the square of the magnitude we wish to minimize must be an eigenvalue of $\mathbf{X}^T\mathbf{X}$, so the $\mathbf{v}$ that produces the $\mathbf{a}$ with minimum length must be the eigenvector of $\mathbf{X}^T\mathbf{X}$ that corresponds to the smallest eigenvalue. Clearly, the minimum length $\|\mathbf{a}\|$ is simply the positive square root of this smallest eigenvalue. This is what prompted Kendall, as noted in the previous chapter, to suggest using small eigenvalues as evidence of collinearity.

Let us call the $\mathbf{v}$, λ, and $\mathbf{a}$ just derived $\mathbf{v}_1$, λ_1, and $\mathbf{a}_1$, respectively, so we can proceed to ask for the $\mathbf{v}_2$ producing the linear combination of $\mathbf{X}$ having the next smallest length $\|\mathbf{a}_2\|$ subject to (3.2) and the additional constraint that it be orthogonal to $\mathbf{v}_1$. This is clearly the eigenvector of $\mathbf{X}^T\mathbf{X}$ corresponding to the next smallest eigenvalue, λ_2, and having length $\|\mathbf{a}_2\| = \lambda_2^{1/2}$. Continuing in this way, we can find a set of p mutually orthogonal eigenvectors $\mathbf{v}_1, \ldots, \mathbf{v}_p$ of $\mathbf{X}^T\mathbf{X}$, each of unit length, and the corresponding set of p eigenvalues $\lambda_1, \ldots, \lambda_p$. The p eigenvectors can be joined into the $p \times p$ orthonormal matrix $\mathbf{V} \equiv [\mathbf{v}_1 \cdots \mathbf{v}_p]$, and constructing the diagonal matrix $\mathbf{\Lambda} \equiv \text{diag}(\lambda_1, \ldots, \lambda_p)$, we have

$$\mathbf{X}^T\mathbf{X}\mathbf{V} = \mathbf{V}\mathbf{\Lambda}. \tag{3.6}$$

It should be noted that the eigenvectors are not uniquely determined if there are multiplicities in the eigenvalues, that is, if several of the p eigenvalues have the same magnitude. When $\mathbf{X}$ is a data matrix whose elements have been generated by a stochastic mechanism that is nondegenerately distributed, equal eigenvalues will occur with probability zero, but it is still possible for eigenvalues to be close. We shall see that this can add some minor complexities (competing near dependencies) in interpreting diagnostic information gotten from the eigenvectors.

The Singular-Value Decomposition of X

A notion very closely allied to that of the eigensystem of $\mathbf{X}^T\mathbf{X}$ is the singular-value decomposition (SVD) of the matrix $\mathbf{X}$. We shall use this decomposition extensively in this work and so review it here.

Any $n \times p$ matrix $\mathbf{X}$ $(n \geqslant p)$ may be decomposed as

$$\mathbf{X} = \mathbf{U}\mathbf{D}\mathbf{V}^T, \tag{3.7}$$

where $\mathbf{U}^T\mathbf{U} = \mathbf{V}^T\mathbf{V} = I_p$ and $\mathbf{D}$ is diagonal with nonnegative diagonal elements

[2]Johnston (1984).

$\mu_1, \ldots, \mu_p$, called the *singular values* of **X**.[3] This is a very powerful decomposition, for its three components are matrices with very special, highly exploitable properties: **U** is column orthogonal, **V** is both row and column orthogonal so $\mathbf{V}^T\mathbf{V} = \mathbf{V}\mathbf{V}^T = \mathbf{I}_p$, and **D** is nonnegative and diagonal. Here, we assume **U** is $n \times p$ (the same size as **X**), **D** is $p \times p$, and **V** is $p \times p$, but other configurations for this decomposition are possible and may prove more useful in other applications.[4] It is clear from examining (3.7) that the columns of **X** are linear combinations of the columns of **U** and the rows of **X** are linear combinations of the columns of **V**. Thus, the columns of **U** are an orthogonal basis for the column space of **X** (the p-dimensional subspace of $\mathscr{R}^n$ spanned by the columns of **X**), and the columns (and rows) of **V** are an orthogonal basis for the row space of **X**.

The singular values from this decomposition are unique (though not necessarily distinct). The columns of **V** are uniquely determined up to order and sign if the singular values are all distinct. However, when there are singular values with multiplicity greater than 1, the orthonormal columns of **V** corresponding to these singular values may be chosen in numerous ways. For reasons that will become clear shortly, this is precisely the same indeterminacy affecting the eigenvectors of $\mathbf{X}^T\mathbf{X}$ when there are eigenvalues with multiplicities greater than 1.

The Relation of the Singular-Value Decomposition of X *to the Eigensystem of* $\mathbf{X}^T\mathbf{X}$

It is clear that the singular-value decomposition of **X** is very closely related to the eigensystem of $\mathbf{X}^T\mathbf{X}$. From (3.7) we see that $\mathbf{X}^T\mathbf{X} = \mathbf{V}\mathbf{D}\mathbf{U}^T\mathbf{U}\mathbf{D}\mathbf{V}^T = \mathbf{V}\mathbf{D}^2\mathbf{V}^T$ or, postmultiplying by **V**, that $\mathbf{X}^T\mathbf{X}\mathbf{V} = \mathbf{V}\mathbf{D}^2$. Comparing this to (3.6), we recognize that the orthonormal matrix **V** from the singular-value decomposition of **X** is a matrix whose columns are eigenvectors of $\mathbf{X}^T\mathbf{X}$—this is the reason that the symbol **V** is used in both (3.6) and (3.7)—and that the diagonal matrix **D** of singular values from the singular-value decomposition of **X** is a matrix whose diagonal elements are the positive square roots of the eigenvalues of $\mathbf{X}^T\mathbf{X}$. It is also straightforward to show that the columns of **U** comprise a subset of p of the n eigenvectors of $\mathbf{X}\mathbf{X}^T$ that correspond to p of its n eigenvalues, including all of the nonzero ones.

The singular-value decomposition of the matrix **X**, therefore, provides information that encompasses that given by the eigensystem of $\mathbf{X}^T\mathbf{X}$. As a practical matter, however, there are reasons for preferring the use of the singular-value decomposition. First, it applies directly to the data matrix **X** that

[3]See, e.g., Golub (1969), Golub and Reinsch (1970), Stewart (1973), and Becker et al. (1974). On some related computational issues see Laub and Klema (1980).

[4]Thus, one may have

$$\underset{n\times p}{\mathbf{X}} = \underset{n\times n}{\mathbf{U}}\ \underset{n\times p}{\mathbf{D}}\ \underset{p\times p}{\mathbf{V}^T} \quad \text{or} \quad \underset{n\times p}{\mathbf{X}} = \underset{n\times r}{\mathbf{U}}\ \underset{r\times r}{\mathbf{D}}\ \underset{r\times p}{\mathbf{V}^T}$$

where $r = \text{rank}(\mathbf{X})$. In this latter formulation, **D** is always of full rank, even when **X** is not.

is the focus of our concern and not to the cross-product matrix $\mathbf{X}^T\mathbf{X}$. Second, the lengths $\|\mathbf{a}\|$ of the linear combinations (3.1) of $\mathbf{X}$ that we sought to minimize above—and the notion of the condition number that will shortly be used to give meaning to these lengths—are both properly defined in terms of the square roots of the eigenvalues of $\mathbf{X}^T\mathbf{X}$, that is to say, the singular values of $\mathbf{X}$. Third, the concept of the singular-value decomposition has many practical and analytical uses in statistics, econometrics, and data analysis, so it is worthwhile taking this opportunity to extend knowledge of it to a wider audience than numerical analysts. And fourth, whereas the eigensystem and the singular-value decomposition seem, for our purposes, to be mathematically equivalent, computationally they are not. The algorithms that exist for computing the singular-value decomposition are numerically far more stable than those for computing the eigensystem of $\mathbf{X}^T\mathbf{X}$. This is true for the case that is of central interest here, where $\mathbf{X}$ is ill conditioned. Furthermore, in operating directly on the $n \times p$ data matrix $\mathbf{X}$, the singular-value decomposition avoids the additional computational burden of forming $\mathbf{X}^T\mathbf{X}$, an operation involving np^2 unneeded sums and products and providing an unnecessary source of truncation error.

Singular-Value Decomposition Algorithms

As a practical matter, then, the collinearity diagnostics we discuss should be carried out using the computationally stable algorithm for the singular-value decomposition of $\mathbf{X}$ rather than an algorithm for determining the eigenvalues and eigenvectors of $\mathbf{X}^T\mathbf{X}$. FORTRAN routines that perform the singular-value decomposition are readily available, most allowing for tolerances to be set to operate optimally on a wide variety of machines. These include subroutines to be found in EISPACK II (Smith et al., 1976, Garbow et al., 1977) and LINPACK (Dongarra et al., 1979). Furthermore, the *Mathematica* application available for the Macintosh, the Sun, and various other personal computers incorporates a flexible singular-value decomposition routine. The C source code for a singular-value decomposition routine is available in Vetterling and Press (1988) or from the author. Whereas, in a pinch, an eigensystem algorithm is better than nothing, we continue to use the singular-value decomposition as our main tool to remind us that there are better ways to proceed.

***The Nature of* X**

Not surprisingly, the preceding has made little mention of the nature of the matrix $\mathbf{X}$ since the singular-value decomposition can be applied to any matrix. In our applications, however, $\mathbf{X}$ will be a data matrix, presumably relevant to a least-squares analysis, and the singular-value decomposition will be employed for the purpose of producing meaningful collinearity diagnostics. These circumstances typically require certain restrictions on the form of $\mathbf{X}$, some of which we mention now. This issue will be dealt with much more fully in Chapter 6.

First, the columns of $\mathbf{X}$ should be column equilibrated; that is, they should be scaled to have equal length. This is usually effected by scaling each column of $\mathbf{X}$ to have unit length, although there is nothing special about unity here and any positive value would serve as well. Second, the columns of **X** should not be mean centered, and if the data are relevant to a model with a constant term, **X** should contain a column of ones (before column equilibration) to represent the intercept.[5] This is true whether or not mean-centered data have been otherwise employed, say, for estimation purposes. We shall see that the use of mean-centered data can mask the role played by the constant term in any underlying near dependencies and, despite the myriad opinions you may have heard to the contrary, produce misleading diagnostics. Third, the data matrix **X** should be relevant to a truly linear regression model, that is, one linear in both parameters and variates. Diagnostics relevant to nonlinear models and estimators other than least squares are treated in Chapters 9 and 11. In at least one important instance, that of models with logarithms, collinearity is readily diagnosed with minor modifications to the diagnostics developed in this chapter.

There is no restriction that the columns of **X** represent only exogenous variates in a regression analysis. Thus, **X** may contain lagged values of the response variate **y** as well as other predetermined variates.

Exact Linear Dependencies: Rank Deficiency

In the first instance, let us make use of the singular-value decomposition to reexamine a matrix **X** (which may contain a constant column of ones) that has $p - r$ exact linear dependencies among its columns so that $\text{rank}(\mathbf{X}) = r < p$. Since in the singular-value decomposition (3.7) of **X** the matrices **U** and **V** are each column orthogonal (and hence are necessarily of full rank), we must have $\text{rank}(\mathbf{X}) = \text{rank}(\mathbf{D})$. Thus, there will be exactly as many zero elements along the diagonal of **D** as the nullity of **X**, namely, $p - r$, and the singular-value decomposition may be partitioned as

$$\mathbf{X} = \mathbf{U}\mathbf{D}\mathbf{V}^{\mathrm{T}} = \mathbf{U}\begin{bmatrix} \mathbf{D}_{11} & \mathbf{0} \\ \mathbf{0} & \mathbf{0} \end{bmatrix}\mathbf{V}^{\mathrm{T}}, \tag{3.8}$$

where $\mathbf{D}_{11}$ is $r \times r$ and nonsingular. Postmultiplying by **V** and further partitioning, we obtain

$$\mathbf{X}[\mathbf{V}_1 \;\; \mathbf{V}_2] = [\mathbf{U}_1 \;\; \mathbf{U}_2]\begin{bmatrix} \mathbf{D}_{11} & \mathbf{0} \\ \mathbf{0} & \mathbf{0} \end{bmatrix}, \tag{3.9}$$

where $\mathbf{V}_1$ is $p \times r$, $\mathbf{V}_2$ is $p \times (p - r)$, $\mathbf{U}_1$ is $n \times r$, and $\mathbf{U}_2$ is $n \times (p - r)$. Equation (3.9) results in two matrix equations:

$$\mathbf{X}\mathbf{V}_1 = \mathbf{U}_1\mathbf{D}_{11}, \tag{3.10}$$

$$\mathbf{X}\mathbf{V}_2 = \mathbf{0}. \tag{3.11}$$

[5]That is, the data matrix **X** should have a column of ones or *its equivalent*. If, e.g., **X** contains a set of dummy or other variates that add up to a column of ones, this is equivalent to having a constant term in the model, and it is neither necessary nor proper to have a constant column as well.

Interest centers on (3.11), for it displays all of the exact linear dependencies among the columns of $\mathbf{X}$. The $p \times (p - r)$ matrix $\mathbf{V}_2$ provides an orthonormal basis for the null space associated with the columns of $\mathbf{X}$.

If, then, $\mathbf{X}$ possessed $p - r$ exact linear relations among its columns and if computers possessed exact arithmetic, there would be exactly $p - r$ zero singular values along the diagonal of $\mathbf{D}$, and the variates involved in each of these dependencies would be determined by the nonzero elements of $\mathbf{V}_2$ in (3.11).

Needless to say, in most statistical applications, the interrelations among the columns of $\mathbf{X}$ are not exact dependencies, and computers deal in finite, not exact, arithmetic. Exact zeros for the singular values or for the elements of $\mathbf{V}_2$ will therefore rarely, if ever, occur. In general, then, it will be difficult to determine the nullity of $\mathbf{X}$ (as determined by zero μ's) or those columns of $\mathbf{X}$ that do not enter into specific linear relationships (as determined by the zeros of $\mathbf{V}_2$). This difficulty was amply illustrated in Chapter 2 in the exercise surrounding the data in Exhibit 2.5. Nevertheless, as we have seen, each linear dependency among the columns of $\mathbf{X}$ will manifest itself in a small singular value μ (or eigenvalue λ), justifying the Kendall–Silvey notion that the presence of collinearity is revealed by the existence of a small eigenvalue. The question is to determine what is small.

Determining Small Singular Values

We begin this discussion with some geometric considerations that, while not rigorous, strongly sharpen our intuition about what is needed to measure a small singular value. We then provide a rigorous underpinning for this notion with the condition number of a matrix.

Two Geometric Aids

There are two geometric notions associated with a matrix $\mathbf{X}$ that help to demonstrate that singular values should be judged to be small only relative to other singular values and not relative to zero. The first of these is the volume of the parallelepiped associated with $\mathbf{X}$—or, more simply, the volume of $\mathbf{X}$—, and the second is the ellipse associated with $(\mathbf{X}^T\mathbf{X})^{-1}$.

The Volume of $\mathbf{X}$

Since the p columns of $\mathbf{X}$ are each vectors in an n-dimensional space, we can plot them and use them to generate a p-sided parallelepiped. This is exemplified in Exhibit 3.1 for an $n \times 3$ matrix $\mathbf{X} = [\mathbf{X}_1 \;\; \mathbf{X}_2 \;\; \mathbf{X}_3]$. Thus, this parallelepiped and its volume, which we shall denote by $\mathrm{Vol}(\mathbf{X})$, are one means for interpreting the matrix $\mathbf{X}$ geometrically. While we cannot draw such volumes for $p > 3$, we can certainly conceive of such figures quite generally.

The length of the ith side of this parallelepiped is just $\|\mathbf{X}_i\|$, the length of the ith column of $\mathbf{X}$. For a given orientation of the $\mathbf{X}_i$'s (given angles between them), the volume may be made large or small simply by rescaling some $\mathbf{X}_i$ to have longer or shorter lengths. But we can standardize the problem by scaling all the

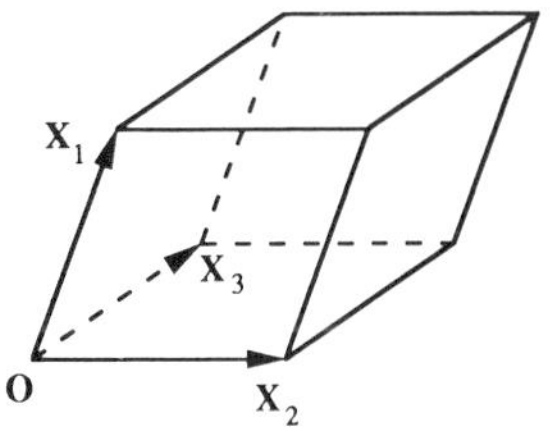

Exhibit 3.1 Parallelpiped corresponding to $\mathbf{X} = [\mathbf{X}_1\ \mathbf{X}_2\ \mathbf{X}_3]$.

columns of **X** to have the same length, say, unity. When this is done, the maximum volume that can be associated with a matrix **X** is 1, and this clearly occurs when the columns of **X** are mutually orthogonal, so that the geometric figure is a rectangular parallelepiped. This is shown in Exhibit 3.2*a*. Intuitively, we feel such a matrix is well conditioned; each column is allowing its weight fully to be felt.

By contrast, a matrix whose parallelepiped is like that shown in Exhibit 3.2*b* has small volume (even though its sides are still individually of unit length) because one of the sides is squeezed toward the other two. Such a matrix is less well conditioned. And in Exhibit 3.2*c*, we see a worse case yet, where two sides are each close to the third. Both of the matrices depicted in Exhibits 3.2*b*, *c* are ill conditioned because their volumes are small, but a distinction can certainly be made between these two cases since there are two sources to this smallness in the case of Exhibit 3.2*c*, whereas there is only one in Exhibit 3.2*b*. It should be noted that the terms *ill conditioned* and *well conditioned* have not yet been well defined, and their use here is loose, in keeping with the intuitive spirit of this discussion.

We can use the singular-value decomposition to reinterpret the volume of **X** in an illuminating way. First, we note that an orthogonal matrix corresponds to a linear transformation that preserves lengths and angles. Thus, let **C** be a $q \times q$ orthogonal matrix (so $\mathbf{C}^T\mathbf{C} = \mathbf{I}$), and let **x** and **z** be q-vectors with $\mathbf{z} \equiv \mathbf{Cx}$. Then $\|\mathbf{z}\| = (\mathbf{z}^T\mathbf{z})^{-1/2} = (\mathbf{x}^T\mathbf{C}^T\mathbf{C}\mathbf{x})^{-1/2} = (\mathbf{x}^T\mathbf{x})^{-1/2} = \|\mathbf{x}\|$. A similar exercise using (1.13) will convince the reader that the angle ϕ between two vectors **x** and **y** is also invariant to an orthogonal transformation of both vectors. Orthogonal

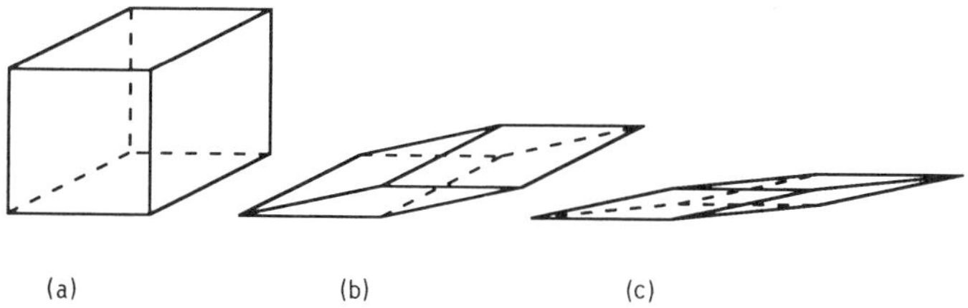

Exhibit 3.2 Parallelepipeds of **X** matrices: (*a*) well-conditioned matrix; (*b*) matrix with one deficient dimension; (*c*) matrix with two deficient dimensions.

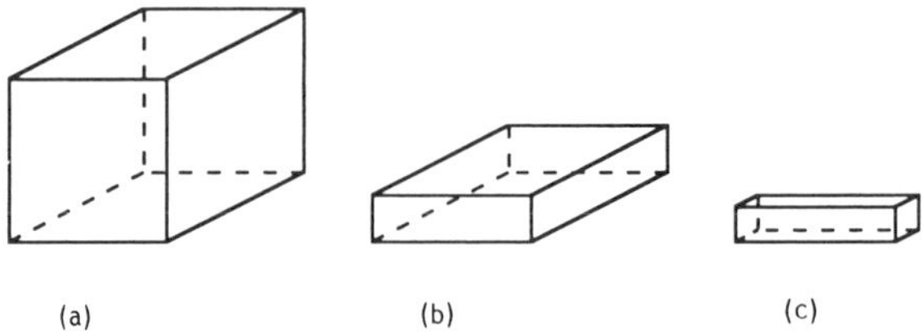

Exhibit 3.3 Rectangular parallelepipeds of corresponding **D** matrices.

transformations, then, rotate vectors but leave their relative lengths and orientations unchanged. We use this fact to note from (3.7) that $\text{Vol}(\mathbf{X}) = \text{Vol}(\mathbf{UDV}^T) = \text{Vol}(\mathbf{D})$, this last equality holding since the orthogonal transforms **U** and **V** merely rotate the parallelepiped of **D** and therefore do not change its volume.

Thus, the volume of **X** is the same as that of **D**. But Vol(**D**) is particularly easy to determine since **D** is a diagonal matrix whose columns are necessarily mutually orthogonal and have lengths that are simply the singular values of **X**. The matrix **D**, therefore, forms a rectangular parallelepiped, and its volume, as well as that of **X**, is just the product of the lengths of its sides, that is, the product of the singular values.[6] Exhibit 3.3 shows the parallelepipeds of the **D**'s corresponding to the **X**'s of Exhibit 3.2. Clearly, the **X** corresponding to the **D** in Exhibit 3.3a shows good measure in all dimensions. That for Exhibit 3.3*b*, however, is deficient in one dimension and that for Exhibit 3.3*c* in two.

Thus, this exercise in geometric intuition strongly suggests considering a matrix **X** to be ill conditioned when one side of its corresponding **D** is short relative to another, say, the longest, or, equivalently, when one of its singular values becomes small relative to its largest singular value. This in turn suggests something even more important for our needs, namely, that the appropriate means for assessing the smallness of a singular value is not its smallness absolutely but rather its smallness relatively—in relation to the largest singular value.

We can see the virtue in this suggestion by recalling that, in determining the volume of **X**, each of its columns is scaled to have unit length. This allows us to see that the volume of the resulting parallelepiped can become small even though its generating sides are all of equal length and thus to associate this reduced volume with ill conditioning. But this scaling is completely arbitrary; we could have chosen any equal length. And were we to halve all the column lengths, it is clear that the shape of the resulting parallelepiped would remain unchanged but all its dimensions would halve. Thus, the ill conditioning of **X** as

[6] This product is also det(**D**). Indeed this exercise can be generalized very easily to show that the absolute value of the determinant of any square matrix is the volume of the parallelepiped formed by the columns (or rows) of that matrix, and to show that, for any $n \times p$ ($n \geqslant p$) matrix **X**, $\det(\mathbf{X}^T\mathbf{X})$ is the square of the volume of the parallelepiped formed by the columns of **X**.

judged by the shape of its parallelepiped has nothing to do with the absolute dimensions of the parallelepiped but rather with its relative dimensions, and so a small singular value must be judged to be so relative to a large one.

The Ellipse of $\mathbf{X}^T\mathbf{X}$

The preceding conclusion also arises if we consider the ellipse associated with **X**. Let us assume that the $n \times p$ matrix **X** is of full rank so that $\mathbf{X}^T\mathbf{X}$ and $(\mathbf{X}^T\mathbf{X})^{-1}$ are positive definite. Then the locus of all ξ that satisfy

$$\xi^T(\mathbf{X}^T\mathbf{X})^{-1}\xi = 1 \tag{3.12}$$

is an ellipse whose shape is directly associated with **X**.

We will again suppose that the columns of **X** have been scaled to have unit length. With this normalization, if the columns of **X** were mutually orthogonal, then $(\mathbf{X}^T\mathbf{X}) = \mathbf{I}$, and (3.12) is simply the equation of a p-dimensional unit sphere; all of its principal axes will be of the same length, as shown in Exhibit 3.4*a*. This situation again accords with our intuition of a well-conditioned matrix. If, however, the columns are not orthogonal, an ellipse will result with principal axes of different lengths. And if some of these axes are very small in relation to the major axis, the matrix would appear ill conditioned. This is exemplified in Exhibit 3.4*b*, which depicts the ellipse of a matrix **X** that has a deficiency in the direction of one principal axis, and in Exhibit 3.4*c*, depicting an **X** with deficiencies in two principal directions. To formalize these thoughts, we need to determine the lengths of the principal axes of the ellipse (3.12).

The first principal axis is clearly that ξ with longest length $\|\xi\|$. This can be found by maximizing $\xi^T\xi$ subject to (3.12). An argument closely following that previously given for the eigenvectors and eigenvalues of $\mathbf{X}^T\mathbf{X}$ readily shows that the principal axis ξ is in the direction of the eigenvector corresponding to the largest eigenvalue of $\mathbf{X}^T\mathbf{X}$ and that its length is the square root of this largest eigenvalue, that is, the largest singular value of **X**. Similarly, the next longest principal axis is in the direction of the eigenvector corresponding to the next largest eigenvalue and has length equal to the next largest singular value of **X**. The shortest principal axes have lengths that are the smallest singular values. So we are once again led to associate the ill conditioning of a matrix with the

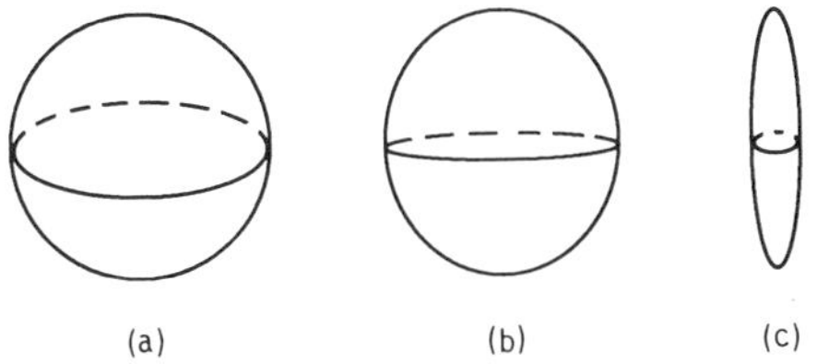

Exhibit 3.4 Ellipses of $\mathbf{X}^T\mathbf{X}$ matrices: (*a*) well-conditioned **X** matrix; (*b*) **X** matrix with one deficient dimension; (*c*) **X** matrix with two deficient dimensions.

situation in which one or more of its singular values is small in comparison to its largest singular value.

The preceding geometric considerations are illuminating but only suggestive. To provide a theoretical underpinning to these conclusions regarding assessing a small singular value, we turn to the concept of the condition number.

The Condition Number

Intuition is hard pressed to define a notion of an ill-conditioned matrix. The geometric views given in the preceding are gallant efforts, but like so many intuitive aids to understanding, they presuppose that which they illustrate and, hence, cannot serve as a definition. Verbally, one might be tempted to say that a matrix is ill conditioned if it is "almost not of full rank" or if its inverse "almost does not exist"—two obviously absurd statements. Yet, this is what is meant, in effect, when it is said that a square matrix is ill conditioned if it has a small determinant [or a rectangular matrix is ill conditioned if it has small $\det(\mathbf{X}^T\mathbf{X})$]. A small determinant, of course, has nothing to do with the invertibility of a matrix, for the matrix $\mathbf{A} = \alpha\mathbf{I}$ has determinant α^n, which can be made arbitrarily small by letting $\alpha \to 0$; yet it is clear that $\mathbf{A}^{-1}$ always exists for $\alpha \neq 0$ and is readily calculated as $\alpha^{-1}\mathbf{I}$. Rather more generally, we certainly know that if $\mathbf{A}$ is invertible, so is its inverse. Yet if $\det(\mathbf{A})$ is small, then $\det(\mathbf{A}^{-1})$ is large, so that, if we measure the invertibility or conditioning of a matrix by its determinant, the same information that tells us that $\mathbf{A}$ is not invertible tells us that $\mathbf{A}^{-1}$ is.[7]

A means for defining the conditioning of a matrix that accords with our growing intuition but avoids such pitfalls as those just mentioned is afforded by the singular-value decomposition. The motivation behind this definition derives from a more meaningful method for determining when an inverse of a given matrix "blows up." As we shall see, it becomes reasonable to consider a matrix $\mathbf{A}$ to be ill conditioned if the product of its spectral norm (defined in what follows) with that of its inverse $\mathbf{A}^{-1}$ is large. This measure, called the *condition number* of $\mathbf{A}$, provides summary information on the potential difficulties to be encountered in various calculations based on $\mathbf{A}$; the larger the condition number, the more ill conditioned the given matrix.

[7]It is also sometimes thought that the ill conditioning of a given matrix can be discovered by the presence of small diagonal elements in a triangular factorization of the matrix. This too is not true. Two examples from Golub and Reinsch (1970) and Wilkinson (1965) illustrate this point. Consider

$$\begin{bmatrix} 0.501 & -1 & & & \\ & 0.502 & -1 & & \\ & & \ddots & \ddots & \\ & & & 0.559 & -1 \\ & & & & 0.600 \end{bmatrix} \quad \text{and} \quad \begin{bmatrix} 1 & -1 & -1 & \cdots & -1 \\ & 1 & -1 & \cdots & -1 \\ & & \ddots & & \\ & & & & 1 \end{bmatrix}.$$

Each of these matrices can be shown by the singular-value decomposition, in a way described later, to be quite ill conditioned even though neither possesses a small diagonal element.

In particular, we shall show that the condition number provides (1) a measure of the potential sensitivity of the solution vector $\mathbf{z}$ of the linear system $\mathbf{Az} = \mathbf{c}$ to small changes (perturbations) in the elements of $\mathbf{c}$ and $\mathbf{A}$ and (2) a measure of the distance of a matrix from singularity (or exact collinearity). Furthermore, it is shown in Section 3.5 that the condition number meaningfully measures the difficulty of inverting a matrix and that the condition number of a data matrix $\mathbf{X}$ provides an upper bound on the elasticity (a measure of sensitivity frequently employed in economics) of the diagonal elements of the matrix $(\mathbf{X}^T\mathbf{X})^{-1}$ with respect to any element of the data matrix $\mathbf{X}$. These diagonal elements are, of course, proportional to the estimated least-squares coefficient variances and so the relevance of this result to least-squares estimation is obvious.

As an aid to developing and understanding the condition number, we begin by summarizing several related concepts—generalized inverses and spectral norms—and several important properties of the condition number in the context of solutions to linear systems of equations. We then provide two illuminating examples. For a more detailed treatment of what follows, along with the proofs and derivations that are omitted here, the reader is referred to Faddeeva (1959), Wilkinson (1965), Forsythe and Moler (1967), Hanson and Lawson (1969), Golub and Van Loan (1983), and Stewart (1987).

The Generalized Inverse

The generalized inverse (also known as the Moore–Penrose inverse) of any $m \times n$ matrix $\mathbf{A}$ is any $n \times m$ matrix, denoted $\mathbf{A}^+$, having the properties

(i) $\mathbf{AA}^+\mathbf{A} = \mathbf{A}$,

(ii) $\mathbf{A}^+\mathbf{AA}^+ = \mathbf{A}^+$,

(iii) $\mathbf{A}^+\mathbf{A} = (\mathbf{A}^+\mathbf{A})^T$,

(iv) $\mathbf{AA}^+ = (\mathbf{AA}^+)^T$.

The generalized inverse can be shown to be unique, and it is clearly a true generalization of the regular inverse. If, for example, $\mathbf{A}$ is square and invertible, it is easy to show that $\mathbf{A}^+ = \mathbf{A}^{-1}$ and, somewhat more generally, if $\mathbf{A}$ has full column rank, to demonstrate that $\mathbf{A}^+ = (\mathbf{A}^T\mathbf{A})^{-1}\mathbf{A}^T$, a result clearly relevant to least squares. Quite generally, from the singular-value decomposition of $\mathbf{A} = \mathbf{UDV}^T$, it is straightforward to show that $\mathbf{A}^+ = \mathbf{VD}^+\mathbf{U}^T$, where $\mathbf{D}^+$ is the generalized inverse of $\mathbf{D}$ and is simply $\mathbf{D}$ with its nonzero diagonal elements inverted. Hence, the singular values of $\mathbf{A}^+$ are merely the reciprocals of those of $\mathbf{A}$. Furthermore, $\mathbf{A}^+$ plays an analogous role to $\mathbf{A}^{-1}$ in solutions to rectangular systems of linear equations. Thus, consider a consistent equation system $\mathbf{Az} = \mathbf{c}$, where $\mathbf{z} \in \mathscr{R}^n$ and $\mathbf{c} \in \mathscr{R}^m$. Then it can be shown that $\mathbf{z} = \mathbf{A}^+\mathbf{c}$ is a solution to this system of equations.

The Spectral Norm

In Chapter 1 we defined the Euclidean norm of the n-vector $\mathbf{z}$ as $\|\mathbf{z}\| = (\mathbf{z}^T\mathbf{z})^{-1/2}$.

An important generalization of this to an $m \times n$ matrix $\mathbf{A}$ is the *spectral norm*, denoted $\|\mathbf{A}\|$ and defined as

$$\|\mathbf{A}\| \equiv \sup_{\|\mathbf{z}\| = 1} \|\mathbf{Az}\|. \tag{3.13}$$

It is readily shown (again using an argument similar to that used in our discussion of eigenvalues and eigenvectors) that $\|\mathbf{A}\| = \mu_{\max}$; that is, the spectral norm of the matrix $\mathbf{A}$ is simply the maximal singular value of $\mathbf{A}$, and hence, recalling that the singular values of $\mathbf{A}^+$ are the inverses of those of $\mathbf{A}$, we must also have $\|\mathbf{A}^+\| = 1/\mu_{\min}$. Further, like the Euclidean norm, the spectral norm is a true norm; that is, it possesses the following properties:

1. $\|\lambda\mathbf{A}\| = |\lambda| \cdot \|\mathbf{A}\|$ for all real λ and all $\mathbf{A}$.
2. $\|\mathbf{A}\| = 0$ if and only if $\mathbf{A} = \mathbf{0}$, the matrix of zeros.
3. $\|\mathbf{A} + \mathbf{B}\| \leqslant \|\mathbf{A}\| + \|\mathbf{B}\|$ for all $m \times n$ matrices $\mathbf{A}$ and $\mathbf{B}$.

In addition, the spectral norm obeys the following relations:

4. $\|\mathbf{Az}\| \leqslant \|\mathbf{A}\| \cdot \|\mathbf{z}\|$ (which follows directly from the definition).
5. $\|\mathbf{AB}\| \leqslant \|\mathbf{A}\| \cdot \|\mathbf{B}\|$ for all commensurate $\mathbf{A}$ and $\mathbf{B}$.

Sensitivity of Exact Linear Systems

We shall now use the preceding concepts in analyzing the conditioning of a linear system of equations. We begin with a consistent system of exact equations $\mathbf{Az} = \mathbf{c}$, where, as above, $\mathbf{A}$ is $m \times n$, $\mathbf{z}$ is an n-vector, and $\mathbf{c}$ is an m-vector. A solution to this system, which is unique if $\mathbf{A}$ has full column rank, is $\mathbf{z} = \mathbf{A}^+\mathbf{c}$.[8] We may now ask how much the solution vector $\mathbf{z}$ would change ($\delta\mathbf{z}$) if there were small changes, or perturbations, in the elements of $\mathbf{c}$ or $\mathbf{A}$, denoted $\delta\mathbf{c}$ and $\delta\mathbf{A}$. In the event that $\mathbf{A}$ is fixed but $\mathbf{c}$ changes by $\delta\mathbf{c}$, we have $\delta\mathbf{z} = \mathbf{A}^+\delta\mathbf{c}$, or using the preceding property 4,

$$\|\delta\mathbf{z}\| \leqslant \|\mathbf{A}^+\| \cdot \|\delta\mathbf{c}\|. \tag{3.14}$$

Further, employing the same property with the equation system $\mathbf{Az} = \mathbf{c}$, we have

$$\|\mathbf{c}\| \leqslant \|\mathbf{A}\| \cdot \|\mathbf{z}\|; \tag{3.15}$$

and by multiplying these two expressions, we obtain

$$\frac{\|\delta\mathbf{z}\|}{\|\mathbf{z}\|} \leqslant \|\mathbf{A}\| \cdot \|\mathbf{A}^+\| \cdot \frac{\|\delta\mathbf{c}\|}{\|\mathbf{c}\|}. \tag{3.16}$$

[8]Actually, this solution has some uniqueness characteristics even if $\mathbf{A}$ has less than full column rank. In this case, although there is an infinity of solutions that satisfy $\mathbf{Az} = \mathbf{c}$, the solution $\mathbf{z} = \mathbf{A}^+\mathbf{c}$ can be shown to be the one with the smallest norm $\|\mathbf{z}\|$. This interesting fact has little relevance to statistical applications since estimators with least norm are rarely of a priori interest.

From this, we see that the magnitude $\|\mathbf{A}\| \cdot \|\mathbf{A}^+\|$ provides a bound for the relative change in the length of the solution vector $\mathbf{z}$ that can result from a given relative change in the length of $\mathbf{c}$. A similar result holds for perturbations $\delta\mathbf{A}$ in the elements of the matrix $\mathbf{A}$ as long as they are small enough so that $\mathbf{A} + \delta\mathbf{A}$ retains full column rank (see what follows for the conditions for which this is true). Here it can be shown that

$$\frac{\|\delta\mathbf{z}\|}{\|\mathbf{z}\|} \leqslant 2\|\mathbf{A}\| \cdot \|\mathbf{A}^+\| \cdot \frac{\|\delta\mathbf{A}\|}{\|\mathbf{A}\|} + O, \tag{3.17}$$

where O denotes a term of the second order of smalls in $\|\delta\mathbf{A}\|/\|\mathbf{A}\|$.[9] Again, the magnitude $\|\mathbf{A}\| \cdot \|\mathbf{A}^+\|$ provides the relevant bound.

Because of its usefulness in this context, the magnitude $\|\mathbf{A}\| \cdot \|\mathbf{A}^+\|$ is given the name the condition number of the matrix $\mathbf{A}$ and is denoted by $\kappa(\mathbf{A})$. Recalling that $\|\mathbf{A}\| = \mu_{\max}$ and $\|\mathbf{A}^+\| = 1/\mu_{\min}$, we can also define the condition number for any matrix $\mathbf{A}$ to be

$$\kappa(\mathbf{A}) = \frac{\mu_{\max}}{\mu_{\min}} \geqslant 1. \tag{3.18}$$

In this formulation, we see immediately the justification for assessing the smallness of a singular value relative to the largest singular value. Indeed, this measure of conditioning has many desirable properties. It is readily shown that the condition number of a matrix with orthonormal columns is unity, so that $\kappa(\mathbf{A})$ reaches its lowest possible value in this cleanest of all possible cases. Furthermore, it is clear that $\kappa(\mathbf{A}) = \kappa(\mathbf{A}^+)$, so the condition number tells us the same story about a matrix $\mathbf{A}$ whether we are dealing with it or its (generalized) inverse.

The condition number $\kappa(\mathbf{A})$, then, provides a measure of the potential sensitivity of the solution of a linear system of exact equations to changes in the elements of $\mathbf{c}$ and $\mathbf{A}$. We shall see in a moment that a similar result is true for a solution to an inexact system of equations, such as a regression equation, but first let us examine two examples that will help us to understand better the behavior of the condition number.

Consider first the matrix

$$\mathbf{A} = \begin{bmatrix} 1 & \alpha \\ \alpha & 1 \end{bmatrix}.$$

Clearly, as $\alpha \to 1$, this matrix tends toward perfect singularity. The singular values of $\mathbf{A}$ are readily shown[10] to be $1 \pm \alpha$ and those of $\mathbf{A}^+$ (which here

[9]In fact, only the component of $\|\delta\mathbf{A}\|$ in the space orthogonal to that spanned by the columns of $\mathbf{A}$ is relevant here. Thus, if $\mathbf{A}$ is perturbed so as to leave its column space unchanged, O is zero in (3.17). See, e.g., Stewart (1973).

[10]Applying the singular-value decomposition (3.7) to a real symmetric matrix, such as is $\mathbf{A}$ here, quickly shows that the singular values of $\mathbf{A}$ are also its eigenvalues.

is $\mathbf{A}^{-1}$) to be $(1 \pm \alpha)^{-1}$. Hence, as $\alpha \to 1$, the condition number $\kappa(\mathbf{A}) = \|\mathbf{A}\| \cdot \|\mathbf{A}^{+}\| = (1+\alpha)(1-\alpha)^{-1}$ explodes; and as expected, $\mathbf{A}$ is ill conditioned for α near unity.

By way of contrast, consider the admittedly well-conditioned matrix

$$\mathbf{B} = \begin{bmatrix} 0 & \alpha \\ \alpha & 0 \end{bmatrix}.$$

As we have seen, the often-held belief that $\mathbf{B}$ becomes ill conditioned as $\alpha \to 0$ is incorrect, and this is correctly reflected in the condition number. Here we have $\|\mathbf{B}\| = \alpha$ and $\|\mathbf{B}^{+}\| = \alpha^{-1}$, so the condition number $\kappa(\mathbf{B}) = \|\mathbf{B}\| \cdot \|\mathbf{B}^{+}\| = \alpha\alpha^{-1} = 1$ as $\alpha \to 0$. In this case, then, the condition number does not blow up, and $\mathbf{B}$ is well conditioned for all $\alpha \neq 0$.

Sensitivity of Inexact Linear Systems (Least Squares)

Things are somewhat more complicated when examining the sensitivity of an inexact system of linear equations: $\mathbf{y} \approx \mathbf{Xb}$; that is, where we have $\mathbf{y} - \mathbf{Xb} \neq \mathbf{0}$. I have switched notation from the $\mathbf{z}$, $\mathbf{A}$, and $\mathbf{c}$ used above only because I want to emphasize the relation of this analysis to the least-squares regression context. Here, then, $\mathbf{y}$, is an n-vector, $\mathbf{b}$ is a p-vector, and $\mathbf{X}$ is an $n \times p$ matrix. The least-squares solution to this equation, that is, the $\mathbf{b}$ that minimizes $\|\mathbf{y} - \mathbf{Xb}\|^2$, is well known to be $\mathbf{b} = \mathbf{X}^{+}\mathbf{y}$, which in the case that $\mathbf{X}$ has full column rank takes the familiar form $\mathbf{b} = (\mathbf{X}^{\mathrm{T}}\mathbf{X})^{-1}\mathbf{X}^{\mathrm{T}}\mathbf{y}$. Now let $\hat{R}^2$ be the uncentered squared multiple-correlation coefficient (2.4a) of $\mathbf{y}$ regressed on $\mathbf{X}$, and consider perturbations $\delta\mathbf{X}$, $\delta\mathbf{b}$, and $\delta\mathbf{y}$ in $\mathbf{X}$, $\mathbf{b}$, and $\mathbf{y}$, respectively. Here we assume that both $\mathbf{X}$ and $\mathbf{X} + \delta\mathbf{X}$ are of full column rank. Given that $\mathbf{X}$ is of full rank, it is shown in Hanson and Lawson (1969) that $\mathbf{X} + \delta\mathbf{X}$ is also if $\|\delta\mathbf{X}\|/\|\mathbf{X}\| < \kappa^{-1}(\mathbf{X})$.[11] Then as can be derived from Golub and Van Loan (1983),

$$\frac{\|\delta\mathbf{b}\|}{\|\mathbf{b}\|} \leqslant \kappa(\mathbf{X})\hat{R}^{-1}[2 + (1 - \hat{R}^2)^{1/2}\kappa(\mathbf{X})]v + O(v^2), \tag{3.19}$$

where $v = \max(\|\delta\mathbf{y}\|/\|\mathbf{y}\|, \|\delta\mathbf{X}\|/\|\mathbf{X}\|)$. Exact equality here is possible and does not depend upon v.

The sensitivity of the solution of an inexact system of equations (such as the least-squares estimator) to perturbations in either the $\mathbf{y}$ or $\mathbf{X}$ data, then, is also seen to be bounded by a relation that depends on the condition number $\kappa(\mathbf{X})$. The worse the conditioning of the $\mathbf{X}$ data, the greater the potential for large relative changes in the solution to result from small relative changes in the data. In the case of inexact equations, however, we note that this sensitivity is further

[11]This result warrants a moment's interpretation, for it says that the condition number $\kappa(\mathbf{X})$, among its many other uses, also tells us the amount by which we can "move" a matrix $\mathbf{X}$ of full rank so that it remains of full rank. Clearly, the more ill conditioned is a matrix $\mathbf{X}$, the less it can be perturbed and remain of full rank, or equivalently, the closer it is (in the sense of a matrix norm) to a matrix of less than full rank. This result is discussed in greater detail shortly.

affected by the strength of the linear relation between **y** and **X** as measured by $\hat{R}^2$. If this relation is perfect, so $\hat{R}^2 = 1$, (3.19) imposes its tightest bounds, which are seen to be similar to those for exact dependencies given in (3.16) and (3.17), and the conditioning is measured by $\kappa(\mathbf{X})$. As the relation becomes looser, however, the potential for sensitivity increases regardless of $\kappa(\mathbf{X})$, and for small $\hat{R}^2$, the conditioning is measured more nearly by $\kappa^2(\mathbf{X})$. Thus, we see that a high condition number is a conservative indicator of the potential sensitivity of the solution of inexact equations. It is possible for an ill-fitting equation to show high sensitivity even with well-conditioned data. But it is impossible to avoid the potential sensitivity when the data are ill conditioned regardless of the quality of the fit. And of course, when the fit is good, the condition number $\kappa(\mathbf{X})$ behaves much as it would for exact equations.

Another Interpretation of the Condition Number

We have just seen that the condition number can be interpreted as a measure of the potential sensitivity of a system of linear equations to perturbations in its data. There is another interpretation of $\kappa(\mathbf{X})$ that also helps to shed light on its meaning for assessing collinearity. Suppose we were to perturb a matrix **X** of full rank by **E** to get **X** + **E** and then ask what the smallest **E** would be that makes **X** + **E** exactly collinear. It can be shown that the **E** with the smallest spectral norm that does this task has as its spectral norm $\mu_{\text{min}}(\mathbf{X})$, the smallest singular value of **X**.[12] Therefore, $\mu_{\text{min}}(\mathbf{X})$ provides a measure of the absolute distance of **X** from rank deficiency (exact collinearity), and hence, $\mu_{\text{min}}(\mathbf{X})/\mu_{\text{max}}(\mathbf{X})$ provides a relative measure of **X** from exact collinearity, which by (3.18) is seen to be $\kappa^{-1}(\mathbf{X})$. Thus, the larger the condition number, the stronger the degree of collinearity among the columns of **X**.

Condition Indexes

By way of brief summary, we have seen that for each exact linear dependency among the columns of **X** there is one zero singular value. Extending this property to near dependencies leads one to suggest, as did Kendall (1975) and Silvey (1969), that for each near linear dependency there will be a "small" singular value of **X** (or eigenvalue of $\mathbf{X}^{\mathrm{T}}\mathbf{X}$). But this suggestion does not include a means for determining what small is. The preceding geometric and analytic discussions do, however, provide a basis for assessing smallness. Specifically, the degree of ill conditioning depends on how small the minimum singular value is relative to the maximal singular value; that is, μ_{max} provides a yardstick against which smallness can be measured. We therefore rephrase this suggestion to state that there will be a near dependency for each singular value that is small relative to μ_{max}. In this context, it is clearly useful to define the set of *condition indexes* of the $n \times p$ matrix **X**

$$\eta_k \equiv \frac{\mu_{\text{max}}}{\mu_k}, \qquad k = 1, \ldots, p. \tag{3.20}$$

[12]See Stewart (1987). These results are due to Eckart and Young (1936) and Mirsky (1960).

Of course, $\eta_k \geqslant 1$ for all k, and the lower bound must occur for some k. The largest value for η_k is also the condition number of **X**. A singular value that is small relative to its yardstick μ_{max}, then, has a high condition index.

We may therefore extend the Kendall–Silvey suggestion as follows: there are as many near dependencies among the columns of a data matrix **X** as there are high condition indexes (singular values small relative to μ_{max}). Two points regarding this extension must be emphasized.

First, we have not merely redirected the problem from one of determining when small is small to one of determining when large is large. As we have seen, taken alone, the singular values (or eigenvalues) shed no light on the conditioning of the data matrix: equally well-conditioned data matrices can have arbitrarily low singular values. Thus, determining that a singular value is absolutely small has no direct relevance to determining the presence of a near dependency causing the data matrix to be ill conditioned. We did see, however, in our discussion surrounding the condition number, that determining that a singular value is small relative to μ_{max}, or equivalently that determining that a condition index is high, is directly related to this problem. The meaningfulness of the condition indexes in this context is verified empirically in the experiments of Chapter 4.[13]

Second, even if there is measurable content to the term *large* in connection with the condition indexes, there is no a priori basis for determining how large a condition index must be before there is evidence of collinear data or, even more importantly, evidence of data so collinear that its presence is degrading or harming regression estimates. Just what is to be considered a large condition index is a matter to be determined empirically, and the experiments of Chapter 4 are aimed at aiding such an understanding. There we learn that weak dependencies are associated with condition indexes around 5–10, whereas moderate to strong relations are associated with condition indexes of 30–100.

The use of the condition indexes, then, extends the Kendall–Silvey suggestion in two ways. First, practical experience will allow an answer to the question of when small is small (or large is large), and second, the simultaneous occurrence of several large η's keys the simultaneous presence of more than one near dependency. The condition indexes, then, are a means for determining when there are collinear relations among the columns of a data matrix **X** and for determining how many of them there are.

3.2 THE VARIANCE-DECOMPOSITION PROPORTIONS: DETERMINING VARIATE INVOLVEMENT

Having found a means for determining the presence and number of near dependencies, we turn now to means for determining which variates are involved in them. We recall from our survey of collinearity diagnostics in

[13]The closely related problem of determining when, as a practical matter, a matrix may be considered rank deficient (i.e., when a singular value is effectively zero relative to μ_{max}) is treated in Golub et al. (1976).

Chapter 2 that the direct means for revealing variate involvement have significant drawbacks. The term *direct* here refers to those methods that directly examine the magnitude of the coefficient (or loading) of a given variate in a linear relation with the others. We found, for example, that examining the elements of the eigenvectors is not very useful for this purpose because it is possible for some variates known to be involved in a relation nevertheless to have arbitrarily small eigenvector elements. Likewise, all-subsets regression, while possibly providing the desired information, does so only at the expense of much effort in digesting the results, even for simple cases, and with the great risk that many of these regressions would be misleading precisely in the cases of interest, where there are several coexisting near dependencies.

To find a successful diagnostic of variate involvement, then, it appears we must eschew direct indicators and turn to indirect ones. The VIF, for example, provides indirect evidence of the involvement of a given column of **X** in a linear relation by looking at the magnitude of the R^2 (preferably uncentered) that results when that column is regressed on the others. High VIFs are indeed indicators of variate involvement in at least some linear relation. But we have also seen that VIFs cannot be used to determine the number of relations, and so, when there are multiple relations, high VIFs cannot be used to indicate the relation in which the given variates is involved. VIFs, then, go part way toward our objective—but we would like something better.

Another well-known indirect consequence of the involvement of a column of **X** in a collinear relation is that the variance of its estimated coefficient "blows up" when **X** is used in a least-squares regression. Thus, it would seem that there is diagnostic information about the involvement of the columns of **X** in collinear relations to be gotten from these variances. Indeed, we show here that we can extend and reinterpret the work of Silvey (1969) to provide a decomposition of the estimated variance of each regression coefficient into a sum of terms, each of which is associated with a condition index, thereby providing a means for determining the extent to which each near dependency (high condition index) becomes a dominant part of (degrades) each variance. Since this variance inflation occurs for each variate involved in each near dependency, this provides a means for diagnosing variate involvement in separate coexisting near dependencies.

It is worth emphasizing that, viewed strictly diagnostically, this decomposition need not be relevant to an actual regression to be useful in diagnosing variate involvement in a collinear relation. If there are collinear relations among the columns of **X**, then there will be inflated variances when **X** is used in a regression, whether or not it is actually intended to use **X** in a regression. As indirect evidence of the involvement of the columns of **X** in collinear relations, then, this information is of general interest as a collinearity diagnostic regardless of the use to which **X** is to be put. However, when **X** is ultimately to be used in a regression analysis, this decomposition further provides a link between the numerical analysis of a data matrix **X**, as embodied in its singular-value decomposition, and the statistical quality of the subsequent regression analysis using $\mathbf{X}$ as a data matrix, as embodied in the variance–covariance matrix of $\mathbf{b}$. This link is of obvious interest, for the resulting diagnostics clearly acquire a more profound meaning in a regression context because of it; the collinearity diagnostics now also become regression diagnostics. By way of contrast, we saw that such a link was missing in the Farrar and Glauber technique described in Section 2.3.

The Variance Decomposition

Under the usual assumptions, the variance–covariance matrix $\mathbf{V}(\mathbf{b})$ of the least-squares estimator $\mathbf{b} = (\mathbf{X}^{\mathrm{T}}\mathbf{X})^{-1}\mathbf{X}^{\mathrm{T}}\mathbf{y}$ is, of course, $\sigma^2(\mathbf{X}^{\mathrm{T}}\mathbf{X})^{-1}$, where σ^2 is the common variance of the components of $\boldsymbol{\varepsilon}$ in the linear model $\mathbf{y} = \mathbf{X}\boldsymbol{\beta} + \boldsymbol{\varepsilon}$. Using the singular-value decomposition, $\mathbf{X} = \mathbf{U}\mathbf{D}\mathbf{V}^{\mathrm{T}}$, this variance–covariance matrix may be written as

$$\mathbf{V}(\mathbf{b}) = \sigma^2(\mathbf{X}^{\mathrm{T}}\mathbf{X})^{-1} = \sigma^2\mathbf{V}\mathbf{D}^{-2}\mathbf{V}^{\mathrm{T}}. \tag{3.21}$$

Thus, the variance of the kth regression coefficient, b_k, which is the kth diagonal element of (3.21), is simply

$$\operatorname{var}(b_k) = \sigma^2 \sum_j \frac{v_{kj}^2}{\mu_j^2}, \tag{3.22}$$

where the μ_j's are the singular values of $\mathbf{X}$ and $\mathbf{V} \equiv (v_{ij})$.

Note that (3.22) decomposes $\operatorname{var}(b_k)$ into a sum of components, each associated with one and only one of the p singular values μ_j of the $n \times p$ matrix $\mathbf{X}$ (or eigenvalues $\lambda_j = \mu_j^2$ of $\mathbf{X}^{\mathrm{T}}\mathbf{X}$). Since these μ_j^2 appear in the denominator, other things being equal, those components associated with near dependencies—that is, with small μ_j—will be large relative to the other components. This suggests, then, that an unusually high *proportion* of the variance of two or more coefficients concentrated in components associated with the same *small* singular value provides evidence that the variates (columns of $\mathbf{X}$) corresponding to those coefficients are involved in the near dependency corresponding to that small singular value. The preceding says "two or more" because it is clear that it takes at least two variates to make a collinear relation.[14] So let us pursue this suggestion.

Define the (k, j)th *variance–decomposition proportion* as the proportion of the variance of the kth regression coefficient associated with the jth component of its decomposition in (3.22). These proportions are readily calculated as follows:

First, let

$$\phi_{kj} \equiv \frac{v_{kj}^2}{\mu_j^2} \quad \text{and} \quad \phi_k \equiv \sum_{j=1}^{p} \phi_{kj}, \qquad k = 1, \ldots, p. \tag{3.23}$$

[14]We shall see in Chapter 7, however, that a closely allied problem, short data, can affect a single variate.

Then the variance–decomposition proportions are

$$\pi_{jk} \equiv \frac{\phi_{kj}}{\phi_k}, \qquad k, j, = 1, \ldots, p. \tag{3.24}$$

These variance–decomposition proportions are most easily digested when summarized in a **Π** matrix, that is, a table like that in Exhibit 3.5. Here, each row corresponds to a given singular value, μ_j, or equivalently, the associated condition index $\eta_j \equiv \mu_{\max}/\mu_j$. These rows can be ordered so that the condition indexes are in increasing (or decreasing) order. Naturally, the columns of π's should sum to 1. Interest centers on patterns where two or more variates have large values associated with the same high condition index. We shall have a good deal more to say about how to view and interpret these tables as we proceed, specifically in what follows and again in Section 5.3.

The reader is warned here of the switch in the order of the subscripts between the left- and right-hand sides of (3.24). This convention has been adopted solely for the convenience it offers in displaying tables like Exhibit 3.5 in this book. However, in the output of a computer program to effect these diagnostics, it often proves desirable to transpose the **Π** matrix, printing the condition indexes in decending order as column headings along the first row and the variance–decomposition proportions for each variate occupying each succeeding row. This format is particularly suitable when p is large and is discussed in greater detail in Section 5.3.

Two Interpretative Considerations

The next chapter reports detailed experiments using the two tools developed above: the set of condition indexes and the associated **Π** matrix of variance–decomposition proportions. These experiments are designed to give empirical content to the behavior of these tools when used for analyzing collinearity, detecting it, and assessing the potential damage its presence may cause least-squares estimates. Before this, however, it is useful to develop two fundamental interpretive properties of the **Π** matrix of variance–decomposition proportions.

Exhibit 3.5 Π Matrix of Variance–Decomposition Proportions

Condition Index	Proportions of $\mathrm{var}(b_1)$	$\mathrm{var}(b_2)$	$\cdots$	$\mathrm{var}(b_p)$
η_1	π_{11}	π_{12}	$\cdots$	π_{1p}
η_2	π_{21}	π_{22}	$\cdots$	π_{2p}
$\vdots$	$\vdots$	$\vdots$		$\vdots$
η_p	π_{p1}	π_{p2}	$\cdots$	π_{pp}

Near Collinearity Nullified by Near Orthogonality
In the variance decomposition (3.22), small μ_j, other things equal, lead to large components of var(b_k). However, not all var(b_k) need be adversely affected by a small μ_j, for the v_{kj}^2 in the numerator may be smaller yet. In the extreme case where $v_{kj} = 0$, var(b_k) would be unaffected by any near dependency among the columns of **X** that would cause μ_j to become even very small. This can be shown to occur exactly when the kth and jth columns of **X** belong to separate orthogonal blocks. Thus, suppose $\mathbf{X} \equiv [\mathbf{X}_1 \;\; \mathbf{X}_2]$ with $\mathbf{X}_1^T\mathbf{X}_2 = \mathbf{0}$, and let the singular-value decompositions of $\mathbf{X}_1$ and $\mathbf{X}_2$ be given, respectively, as $\mathbf{X}_1 = \mathbf{U}_1\mathbf{D}_{11}\mathbf{V}_{11}^T$ and $\mathbf{X}_2 = \mathbf{U}_2\mathbf{D}_{22}\mathbf{V}_{22}^T$. Now, since $\mathbf{U}_1$ is an orthogonal basis for the space spanned by the columns $\mathbf{X}_1$, and $\mathbf{U}_2$ is an orthogonal basis for the space spanned by the columns of $\mathbf{X}_2$, it must be that $\mathbf{X}_1^T\mathbf{X}_2 = \mathbf{0}$ implies $\mathbf{U}_1^T\mathbf{U}_2 = \mathbf{0}$, so $\mathbf{U} \equiv [\mathbf{U}_1 \;\; \mathbf{U}_2]$ is column orthogonal. It is now straightforward to verify that the singular-value decomposition of **X** is simply $\mathbf{X} = \mathbf{UDV}^T$, with

$$\mathbf{D} \equiv \begin{bmatrix} \mathbf{D}_{11} & \mathbf{0} \\ \mathbf{0} & \mathbf{D}_{22} \end{bmatrix}$$

and

$$\mathbf{V} \equiv \begin{bmatrix} \mathbf{V}_{11} & \mathbf{0} \\ \mathbf{0} & \mathbf{V}_{22} \end{bmatrix}.$$

Thus, $\mathbf{V}_{12} = \mathbf{0}$.

One important implication of this is a fact well known to users of least squares, namely, that the introduction into a regression equation of variates that are orthogonal to all preceding variates will not affect the ordinary least-squares estimates of the coefficients or their true standard errors.[15] Thus, consider introducing into a regression equation two variates that are very highly collinear with one another but are also orthogonal to all the previous variates. Then the estimates of the coefficients of all the previous variates and their true variances must be unaffected. In terms of the variance decomposition (3.22), this is a situation in which at least one μ_j is very small (that corresponding to the two closely collinear variates) but has no weight in determining any of the var(b_k) for k corresponding to the previously included variates. Clearly, this can only occur if the v_{kj} between the previously included variates and the two additional variates are zero.

Hence, we see that, in the singular-value decomposition of $\mathbf{X} \equiv [\mathbf{X}_1 \;\; \mathbf{X}_2]$

[15]The term *true standard errors* refers to the standard errors or variances taken from the diagonal elements of $\sigma^2(\mathbf{X}^T\mathbf{X})^{-1}$ and not to the estimated standard errors based on $s^2(\mathbf{X}^T\mathbf{X})^{-1}$, which will not necessarily remain unaltered when a variate (even one orthogonal to all other variates) is added to the equation. Of course, all that is really needed here is the invariance of the diagonal elements of $(\mathbf{X}^T\mathbf{X})^{-1}$ to the inclusion of such a variate.

with $\mathbf{X}_1^T\mathbf{X}_2 = \mathbf{0}$, the columns of the $\mathbf{V}$ matrix may always be ordered so that it takes the form

$$\mathbf{V} = \begin{bmatrix} \mathbf{V}_{11} & \mathbf{0} \\ \mathbf{0} & \mathbf{V}_{22} \end{bmatrix}. \tag{3.25}$$

An analogous result clearly applies to any number of mutually orthogonal subgroups. Thus, we note that for some coefficients the deleterious effects of collinearity, associated with relatively small μ's, may be mitigated by the isolation that is afforded by the near orthogonality of subgroups of the columns of $\mathbf{X}$, resulting in small v_{kj}'s.

At Least Two Variates Must Be Involved

At first it would seem that the concentration of the variance of any one regression coefficient in any one of its components would signal that collinearity may be present. However, since two or more variates are required to create a near dependency, it must be that the variances of two or more regression coefficients are together adversely affected, each showing high variance-decomposition proportions associated with a single condition index (i.e., a single near dependency), before a collinear relation may be said to be present.

To illustrate this, consider a data matrix $\mathbf{X}$ with mutually orthogonal columns—the best possible experimental data. Our previous result immediately implies that the $\mathbf{V}$ matrix of the singular-value decomposition of $\mathbf{X}$ is diagonal since each $v_{ij} = 0$, $i \neq j$. Hence, the associated $\mathbf{\Pi}$ matrix of variance-decomposition proportions must take the form

Condition Index	Proportions of			
	$\text{var}(b_1)$	$\text{var}(b_2)$	$\cdots$	$\text{var}(b_p)$
η_1	1			
η_2		1		$\mathbf{0}$
$\vdots$			$\ddots$	
η_p		$\mathbf{0}$		1

It is clear that a high proportion of any variance associated with a *single* condition index is hardly indicative of collinearity, for the variance-decomposition proportions here are those for orthogonal data. Rather, reflecting the fact that there must be two or more columns of $\mathbf{X}$ involved to make a near dependency, the degradation of a regression estimate or the involvement of a variate in a collinear relation can be observed only when a high condition index is associated with a large proportion of the variance of *two or more* coefficients. If, for example, in a case with $p = 5$, columns 4 and 5 of **X** are highly collinear and all columns are otherwise mutually orthogonal, we would expect a **Π** matrix of variance–decomposition proportions something like

Condition Index	Proportions of var(b_1)	var(b_2)	var(b_3)	var(b_4)	var(b_5)
η_1	1.0	0	0	0	0
η_2	0	1.0	0	0	0
η_3	0	0	1.0	0	0
η_4	0	0	0	0.05	0.03
η_5	0	0	0	0.95	0.97

Here η_5, which is assumed to be large (so μ_5 is small relative to μ_{max}), plays a large role in both var(b_4) and var(b_5). Viewed as a collinearity diagnostic, this information indicates the involvement of $\mathbf{X}_4$ and $\mathbf{X}_5$ in a collinear relation (regardless of whether **X** is being used in a regression). Viewed as a regression diagnostic, this information also indicates the degradation of the least-squares estimates b_4 and b_5 on account of the collinearity.

An Example

An example of the preceding two interpretive considerations is useful at this point. Consider the 6×5 data matrix given in Exhibit 3.6. The first four columns

$$\mathbf{X} = [\mathbf{X}_1 \ \mathbf{X}_2] = \begin{bmatrix} -74 & 80 & 18 & -56 & -112 \\ 14 & -69 & 21 & 52 & 104 \\ 66 & -72 & -5 & 764 & 1{,}528 \\ -12 & 66 & -30 & 4{,}096 & 8{,}192 \\ 3 & 8 & -7 & -13{,}276 & -26{,}552 \\ 4 & -12 & 4 & 8{,}421 & 16{,}842 \end{bmatrix}$$

Exhibit 3.6 The Modified Bauer matrix.

of this matrix are due to Bauer (1971) and have the property that the fourth column is exactly orthogonal to the first three (which, however, are not orthogonal to each other). It has been modified by the addition of a fifth column that is exactly twice the fourth. Thus, $\mathbf{X}_2$ is of rank 1 and $\mathbf{X}_1^T\mathbf{X}_2 = \mathbf{0}$. We therefore know from the preceding discussion that, in the singular-value decomposition of X, (i) one of the singular values associated with $\mathbf{X}_2$ will be zero (i.e., within the machine tolerance of zero[16]) and (ii) in

$$\mathbf{V} = \begin{bmatrix} \mathbf{V}_{11} & \mathbf{V}_{12} \\ \mathbf{V}_{21} & \mathbf{V}_{22} \end{bmatrix}, \qquad \mathbf{V}_{12} = \mathbf{V}_{21}^T = \mathbf{0}.$$

[16]Approximately 10^{-14} on the IBM 370 in double precision; $\approx 10^{-19}$ on a Macintosh using the SANE routines or the floating-point processor, which was used for these calculations.

Indeed, application of a singular-value decomposition routine (MINFIT in this case—see Section 3.1 for references to software) to **X** gives, for **V** (partitioned),

$$\begin{bmatrix} -0.548 & 0.625 & -0.556 & 0.71\times10^{-23} & 0.54\times10^{-40} \\ 0.836 & 0.383 & -0.393 & 0.89\times10^{-22} & -0.24\times10^{-37} \\ -0.033 & -0.680 & -0.733 & 0.22\times10^{-23} & -0.99\times10^{-39} \\ & & & & \\ -0.31\times10^{-22} & -0.16\times10^{-22} & 0.18\times10^{-22} & 0.447 & 0.894 \\ -0.63\times10^{-22} & -0.36\times10^{-22} & 0.36\times10^{-22} & 0.894 & -0.447 \end{bmatrix}$$

with the following singular values (in descending order) and corresponding condition indexes:

Singular Values	Condition Indexes
$\mu_1 = 36368.4$	$\eta_1 = 1$
$\mu_2 = 170.7$	$\eta_2 = 213$
$\mu_3 = 60.5$	$\eta_3 = 601$
$\mu_4 = 7.6$	$\eta_4 = 4785$
$\mu_5 = 8.5\times10^{-16}$	$\eta_5 = 4\times10^{19}$

A glance at **V** verifies that the off-diagonal blocks are small—all are of the order of 10^{-20} or smaller—and well within the effective zero tolerance 10^{-19} of the computational precision of the Macintosh used for these calculations. Only somewhat less obvious is the fact that the smallest singular value μ_5 is also zero. Since μ_5 is of the order 10^{-16}, it would seem to be nonzero relative to the machine tolerance, but as we have seen, the magnitude of each μ_j has meaning only relative to μ_{max}, and in this case $\mu_5/\mu_{max} = \eta_5^{-1} = 2.5\times10^{-20}$, well within the machine tolerance of zero. Since this number is effectively zero, its inverse, $\eta_5 = 4\times10^{19}$, the largest condition index and thus the condition number of the modified Bauer matrix, is really a junk number that must be viewed as "infinity" relative to machine tolerance.

The reader is warned against attempting too fine an interpretation of these condition indexes at this point. For reasons that will be explained, the data should first be scaled to have equal column lengths before the resulting condition indexes acquire meaningful diagnostic values other than zero. Since it is this latter value that is of interest here, however, such scaling is unneeded for the current analysis. These data will be examined again more completely in Section 5.4.

The **Π** matrix of variance–decomposition proportions for this data matrix is given in Exhibit 3.7. Several of its properties are noteworthy. First, we would expect that the large condition index η_5 associated with the exact linear

Exhibit 3.7 Variance-Decomposition Proportions: *Modified Bauer Matrix*

Condition Index, η	Proportions of var(b_1)	var(b_2)	var(b_3)	var(b_4)	var(b_5)
1	.000	.000	.000	.000	.000
213	.002	.009	.000	.000	.000
601	.019	.015	.013	.000	.000
4785	.979	.976	.987	.000	.000
4×10^{19}	.000	.000	.000	1.000	1.000

dependency between columns 4 and 5 (**C4** = 0.5**C5**) would dominate several variances—at least those of the two variates involved—, and this is seen to be the case; the component associated with the essentially infinite $\eta_5 = 4 \times 10^{19}$ accounts for virtually all of the variance of both b_4 and b_5.

Second, we would expect the fact that the first three columns of **X** are orthogonal to the two involved in the strong linear dependency would be evident: viewing these figures as a collinearity diagnostic, there should be no indication of any involvement of these first three columns in the linear dependency with columns 4 and 5, and viewing them as a regression diagnostic, there should be evidence that the estimated coefficients of these three variates are isolated from the deleterious effects of the collinearity affecting the other two. This too is seen to be the case; the variance–decomposition proportions of these three variances associated with η_5 are very small indeed, each being zero to at least three places.[17] This serves to exemplify the point that the collinearity diagnostics suggested here can aid the user in determining not only those regression estimates that are degraded by the presence of collinearity but also those that are not—and so may be salvaged despite the presence of collinearity.

Third, a somewhat unexpected result appears. The fourth condition index ($\eta_4 = 4785$) accounts for 97% or more of var(b_1), var(b_2), and var(b_3). This suggests that a second near dependency is present in **X**, one associated with η_4 and involving the first three columns. This, in fact, turns out to be the case. We reexamine this example in Section 5.4 once we have gained further experience in interpreting the magnitudes of condition indexes and variance–decomposition proportions.

Fourth, to the extent that there are two separate near dependencies in **X** (one among its first three columns, one between the last two), the **Π** matrix does indeed provide a simple means for determining which variates are involved in each.

The simplicity with which this diagnostic technique discerns variate involve-

[17] That these values might be nonzero beyond the third decimal place is due only to the finite arithmetic of the machine. In theory these components are an undefined ratio of zeros that would be defined to be zero in this application.

ment in separate dependencies is important because it is not true of any of the alternative means for analyzing near dependencies among the columns of **X** that we have encountered. We have seen, for example, that one could hope to investigate such near dependencies by regressing selected columns of **X** on other columns or by examining partial correlations. But to do this in anything other than a shotgun manner, such as all-subsets regression, would require prior knowledge about which columns of **X** are best preselected to regress on the others, and to do so when there are several coexisting near dependencies would prove a terrific burden, if only because any collinearity that may exist among the subsets chosen for the right-hand side could distort the results. Typically, the user of linear regression, when presented with a specific data matrix, has no rational means for preselecting offending variates. Fortunately, the **Π** matrix of variance–decomposition proportions avoids this problem altogether, for it displays all such near dependencies right from the outset, treating the columns of **X** symmetrically and requiring no prior information on the numbers of near dependencies or their composition.

3.3 A SUGGESTED COLLINEARITY DIAGNOSTIC

The preceding material puts us but one short step away from being able to formalize both a definition of collinearity and an appropriate procedure for diagnosing its presence: the step of column scaling. After a brief discussion of the need for column scaling, we introduce the associated notions of scaled condition numbers and scaled condition indexes and then turn to the suggested diagnostic procedure.

The Need for Column Scaling

Data matrices that differ from one another only by the scale assigned to their columns (matrices of the form **XB**, where **B** is a nonsingular diagonal matrix with positive diagonal elements) represent essentially equivalent information; it does not matter, for example, whether one specifies monetary data in dollars, cents, or billions of dollars or work in foot-pounds or joules. It is very clear from above, however, that such scale changes do affect the numerical properties of the data matrix and result in very different variance–decomposition proportions and condition indexes.[18] Without further adjustment, then, we have a situation in which near dependencies among structurally equivalent variates (differing only in their units of measurement) can result in greatly differing condition indexes. Clearly, the condition indexes can provide no stable information to the user of linear regression on the degree of collinearity among the **X** variates in such a case.

[18]Such scale changes do not, however, affect the presence of exact linear dependencies among the columns of **X** since for any nonsingular matrix **B** there exists a nonzero **c** such that $\mathbf{Xc} = \mathbf{0}$ if and only if $[\mathbf{XB}][\mathbf{B}^{-1}\mathbf{c}] \equiv \bar{\mathbf{X}}\bar{\mathbf{c}} = \mathbf{0}$, where $\bar{\mathbf{X}} = \mathbf{XB}$ and $\bar{\mathbf{c}} = \mathbf{B}^{-1}\mathbf{c}$. For a more general discussion of the effects of column scaling, see Sections 6.1 and 6.2.

It is necessary, therefore, to standardize data matrices that correspond to equivalent model structures in a way that makes comparisons of condition indexes meaningful. A natural standardization is to scale each column to have equal length—column equilibration. This scaling is natural because it transforms a data matrix $\mathbf{X}$ with mutually orthogonal columns, seemingly ideal data, into one whose condition indexes would all be unity, the smallest and most ideal possible. Any other scaling would fail to reflect this desirable property. And an important converse is true with column-equilibrated data; namely, when all condition indexes of a data matrix are equal to unity, the columns are mutually orthogonal. This is readily proved by noting that, if all condition indexes are equal to 1, then, in the singular-value decomposition of $\mathbf{X}$, the $\mathbf{D}$ matrix must take the form $\mathbf{D} = \lambda\mathbf{I}$ for some λ. Hence, we have $\mathbf{X} = \mathbf{UDV}^{\mathrm{T}} = \lambda\mathbf{UV}^{\mathrm{T}}$, or $\mathbf{X}^{\mathrm{T}}\mathbf{X} = \lambda^2\mathbf{VU}^{\mathrm{T}}\mathbf{UV}^{\mathrm{T}} = \lambda^2\mathbf{I}$, due to the orthogonality of $\mathbf{U}$ and $\mathbf{V}$. This result is important because it rules out the possibility that several high variance–decomposition proportions could be associated with a very low (near unit) condition index. Further, it is shown in Section 6.2 that column equilibration results in condition numbers that are most meaningful as a collinearity diagnostic.

The exact length to which the columns are scaled is unimportant, just so long as they are equal, since the condition indexes are readily seen to be invariant to scale changes that affect all columns equally. But as a matter of practice, we effect column equilibration by scaling each column of $\mathbf{X}$ to have unit length. This scaling is similar to that used to transform the cross-product matrix $\mathbf{X}^{\mathrm{T}}\mathbf{X}$ into a correlation matrix, which appears to cause some to confuse these two issues. Thus, it is worth reemphasizing that (1) any equal length, not just unit length, would suffice for our purposes here and (2) whereas the columns have been scaled, they have not been centered to have zero means, as would be needed for correlations. Indeed, we shall see that centering, while useful for some purposes, is almost always inappropriate for analyzing collinearity and indeed, if done, tends to produce misleading conditioning diagnostics. Full justifications for column equilibration and the need for avoiding centering are given in Chapter 6.

By way of terminology, then, we continue to define the condition number $\kappa(\mathbf{X})$ and condition indexes $\eta_i(\mathbf{X})$ of a matrix $\mathbf{X}$ by (3.18) and (3.20), respectively, where the singular values are those of $\mathbf{X}$. But we now introduce the *scaled condition number* $\tilde{\kappa}(\mathbf{X})$ and the *scaled condition indexes* $\tilde{\eta}_i(\mathbf{X})$ of the matrix $\mathbf{X}$, which are similarly defined, except that the singular values are those that result from applying the singular-value decomposition to $\mathbf{X}$ after its columns have first been scaled to have equal (unit) length.

That is, if $\mathbf{X} = [\mathbf{X}_1 \cdots \mathbf{X}_p]$, $s_i \equiv (\mathbf{X}_i^{\mathrm{T}}\mathbf{X}_i)^{-1/2}$, and $\mathbf{S} \equiv \mathrm{diag}(s_1, \ldots, s_p)$, then

$$\tilde{\kappa}(\mathbf{X}) \equiv \kappa(\mathbf{XS}) \tag{3.26}$$

and

$$\tilde{\eta}_i(\mathbf{X}) \equiv \eta_i(\mathbf{XS}), \qquad i = 1, \ldots, p. \tag{3.27}$$

We can similarly refer to the scaled variance–decomposition proportions as those relevant to the scaled matrix **XS**.

This terminology may at first appear confusing since a scaled condition index is not a condition index that has been scaled but rather is a condition index of a matrix that has been scaled (column equilibrated). However, this confusion is slight and does not outweigh the economy of terminology that otherwise results.

The Diagnostic Procedure

The preceding discussions can be summarized to suggest the following practical procedure for (1) determining the presence of one or more near dependencies among the columns of a data matrix **X**, (2) determining which variates are involved in each such near dependency, and (3) in the event that the data are being used for the estimation of a linear regression equation by least squares, assessing the degree to which each estimated regression coefficient is being degraded by the presence of each such near dependency.

We suggest, then, the following double condition for diagnosing the presence of degrading collinearity:

I A scaled condition index judged to be high is associated with

II High scaled variance–decomposition proportions for *two or more* estimated regression coefficient variances.

The number of scaled condition indexes deemed large (say, greater than 30) in I identifies the number of near dependencies among the columns of the data matrix **X**, and the magnitudes of these high scaled condition indexes provide a measure of their relative "tightness." Furthermore, the determination in II of large variance–decomposition proportions (say, greater than 0.5) associated with each high scaled condition index identifies those variates that are involved in that near dependency, and the magnitude of these proportions in conjunction with the high scaled condition index provides a measure of the degree to which the corresponding regression estimate has been degraded by the presence of collinearity. When there are several coexisting or simultaneous near dependencies, additional considerations must be taken into account in interpreting the size of the variance–decomposition proportions. These issues are developed in the next chapter along with experiments that help us put meaning to *high* and *large*, two terms whose meaning in this context can only be determined empirically and that must necessarily be used loosely here.

Examining the Near Dependencies: Auxiliary Regressions

Once the variates involved in each near dependency have been identified by their high variance–decomposition proportions in II, the near dependency itself can be examined—for example, by regressing one of the variates involved on the others. Another procedure is suggested by (3.11). In the event that there are $p-r$

exact dependencies among the columns of $\mathbf{X}$, $\mathbf{V}_2$ in (3.11) has rank $p - r$, and we may partition $\mathbf{X}$ and $\mathbf{V}_2$ to obtain

$$[\mathbf{X}_1 \ \ \mathbf{X}_2]\begin{bmatrix}\mathbf{V}_{21}\\ \mathbf{V}_{22}\end{bmatrix} = \mathbf{X}_1\mathbf{V}_{21} + \mathbf{X}_2\mathbf{V}_{22} = \mathbf{0},$$

where $\mathbf{V}_{21}$ is chosen square and nonsingular. Hence, the dependencies among the columns of $\mathbf{X}$ are displayed as

$$\mathbf{X}_1 = -\mathbf{X}_2\mathbf{V}_{22}\mathbf{V}_{21}^{-1} \equiv \mathbf{X}_2\mathbf{G}. \tag{3.29}$$

The elements of $\mathbf{G} \equiv -\mathbf{V}_{22}\mathbf{V}_{21}^{-1}$ can be calculated directly from those of $\mathbf{V}$ resulting from the singular-value decomposition and provide one means for displaying the linear relations between the elements in $\mathbf{X}_1$ and those in $\mathbf{X}_2$.

Of course, (3.28) holds exactly only in the event that the linear dependencies in $\mathbf{X}$ are exact. It is also straightforward in this event to show that the $\mathbf{G}$ in (3.29) is exactly the matrix of least-squares estimates that results when $\mathbf{X}_1$ is regressed on $\mathbf{X}_2$. It seems reasonable, therefore, because of the relative simplicity involved, to employ least-squares regression as the descriptive mechanism for displaying the linear dependencies once the variates are discerned in II. It then also becomes possible to use the accompanying t-statistics as descriptive measures of the strength of individual variate involvement. These descriptive regressions are called the *auxiliary regressions*, and the way in which they are to be formed will be discussed in detail in Section 5.3.

It is important to reiterate the point made earlier that least-squares regression applied blindly to the columns of $\mathbf{X}$ does not and cannot substitute for the diagnostic procedure suggested here. Descriptive regressions can be applied rationally only after it has first somehow been determined how many dependencies there are among the columns of $\mathbf{X}$ and which variates are involved. The diagnostic procedure suggested here, by contrast, requires no prior knowledge of the numbers of near dependencies involved or of the variates involved in each; it discovers this information—treating all columns of $\mathbf{X}$ symmetrically and requiring that none be chosen to become the "dependent" variate. This information can then in turn be used directly to indicate the relevant auxiliary regressions.

It should also be asked just what these auxiliary regressions do and do not mean. Most importantly, the auxiliary regressions are not intended to specify and estimate any process supposed to generate the exogenous or predetermined variates $\mathbf{X}$. Rather, they are merely one possible descriptive summary of the near dependencies that occur in a specific set of data, and as such, they are not intended to be viewed inferentially. Clearly, any linear combination of these relations would also suffice for this purpose, and many compositions for $\mathbf{X}_1$ and $\mathbf{X}_2$ in (3.29) are possible. A mechanical means for selecting $\mathbf{X}_1$ and $\mathbf{X}_2$ that works effectively for many cases will be described in Section 5.3.

The auxiliary regressions do not, therefore, determine whether the process generating the **X** data is itself ill conditioned, only whether the specific result of that process currently being used, the **X** matrix, is ill conditioned. The broader question of whether the **X**-generating process is itself ill conditioned is an interesting one, but one that extends far beyond diagnostic issues considered here and one that, for diagnostic purposes, would seem often to be of limited relevance. If, for example, this process were determined to be well conditioned, this does not preclude the possibility (indeed, the probability) that it could nevertheless generate a specific **X** matrix that was ill conditioned. And for the user of such ill-conditioned data in a least-squares regression analysis, there would be little comfort in knowing that nature had played him a cruel trick: the given data would still be ill conditioned, and this is the diagnostic information that would be important to know. Suggestions for appropriate corrective measures, however, would benefit from knowing if additional data generated from this process are likely to be better conditioned or whether they too are likely to possess problems similar to the given sample.

An Alternative Definition of Collinearity

The scaled condition number of a matrix **X** along with the variance–decomposition proportions give us the basis for defining collinearity that eluded the various attempts discussed in Section 2.3. Collinearity exists when I and II are met. We arrived at this definition by pursuing means for determining when $\|\mathbf{a}\|$ in (3.1) is small, a pursuit that led to the use of scaled condition indexes. Thus, in this context, it is also of interest briefly to discuss another means of defining small and its corresponding definition of collinearity as suggested by Gunst (1983). Although initially this definition seems different from the one previously given, it is in fact seen to be equivalent and therefore provides useful perspective.

Gunst defines the $n \times p$ matrix **X** to be collinear if, for γ chosen suitably small, there exists $\mathbf{c} \neq \mathbf{0}$ such that, for $\mathbf{a} \equiv \mathbf{Xc}$,

$$\|\mathbf{a}\| < \gamma\|\mathbf{c}\|. \tag{3.30}$$

Note the relation between this and (3.1) and the means for defining small. Here, it is assumed that **X** is of full rank p and has its columns scaled to unit length.[19] To see the equivalence between this definition of collinearity and one based on the scaled condition number $\tilde{\kappa}(\mathbf{X})$, let **V** be the orthogonal matrix from (3.6) giving $\mathbf{V}^T\mathbf{X}^T\mathbf{XV} = \mathbf{\Lambda}$, the diagonal matrix of the eigenvalues of $\mathbf{X}^T\mathbf{X}$, which we assume here are ordered so $\lambda_1 \geqslant \cdots \geqslant \lambda_p > 0$. Now, make the orthonormal linear substitution

$$\mathbf{c} \equiv \mathbf{Vh}, \tag{3.31}$$

[19]The treatment given in Gunst (1983) fails to include this essential column equilibration.

so that $\|\mathbf{c}\| = \|\mathbf{h}\|$. Then we have

$$\|\mathbf{a}\| \equiv \|\mathbf{Xc}\| = \|\mathbf{\Lambda}^{1/2}\mathbf{h}\| = (h_1^2\lambda_1 + \cdots + h_p^2\lambda_p)^{1/2}. \tag{3.32}$$

Furthermore, since the columns of $\mathbf{X}$ have unit length,

$$p = \operatorname{tr}(\mathbf{X}^T\mathbf{X}) = \operatorname{tr}(\mathbf{\Lambda}) = \sum_{i=1}^{p} \lambda_i. \tag{3.33}$$

Now, when $\mathbf{X}$ is well conditioned, $\lambda_1 \approx \cdots \approx \lambda_p$, and (3.33) implies $\lambda_i \approx 1$ for $i = 1, \ldots, p$. From (3.32) and (3.31), then, we have $\|\mathbf{a}\| \approx (\Sigma\, h_i^2)^{1/2} = \|\mathbf{h}\| = \|\mathbf{c}\|$. Thus, (3.30) can hold only if $\mathbf{X}$ is ill conditioned.

Conversely, suppose $\mathbf{X}$ is well conditioned. From (3.33) we see that $\lambda_1 \leqslant p$, and hence that $\tilde{\kappa}(\mathbf{X}) \equiv (\lambda_1/\lambda_p)^{1/2} \leqslant (p/\lambda_p)^{1/2}$, or

$$\lambda_p^{1/2} \leqslant p^{1/2}\tilde{\kappa}^{-1}(\mathbf{X}). \tag{3.34}$$

Now, choose $\mathbf{h} = (0, \ldots, 0, h_p)^T \neq \mathbf{0}$ and $\gamma = p^{1/2}\tilde{\kappa}^{-1}(\mathbf{X})$. Then from (3.32), (3.31), and (3.34),

$$\|\mathbf{a}\| = h_p\lambda_p^{1/2} = \lambda_p^{1/2}\|\mathbf{h}\| = \lambda_p^{1/2}\|\mathbf{c}\| \leqslant \frac{p^{1/2}}{\tilde{\kappa}(\mathbf{X})}\|\mathbf{c}\| = \gamma\|\mathbf{c}\|. \tag{3.35}$$

Thus, Gunst's defining relation (3.30) for collinearity holds if and only if $\mathbf{X}$ is ill conditioned, that is, if and only if $\tilde{\kappa}(\mathbf{X})$ is sufficiently large.

What is "Large" or "High"?

Just what constitutes a large scaled condition index or a high variance–decomposition proportion is a matter that can only be decided empirically. And in the next chapter we turn to a set of systematic experiments designed to shed light on this matter.[20] To provide a meaningful background against which to interpret those empirical results, however, we first give a more specific idea of what it means for collinearity to harm or to degrade a regression estimate.

3.4 THE ILL EFFECTS OF COLLINEARITY ON LINEAR REGRESSION ESTIMATED BY LEAST SQUARES

The ill effects that collinearity can cause the least-squares estimation of a linear regression model are two: one computational and one statistical.

Computational Problems

When we introduced the condition number earlier in this chapter, we showed that it provided a measure of the potential sensitivity of a solution of a linear

[20]It is of interest to note that, once we have determined when $\tilde{\kappa}(\mathbf{X})$ is to be considered large, we can use this information along with the relation $\gamma = p^{1/2}\tilde{\kappa}^{-1}(\mathbf{X})$ used following (3.34) to help put meaning to the otherwise undefined phrase *suitably small* used in Gunst's definition (3.30).

system of equations (exact or inexact) to changes in the elements of the system. Computationally this means that the solution to a least-squares problem contains a number of digits whose meaningfulness is limited by the conditioning of the data in a manner directly related to the condition number.[21] Indeed, the condition number gives a magnification factor by which imprecision in the data can be blown up to produce even greater imprecision in the solution to a linear system of equations. Somewhat loosely, if the data are known to d significant digits and the condition number of the matrix $\mathbf{A}$ of a linear system $\mathbf{Az} = \mathbf{c}$ is of the order of magnitude 10^r, then a small change in the data in its last place can (but need not) affect the solution $\mathbf{z} = \mathbf{A}^{-1}\mathbf{c}$ in the $(d - r)$th place.

Thus, if GNP data are trusted to four digits and the condition number of the data matrix $\mathbf{X}$ is 10^3, then a shift in the fifth place of GNP (which, since only the first four digits count, results in what must be considered an *observationally equivalent* data matrix) could affect the least-squares solution in its second $(5 - 3)$ significant digit. Only the first digit of the estimated regression coefficients is therefore trustworthy, the others potentially being worthless, arbitrarily alterable by modifications in $\mathbf{X}$ that do not affect the degree of accuracy to which the data are known. Needless to say, had the condition number of $\mathbf{X}$ been 10^4 or 10^5 in this case, one could trust none of the significant digits of the least-squares solution $\mathbf{b} = \mathbf{X}^+\mathbf{y}$.

This computational problem in the calculation of the least-squares estimates can be minimized but never removed.[22] The intuitive distrust held by users of least squares of estimates based on ill-conditioned data is therefore justified. A discussion of this topic in the context of the Longley (1967) data is to be found in Beaton, Rubin, and Barone (1976).

Statistical Problems

Statistically, it is well known that the problem introduced by the presence of collinearity in the data matrix is the decreased precision with which statistical estimates conditional[23] upon those data may be known; that is, collinearity can cause the conditional variances of the estimates to be high. The reason for this may be seen directly from the variance decomposition (3.22), which shows how a small singular value resulting from a collinear relation causes a term in the

[21] For a more detailed discussion of this topic, see Stewart (1973), Belsley and Klema (1974), and Belsley (1976).

[22] One is better off, for example, in calculating the least-squares estimates with the pseudo-inverse as $\mathbf{b} = \mathbf{X}^+\mathbf{y}$ than with the normal equations $\mathbf{b} = (\mathbf{X}^T\mathbf{X})^{-1}\mathbf{X}^T\mathbf{y}$. This is because the conditioning of $\mathbf{X}^+$ is $\kappa(\mathbf{X})$, while that of $(\mathbf{X}^T\mathbf{X})^{-1}$ is $\kappa(\mathbf{X}^T\mathbf{X}) = \kappa^2(\mathbf{X})$. This last equality is seen by using the singular-value decomposition of $\mathbf{X} = \mathbf{UDV}^T$ to show that $\mathbf{X}^T\mathbf{X} = \mathbf{VD}^2\mathbf{V}^T$, which, by definition, must be the SVD of $\mathbf{X}^T\mathbf{X}$. Clearly, then, $\kappa(\mathbf{X}^T\mathbf{X}) = \mu^2_{max}/\mu^2_{min} = \kappa^2(\mathbf{X})$. Hence, any ill conditioning in $\mathbf{X}$ can be greatly compounded in its ill effects on the least-squares solution calculated as $\mathbf{b} = (\mathbf{X}^T\mathbf{X})^{-1}\mathbf{X}^T\mathbf{y}$. Procedures like the singular-value decomposition or the QR decomposition that avoid forming $\mathbf{X}^T\mathbf{X}$ and its inverse, then, are greatly to be preferred. See Golub (1969) or Belsley (1974).

[23] To avoid any possible confusion, we note that this statistical use of the word *conditional* has nothing to do with (and thus is to be contrasted with) the numerical-analytic notion of conditioning or ill-conditioned data.

variance decomposition to blow up. Intuitively, this problem reflects the fact that, when data are ill conditioned, some data series are nearly linear combinations of others and hence add very little new, independent information from which additional statistical information may be gleaned.

Needless to say, inflated variances are quite harmful to the use of regression as a basis for hypothesis testing, estimation, and forecasting. All users of linear regression have had the suspicion that an important test of significance has been rendered inconclusive through a needlessly high error variance induced by collinear data or that a confidence interval or forecast interval is uselessly large, reflecting the lack of properly conditioned data from which appropriately refined intervals could conceivably have been estimated.

Both of the preceding ill effects of collinear data are most directly removed through the introduction of new and well-conditioned data. In many applications, however, new data are either altogether unavailable or available only at great cost in time and effort. The next best substitute for new data is good prior information. We discuss and exemplify this later in Chapter 10. Prior information can often reduce the statistical (but not always the computational) problem due to collinearity. The usefulness of having diagnostic tools that signal the presence of collinearity and even indicate the variates involved is therefore apparent, for with them the investigator can at least determine whether the additional effort (costs) to correct for collinearity, either by collecting new data or developing and introducing prior information, is potentially worthwhile—and perhaps he can learn a great deal more. But just how much can be learned? To what extent can diagnostics tell the degree to which collinearity has caused harm?

Harmful versus Degrading Collinearity

At the outset, it should be noted that not all collinearity need be harmful. We have already seen in the example of the Bauer matrix given in Exhibit 3.6 and its $\mathbf{\Pi}$ matrix of Exhibit 3.7 that near orthogonality can isolate regression estimates from the presence of even extreme collinearity. Also, it is well known that specific linear combinations of estimated regression coefficients may be well determined even if the individual coefficients are not.[24] If by chance, then, the investigator's interest centers only on unaffected parameter estimates or on well-determined linear combinations of the estimated coefficients, clearly, no problem exists.[25] The estimates of the constant term in Exhibit 2.2*d* or of β_3 in Exhibit 2.2*e* provide simple examples of this.

[24]See Silvey (1969) or Theil (1971, pp. 152–154). Further, the effect on the collinearity diagnostics of linear transformations of the data matrix $\mathbf{X}$ $(=\mathbf{ZG}$ for $\mathbf{G}$ nonsingular) or, equivalently, of reparameterizations of the model $\mathbf{y} = \mathbf{X}\boldsymbol{\beta} + \boldsymbol{\varepsilon}$ into $\mathbf{y} = \mathbf{Z}\boldsymbol{\delta} + \boldsymbol{\varepsilon} = (\mathbf{X}\mathbf{G}^{-1})\mathbf{G}\boldsymbol{\beta} + \boldsymbol{\varepsilon}$ is treated in Section 6.1.

[25]No problem, that is, as long as a regression algorithm is used that does not blow up in the presence of highly collinear data. We have already noted (footnote 22) that routines that solve the normal equations as $\mathbf{b} = (\mathbf{X}^T\mathbf{X})^{-1}\mathbf{X}^T\mathbf{y}$ are unduly sensitive to ill-conditioned $\mathbf{X}$. This problem is lessened by using regression routines based on the SVD or QR decomposition.

In a less extreme and therefore practically more useful example, we note that the problems caused by collinearity in **X** may also be mitigated by low noise in the generation of the response variate **y**. We recall from (3.22) that the variance of the kth regression coefficient, var(b_k), is $\sigma^2 \Sigma_j v_{kj}^2/\mu_j^2$. If σ^2 is sufficiently small, it may be that particular var(b_k)'s are small enough for specific testing purposes despite large components in the v_{kj}^2/μ_j^2 terms resulting from near dependencies. One can see from Exhibit 2.2*c* that, if the height of the cloud, which is determined by σ^2, is made smaller relative to its "width," the regression plane becomes better defined even though the conditioning of the **X** data remains unchanged. This is seen even more clearly in the n-dimensional geometry of Exhibit 2.3, where the radius of the circle is directly related to σ^2. Clearly, even the extreme collinearity in Exhibit 2.3*c* can be overcome by a sufficiently small σ^2. Thus, if an investigator is only interested in whether a given coefficient is significantly positive and is able, even in the presence of collinearity, to accept that hypothesis on the basis of the relevant t-test, then collinearity has caused no problem. Of course, the resulting forecasts or estimates may have wider confidence intervals than would be needed to satisfy a more ambitious researcher, but for the limited purpose of the test of significance initially proposed, collinearity has caused no practical harm. These cases serve to exemplify the pleasantly pragmatic philosophy that collinearity doesn't hurt so long as it doesn't bite.

How, then, can one know when collinearity starts to hurt? To help answer this question, we make a distinction between degrading collinearity and harmful collinearity. We have seen that the mere presence of collinearity does not mean it is harmful. But if the researcher were provided with information that (1) there are strong near dependencies among the data, so that collinearity is *potentially a problem*, and (2) variances of parameters (or confidence intervals based on them) that are of interest have large proportions of their magnitude associated with the presence of the collinear relation(s), so that collinearity is *potentially harmful*, then he would be a long way toward deciding whether the costs of corrective action were worth undertaking. In addition, such information would help to indicate when variances of interest were not being adversely affected and so could be relied upon without further action. This information is, of course, precisely that provided by the scaled condition indexes and variance–decomposition proportions used in the two-pronged diagnostic procedure I and II suggested earlier. Thus, we say that, when this joint condition is met, the affected regression coefficients have been *degraded* (but not necessarily harmed) by the presence of collinearity—degraded in the sense that the magnitude of the estimated variance is being determined primarily by the presence of a collinear relation. Therefore, there is a presumption that confidence intervals, prediction intervals, and point estimates based on this estimate could be refined, if need be, by introducing better-conditioned data or relevant prior information.

Providing evidence that collinearity has actually harmed estimation, however, is a significantly more difficult task. To do this, one must show, for example, that a prediction interval that is too wide for a given purpose could be appropriately narrowed if made statistically conditional on a better-conditioned set of data, or that a confidence interval could be appropriately narrowed, or that the computational precision of a point estimator could be appropriately increased. Tools that would allow this clearly require knowledge of the response variate $\mathbf{y}$ and the error noise σ^2 (or its estimate s^2) and go beyond the two-step diagnostic test I and II presented previously, which uses only information on the $\mathbf{X}$ matrix. Eventually we shall be able to develop such a test for *harmful* collinearity, but that must await Chapter 7. First we must develop and interpret fully the two-pronged test for degrading collinearity that is a part of it.

At what point, then, do estimates become degraded? We turn in the next chapter to a series of experiments to help answer this question. These result in a rule of thumb that estimates are degraded when two or more variances have at least half of their magnitude associated with a scaled condition index of 30 or more.

3.5 APPENDIX: THE CONDITION NUMBER AND INVERTIBILITY

In this appendix we examine a means for interpreting the relation that exists between the condition number of a matrix and the "invertibility" of that matrix. This material is not required for an effective understanding of the collinearity diagnostics, but those wishing to skip it for the first time round may find the concluding paragraph of interest.

Here we show that the higher is the condition number of a matrix, the greater is the potential sensitivity of the elements of its inverse to small changes in the elements of the matrix itself. Sensitivity is measured here by the economist's familiar notion of elasticity. In particular, we see that $2\kappa(\mathbf{A})$, twice the condition number of a real symmetric matrix $\mathbf{A}$, provides an upper bound for the elasticity of the diagonal elements of $\mathbf{A}^{-1}$ with respect to the elements of $\mathbf{A}$. This result is then particularized to the special case where $\mathbf{A} = \mathbf{X}^T\mathbf{X}$, and it is shown that $2\kappa(\mathbf{X})$ plays a similar role for the elasticity of the diagonal elements of $(\mathbf{X}^T\mathbf{X})^{-1}$ with respect to elements of $\mathbf{X}$. This latter result shows how the condition number of the data matrix $\mathbf{X}$ provides a measure of the potential sensitivity of the estimated standard errors of regression coefficients to small changes in the data, a result of direct interest to users of least-squares estimation.

The elements of a matrix $\mathbf{A}$ are denoted by $\mathbf{A} = (a_{rs})$ and those of the inverse (when $\mathbf{A}$ is square and invertible) by $\mathbf{A}^{-1} = (a^{ij})$. The m rows of an $m \times n$ matrix are denoted according to

$$\mathbf{A} = \begin{bmatrix} \mathbf{a}_1^T \\ \vdots \\ \mathbf{a}_m^T \end{bmatrix}.$$

The notation $|\cdot|$ indicates absolute value and $\|\cdot\|$ denotes Euclidean length (1.10).

We first show a result applicable to any nonsingular matrix **A**, which gains strength when applied to the case where **A** is a real symmetric matrix. Here we employ the elasticity notation $\xi_{rs}^{ij} \equiv (\partial a^{ij}/\partial a_{rs})(a_{rs}/a^{ij})$.

Theorem 3.1. Let **A** be a nonsingular matrix with condition number $\kappa(\mathbf{A})$. Then $|\xi_{rs}^{ij}| \leqslant c_{ij}\kappa(\mathbf{A})$, where $c_{ij} \geqslant 1$ for all i and j.

Proof. First, we recall[26] that $\partial a^{ij}/\partial a_{rs} = -a^{ir}a^{sj}$, and from the singular-value decomposition of $\mathbf{A} = \mathbf{UDV}^{\mathrm{T}}$, we note $a_{rs} = \mathbf{u}_r^{\mathrm{T}}\mathbf{D}\mathbf{v}_s$ and $a^{hk} = \mathbf{v}_h^{\mathrm{T}}\mathbf{D}^{-1}\mathbf{u}_k$. Hence, we may write

$$\xi_{rs}^{ij} \equiv \frac{\partial a^{ij}}{\partial a_{rs}}\frac{a_{rs}}{a^{ij}} = -\frac{(\mathbf{v}_i^{\mathrm{T}}\mathbf{D}^{-1}\mathbf{u}_r)(\mathbf{v}_s^{\mathrm{T}}\mathbf{D}^{-1}\mathbf{u}_j)(\mathbf{v}_s^{\mathrm{T}}\mathbf{D}\mathbf{u}_r)}{(\mathbf{v}_i^{\mathrm{T}}\mathbf{D}^{-1}\mathbf{u}_j)}. \tag{3.36}$$

Taking absolute values and applying the Cauchy–Schwarz inequality to the numerator produces

$$|\xi_{rs}^{ij}| = \frac{|\mathbf{v}_i^{\mathrm{T}}\mathbf{D}^{-1}\mathbf{u}_r||\mathbf{v}_s^{\mathrm{T}}\mathbf{D}^{-1}\mathbf{u}_j||\mathbf{v}_s^{\mathrm{T}}\mathbf{D}\mathbf{u}_r|}{|\mathbf{v}_i^{\mathrm{T}}\mathbf{D}^{-1}\mathbf{u}_j|} \leqslant \frac{(\mathbf{v}_i^{\mathrm{T}}\mathbf{D}^{-1}\mathbf{v}_i)^{1/2}(\mathbf{u}_j^{\mathrm{T}}\mathbf{D}^{-1}\mathbf{u}_j)^{1/2}}{|\mathbf{v}_i^{\mathrm{T}}\mathbf{D}^{-1}\mathbf{u}_j|}$$
$$\times (\mathbf{u}_r^{\mathrm{T}}\mathbf{D}^{-1}\mathbf{u}_r)^{1/2}(\mathbf{u}_r^{\mathrm{T}}\mathbf{D}\mathbf{u}_r)^{1/2}(\mathbf{v}_s^{\mathrm{T}}\mathbf{D}^{-1}\mathbf{v}_s)^{1/2}(\mathbf{v}_s^{\mathrm{T}}\mathbf{D}\mathbf{v}_s)^{1/2}. \tag{3.37}$$

Recalling that $\|\mathbf{u}_k\| = \|\mathbf{v}_k\| = 1$ for all k and that **D** is diagonal, we employ the fact[27] that for any symmetric matrix **B** and vector $\boldsymbol{\zeta}$ with $\|\boldsymbol{\zeta}\| = 1$, $\boldsymbol{\zeta}^{\mathrm{T}}\mathbf{B}\boldsymbol{\zeta} \leqslant \lambda_{\max}$, where $\lambda_{\max}$ is the maximal eigenvalue of **B**. Hence, we have

$$|\xi_{rs}^{ij}| \leqslant c_{ij}\kappa(\mathbf{A}), \tag{3.38}$$

where

$$c_{ij} \equiv \frac{(\mathbf{v}_i^{\mathrm{T}}\mathbf{D}^{-1}\mathbf{v}_i)^{1/2}(\mathbf{u}_j^{\mathrm{T}}\mathbf{D}^{-1}\mathbf{u}_j)^{1/2}}{|\mathbf{v}_i^{\mathrm{T}}\mathbf{D}^{-1}\mathbf{u}_j|} \geqslant 1,$$

by the Cauchy–Schwarz inequality. □

We next examine the case where **A** is a nonsingular real symmetric matrix. Recognizing that a change in a_{rs} is now also a change in a_{sr}, the elasticity of an element of $\mathbf{A}^{-1}$ with respect to a_{rs} becomes

$$\tilde{\xi}_{rs}^{ij} = \begin{cases} \left(\dfrac{\partial a^{ij}}{\partial a_{rs}} + \dfrac{\partial a^{ij}}{\partial a_{sr}}\right)\dfrac{a_{rs}}{a^{ij}} & \text{for } r \neq s, \\[2ex] \dfrac{\partial a^{ij}}{\partial a_{rr}}\dfrac{a_{rr}}{a^{ij}} & \text{for } r = s. \end{cases} \tag{3.39}$$

[26]Theil (1971, p. 33).
[27]Rao (1973).

Furthermore, along the diagonal of $\mathbf{A}^{-1}$, where $i = j$, we have

$$\tilde{\xi}_{rs}^{ii} = \begin{cases} 2\dfrac{\partial a^{ii}}{\partial a_{rs}}\dfrac{a_{rs}}{a^{ii}} = 2\xi_{rs}^{ii} & \text{for } r \neq s, \\ \dfrac{\partial a^{ii}}{\partial a_{rr}}\dfrac{a_{rr}}{a^{ii}} = \xi_{rr}^{ii} & \text{for } r = s, \end{cases} \tag{3.40}$$

where ξ_{rs}^{ij} is defined as (3.36).

Finally, we note that the symmetry of $\mathbf{A}$ implies that its singular-value decomposition takes the form $\mathbf{VDV}^{\mathrm{T}}$, so for all i, $\mathbf{v}_i = \pm\mathbf{u}_i$. Hence, along the diagonal, $c_{ii} = 1$ for all i in Theorem 3.1. Joining this fact to (3.40), we have just proved Theorem 3.2.

Theorem 3.2. Let $\mathbf{A}$ be a nonsingular real symmetric matrix with condition number $\kappa(\mathbf{A})$. Then for the diagonal elements $a^{ii} \neq 0$ of $\mathbf{A}^{-1}$, $|\tilde{\xi}_{rs}^{ii}| \leqslant 2\kappa(\mathbf{A})$ for $r \neq s$ and $|\tilde{\xi}_{rr}^{ii}| \leqslant \kappa(\mathbf{A})$ for all r.

From the point of view of users of least-squares regression, particular interest is attached to the case where $\mathbf{A} = \mathbf{X}^{\mathrm{T}}\mathbf{X}$, where $\mathbf{X}$ is an $n \times p$ data matrix with condition number $\kappa(\mathbf{X})$. We now prove Theorem 3.3.[28]

Theorem 3.3. Let $\mathbf{X} = (x_{tk})$ be an $n \times p$ data matrix, and let $\mathbf{A} = \mathbf{X}^{\mathrm{T}}\mathbf{X}$. Then

$$|\xi_{tk}^{ii}| \equiv \frac{\partial a^{ii}}{\partial x_{tk}}\frac{x_{tk}}{a^{ii}} \leqslant 2\kappa(\mathbf{X})$$

for $i, k = 1, \ldots, p$ and $t = 1, \ldots, n$.

Proof

$$\begin{aligned} \xi_{tk}^{ii} \equiv \frac{\partial a^{ij}}{\partial x_{tk}}\frac{x_{tk}}{a^{ij}} &= \left(\sum_{r,s}\frac{\partial a^{ij}}{\partial a_{rs}}\frac{\partial a_{rs}}{\partial x_{tk}}\right)\frac{x_{tk}}{a^{ij}} \\ &= 2\left(\sum_{s=1}^{p}\frac{\partial a^{ij}}{\partial a_{ks}}x_{ts}\right)\frac{x_{tk}}{a^{ij}} \\ &= -2\frac{a^{ik}}{a^{ij}}x_{tk}\sum_{s=1}^{p}a^{sj}x_{ts} \\ &= -2\frac{a^{ik}}{a^{ij}}x_{tk}\mathbf{x}_t^{\mathrm{T}}\mathbf{a}^j \\ &= -2\frac{a^{ik}}{a^{ij}}x_{tk}z_{tj}, \end{aligned} \tag{3.41}$$

[28] I am indebted to R. J. O'Brien of the University of Southampton for providing a proof [employed here following Equation (3.41)] that substantially tightens my original bounds. In O'Brien (1975) a study of the sensitivity of ordinary least-squares estimates to perturbations in the data is undertaken.

where $\mathbf{x}_t^{\mathrm{T}}$ is the tth row of $\mathbf{X}$ and $\mathbf{a}^j$ is the jth column of $\mathbf{A}^{-1}$ and where we note that $\Sigma_{s=1}^{p} a^{sj} x_{ts} = \mathbf{x}_t^{\mathrm{T}} \mathbf{a}^j$ is the (j, t) element of the $p \times n$ matrix $\mathbf{Z}^{\mathrm{T}} \equiv (\mathbf{X}^{\mathrm{T}}\mathbf{X})^{-1}\mathbf{X}^{\mathrm{T}}$. Letting $\mathbf{z}_j$ be the jth column of $\mathbf{Z}$, we note that $|z_{tj}| \leqslant \|\mathbf{z}_j\| = (a^{jj})^{1/2}$, this latter since $\mathbf{Z}^{\mathrm{T}}\mathbf{Z} = (\mathbf{X}^{\mathrm{T}}\mathbf{X})^{-1}$. Likewise, $|x_{tk}| \leqslant \|\mathbf{x}_k\| = (a_{kk})^{1/2}$. Hence, for $i = j$, (3.41) becomes

$$|\xi_{tk}^{ii}| = 2 \frac{|a^{ik}|}{a^{ii}} |x_{tk}||z_{ti}| \leqslant 2 \frac{|a^{ik}|}{a^{ii}} (a_{kk})^{1/2}(a^{ii})^{1/2}. \tag{3.42}$$

Further, $\mathbf{A}$ positive definite implies $|a^{ik}| \leqslant (a^{ii})^{1/2}(a^{kk})^{1/2}$, resulting in

$$|\xi_{tk}^{ii}| \leqslant 2(a_{kk})^{1/2}(a^{kk})^{1/2}. \tag{3.43}$$

Now, for any p-vector $\boldsymbol{\alpha}$ such that $\|\boldsymbol{\alpha}\| = 1$, $\boldsymbol{\alpha}^{\mathrm{T}}\mathbf{A}\boldsymbol{\alpha} = \boldsymbol{\alpha}^{\mathrm{T}}\mathbf{V}\mathbf{D}^2\mathbf{V}^{\mathrm{T}}\boldsymbol{\alpha} \equiv \boldsymbol{\zeta}^{\mathrm{T}}\mathbf{D}^2\boldsymbol{\zeta}$, where $\|\boldsymbol{\zeta}\| \equiv \|\mathbf{V}^{\mathrm{T}}\boldsymbol{\alpha}\| = 1$, and hence,

$$\boldsymbol{\alpha}^{\mathrm{T}}\mathbf{A}\boldsymbol{\alpha} \leqslant \mu_{\max}^2. \tag{3.44}$$

In particular, letting $\boldsymbol{\alpha}$ be the kth-component unit vector, (3.44) becomes

$$a_{kk} \leqslant \mu_{\max}^2. \tag{3.45}$$

Using $\mathbf{A}^{-1}$ and $\mathbf{D}^{-2}$ in the preceding produces

$$a^{kk} \leqslant \frac{1}{\mu_{\min}^2}, \tag{3.46}$$

which implies, in conjunction with (3.45), that

$$(a_{kk})^{1/2}(a^{kk})^{1/2} \leqslant \frac{\mu_{\max}}{\mu_{\min}} = \kappa(\mathbf{X}). \tag{3.47}$$

Hence, (3.43) becomes

$$|\xi_{tk}^{ii}| \leqslant 2\kappa(\mathbf{X}). \tag{3.48}$$

□

This last result is directly interpretable in a least-squares context, for it says that the elasticity of the variance of any least-squares estimate with respect to any element in the data matrix $\mathbf{X}$ is bounded by twice the condition number of $\mathbf{X}$. That is, $2\kappa(\mathbf{X})$ provides an upper bound to the possible sensitivity of the parameter variances to changes in $\mathbf{X}$. Since condition numbers in excess of 100 are not uncommon for nonexperimental data matrices (the consumption function data, we shall see, have a $\tilde{\kappa}$ of 376), a 1% change in any element of $\mathbf{X}$

could result in a $2 \times 100\%$ change in the variance of any estimate or, roughly, a 14% change in its standard error. It is to be emphasized that this result is an inequality and hence shows the maximum potential sensitivity; it is not an immutable and incontrovertible fact of life. Ill-conditioned data, however, obviously have the potential for causing troubles, and the condition index provides a quick measure of the extent of this potential.

CHAPTER 4

The Experimental Experience

The test for the presence of degrading collinearity suggested in the previous chapter requires the joint occurrence of high variance–decomposition proportions for two or more coefficients associated with a single scaled condition index deemed to be large. Knowledge of what constitutes a high variance–decomposition proportion and a large scaled condition index must be determined empirically, and it is the purpose of this chapter to describe and report the results of a set of experiments designed to provide such experience.

4.1 THE EXPERIMENTAL PROCEDURE

Each of the six sets of experiments reported in this chapter examines the behavior of the scaled condition indexes and variance–decomposition proportions of a series of data matrices that are made to become systematically more and more ill conditioned by the presence of one or more near dependencies constructed to become more nearly exact.

Each experiment begins with a "basic" data set $\mathbf{X}$ of n observations on p_1 variates. The number of observations n, which is unimportant, varies between 24 and 30, and the number of basic variates p_1 varies between 3 and 5, depending upon the experiment. The basic data series are chosen either as actual economic time series, as constructs that mimic actual economic time series, or as random series.

These basic data series are used to construct additional collinear data series displaying increasingly tighter linear near dependencies with the basic series as follows. Let $\mathbf{c}$ be a p_1-vector of constants, and construct

$$\mathbf{w}_i = \mathbf{Xc} + \mathbf{e}_i, \tag{4.1}$$

where the components of $\mathbf{e}_i$ are generated iid normal with mean zero and variance $\sigma_i^2 = 10^{-i} s_{\mathbf{Xc}}^2$, where $s_{\mathbf{Xc}}^2 \equiv \text{var}(\mathbf{Xc})$, $i = 0, \ldots, 4$. Each $\mathbf{w}_i$, then, is constructed to be a known linear combination $\mathbf{Xc}$ of the basic data series plus a zero-mean random error term $\mathbf{e}_i$ whose variance becomes smaller and smaller (i.e., the dependency becomes tighter and tighter) in five stages with increasing i. In the $i = 0$ case, the variance in $\mathbf{w}_i$ due to the error term $\mathbf{e}_i$ is seen to be equal to that due to the systematic part $\mathbf{Xc}$. This produces an imposed linear dependency that is weak, having a correlation between $\mathbf{w}_i$ and $\mathbf{Xc}$ that tends to be between .4 and .6. By the time $i = 4$, however, only 1/10,000 of $\mathbf{w}_i$'s variance is due to the additive noise, and the dependency between $\mathbf{w}_i$ and $\mathbf{Xc}$ is tight, displaying a correlation very close to unity. A set of data matrices that become systematically more ill conditioned may therefore be constructed by augmenting the basic data matrix $\mathbf{X}$ with each $\mathbf{w}_i$, that is, by constructing the set

$$\mathbf{X}\{i\} \equiv [\mathbf{X}\ \ \mathbf{w}_i], \qquad i = 0, \ldots, 4. \tag{4.2}$$

The experiments are readily extended to the analysis of matrices possessing two or more coexisting or simultaneous near dependencies by the addition of more data series similarly constructed from the basic series. Thus, for a given p_1-vector $\mathbf{a}$, let

$$\mathbf{z}_j = \mathbf{Xa} + \mathbf{e}_j, \tag{4.3}$$

where the components of $\mathbf{e}_j$ are iid normal with mean zero and variance $\sigma_j^2 = 10^{-j} s_{\mathbf{Xa}}^2, j = 0, \ldots, 4$. Experimental matrices with two linear dependencies of varying strengths are constructed as

$$\mathbf{X}\{i, j\} \equiv [\mathbf{X}\ \ \mathbf{w}_i\ \ \mathbf{z}_j], \qquad i, j = 0, \ldots, 4. \tag{4.4}$$

In the third set of experiments that follows, three coexisting dependencies are examined.

In the first three sets of experiments, called the **X**, **Y**, and **Z** series, all multiple dependencies are coexisting; that is, they involve nonoverlapping subsets of the basic variates. In the last three sets of experiments, the **U**, **V**, and **W** series, truly simultaneous relations are examined in which some variates exist in several near dependencies.

The Choice of X's

For the first three sets of experiments, the data series chosen for the basic matrices **X** are either actual economic time series or variates constructed to mimic actual economic time series. In the last three sets of experiments, the basic data are simply chosen randomly. The principle of selection in all cases is to produce a basic data matrix that is reasonably well conditioned, so that all significant ill conditioning can be controlled through the introduced near dependencies such as (4.1).[1]

[1]As we shall see in experiment 2, this objective is only partially achieved, leading to an unexpected set of dependencies that nevertheless provides a further successful test of the usefulness of this diagnostic procedure.

The various series of matrices that comprise any one experiment all have the same basic data matrix and differ only in the constructed near dependencies used to augment them. Within any one such series, the augmenting near dependencies become systematically tighter with increased i, j, or k, and it is in this sense that we can speak meaningfully of what happens to scaled condition indexes and variance–decomposition proportions as the data matrix becomes "more ill conditioned" or "more nearly singular" or as "the near dependencies get tighter" or "the degree of collinearity increases."

Experimental Shortcomings

The experiments given here, while not Monte Carlo experiments,[2] are nevertheless subject to similar weaknesses; namely, the results depend on the specific experimental matrices chosen and cannot be generalized to different situations with complete assurance. It has been attempted, therefore, within the necessarily small number of experiments reported here, to choose basic data matrices using data series and combinations of data series representing a wide variety of actual circumstances. Needless to say, not all meaningful cases can be considered, and the reader will no doubt think of cases he would rather have seen analyzed.

The results also depend on the particular sample realizations that occur for the various $\mathbf{e}_i$ and $\mathbf{e}_j$ series that are used to generate the dependencies such as (4.1) or (4.3). However, the cases offered here are sufficiently varied that any systematic patterns that emerge from them are worthy of being reported and will certainly provide an excellent basis for any refinements that subsequent experience may suggest. In fact, the results are seen to be extremely stable across all cases and experiments, an outcome that is very reassuring.

The Experimental Report

Selected tables displaying the scaled condition indexes and the variance–decomposition proportions ($\mathbf{\Pi}$ matrices) are reported for each experiment in order to show how these two principal pieces of diagnostic information change as the near dependencies get tighter. The term *scaled* merely reminds us that the data are first scaled to have equal (unit) column length before being subjected to the singular-value decomposition, from which the condition indexes and variance–decomposition proportions are calculated. In the event that the linear relations between the variates are displayed through supplementary regressions, however, the regression coefficients have been rescaled to their original units.

Additional statistics, such as the simple correlations of the contrived dependencies and their uncentered $\hat{R}^2$s as measured from relevant regressions,[3] are also reported to provide a link between the magnitudes of scaled condition

[2]No attempt is made here to infer any distributional properties through repeated samplings.

[3]Beyond the fact that there may be no constant term in these regressions, the reasons for preferring the use of uncentered $\hat{R}^2$ in this context are made clear in Chapters 2 and 6.

indexes and these more familiar notions. It is clear, however, that these additional statistics cannot substitute for the information provided by the scaled condition indexes and the variance–decomposition proportions. In the experiments that follow, we know a priori which variates are involved in which relations and what the generating constants, such as **c** in (4.1), are. It is therefore possible to compute simple correlations between $\mathbf{w}_i$ and **Xc** and to run regressions of $\mathbf{w}_i$ on **X**. In practice, of course, **c** is unknown and one does not know which columns of the data matrix are involved in which dependencies, if any at all. These supplementary statistics are, therefore, not available to the investigator as independent analytic or diagnostic tools. However, one can learn from the variance–decomposition proportions which variates are involved in which relations, and regressions may then be run among these variates to display the near dependencies. Furthermore, the t-statistics that result from these regressions can be used in the standard way for providing additional *diagnostic* evidence of the "significance" (strength of involvement) of each variate in the specific linear dependency. Thus, a collinearity analysis using scaled condition indexes and variance–decomposition proportions can suggest useful auxiliary regressions as a means for exhibiting the near dependencies, but regression, by itself, cannot provide similar information, particularly if there are two or more simultaneous near dependencies.[4]

4.2 THE INDIVIDUAL EXPERIMENTS

Six sets of experiments are conducted, each using a separate series of data matrices designed to represent different types of data and different types of collinearity. Thus, "levels" data (manufacturing sales), "trended" data (GNP), "rate-of-change" data (inventory investment), "rates" data (unemployment), and random data are all represented. Similarly, the types of collinear relations that are generated include simple relations between two variates, relations involving more than two variates, coexisting and simultaneous near dependencies, and near dependencies among variates with essential scaling problems. In the first three sets of experiments, the different cases of relations involving more than two variates have been chosen to involve different mixes of the various types of economic variates listed above. In these cases, the near dependencies are generated from the unscaled (natural) economic data, and hence the various test data sets represent as closely as possible economic data with natural near dependencies.

[4]Presumably, a time-consuming and relatively costly all-subsets regression could reveal some or all of this information. But in choosing among the various ways to skin a cat, the hair-by-hair method might well be avoided.

Experiment 1: The X Series

The basic data set employed here is

$$\mathbf{X} \equiv [\mathbf{MFGS}, \mathbf{IVM}^*, \mathbf{MV}],$$

where

MFGS is total manufacturers' shipments,
IVM* is total manufacturers' inventories,
MV is total manufacturers' unfilled orders.

Each series is in millions of U.S. dollars, annual 1947–1970 ($n = 24$). This basic data set provides the type of series that would be relevant, for example, to an econometric study of inventory investment.[5] The asterisk after a series name indicates that it has been constructed to retain the same mean and variance as the given series but modified to provide a well-conditioned basic data matrix.

Two sets of additional dependency series are generated from **X** as follows:

$$\mathbf{w}_i = \mathbf{MV} + \mathbf{e}_i, \qquad i = 0, \ldots, 4, \tag{4.5a}$$

with $\mathbf{e}_i$ generated as normal with mean zero and variance–covariance matrix $\sigma_i^2\mathbf{I} = 10^{-i}s_{\mathbf{MV}}^2\mathbf{I}$ [denoted $\mathbf{e}_i \sim N(\mathbf{0}, 10^{-i}s_{\mathbf{MV}}^2\mathbf{I})$], $s_{\mathbf{MV}}^2$ being the sample variance of the **MV** series, and

$$\begin{aligned} \mathbf{z}_j &= 0.8\mathbf{MFGS} + 0.2\mathbf{IVM}^* + \mathbf{e}_j, \\ \mathbf{e}_j &\sim N(\mathbf{0}, 10^{-j}s_{\mathbf{z}}^2\mathbf{I}), \end{aligned} \tag{4.5b}$$

$s_{\mathbf{z}}^2$ being the sample variance of $0.8\mathbf{MFGS} + 0.2\mathbf{IVM}^*$.

The $\mathbf{w}_i$ and $\mathbf{z}_j$ series are used to augment the basic data set to produce three sequences of matrices:

$$\begin{aligned} \mathbf{X1}\{i\} &\equiv [\mathbf{X}\ \ \mathbf{w}_i] & i &= 0, \ldots, 4, \\ \mathbf{X2}\{j\} &\equiv [\mathbf{X}\ \ \mathbf{z}_j] & j &= 0, \ldots, 4, \\ \mathbf{X3}\{i, j\} &\equiv [\mathbf{X}\ \ \mathbf{w}_i\ \ \mathbf{z}_j] & i, j &= 0, \ldots, 4, \end{aligned} \tag{4.6}$$

each of which is subjected to analysis.

The dependency (4.5a) is a commonly encountered simple relation between two variates. Unlike more complex relations, it is a dependency whose presence can be discovered through examination of the simple correlation matrix of the columns of the $\mathbf{X1}\{i\}$ or the $\mathbf{X3}\{i, j\}$. Its inclusion therefore, allows us to learn

[5]See, e.g., Belsley (1969a) or the case study given in Section 10.3.

how scaled condition indexes and simple correlations compare with one another.

The dependency (4.5b) involves three variates and hence would not generally be discoverable through an analysis of the simple correlation matrix. This equation is designed to present no difficult scaling problems; that is, the two basic data series **MFGS** and **IVM*** have roughly similar magnitudes and variations, and the coefficients (0.8 and 0.2) are of the same order of magnitude. No one variate, therefore, dominates the near dependency, masking the effects of the others. This situation should allow the involvement of the three variates and the relation among them to be determined with relative ease.

Experiment 2: The Y Series

The basic data set employed here is

$$\mathbf{Y} \equiv [\mathbf{GNP58^*}, \mathbf{GAVM^*}, \mathbf{LHTUR^*}, \mathbf{GV58^*}],$$

where

GNP58* is GNP in 1958 dollars,
GAVM* is net corporate dividend payments,
LHTUR* is the unemployment rate, and
GV58* is annual change in total inventories, 1958 dollars.

Each basic series here has been constructed from the indicated series to have similar means and variances but modified to produce a reasonably well conditioned **Y** matrix. The data are annual, 1948–1974 ($n = 27$). The variates included here, then, represent "levels" variates (**GNP58***), rates of change (**GV58***), and "rates" (**LHTUR***).

Three additional dependency series are constructed as ($i, j, k = 0, \ldots, 4$)

$$\begin{aligned} \mathbf{v}_i &= \mathbf{GNP58^*} + \mathbf{GAVM^*} + \mathbf{e}_i, \\ &\quad \mathbf{e}_i \sim N(\mathbf{0}, 10^{-i} s_{\hat{\mathbf{v}}}^2 \mathbf{I}), \\ &\quad s_{\hat{\mathbf{v}}}^2 = \text{var}(\mathbf{GNP58^*} + \mathbf{GAVM^*}), \end{aligned} \tag{4.7a}$$

$$\begin{aligned} \mathbf{w}_j &= 0.1\mathbf{GNP58^*} + \mathbf{GAVM^*} + \mathbf{e}_j, \\ &\quad \mathbf{e}_j \sim N(\mathbf{0}, 10^{-j} s_{\hat{\mathbf{w}}}^2 \mathbf{I}), \\ &\quad s_{\hat{\mathbf{w}}}^2 = \text{var}(0.1\mathbf{GNP58^*} + \mathbf{GAVM^*}), \end{aligned} \tag{4.7b}$$

$$\begin{aligned} \mathbf{z}_k &= \mathbf{GV58^*} + \mathbf{e}_k, \\ &\quad \mathbf{e}_k \sim N(\mathbf{0}, 10^{-k} s_{\mathbf{GV58^*}}^2 \mathbf{I}). \end{aligned} \tag{4.7c}$$

These data sets are used to augment **Y** to produce four series of test matrices:

$$\begin{aligned} \mathbf{Y1}\{i\} &\equiv [\mathbf{Y}\ \ \mathbf{v}_i] && i = 0, \ldots, 4, \\ \mathbf{Y2}\{j\} &\equiv [\mathbf{Y}\ \ \mathbf{w}_j] && j = 0, \ldots, 4, \\ \mathbf{Y3}\{k\} &\equiv [\mathbf{Y}\ \ \mathbf{z}_k] && k = 0, \ldots, 4, \\ \mathbf{Y4}\{i, k\} &\equiv [\mathbf{Y}\ \ \mathbf{v}_i\ \ \mathbf{z}_k] && i, k = 0, \ldots, 4. \end{aligned} \tag{4.8}$$

Dependency (4.7a) presents a relation among three variates with an essential scaling problem; namely, in the units of the basic data, the variation introduced by **GNP58*** is less than 1% of that introduced by **GAVM***. The inclusion of **GNP58*** is therefore dominated by **GAVM***, and its effects should be somewhat masked and difficult to discern. Dependency (4.7b) is of a similar nature except that the scaling problem has been made even more extreme. These are *essential scaling problems* in that the relative weakness of the dominated variate is "locked in" the linear combination and cannot be undone through simple column scalings or more general linear transformations of the data. Dependency (4.7c) is a simple relation between two variates, except in this case, the variate is a rate of change, exhibiting numerous shifts in sign.

Experiment 3: The Z Series

The basic data matrix here is an expanded version of that in the previous experiment:

$$\mathbf{Z} = [\mathbf{GNP58^*}, \mathbf{GAVM^*}, \mathbf{LHTUR^*}, \mathbf{DUM1}, \mathbf{DUM2}],$$

where **DUM1** is generated similar to **GV58***, and **DUM2** is similar to **GNP58***, except that they are generated to have very low intercorrelation with the first three variates. This configuration allows examination of the case described in Section 3.2 when some variates are isolated by near orthogonality from dependencies among others.

The additional dependency series are ($i, j, k, m = 0, \ldots, 4$):

$$\begin{aligned} \mathbf{u}_i &= \mathbf{DUM1} + \mathbf{e}_i, \\ &\quad \mathbf{e}_i \sim N(\mathbf{0}, 10^{-i} s^2_{\mathbf{DUM1}} \mathbf{I}), \end{aligned} \tag{4.9a}$$

$$\begin{aligned} \mathbf{v}_j &= \mathbf{DUM2} - \mathbf{DUM1} + \mathbf{e}_j, \\ &\quad \mathbf{e}_j \sim N(\mathbf{0}, 10^{-j} s^2_{\hat{\mathbf{v}}} \mathbf{I}), \\ &\quad s^2_{\hat{\mathbf{v}}} = \mathrm{var}(\mathbf{DUM2} - \mathbf{DUM1}), \end{aligned} \tag{4.9b}$$

$$\begin{aligned} \mathbf{w}_k &= 3\mathbf{GNP58^*} + 1.5\mathbf{LHTUR^*} + \mathbf{e}_k, \\ &\quad \mathbf{e}_k \sim N(\mathbf{0}, 10^{-k} s^2_{\hat{\mathbf{w}}} \mathbf{I}), \\ &\quad s^2_{\hat{\mathbf{w}}} = \mathrm{var}(3\mathbf{GNP58^*} + 1.5\mathbf{LHTUR^*}), \end{aligned} \tag{4.9c}$$

$$\mathbf{z}_m = \mathbf{GAVM^*} + 0.7\mathbf{DUM2} + \mathbf{e}_m,$$
$$\mathbf{e}_m \sim N(\mathbf{0},\ 10^{-m} s_{\hat{\mathbf{z}}}^2 \mathbf{I}), \qquad (4.9\text{d})$$
$$s_{\hat{\mathbf{z}}}^2 = \text{var}(\mathbf{GAVM^*} + 0.7\mathbf{DUM2}).$$

These data are used to augment $\mathbf{Z}$ to produce seven series of test matrices:

$$\begin{aligned}
\mathbf{Z1}\{i\} &\equiv [\mathbf{Z}\ \ \mathbf{u}_i] & i &= 0, \ldots, 4,\\
\mathbf{Z2}\{j\} &\equiv [\mathbf{Z}\ \ \mathbf{v}_j] & j &= 0, \ldots, 4,\\
\mathbf{Z3}\{k\} &\equiv [\mathbf{Z}\ \ \mathbf{w}_k] & k &= 0, \ldots, 4,\\
\mathbf{Z4}\{m\} &\equiv [\mathbf{Z}\ \ \mathbf{z}_m] & m &= 0, \ldots, 4, \qquad (4.10)\\
\mathbf{Z5}\{j, k\} &\equiv [\mathbf{Y}\ \ \mathbf{v}_j\ \ \mathbf{w}_k] & j, k &= 0, \ldots, 4,\\
\mathbf{Z6}\{i, m\} &\equiv [\mathbf{Y}\ \ \mathbf{u}_i\ \ \mathbf{z}_m] & i, m &= 0, \ldots, 4,\\
\mathbf{Z7}\{i, k, m\} &\equiv [\mathbf{Y}\ \ \mathbf{u}_i\ \ \mathbf{w}_k\ \ \mathbf{z}_m] & i, k, m &= 0, \ldots, 4.
\end{aligned}$$

Each of the first three dependencies (4.9a–c) possesses essential scaling problems with **DUM2**, 3**GNP58***, and **GAVM***, respectively, being the dominant terms. The problem is extreme in the relation with **DUM1** and **DUM2**, where **DUM1** introduces much less than 0.1% of the total variation, and difficult in the other two cases. The relation defined by (4.9b) is isolated by near orthogonality from the one defined by (4.9c). These relations occur separately in the **Z2** and **Z3** test series and together in the **Z5** series. Relation (4.9d) bridges these two situations.

Experiment 4: The U Series

The next three sets of experiments differ somewhat from those just described. It will be noted in the previous experiments that, when there are multiple near dependencies, the variates involved in each are nonoverlapping, and as such, they are coexisting but not truly simultaneous near dependencies. These final experiments examine truly simultaneous near dependencies. The basic data matrices here do not attempt to mimic actual data series but are simply chosen randomly to assure good conditioning of the basic data matrix.

The basic data matrix employed for the **U** series is

$$\mathbf{U} \equiv [\mathbf{D1},\ \mathbf{D2},\ \mathbf{D3}],$$

where each **D** series is chosen from a uniform distribution on the interval 0–1.

Three additional dependency series are constructed as $(i, j, k = 0, \ldots, 4)$

$$\mathbf{u}_i = 0.3\mathbf{D2} - 0.6\mathbf{D3} + \mathbf{e}_i,$$
$$\mathbf{e}_i \sim N(\mathbf{0},\ 10^{-i} s_{\hat{\mathbf{u}}}^2 \mathbf{I}), \qquad (4.11\text{a})$$
$$s_{\hat{\mathbf{u}}}^2 = \text{var}(0.3\mathbf{D2} - 0.6\mathbf{D3}),$$

$$\begin{aligned} \mathbf{v}_j &= 0.7\mathbf{D1} + 0.4\mathbf{D2} + \mathbf{e}_j, \\ &\quad \mathbf{e}_j \sim N(\mathbf{0}, 10^{-j}s_{\hat{\mathbf{v}}}^2\mathbf{I}), \\ &\quad s_{\hat{\mathbf{v}}}^2 = \text{var}(0.7\mathbf{D1} + 0.4\mathbf{D2}), \end{aligned} \tag{4.11b}$$

$$\begin{aligned} \mathbf{w}_k &= 0.03\mathbf{D2} - 0.6\mathbf{D3} + \mathbf{e}_k, \\ &\quad \mathbf{e}_k \sim N(\mathbf{0}, 10^{-k}s_{\hat{\mathbf{w}}}^2\mathbf{I}), \\ &\quad s_{\hat{\mathbf{w}}}^2 = \text{var}(0.03\mathbf{D2} - 0.6\mathbf{D3}). \end{aligned} \tag{4.11c}$$

These data are used to augment **U** to produce two series of test matrices:

$$\begin{aligned} \mathbf{U1}\{i, j\} &\equiv [\mathbf{U}\ \ \mathbf{u}_i\ \ \mathbf{v}_j], \qquad i, j = 0, \ldots, 4, \\ \mathbf{U2}\{j, k\} &\equiv [\mathbf{U}\ \ \mathbf{w}_k\ \ \mathbf{v}_j], \qquad j, k = 0, \ldots, 4. \end{aligned} \tag{4.12}$$

The object here is to see if true simultaneity increases the problems in detecting near dependencies that involve more than simple correlations. The **U2** series simply adds a scaling problem to the well-scaled case in **U1**.

Experiment 5: The V Series

The basic data matrix employed here is

$$\mathbf{V} \equiv [\mathbf{E1}, \mathbf{E2}, \mathbf{E3}],$$

where each **E** series is chosen from a uniform distribution on the interval 0–1.

Two additional dependency series are constructed as $(i, j = 0, \ldots, 4)$

$$\begin{aligned} \mathbf{u}_i &= 0.4\mathbf{E1} - 0.3\mathbf{E2} + 0.6\mathbf{E3} + \mathbf{e}_i, \\ &\quad \mathbf{e}_i \sim N(\mathbf{0}, 10^{-i}s_{\hat{\mathbf{u}}}^2\mathbf{I}), \\ &\quad s_{\hat{\mathbf{u}}}^2 = \text{var}(0.4\mathbf{E1} - 0.3\mathbf{E2} + 0.6\mathbf{E3}), \end{aligned} \tag{4.13a}$$

$$\begin{aligned} \mathbf{v}_j &= 0.7\mathbf{E1} + 0.4\mathbf{E2} + \mathbf{e}_j, \\ &\quad \mathbf{e}_j \sim N(\mathbf{0}, 10^{-j}s_{\hat{\mathbf{v}}}^2\mathbf{I}), \\ &\quad s_{\hat{\mathbf{v}}}^2 = \text{var}(0.7\mathbf{E1} + 0.4\mathbf{E2}). \end{aligned} \tag{4.13b}$$

These data are used to augment **V** to produce one series of test matrices:

$$\mathbf{V1}\{i, j\} \equiv [\mathbf{V}\ \ \mathbf{u}_i\ \ \mathbf{v}_j], \qquad i, j = 0, \ldots, 4. \tag{4.14}$$

The object here is to examine the behavior of the collinearity diagnostics when a truly simultaneous set of near dependencies has more than one variate in common. Here, both **E1** and **E2** enter both near dependencies.

Experiment 6: The W Series

The basic data matrix employed here is

$$\mathbf{W} \equiv [\mathbf{F1}, \mathbf{F2}, \mathbf{F3}],$$

where each **F** series is chosen from a uniform distribution on the interval 0–1.

Two additional dependency series are constructed as $(i, j = 0, \ldots, 4)$

$$\mathbf{u}_i = \mathbf{F1} + \mathbf{e}_i,$$

$$\mathbf{e}_i \sim N(\mathbf{0}, 10^{-i} s_{\mathbf{F1}}^2 \mathbf{I}), \tag{4.15a}$$

$$\mathbf{v}_j = 0.7\mathbf{F1} + 0.4\mathbf{F2} - 0.3\mathbf{F3} + 0.6\mathbf{F4} + \mathbf{e}_j,$$

$$\mathbf{e}_j \sim N(\mathbf{0}, 10^{-j} s_{\mathbf{v}}^2 \mathbf{I}), \tag{4.15b}$$

$$s_{\mathbf{v}}^2 = \text{var}(0.7\mathbf{F1} + 0.4\mathbf{F2} - 0.3\mathbf{F3} + 0.6\mathbf{F4}).$$

These data are used to augment **W** to produce one series of test matrices:

$$\mathbf{W1}\{i, j\} \equiv [\mathbf{W} \ \ \mathbf{u}_i \ \ \mathbf{v}_j], \qquad i, j = 0, \ldots, 4. \tag{4.16}$$

The object here is to examine the behavior of the collinearity diagnostics when there are two simultaneous near dependencies, one involving all of the variates and the other a subset of them. The reader will recall that this is the case mentioned in Section 2.3 for which VIFs would be unable to diagnose the number of near dependencies (although VIFs would be able to determine overall variate involvement), and interest clearly centers on whether the scaled condition indexes and variance–decomposition proportions are able to do so.

4.3 THE EXPERIMENTAL RESULTS

Space limitations obviously prevent reporting the full set of experimental results. Fortunately, after reporting the first experiment in some detail, it is possible to select samples of output from the remaining experiments to convey what further generalizations are possible. This is because the experimental results are so pleasantly stable across all variations that the essential behavior of the diagnostic procedure can be substantively appreciated from the first experiment alone. Even when so abbreviated, the experimental report that follows is necessarily lengthy and may be difficult to digest fully on a first reading. Thus, while it is strongly advised that the reader invest the time and patience to study all the experiments, those in need of a faster pace might wish to skip all but the first experiment first time through and go directly thereafter to the summary given at the beginning of Chapter 5. Ultimately, however, these other experiments are necessary for introducing the important concepts of dominant and competing near dependencies, providing other refinements in interpretation, and demonstrating the empirical stability of the diagnostic procedure over a wide set of circumstances.

Experiment 1: The X Matrices

X1

Let us begin with the simplest set of experimental matrices, the $\mathbf{X1}\{i\}$, $i = 0, \ldots, 4$. Here the data series in column 4, which we denote as **C4**, is related to that of column 3, **C3**, by (4.5a), that is, $\mathbf{C4} = \mathbf{C3} + \mathbf{e}_i$, $i = 0, \ldots, 4$; and this is the only contrived near dependency among the four columns of **X1**. We would therefore expect one high scaled condition index and a large proportion of var(b_3) and var(b_4) to be associated with it. Exhibit 4.1*a* presents the scaled condition indexes and variance–decomposition proportions for this series as the tightness of the relation increases in five steps with i going from 0 to 4.

Exhibit 4.1*a* Scaled Condition Indexes and Variance–Decomposition Proportions: X1 Series. One Constructed Near Dependency (4.5a): $\mathbf{C4} = \mathbf{C3} + \mathbf{e}_i$

Scaled Condition Index, $\tilde{\eta}$	Proportions of			
	var(b_1)	var(b_2)	var(b_3)	var(b_4)
		X1{0}		
1	.005	.012	.001	.003
5	.044	.799	.004	.032
8	.906	.002	.041	.238
14	.045	.187	.954	.727
		X1{1}		
1	.005	.011	.001	.001
5	.094	.834	.003	.002
9	.899	.117	.048	.035
27	.002	.038	.948	.962
		X1{2}		
1	.005	.012	.000	.000
5	.087	.889	.000	.000
9	.901	.083	.003	.003
95	.007	.016	.997	.997

Exhibit 4.1*a* Continued

Scaled Condition Index, $\tilde{\eta}$	Proportions of var(b_1)	var(b_2)	var(b_3)	var(b_4)
		X1{3}		
1	.005	.012	.000	.000
5	.079	.903	.000	.000
9	.855	.079	.001	.001
461	.061	.006	.999	.999
		X1{4}		
1	.005	.010	.000	.000
5	.084	.793	.000	.000
9	.906	.070	.000	.000
976	.005	.127	1.000	1.000

A glance at these results confirms our expectations. In each case, there is a highest scaled condition index that accounts for a high proportion of the variance for two or more of the coefficients, and these are var(b_3) and var(b_4). Furthermore, the pattern is observable in the weakest case **X1**{0} and becomes increasingly clearer as the near dependency becomes tighter: all scaled condition indexes save one remain virtually unchanged, while the scaled condition index corresponding to the imposed near dependency increases strongly with each

Exhibit 4.1*b* Supplementary Correlations and Regressions of C4 on C1, C2, and C3: X1 Series[a]

Data Matrix	*r*(**C3**, **C4**)	**C1**	**C2**	**C3**	$\hat{R}^2$
X1{0}	.766	0.391 [1.11]	−0.135 [−0.91]	0.938 [4.76]	.6229
X1{1}	.931	0.148 [0.97]	0.092 [1.38]	0.885 [10.10]	.8765
X1{2}	.995	−0.008 [−0.17]	0.014 [0.72]	0.998 [38.80]	.9893
X1{3}	.999	0.011 [1.22]	0.002 [0.37]	0.990 [188.96]	.9996
X1{4}	1.000	−0.001 [−0.28]	0.003 [1.76]	0.998 [400.56]	.9999

[a]Consonant with the discussion in Section 1.1, the figures in square brackets here and in other exhibits are *t*-statistics since interest in these results centers wholly on "significance."

jump in i; the variance–decomposition proportions of the two "involved" variates **C3** and **C4** become larger and larger, eventually becoming unity.

To help interpret the scaled condition indexes in Exhibit 4.1*a*, we present in Exhibit 4.1*b* the simple correlation between **C3** and **C4** for each of the $\mathbf{X1}\{i\}$ matrices and also the multiple regressions of **C4** on **C1**, **C2**, and **C3**.

In addition to observing the general pattern that was expected, the following points are noteworthy:

1. The relation between **C3** and **C4** in case $\mathbf{X1}\{0\}$ has a simple correlation of .766 and a regression $\hat{R}^2$ of .6229, which is rather modest in comparison with simple correlations present in many real-life data matrices. It nevertheless shows itself in a scaled condition number of 14 and substantively large proportions of the variances of the affected coefficients, var(b_3) and var(b_4).

2. Also at this lowest level ($i = 0$), the proposed collinearity diagnostic correctly indicates the existence of the one near dependency and correctly points to the variates involved.

3. In light of point 2, the supplementary regressions of Exhibit 4.1*b* that were run for comparative purposes are also the auxiliary regressions that would be suggested for displaying the near dependencies. Exhibit 4.1*b* confirms that, even in the weakest case of $\mathbf{X1}\{0\}$ with a scaled condition index $\tilde{\eta}$ of 14, the proper near dependency among the columns of $\mathbf{X1}\{0\}$ is being clearly observed.

4. With each increase in i (corresponding to a 10-fold reduction in the variance of the noise in the near dependency), the simple correlations and the $\hat{R}^2$s increase one step, roughly adding another 9 in the series .9, .99, .999, and so on, and the scaled condition index increases roughly along the progression 1, 3, 10, 30, 100, 300, 1000, and so on—a pattern we consistently observe in future examples.[6] We shall call the first sequence the "progression of nines" and the second the "progression of 10/30." This relation suggests a means for measuring the order of magnitude of the "tightness" of a near dependency.

5. Also with each increase in i, the variance–decomposition proportions of the affected coefficients associated with the highest $\tilde{\eta}$ increase markedly, again roughly adding one more 9 with each step.

6. In the case of $\mathbf{X1}\{0\}$, the second highest scaled condition index is 8 and is not too different from the highest. But we recall from Section 3.2 that it is the joint condition of high variance–decomposition proportions for two or more coefficients associated with a high scaled condition index that signals the presence of degrading collinearity. These conditions are met for the largest scaled condition index of 14 but not for the next largest 8, which is a dominant component in only one variance, that of var(b_1). Here, then, we note that a scaled condition index of 14 (roughly 10 in the progression) appears to be "high enough" for the presence of collinearity to begin to be observed.

[6]This progression corresponds to roughly equal increments in $\log_{10}\tilde{\eta}_i$ of $\frac{1}{2}$; i.e., $\log_{10}\tilde{\eta}_i = 1 + (i/2)$. Alternatively, and somewhat more directly, this progression results from the sequence $10^{i/2}$, $i \geqslant 0$, and corresponds to $[1/(1 - R^2)]^{1/2}$ for $R^2 = 0, .9, .99, .999, \ldots$.

X2

The **X2**$\{i\}$ series also possesses only one constructed near dependency, (4.5b), but involving three variates, columns 1, 2, and 4 in the form **C4** = 0.8**C1** + 0.2**C2** + $\mathbf{e}_i$. We expect, then, high variance–decomposition proportions for these three variates to be associated with a single high scaled condition index. Exhibit 4.2*a* presents the **Π** matrices of the variance–decomposition proportions and the scaled condition indexes for the **X2**$\{i\}$ data

Exhibit 4.2*a* Scaled Condition Indexes and Variance–Decomposition Proportions: X2 Series. One Constructed Near Dependency (4.5b): C4 = 0.8C1 + 0.2C2 + $\mathbf{e}_i$.

Scaled Condition Index, $\tilde{\eta}$	Proportions of			
	var(b_1)	var(b_2)	var(b_3)	var(b_4)
		X2{0}		
1	.003	.012	.004	.005
4	.028	.735	.001	.068
9	.009	.223	.636	.526
11	.960	.030	.359	.401
		X2{1}		
1	.001	.004	.003	.000
5	.021	.297	.011	.001
10	.091	.026	.767	.006
31	.887	.673	.219	.993
		X2{2}		
1	.000	.001	.004	.000
5	.001	.039	.012	.000
9	.004	.002	.983	.002
102	.995	.958	.001	.998
		X2{3}		
1	.000	.000	.004	.000
5	.000	.003	.014	.000
9	.000	.000	.976	.000
381	1.000	.997	.006	1.000
		X2{4}		
1	.000	.000	.003	.000
5	.000	.000	.013	.000
9	.000	.000	.938	.000
1003	1.000	1.000	.046	1.000

series, and Exhibit 4.2*b* gives the corresponding simple correlations and regressions. In this case, the correlations are between **C4** in (4.5b) and $\widehat{\mathbf{C4}} = 0.8\mathbf{C1} + 0.2\mathbf{C2}$. The regressions are **C4** regressed on **C1**, **C2**, and **C3**.

The following points are noteworthy:

1. Once again, the expected results are clearly observed, at least for $i > 0$.

2. In the case of **X2**{0}, the constructed dependency is weak, having a simple correlation of less than .5. The resulting scaled condition index, 11, is effectively the same as the second highest scaled condition index, 9, and hence this dependency is no tighter than the general background conditioning of the basic data matrix. We see as we proceed that, where *several scaled condition indexes are effectively the same*, the diagnostic procedure can have trouble distinguishing among them, and the variance–decomposition proportions of the variates involved can be arbitrarily distributed among the nearly equal scaled condition indexes, although the total weight distributed among them remains roughly fixed. In **X2**{0}, the two scaled condition indexes 9 and 11 account for over 90% of the variance in b_1, b_3, and b_4. We can explain b_3's presence in this group by the fact that the simple correlation between **C1** and **C3** is .58, greater than the constructed correlation between **C4** and $\widehat{\mathbf{C4}}$. The absence of b_2 is explained by a minor scaling problem: **C2** accounts for only one-half of the variance of the constructed variate $\mathbf{C4} = 0.8\mathbf{C1} + 0.2\mathbf{C2} + \mathbf{e}_i$, and in this case, its influence is being dominated by other background correlations.

3. However, by the time the simple correlation between **C4** and $\widehat{\mathbf{C4}}$ becomes .934, in the case of **X2**{1}, the preceding problems completely disappear. The contrived relation involving columns 1, 2, and 4 now dominates the background

Exhibit 4.2*b* Supplementary Correlations and Regressions of C4 on C1, C2, and C3: X2 Series

Data Matrix	$r(\mathbf{C4}, \widehat{\mathbf{C4}})$, $\widehat{\mathbf{C4}} = 0.8\mathbf{C1} + 0.2\mathbf{C2}$	**C1**	**C2**	**C3**	$\hat{R}^2$
X2{0}	.477	0.827	−0.007	0.109	.2864
		[3.84]	[−0.07]	[0.88]	
X2{1}	.934	0.634	0.178	0.103	.8976
		[10.14]	[6.49]	[2.87]	
X2{2}	.995	0.819	0.188	0.001	.9911
		[40.90]	[21.44]	[0.06]	
X2{3}	.999	0.794	0.202	0.001	.9993
		[149.03]	[86.69]	[0.40]	
X2{4}	1.000	0.799	0.199	0.001	.9999
		[393.94]	[224.32]	[1.02]	

correlations among the columns of the basic data matrix, and the variance–decomposition proportions of these variates associated with the largest scaled condition index, 31, are all greater than .5.

4. We again observe that with each increase in i the scaled condition index corresponding to this ever-tightening relation jumps in the same progression of 10/30 previously noted, namely 1, 3, 10, 30, 100, ... In this regard, it is of interest to observe that the contrived relation among the columns 1, 2, and 4 becomes clearly distinguishable from the background in case **X2**{1}, when its scaled condition index becomes one step in this progression above that of the "noise," that is, when it becomes 31 versus the 10 associated with the background dependencies.

5. Once again, the presence of collinearity begins to be observable with scaled condition indexes around 10. In this instance, however, an unintended relation (the .58 correlation between **C1** and **C3**) also shows itself, confounding clear identification of the intended dependency among **C4**, **C1**, and **C2**.

6. In both this and the **X1** series, a scaled condition index of 30 signals clear evidence of the presence of the linear dependency and degraded regression estimates.

X3

This series of matrices combines the two dependencies (4.5a) and (4.5b) just examined individually into a single five-column matrix with $\mathbf{C4} = \mathbf{C3} + \mathbf{e}_i$ and $\mathbf{C5} = 0.8\mathbf{C1} + 0.2\mathbf{C2} + \mathbf{e}_j$. This, then, offers the first constructed example of coexisting near dependencies.[7] We expect that there should be two high scaled condition indexes, one associated with high variance–decomposition proportions in var(b_3) and var(b_4)—due to dependency (4.5a)—and one associated with high variance–decomposition proportions in var(b_1), var(b_2), and var(b_5)—due to dependency (4.5b).

In an effort to keep the report of the increased number of $\boldsymbol{\Pi}$ matrices relevant to this series to a manageable size, we concentrate our results in two ways. First, we report only representative cases from among the 25 possibilities of **X3**$\{i, j\}$, i, $j = 0, \ldots, 4$, and second, where possible, we report only the rows of the variance–decomposition proportions that correspond to the high scaled condition indexes of interest. We can see from the previous two series that many of the rows corresponding to the smaller scaled condition indexes are effectively unchanging as i varies and convey no useful additional information for the analysis at hand.

Let us begin by holding i constant at 2 and varying $j = 0, \ldots, 4$; dependency (4.5a) is therefore fixed and moderately tight while (4.5b) varies. Exhibit 4.3 presents the results for this subseries. The supplementary correlation and regression statistics need not, of course, be repeated, for they are the same as the relevant portions of Exhibits 4.1*b* and 4.2*b*. In particular, the correlations and regressions for the unchanging **X3**$\{2, j\}$ relation between **C4** and **C3** are the same as those for **X1**{2} in Exhibit 4.1*b*, and the regressions for **C5** on the basic

[7]Although, we have already inadvertently seen something like it in case **X2**{0}.

columns 1, 2, and 3 for $j = 0, \ldots, 4$ are the same as those given in Exhibit 4.2*b* for **X2**$\{j\}, j = 0, \ldots, 4$.

The following points are noteworthy:

1. The unchanging "tight" relation between columns 3 and 4 is observable throughout, having a correlation of .995 and a large scaled condition index in the neighborhood of 100.
2. The relation with varying intensity among **C5**, **C1**, and **C2** begins weakly

Exhibit 4.3 Scaled Condition Indexes and Variance-Decomposition Proportions: X3 Series. Two Constructed Near Dependencies (4.5a) and (4.5b): $C4 = C3 + e_2$ (unchanging); $C5 = 0.8C1 + 0.2C2 + e_j$ ($j = 0, \ldots, 4$).

Scaled Condition Index, $\tilde{\eta}$	Proportions of var(b_1)	var(b_2)	var(b_3)	var(b_4)	var(b_5)
		X3{2, 0}			
1	.002	.007	.000	.000	.003
5	.025	.734	.000	.000	.064
8	.025	.241	.003	.003	.302
12	.941	.002	.001	.001	.630
106	.007	.016	.996	.996	.001
		X3{2, 1}[a]			
9	.076	.008	.003	.003	.009
34	.765	.537	.000	.004	.792
118	.147	.192	.997	.992	.199
		X3{2, 2}[a]			
8	.004	.001	.003	.003	.002
127	.746	.750	.459	.464	.757
99	.249	.211	.538	.533	.241
		X3{2, 3}[a]			
8	.002	.000	.003	.003	.000
469	.998	.997	.173	.182	.999
83	.000	.001	.824	.815	.001
		X3{2, 4}[a]			
8	.000	.000	.003	.003	.000
1124	1.000	1.000	.003	.001	1.000
106	.000	.000	.994	.996	.000

[a]Only the last several rows of the **Π** matrix are shown here; the relatively unchanging and inconsequential earlier rows have not been repeated. See text.

for the **X3**{2, 0} case and is, as before, somewhat lost in the background. Still, the involvement of var(b_1) and var(b_5) associated with the scaled condition index 12 is observable even here, although it is being confounded with the other scaled condition index, 8, of roughly equal value. The unchanging tight relation between **C4** and **C3** with scaled condition index 106 remains unobscured by these other relations.

3. When the varying relation between columns 1, 2, and 5 becomes somewhat tighter than the background, as in the case of **X3**{2, 1}, its effects become separable. This case clearly demonstrates the ability of the suggested collinearity diagnostic correctly to identify two coexisting near dependencies and to indicate the variates involved in each: the $\tilde{\eta}$ of 34 is associated with the high variance–decomposition proportions in var(b_1), var(b_2), and var(b_5), and the $\tilde{\eta}$ of 118 is associated with those of var(b_3) and var(b_4).

4. When the two contrived near dependencies become of roughly equal intensity, as in the case of **X3**{2, 2}, where both have $\tilde{\eta}$'s in the neighborhood of 100, the involvement of the variates in the two relations once again becomes confounded. However, only the information on the separate involvement of the variates is lost through this confounding. It is still possible to determine that there are two near dependencies among the columns of **X3**, and it is still possible to determine which variates are involved in at least one of them; in this case all of them, for the two scaled condition indexes together account for well over 90% of the variance in b_1, b_2, b_3, b_4, and b_5, indicating the involvement of each. The only information being lost here is which variates specifically enter each near dependency.

5. When the relation among the columns 1, 2, and 5 again becomes strong relative to the unchanging relation between columns 3 and 4, as in the cases **X3**{2, 3} and **X3**{2, 4}, their separate identities reemerge.

6. Once again, the order of magnitude of the relative tightness of a near dependency seems to increase with the progression of 10/30 in the scaled condition index, that is, along the scale 1, 3, 10, 30, 100, 300, ... Dependencies of roughly equal magnitude can be confounded; dependencies of differing magnitudes are able to be separately identified.

The preceding analysis examines **X3**{i, j} by varying the second near dependency while holding the first constant at $i = 2$. Let us reverse this order and examine **X3**{i, 1} for $i = 0, \ldots, 4$, holding j constant at 1. As i increases, we would expect there would be two high scaled condition indexes. The one corresponding to the unchanging dependency between columns 1, 2, and 5 will not be too high, since j is held at the relatively low level of 1. The relation between columns 3 and 4 should get tighter and more highly defined as i increases from 0 to 4. Exhibit 4.4 reports these results. Exhibit 4.1*b* and the second row of Exhibit 4.2*b* provide the relevant supplementary correlations and regressions.

Exhibit 4.4 Scaled Condition Indexes and Variance-Decomposition Proportions: X3 Series. Two Constructed Near Dependencies (4.5a) and (4.5b): $C4 = C3 + e_i$ $(i = 0, \ldots, 4)$; $C5 = 0.8C1 + 0.2C2 + e_1$ (unchanging).

Scaled Condition Index, $\tilde{\eta}$	Proportions of				
	var(b_1)	var(b_2)	var(b_3)	var(b_4)	var(b_5)
			X3{0, 1}		
1	.001	.002	.001	.002	.000
5	.008	.274	.003	.032	.000
9	.091	.004	.044	.250	.011
35	.885	.644	.171	.008	.988
15	.015	.076	.781	.708	.001
			X3{1, 1}[a]		
35	.821	.656	.226	.077	.900
30	.073	.018	.720	.880	.089
			X3{2, 1}[a]		
34	.765	.537	.000	.004	.792
118	.147	.192	.997	.992	.199
			X3{3, 1}[a]		
34	.871	.673	.000	.000	.985
519	.028	.010	.999	.999	.005
			X3{4, 1}[a]		
34	.842	.549	.000	.000	.901
1148	.062	.191	1.000	1.000	.089

[a]Only the last several rows of the Π matrix are shown here; the relatively unchanging and inconsequential earlier rows have not been repeated. See text.

The following points are noteworthy:

1. Both relations are observable from the outset.

2. A scaled condition index of 15 (greater than 10) corresponds to a near dependency that is tight enough to be observed.

3. The confounding of the two dependencies is again observable when the scaled condition indexes are close in magnitude—the case of **X3**{1, 1} with $\tilde{\eta}$'s of roughly 30—, but it is not as pronounced here as in the previously described subseries, where such equality takes place at the more intense dependency level, **X3**{2, 2}, with $\tilde{\eta}$'s of roughly 100.

4. The order of the scaled condition indexes along the progression of 10/30 is again observed.

To complete the picture of the behavior of **X3**$\{i, j\}$, we report in Exhibit 4.5

Exhibit 4.5 Scaled Condition Indexes and Variance–Decomposition Proportions: X3 Series. Two Constructed Near Dependencies (4.5a) and (4.5b): C4 = C3 + e_i (selected values); C5 = 0.8C1 + 0.2C2 + e_j (selected values).

Scaled Condition Index, $\tilde{\eta}$	Proportions of				
	var(b_1)	var(b_2)	var(b_3)	var(b_4)	var(b_5)
			X3{0, 0}		
1	.002	.007	.001	.002	.003
5	.016	.750	.000	.009	.041
7	.028	.053	.044	.206	.323
12	.953	.000	.018	.032	.610
15	.001	.190	.937	.751	.023
			X3{1, 0}[a]		
8	.015	.271	.041	.033	.341
12	.952	.009	.012	.006	.585
30	.006	.037	.946	.961	.004
			X3{1, 2}[a]		
123	.995	.962	.094	.117	.998
30	.005	.002	.860	.848	.001
			X3{3, 2}[a]		
115	.967	.936	.000	.000	.979
523	.028	.023	1.000	1.000	.020
			X3{3, 4}[a]		
1129	.999	.999	.015	.013	.999
517	.001	.001	.985	.986	.001
			X3{4, 4}[a]		
1357	.698	.705	.632	.631	.699
960	.302	.294	.368	.369	.301

[a]Only the last several rows of the **Π** matrix are shown here; the relatively unchanging and inconsequential earlier rows have not been repeated. See text.

the scaled condition indexes and variance–decomposition proportions for selected values of i and j increasing together.

The following points are noteworthy:

1. In the **X3**{0, 0} case, three scaled condition indexes are of close magnitude, 7, 12, and 15, and there is some confusion of variate involvement among all three of them.

2. The relation between **C3** and **C4** is freed from the background in the next case, **X3**{1, 0}, but there is still some confusion between the two similar scaled condition indexes, 8 and 12.

3. The two relations become more clearly identified as i and j increase and are strongly separable so long as the scaled condition indexes remain separated by at least one order of magnitude along the progression of 10/30.

4. However, no matter how tight the individual near dependencies, they can be confounded when their scaled condition indexes are of similar magnitude, as is seen in the case of **X3**{4, 4}.

Experiment 2: The Y Matrices

Our interest in examining these new experimental series of data matrices focuses on several questions. First, do totally different sets of data matrices result in similar generalizations on the diagnostic behavior of the scaled condition indexes and the variance–decomposition proportions that were beginning to emerge from the experiment 1 just described? All of the **Y** series help answer this question. Second, do "rates-of-change" and "rates" data behave differently from the "levels" and "trends" data of experiment 1? The **Y3** series is relevant to answering this question. Third, do essential scaling problems cause troubles? Series **Y1** and **Y2** examine this question.

Y1 and Y2

The **Y1**{i} series, we recall from (4.8), consists of a five-column matrix in which **C5** = **C1** + **C2** + $\mathbf{e}_i$, as in (4.7a). The variance of **C5** introduced by **C1** (**GNP58***) is relatively small, less than 1% of that introduced by **C2**. Its influence is therefore easily masked. The **Y2**{i} series is exactly the same, except that the influence of **C1** is made smaller yet. Here **C5** = 0.1**C1** + **C2** + $\mathbf{e}_i$, as is seen in (4.7b). These two experimental series allow us to see how sensitive the procedure for diagnosing collinearity is to strong, and even severe, scaling problems.

For both experimental series, **Y1** and **Y2**, we would expect one high scaled condition index associated with high variance–decomposition proportions in var(b_1), var(b_2), and var(b_5). Exhibits 4.6*a* and 4.7*a* present these results for **Y1** and **Y2**, respectively, with $i = 0, \ldots, 4$. Exhibits 4.6*b* and 4.7*b* present the corresponding supplementary correlations and regressions.

Exhibit 4.6a Scaled Condition Indexes and Variance–Decomposition Proportions: Y1 Series. One Constructed Near Dependency (4.7a): $C5 = C1 + C2 + e_i$ ($i = 0, \ldots, 4$).

Scaled Condition Index, $\tilde{\eta}$	Proportions of var(b_1)	var(b_2)	var(b_3)	var(b_4)	var(b_5)
			Y1{0}		
1	.005	.001	.003	.014	.002
3	.010	.002	.010	.898	.002
7	.782	.036	.001	.010	.081
10	.188	.045	.978	.071	.096
16	.015	.916	.008	.007	.819
			Y1{1}[a]		
8	.607	.019	.031	.008	.003
10	.024	.030	.941	.090	.012
40	.360	.950	.014	.036	.985
			Y1{2}[a]		
7	.394	.001	.003	.010	.000
10	.065	.001	.976	.068	.001
156	.534	.998	.008	.020	.999
			Y1{3}[a]		
7	.099	.000	.003	.010	.000
10	.016	.000	.959	.073	.000
397	.883	.997	.024	.003	1.000
			Y1{4}[a]		
10	.001	.000	.965	.072	.000
1659	.993	1.000	.018	.000	1.000

[a]Only the last several rows of the **Π** matrix are shown here; the relatively unchanging and inconsequential earlier rows have not been repeated. See text.

Exhibit 4.6b Supplementary Correlations and Regressions of C5 on C1, C2, C3, and C4: Y1 Series

Data Matrix	$r(C5, \widehat{C5})$, $\widehat{C5} = C1 + C2$	C1	C2	C3	C4	$\hat{R}^2$
Y1{0}	.776	−0.026	0.926	97.810	28.321	.6186
		[−0.01]	[5.33]	[1.03]	[0.34]	
Y1{1}	.939	2.563	0.824	35.777	35.106	.9116
		[3.78]	[13.20]	[1.05]	[1.17]	
Y1{2}	.997	0.950	1.001	−2.750	−4.911	.9940
		[5.24]	[59.80]	[−0.30]	[−0.61]	
Y1{3}	.999	0.953	0.999	2.875	0.968	.9991
		[13.36]	[151.80]	[0.80]	[0.31]	
Y1{4}	1.000	1.018	0.998	0.563	0.014	.9999
		[59.57]	[633.30]	[0.65]	[0.02]	

Exhibit 4.7a Scaled Condition Indexes and Variance-Decomposition Proportions: Y2 Series. One Constructed Near Dependency (4.7b): C5 = 0.1C1 + C2 + e_i ($i = 0, \ldots, 4$).

Scaled Condition Index, $\tilde{\eta}$	Proportions of				
	var(b_1)	var(b_2)	var(b_3)	var(b_4)	var(b_5)
		Y2{0}			
1	.005	.003	.003	.014	.003
3	.009	.002	.009	.886	.005
7	.804	.023	.006	.002	.205
10	.157	.366	.201	.000	.781
11	.025	.606	.781	.098	.006
		Y2{1}[a]			
7	.736	.009	.000	.022	.012
10	.248	.005	.887	.072	.012
42	.000	.985	.099	.016	.975
		Y2{2}[a]			
7	.747	.001	.000	.013	.000
10	.190	.000	.880	.068	.001
153	.049	.999	.108	.001	.998
		Y2{3}[a]			
7	.648	.000	.000	.012	.000
10	.157	.000	.867	.069	.000
475	.183	1.000	.120	.008	1.000
		Y2{4}[a]			
7	.512	.000	.000	.011	.000
10	.121	.000	.984	.066	.000
1166	.357	1.000	.002	.063	1.000

[a]Only the last several rows of the Π matrix are shown here; the relatively unchanging and inconsequential earlier rows have not been repeated. See text.

Exhibit 4.7b Supplementary Correlations and Regressions of C5 on C1, C2, C3, and C4: Y2 Series

Data Matrix	r(C5, $\widehat{\text{C5}}$), $\widehat{\text{C5}}$ = 0.1C1 + C2	C1	C2	C3	C4	$\hat{R}^2$
Y2{0}	.395	0.379	0.459	260.809	47.965	.1433
		[0.16]	[2.10]	[2.18]	[0.46]	
Y2{1}	.962	0.086	1.061	−35.079	27.010	.9294
		[0.12]	[16.64]	[−1.00]	[0.88]	
Y2{2}	.997	0.208	1.018	−13.861	−0.480	.9943
		[1.15]	[60.94]	[−1.52]	[−0.06]	
Y2{3}	.999	0.133	1.008	−5.067	−1.057	.9994
		[2.28]	[187.80]	[−1.73]	[−0.41]	
Y2{4}	1.000	0.084	1.001	−0.218	1.305	.9999
		[3.58]	[460.06]	[−0.18]	[1.25]	

The following points are noteworthy:

1. The results for both data series are in basic accord with expectations.

2. The essential scaling problems do, however, cause difficulties in identifying the variates involved in the near dependency. In the **Y1** series of Exhibit 4.6*a*, the involvement of the dominated column **C1** is not observed at all in the weakest case, **Y1**{0}, which has a scaled condition index of 10 (and a correlation of .78). The involvement of this column begins to be observed in case **Y1**{1} but does not show itself completely until cases **Y1**{2} and **Y1**{3}. By contrast, the involvement of the dominant column **C2**, along with the generated column **C5**, is observed from the outset. Parallel behavior is exhibited in the corresponding regression coefficients in Exhibit 4.6*b*. The regression parameter of **C1** is insignificant in case **Y1**{0}, becomes significant in case **Y1**{1}, and takes the proper order of magnitude (unity) only in cases **Y1**{2} and beyond. The parameter of the dominant column **C2**, however, is significant and near unity right from the weakest case, **Y1**{0}.

3. The aggravated scaling problem in the **Y2** series (**C1** now accounts for less than 0.01% of the variance in **C5**) has the expected effect, as seen in Exhibit 4.7*a*. Now the involvement of column 1 becomes apparent only in the tightest case, **Y2**{4}, with a scaled condition index of 1166. The regression parameter for **C1**, as seen in Exhibit 4.7*b*, also attains significance only in the tightest cases.

4. The several other general patterns noted in experiment 1 seem still to hold: a near dependency's effects are just beginning to be observed with scaled condition indexes around 10; tightening a near dependency by decreasing the variance of its generating noise by successive factors of 10 causes the scaled condition index to progress roughly as the progression of 10/30, that is, as 1, 3, 10, 30, 100, 300, ...

Y3

The **Y3**{*i*} series consists of a five-column data matrix in which the fifth column is in a simple relation (4.7c) with the fourth column, **C5** = **C4** + $\mathbf{e}_i$. In this case, **C4** is inventory investment, a rate-of-change variate, and interest centers on whether the diagnostics behave similarly with this sort of data as with the levels data considered so far. The results of this series are given in Exhibits 4.8*a* and 4.8*b*.

The following points are noteworthy:

1. An interesting phenomenon emerges in the case of **Y3**{0} that is in need of explanation and provides us with the first real opportunity to apply our diagnostic tools to a near dependency that arises naturally in the data, that is, to one not artificially generated. First, we note that the generated near dependency between **C5** and **C4** is indeed observed—associated with the weak scaled condition index 5. In addition, however, we note that over 80% of var(b_2) and var(b_3) is associated with the larger scaled condition index 11, indicating their involvement in a low-level, unintended, background near dependency. The

Exhibit 4.8a Scaled Condition Indexes and Variance-Decomposition Proportions: Y3 Series. One Constructed Near Dependency (4.7c): C5 = C4 + e_i (i = 0, ..., 4).

Scaled Condition Index, $\tilde{\eta}$	Proportions of var(b_1)	var(b_2)	var(b_3)	var(b_4)	var(b_5)
			Y3{0}		
1	.005	.003	.003	.012	.012
3	.018	.006	.014	.095	.222
8	.953	.180	.112	.003	.001
11	.020	.810	.870	.013	.034
5	.004	.001	.001	.877	.731
			Y3{1}[a]		
8	.706	.133	.118	.027	.019
11	.000	.855	.693	.004	.022
15	.270	.001	.171	.948	.939
			Y3{2}[a]		
8	.953	.171	.102	.000	.000
11	.017	.757	.751	.002	.000
42	.005	.063	.132	.996	.997
			Y3{3}[a]		
8	.956	.180	.115	.000	.000
11	.018	.789	.859	.000	.000
147	.001	.020	.008	.999	.999
			Y3{4}[a]		
8	.890	.179	.107	.000	.000
11	.016	.790	.798	.000	.000
416	.071	.020	.078	1.000	1.000

[a] Only the last several rows of the **Π** matrix are shown here; the relatively unchanging and inconsequential earlier rows have not been repeated. See text.

Exhibit 4.8b Supplementary Correlations and Regressions of C5 on C1, C2, C3, and C4: Y3 Series

Data Matrix	r(C5, C4)	C1	C2	C3	C4	$\hat{R}^2$
Y3{0}	.643	0.002	0.000	−0.201	0.918	.4231
		[0.27]	[0.81]	[−0.74]	[3.82]	
Y3{1}	.950	−0.003	0.000	0.084	0.949	.9188
		[−1.93]	[0.62]	[1.20]	[15.38]	
Y3{2}	.995	0.000	0.000	−0.045	1.005	.9902
		[0.36]	[1.23]	[−1.80]	[45.67]	
Y3{3}	.999	0.000	−0.000	0.003	1.000	.9992
		[0.17]	[−0.65]	[0.40]	[161.11]	
Y3{4}	1.000	−0.000	−0.000	0.004	1.001	.9999
		[−1.31]	[−0.67]	[1.42]	[455.66]	

simple correlation between **C2** and **C3** is very low, .09,[8] so we must look further than a simple dependency between **C2** and **C3**. Further examination of **Y3**{0} in Exhibit 4.8*a* shows that there is really not one, but two, unintended near dependencies of roughly equal intensity in the basic data matrix, associated with the effectively equal scaled condition indexes 11 and 8. Moreover, these two background near dependencies together account for over 95% of var(b_1), var(b_2), and var(b_3), and the three roughly equal scaled condition indexes 5, 8, and 11 account for virtually all of each of the five variances. Applying what we have learned from experiments 1 and 2 so far, we conclude that there are three weak near dependencies of roughly equal intensity whose individual effects cannot be clearly separated, a problem we have seen arise when there are several scaled condition indexes of the same order of magnitude. Since there are three near dependencies among the five variates, and we know the relation between **C5** and **C4** is one of them, we would expect to find two additional low-level near dependencies among the four columns **C1**, **C2**, **C3**, and **C4**.[9] Indeed, regressing **C1** and **C3** separately on **C2** and **C4** gives[10]

$$\mathbf{C1} = \underset{[5.86]}{0.054}\mathbf{C2} + \underset{[1.96]}{16.670}\mathbf{C4} \qquad \hat{R}^2 = .8335$$

$$\mathbf{C3} = \underset{[8.43]}{0.002}\mathbf{C2} + \underset{[2.27]}{0.037}\mathbf{C4} \qquad \hat{R}^2 = .9045,$$

which shows these additional two near dependencies beyond question.

These background near dependencies are, of course, also present in the **Y1** and **Y2** series (the first four columns being the same in all **Y** series), but their effects there are overshadowed by the presence of the relatively stronger contrived near dependencies involving **C1**, **C2**, and **C5**. The experience we have gained from these experiments in the use of these diagnostic techniques, however, has clearly led us very much in the right direction.

2. The previously described phenomenon serves also to emphasize the point that care must be taken in applying the diagnostic test when two or more scaled condition indexes are of equal, or closely equal, magnitude. In such cases, the variance–decomposition proportions can be arbitrarily distributed across the roughly equal scaled condition indexes so as to obscure the involvement of a given variate in any of the competing (nearly equal) near dependencies. In

[8]Indeed, the highest (absolute) simple correlation among the four basic columns is .32.

[9]Even though var(b_4) is not greatly determined by the condition indexes of 8 and 11, **C4**'s involvement in these two near dependencies cannot be ruled out. The variance–decomposition proportions, as we have seen, can be arbitrarily distributed among the $\tilde{\eta}$'s of nearly equal magnitude.

[10]The choice of **C1** and **C3** on **C2** and **C4** is arbitrary. Any two of the four variates with a nonvanishing Jacobian could be selected for this descriptive use of least squares. The figures in the square brackets are t-statistics, not standard deviations, and the R^2s are uncentered, as in (2.4a), since there is no constant term.

Y3{0}, for example, the fact that over 80% of var(b_2) and var(b_3) is associated with the single scaled condition index of 11 need not imply only a simple relation between **C2** and **C3**. Other variates (here **C1** and **C4**) associated with competing scaled condition indexes (8 and 5) can be involved as well. Furthermore, when there are competing scaled condition indexes, the fact that a single scaled condition index (like 8 in **Y3**{0}) is associated with only one high variance–decomposition proportion (95% of var(b_1)) need not imply, as it otherwise could,[11] that the corresponding variate (**C1**) is free from involvement in any near dependency. Its interrelation with the variates involved in the competing dependencies must also be investigated.

In sum, when there are *competing dependencies* (scaled condition indexes of similar value), they must be treated together in the application of the diagnostic test. That is, the variance–decomposition proportions for each coefficient should be summed across the competing scaled condition indexes, and a high total variance–decomposition proportion for two or more variances associated with the set of competing high scaled condition indexes is to be interpreted as evidence of degrading collinearity. The exact involvement of specific variates in specific near dependencies cannot be learned in this case, but it is still possible to learn (1) which variates are degraded (those with high *aggregate* variance–decomposition proportions) and (2) the number of near dependencies present (the number of competing high scaled condition indexes).

3. Another, quite different, form of confounded involvement is also exemplified by the foregoing: the *dominant dependency*. Column 4 is apparently involved simultaneously in several near dependencies: weakly, and with scaling problems in the near dependencies associated with $\tilde{\eta}$'s of 8 and 11; and without scaling problems in the contrived dependency between **C4** and **C5**. This latter near dependency dominates the determination of var(b_4) in all cases, but particularly as this relation becomes tighter, thereby obscuring the weak involvement of **C4** in the other near dependencies. Dominant dependencies (scaled condition indexes of higher magnitude), then, can mask the simultaneous involvement of a single variate in weaker near dependencies. That is, the possibility always exists that a variate having most or all of its variance determined by a near dependency with a high scaled condition index is also involved in near dependencies with lower scaled condition indexes—unless, of course, that variate is known to be buffered from the other near dependencies through near orthogonality.

4. From within the intricacies of the foregoing points, however, one must not lose sight of the fact that the test for potentially damaging collinearity requires the *joint* condition of (I) a single *high* scaled condition index associated with (II) two or more variances with high variance–decomposition proportions[12];

[11]See, however, point 3 following.

[12]As we have just seen in points 2 and 3, condition II requires some modifications when there are either competing or dominating near dependencies. These modifications are treated fully in Section 5.1.

condition II by itself is not enough. It is true in the **Y3**{0} case, for example, that the three scaled condition indexes 5, 8, and 11 account for most of the variance of all five coefficients, but by very rough standards, these scaled condition indexes do not seem to be high, and the data matrix **Y3**{0}, quite likely, would be suitable for many statistical applications.

Let us examine this condition further. In our prior examples we note that contrived dependencies begin to be observed when their "tightness" results in scaled condition indexes around 10. We are also able to calculate the correlations that correspond to these relations, so we can associate the magnitudes of scaled condition indexes with this more familiar measure of tightness. A glance through Exhibits 4.1–4.8 shows that scaled condition indexes of magnitude 10 result from underlying near dependencies whose correlations are in the range of .4 to .6, relatively loose relations by much econometric experience. It is not until scaled condition indexes climb to a level of 15–30 that the underlying relations have correlations around .9, a level that much experience suggests is high.

Further insight is afforded by examining the actual variances whose decomposition proportions are given in Exhibit 4.8*a*. These are presented in Exhibit 4.9. In the case of **Y3**{0}, all variance magnitudes are relatively small, certainly in comparison to the size attained by var(b_4) and var(b_5) in the cases of **Y3**{2} and **Y3**{3}, when the contrived near dependency between them becomes tighter. In short, high variance–decomposition *proportions* surely need not imply large *component values*. This merely restates the notion of Section 3.4 that degraded estimates (capable of being improved if calculated from better-conditioned data), which apparently can result from even low-level dependencies, need not be harmful; harmfulness depends, in addition, on the specific regression model that is being used with the given data matrix, on the variance σ^2, and on the statistical use to which the results are to be put. We will examine the issue of harm in greater detail in Chapter 7.

Exhibit 4.9 Underlying Regression Variances[a]: Y3 Series

Data Matrix	var(b_1)	var(b_2)	var(b_3)	var(b_4)	var(b_5)
Y3{0}	9.87	16.94	16.32	3.71	3.04
Y3{1}	11.43	16.74	16.94	25.58	26.38
Y3{2}	9.90	17.56	18.18	207.76	204.17
Y3{3}	9.86	16.78	16.06	2559.53	2547.85
Y3{4}	10.59	16.79	17.28	20389.78	20457.85

[a]Representing the respective **Y3**{*i*} matrices by **X**, the figures reported here are the diagonal elements of $(\mathbf{X}^T\mathbf{X})^{-1}$—the ϕ_{kk}'s of (3.23)—and do not include the constant factor of s^2, the estimated error variance. Of course, s^2 can only be calculated once a specific **y** has been regressed on **X**.

5. The contrived near dependency between **C5** and **C4**, both rates-of-change variates, seems to behave somewhat differently from previous experience based on "levels" data; namely, its scaled condition index is lower for comparable degrees of tightness in the underlying relation as measured by correlations. Perusal of Exhibits 4.1–4.8 indicates that, quite roughly, the scaled condition index jumps one step along the 1, 3, 10, 30, 100, 300, 1000 progression each time another 9 digit is added to the correlation of the underlying near dependency, with a correlation of .9 corresponding to an $\tilde{\eta}$ of roughly 30. That is, there is a pairing between correlations and scaled condition indexes $r \rightarrow \tilde{\eta}$ that goes roughly as $.5 \rightarrow 10$, $.9 \rightarrow 30$, $.99 \rightarrow 100$, $.999 \rightarrow 300$, and so on. Exhibit 4.9, however, indicates the rate-of-change data begin this sequence one step lower, pairing $.5 \rightarrow 3$, $.9 \rightarrow 10$, $.99 \rightarrow 30$, $.999 \rightarrow 100$, and so on. It seems, therefore, that there is no simple pairing of the level of the strength of a relation as measured by a scaled condition index with that of the same relation as measured by a correlation; it depends on the nature of the data. There does, however, seem to be stability in the relative magnitudes of these two measures along the progression of 10/30 noted previously.

6. In all of the foregoing, one should not lose sight of the fact that, basically, the diagnostic procedure works fully in accord with expectations. The contrived relation between **C5** and **C4** is observed from the outset and is unmistakably evident by case **Y3**{1}, once it becomes a level stronger than the background relations.

Y4

In the **Y4**{i, k} series, the two dependencies of **Y1** and **Y3** are made to coexist. Here $\mathbf{C5}=\mathbf{C1}+\mathbf{C2}+\mathbf{e}_i$, according to (4.7a), and $\mathbf{C6}=\mathbf{C4}+\mathbf{e}_k$, as in (4.7c). What is new to be learned from this experimental series can be seen from a very few selected variance–decomposition proportion **Π** matrices. These are reported in Exhibit 4.10.

Exhibit 4.10 Scaled Condition Indexes and Variance–Decomposition Proportions: Y4 Series. Two Constructed Near Dependencies (4.7a) and (4.7c): $\mathbf{C5}=\mathbf{C1}+\mathbf{C2}+\mathbf{e}_i$ (selected values); $\mathbf{C6}=\mathbf{C4}+\mathbf{e}_k$ (selected values).

Scaled Condition Index, $\tilde{\eta}$	Proportions of var(b_1)	var(b_2)	var(b_3)	var(b_4)	var(b_5)	var(b_6)
			Y4{0, 0}			
1	.003	.001	.002	.008	.001	.008
3	.010	.002	.009	.104	.002	.223
7	.788	.036	.001	.001	.079	.004
11	.181	.051	.981	.004	.100	.044
17	.017	.910	.007	.008	.817	.001
5	.001	.000	.000	.875	.001	.720

Exhibit 4.10 Continued

Scaled Condition Index, $\tilde{\eta}$	Proportions of var(b_1)	var(b_2)	var(b_3)	var(b_4)	var(b_5)	var(b_6)
			Y4{0, 2}			
1	.003	.001	.002	.001	.001	.000
3	.011	.002	.008	.002	.003	.002
8	.779	.036	.001	.000	.077	.000
11	.186	.045	.822	.001	.092	.001
17	.015	.916	.008	.000	.772	.000
47	.006	.000	.159	.996	.055	.997
			Y4{1, 1}[a]			
8	.477	.014	.045	.024	.002	.014
11	.001	.034	.777	.006	.013	.025
43	.346	.947	.010	.000	.984	.005
17	.167	.005	.158	.948	.000	.936
			Y4{1, 2}[a]			
8	.587	.018	.027	.000	.003	.000
11	.023	.028	.847	.001	.011	.000
39	.173	.453	.068	.234	.543	.250
51	.209	.501	.048	.762	.442	.747
			Y4{3, 3}[a]			
11	.016	.000	.954	.000	.000	.000
428	.884	1.000	.025	.003	1.000	.003
162	.000	.000	.007	.997	.000	.997

[a]Only the last several rows of the **Π** matrix are shown here; the relatively unchanging and inconsequential earlier rows have not been repeated. See text.

The following points are noteworthy:

1. Both relations are observable even in the weakest instance of **Y4**{0, 0}, one with scaled condition index 17, the other with scaled condition index 5. The presence of the contrived relation among **C1**, **C2**, and **C5** has somewhat masked the background relation among **C1**, **C2**, and **C3** that was observed in the **Y3** series, although var(b_1) is still being distributed among these relations.

2. Near dependencies with differing scaled condition indexes tend to be separately identified, as in the cases of **Y4**{0, 2}, **Y4**{1, 1}, and **Y4**{3, 3}. When the scaled condition indexes are nearly equal, however, as in the case **Y4**{1, 2},

the involvement of separate variates is confounded between the two near dependencies. This fact, observed frequently before, is particularly interesting in this instance. In earlier experiments, roughly equal scaled condition indexes corresponded to roughly equal underlying correlations. In this case, however, the relation between the rate-of-change variates **C4** and **C6** is .9 while that underlying the relation among **C1**, **C2**, and **C5** is .99, one 9 stronger. Thus, the problem of confounding relations results from near dependencies of nearly equal tightness as judged by scaled condition indexes, not as judged by correlations. This is not surprising since this problem arises, as noted in Section 3.1, as a consequence of the nonuniqueness of the eigenvectors when the singular values (eigenvalues) are equal, a problem that clearly manifests itself computationally when the singular values, and hence the scaled condition indexes, are nearly equal.

3. For the most part, however, the two constructed relations behave together quite independently and much as they do separately. This was true in experiment 1 as well; the individual behavior of the dependencies in the **X1** and **X2** series is carried over to their coexisting behavior in the **X3** series. Thus, with the exception of the minor problem of confounded variance–decomposition proportions that results from the presence of near dependencies with competing or dominating scaled condition indexes, it seems appropriate to conclude that the presence of several coexisting near dependencies poses no critical problems to the diagnostic procedure.

Experiment 3: The Z Matrices

The purposes of experiment 3 are (1) to analyze slightly larger data matrices (up to eight columns) to see if size has any notable effect on the diagnostic procedure; (2) to allow up to three coexisting near dependencies, again to see if new complications arise; (3) to recast some previous experimental series in a slightly different setting to see if their behavior remains stable; and (4) to create cases where near orthogonality among data series exists in order to observe its buffering effect against dependencies within nearly orthogonal subgroups. Toward this last objective, we recall that columns 4 and 5 of the basic data matrix **Z** are generated having correlations with columns 1–3 of no more than .18 in absolute value. Columns 1–3 here are the same as columns 1–3 in experiment 2 with the **Y** matrices. Four dependency relations are contrived according to (4.9a)–(4.9d). Equations (4.9a) and (4.9b) generate near dependencies between the two columns **C4** and **C5**, which are constructed to have low intercorrelations with columns 1–3. Equation (4.9c) generates a near dependency using only **C1**, **C2**, and **C3**, which is thereby buffered from **C4** and **C5**. These two data groups are bridged by (4.9d). There are scaling problems built into the generated near dependencies. The following are dominated: **DUM1** [$<0.1\%$ of the variance in (4.9b)]; **LHTUR*** [$<0.1\%$ of the variance in (4.9c)]; and **DUM2** [$<0.1\%$ of the variance in (4.9d)].

Many of the **Z** series experiments that follow are designed to duplicate previous experiments with different data in order to observe whether the diagnostic procedure exhibits some degree of stability in operation. Quite happily, this stability is in fact observed in all such cases, and hence these experiments are simply reported as such to avoid unnecessarily repetitive tabulations.

Z1

In this series, **C6** = **C4** + $\mathbf{e}_i$. In this basic data matrix **Z**, **C4** (**DUM1**) is generated to have the same mean and variance as the rate-of-change variate (**GV58***), column 4 of the basic data matrix **Y** of experiment 2. Hence the **Z1** series is quite similar to the **Y3** series of experiment 2, and we would hope that the collinearity diagnostics would exhibit similar experimental behavior in dealing with this series. This expectation is met in full.

Z2

In this series, the near dependency is generated by the two "isolated" columns, 4 and 5, by **C6** = **C5** − **C4**. It also mixes a rate-of-change variate, **C4**, and a levels variate, **C5**, and, as noted, has a scaling problem. Exhibit 4.11 presents two $\mathbf{\Pi}$ matrices for cases **Z2**{2} and **Z2**{4}.

Exhibit 4.11 Scaled Condition Indexes and Variance–Decomposition Proportions: Z2 Series. One Constructed Near Dependency (4.9b): C6 = C5 − C4 + $\mathbf{e}_i$ (selected values).

Scaled Condition Index, $\tilde{\eta}$	Proportions of var(b_1)	var(b_2)	var(b_3)	var(b_4)	var(b_5)	var(b_6)
			Z2{2}			
1	.004	.003	.002	.000	.000	.000
2	.000	.000	.000	.825	.000	.000
6	.047	.114	.043	.019	.002	.002
8	.944	.162	.085	.016	.000	.000
11	.001	.721	.836	.000	.000	.000
104	.004	.000	.034	.140	.998	.998
			Z2{4}[a]			
11	.001	.661	.839	.000	.000	.000
1039	.003	.088	.029	.817	1.000	1.000

[a]Only the last several rows of the $\mathbf{\Pi}$ matrix are shown here; the relatively unchanging and inconsequential earlier rows have not been repeated. See text.

The following points are noteworthy:

1. The "background" relations with scaled condition indexes 8 and 11 are still present (the first three columns here are the same as in experiment 2).

2. The generated near dependency is quite observable, but the scaling problem with **C4** is evident. Even in the case **Z2**{2} with a scaled condition index of 104, the involvement of **C4** is not clearly observed and does not become strongly evident until the scaled condition index increases to the very high value of roughly 1000.

3. The isolation of columns 1–3 from columns 4–6 is very evident. Even in the case of **Z2**{4}, the high scaled condition index of 1039 does not add any significant degradation to var(b_1), var(b_2), and var(b_3).

Z3

This series, in which $\mathbf{C6} = 3\mathbf{C1} + 1.5\mathbf{C3} + \mathbf{e}_i$, is very similar to the **Y1** series and shows effectively identical behavior. The scaling problem here is severe, and the involvement of the dominated variate **C3** is not strong even in the **Z3**{4} case, as is seen by the one relevant row of the **Π** matrix:

Z3{4}

Scaled Condition Index, $\tilde{\eta}$	Proportions of var(b_1)	var(b_2)	var(b_3)	var(b_4)	var(b_5)	var(b_6)
980	1.000	.027	.451	.072	.000	1.000

Z4

In this series, there is the single contrived relation $\mathbf{C6} = \mathbf{C2} + 0.7\mathbf{C5} + \mathbf{e}_i$. The behavior completely accords with expectations, paralleling that of the qualitatively similar **Y1** and **X2** series.

Z5

This series possesses two coexisting near dependencies, each isolated from the other by the low intercorrelations that exist between columns 1–3 and columns 4–6 of the basic data matrix **Z**. Here $\mathbf{C6} = \mathbf{C5} - \mathbf{C4} + \mathbf{e}_i$ and $\mathbf{C7} = 3\mathbf{C1} + 1.5\mathbf{C3} + \mathbf{e}_j$. A typical **Π** matrix for this series is given in Exhibit 4.12.

The following points are noteworthy:

1. The presence of the two relations is clear, and the scaling problems that beset the two relations are observed.

Exhibit 4.12 Scaled Condition Indexes and Variance-Decomposition Proportions: Z5 Series. Two Constructed Near Dependencies (4.9b) and (4.9c): $C6 = C5 - C4 + e_i$ ($i = 2$); $C7 = 3C1 + 1.5C3 + e_j$ ($j = 3$).

Scaled Condition Index, $\tilde{\eta}$	Proportions of						
	var(b_1)	var(b_2)	var(b_3)	var(b_4)	var(b_5)	var(b_6)	var(b_7)
			Z5{2, 3}				
1	.000	.002	.001	.000	.000	.000	.000
2	.000	.000	.000	.743	.000	.000	.000
6	.000	.029	.005	.029	.002	.002	.000
8	.001	.243	.091	.005	.000	.000	.000
12	.000	.711	.602	.000	.000	.000	.001
114	.000	.000	.021	.121	.950	.949	.000
368	.999	.015	.280	.102	.048	.049	.999

2. Of principal interest is the verification of the expected isolation of the relation among **C7**, **C1**, and **C3** from that among **C4**, **C5**, and **C6**. The low intercorrelations between these two sets of columns allow the variances within each group to be unaffected by the relation within the other group.

3. Although not shown, it should be noted that the usual confounding of relations occurs in this series when the condition numbers are of equal magnitude.

Z6

This series has two contrived near dependencies: $\mathbf{C6} = \mathbf{C4} + \mathbf{e}_i$ and $\mathbf{C7} = \mathbf{C2} + 0.7\mathbf{C5} + \mathbf{e}_j$. It therefore combines the two cases examined separately in series **Z1** and **Z4**, both of which behave exactly as expected. These results too are fully in accord with expectations.

Z7

This series presents the first occurrence of three coexisting near dependencies: $\mathbf{C6} = \mathbf{C4} + \mathbf{e}_i$, $\mathbf{C7} = 3\mathbf{C1} + 1.5\mathbf{C3} + \mathbf{e}_j$, and $\mathbf{C8} = \mathbf{C2} + 0.7\mathbf{C5} + \mathbf{e}_k$. Exhibit 4.13 displays $\mathbf{\Pi}$ matrices for three selected cases from this series.

The following points are noteworthy:

1. The presence of three coexisting near dependencies causes no special problems, each behaving essentially as it did separately.

Exhibit 4.13 Scaled Condition Indexes and Variance–Decomposition Proportions: Z7 Series. Three Constructed Near Dependencies (4.9a), (4.9c), and (4.9d): $C6 = C4 + e_i$ (selected values); $C7 = 3C1 + 1.5C3 + e_j$ (selected values); $C8 = C2 + 0.7C5 + e_k$ (selected values).

Scaled Condition Index, $\tilde{\eta}$	Proportions of var(b_1)	var(b_2)	var(b_3)	var(b_4)	var(b_5)	var(b_6)	var(b_7)	var(b_8)
				Z7{2, 2, 3}[a]				
11	.000	.000	.657	.001	.020	.001	.002	.000
35	.000	.000	.076	.835	.007	.852	.000	.000
153	.970	.001	.071	.083	.000	.073	.973	.000
455	.028	.999	.177	.078	.829	.072	.025	1.000
				Z7{2, 3, 3}[a]				
35	.000	.000	.071	.886	.007	.900	.000	.000
345	.714	.105	.332	.104	.098	.089	.718	.108
482	.286	.895	.003	.006	.734	.007	.282	.892
				Z7{3, 2, 2}[a]				
221	.401	.370	.017	.818	.065	.815	.380	.388
116	.161	.113	.020	.179	.048	.182	.169	.094
164	.436	.506	.068	.003	.124	.003	.451	.516

[a]Only the last several rows of the **Π** matrix are shown here; the relatively unchanging and inconsequential earlier rows have not been repeated. See text.

2. The **Z7**{2, 2, 3} case illustrates the separate identification of all three relationships, although the severe scaling problem of **C3** is masking its presence in the near dependency associated with $\tilde{\eta}_7 = 153$.

3. The other two cases exemplify the problem of separating the individual near dependencies when the scaled condition indexes are of the same order of magnitude. In **Z7**{2, 3, 3}, the two relations with similar scaled condition indexes, 345 and 482, are confounded; while in **Z7**{3, 2, 2}, all three near dependencies have roughly equal scaled condition indexes, 221, 116, and 164, and the involved variates have the variances of their estimated regression parameter spread out over all of them.

One final conclusion may be drawn rather generally from experiment 3; namely, those **Z** series that are qualitatively similar to previous **X** and **Y** series result in quantitatively similar **Π** matrices and scaled condition indexes, attesting to a degree of stability in the diagnostic procedure.

Experiment 4: The U Series

In the previous experiments, the constructed multiple near dependencies consist of coexisting, not truly simultaneous, near dependencies; that is, the several near dependencies are comprised of nonoverlapping sets of variates. Some truly simultaneous situations have arisen naturally in the "background" dependencies, but these are accidental instances, and they provide no systematic experimental evidence. Here, then, we examine for the first time the systematic behavior of the collinearity diagnostics when applied to a data matrix with truly simultaneous near dependencies. Interest centers on determining whether there is behavior that is notably different from that exhibited with coexisting near dependencies and whether there are means for determining the nature of the simultaneity.

U1

In this series there are two contrived simultaneous near dependencies: $\mathbf{C4}=0.3\mathbf{C2}-0.6\mathbf{C3}+\mathbf{e}_i$, determined from (4.11a), and $\mathbf{C5}=0.7\mathbf{C1}+0.4\mathbf{C2}+\mathbf{e}_j$, determined from (4.11b). These near dependencies have one element, **C2**, in common. Once again, to keep the report as short as possible, only selected subseries are displayed. Exhibit 4.14 shows the results holding the first relation fixed at level $i = 1$ while varying the second through the entire range $j = 0, \ldots, 4$. The corresponding supplementary correlations between **C5** and $\widehat{\mathbf{C5}}$ and regressions for **C5** on **C1**, **C2**, and **C3** are given in Exhibit 4.16*b*. Exhibit 4.15 shows the results holding the second relation fixed at level $j = 3$ while varying the first through its range, $i = 0, \ldots, 4$. The supplementary correlations and regressions for the contrived near dependency with **C4** are given in Exhibit 4.16*a*.

Exhibit 4.14 Scaled Condition Indexes and Variance-Decomposition Proportions: U1 Series. Two Constructed Near Dependencies (4.11a) and (4.11b): $\mathbf{C4} = 0.3\mathbf{C2} - 0.6\mathbf{C3} + \mathbf{e}_1$ (unchanging); $\mathbf{C5} = 0.7\mathbf{C1} + 0.4\mathbf{C2} + \mathbf{e}_j$ ($j = 0, \ldots, 4$).

Scaled Condition Index, $\tilde{\eta}$	Proportions of var(b_1)	var(b_2)	var(b_3)	var(b_4)	var(b_5)
			U1{1, 0}		
1	.011	.004	.001	.002	.007
2	.006	.021	.003	.031	.027
4	.367	.088	.007	.000	.011
17	.000	.779	.986	.955	.013
7	.616	.108	.003	.012	.942

Exhibit 4.14 Continued

Scaled Condition Index, $\tilde{\eta}$	Proportions of var(b_1)	var(b_2)	var(b_3)	var(b_4)	var(b_5)
			U1{1, 1}[a]		
4	.053	.045	.007	.000	.002
17	.107	.110	.862	.859	.094
22	.836	.831	.127	.106	.901
			U1{1, 2}[a]		
4	.005	.008	.007	.000	.000
17	.002	.050	.982	.961	.001
68	.993	.939	.007	.004	.998
			U1{1, 3}[a]		
4	.001	.001	.007	.000	.000
17	.000	.006	.985	.963	.000
215	.999	.993	.003	.002	1.000
			U1{1, 4}[a]		
4	.000	.000	.007	.000	.000
17	.000	.001	.986	.963	.000
679	1.000	.999	.003	.001	1.000

[a]Only the last several rows of the Π matrix are shown here; the relatively unchanging and inconsequential earlier rows have not been repeated. See text.

Exhibit 4.15 Scaled Condition Indexes and Variance-Decomposition Proportions: U1 Series. Two Constructed Near Dependencies (4.11a) and (4.11b): $C4 = 0.3C2 - 0.6C3 + e_i$ $(i = 0, \ldots, 4)$; $C5 = 0.7C1 + 0.4C2 + e_3$ (unchanging).

Scaled Condition Index, $\tilde{\eta}$	Proportions of var(b_1)	var(b_2)	var(b_3)	var(b_4)	var(b_5)
			U1{0, 3}		
1	.000	.000	.008	.010	.000
2	.000	.000	.009	.271	.000
4	.001	.001	.065	.001	.000
7	.000	.002	.913	.717	.000
211	.999	.997	.005	.001	1.000

Exhibit 4.15 Continued

Scaled Condition Index, $\tilde{\eta}$	Proportions of var(b_1)	var(b_2)	var(b_3)	var(b_4)	var(b_5)
			U1{1, 3}[a]		
2	.000	.000	.003	.034	.000
4	.001	.001	.007	.000	.000
17	.000	.006	.985	.963	.000
215	.999	.993	.003	.002	1.000
			U1{2, 3}[a]		
2	.000	.000	.000	.003	.000
4	.001	.001	.001	.000	.000
54	.001	.045	.994	.992	.001
215	.998	.954	.005	.004	.999
			U1{3, 3}[a]		
4	.001	.001	.000	.000	.000
168	.069	.145	.914	.915	.067
219	.930	.854	.086	.085	.933
			U1{4, 3}[a]		
544	.007	.899	.997	.997	.008
213	.993	.101	.003	.003	.992

[a]Only the last several rows of the **Π** matrix are shown here; the relatively unchanging and inconsequential earlier rows have not been repeated. See text.

Exhibit 4.16*a* Supplementary Correlations and Regressions of C4 on C1, C2, and C3: U1 Series

Variate {Level}	$r(\mathbf{C4}, \widehat{\mathbf{C4}})$, $\widehat{\mathbf{C4}} = 0.3\mathbf{C2} - 0.6\mathbf{C3}$	**C1**	**C2**	**C3**	$\hat{R}^2$
C4{0}	.693	−0.128 [−0.77]	0.361 [2.37]	−0.563 [−3.99]	.535
C4{1}	.955	−0.040 [−0.77]	0.319 [6.62]	−0.588 [−13.20]	.918
C4{2}	.995	−0.013 [−0.77]	0.306 [20.10]	[−0.596 [−42.30]	.991
C4{3}	.999	−0.004 [−0.77]	0.302 [62.60]	−0.599 [−134.00]	.999
C4{4}	.9999	−0.001 [−0.77]	0.301 [197.00]	−0.600 [−425.00]	.9999

Exhibit 4.16*b* Supplementary Correlations and Regressions of C5 on C1, C2, and C3: U1 Series

Variate {Level}	$r(\mathbf{C5}, \widehat{\mathbf{C5}})$, $\widehat{\mathbf{C5}} = 0.7\mathbf{C1} + 0.4\mathbf{C2}$	C1	C2	C3	$\hat{R}^2$
C5{0}	.724	0.423	0.714	−0.088	.502
		[1.34]	[2.45]	[−0.33]	
C5{1}	.955	0.612	0.499	−0.028	.911
		[6.11]	[5.42]	[−0.33]	
C5{2}	.995	0.672	0.431	−0.009	.991
		[21.20]	[14.80]	[−0.33]	
C5{3}	.999	0.691	0.410	−0.003	.999
		[69.00]	[44.50]	[−0.33]	
C5{4}	.9999	0.697	0.403	−0.001	.9999
		[220.00]	[138.00]	[−0.33]	

The following points are noteworthy:

1. Looking at either Exhibit 4.14 or 4.15, the presence of both near dependencies is apparent from the outset. The diagnostic procedure has no trouble here in determining the number of simultaneous near dependencies.

2. The role of **C2**, the variate common to the two relations, is of course necessarily confused between them, the stronger of the two near dependencies tending to claim the larger share, as would be expected. This is seen most dramatically from Exhibit 4.15, where the role of the **C4** dependency in the determination of var(b_2) is very minor at the beginning, grows with the increasing scaled condition index $\tilde{\eta}_4$, and, by the **U1**{4, 3} case, has taken the larger share.

3. Even though the dominating near dependency often obscures the individual variate involvement of **C2** in the weaker near dependency (as seen from the variance–decomposition proportions), it is nonetheless clear that the diagnostic information directly and readily leads to the determination of a set of auxiliary regressions that can show the appropriate variate involvement. Thus, suppose one considers case **U1**{1, 1} in Exhibit 4.14. This clearly indicates the involvement of **C4** in the relation with scaled condition index 17 and the involvement of **C5** in that with scaled condition index 22, suggesting as one (but only one) possible pair of auxiliary regressions those of **C4** and **C5** on **C1**, **C2**, and **C3**. These regressions relevant to the **U1**{1, 1} case are given, respectively, in the lines labeled **C4**{1} in Exhibit 4.16*a* and **C5**{1} in Exhibit 4.16*b*. The simultaneous involvement of **C2** in both of these auxiliary regressions is clearly evident. We shall have more to say on the formation of these auxiliary regressions in Section 5.3.

4. The confusion of all variates, not just those entering simultaneously, is still

observed in cases like **U1**{1, 1} or **U1**{3, 3} where the two near dependencies are competing.

5. The progression between the correlations and the scaled condition indexes is as in all previous experiments.

6. Essentially, then, little new is learned here, except that the diagnostic procedure appears to be working in this truly simultaneous case exactly as we have learned to expect from the previous experiments.

U2

This series is essentially the same as **U1**, except that **C2**, the common variate, has been given a scaling problem in the first dependency, which now reads **C4** = 0.03**C2** − 0.6**C3** + $\mathbf{e}_i$, as determined in (4.11c). Exhibit 4.17 shows the **Π** matrices for the subseries **U2**$\{i, 3\}$, $i = 0, \ldots, 4$. This subseries can be directly compared with that for the **U1**$\{i, 3\}$ subseries given in Exhibit 4.15.

Exhibit 4.17 Scaled Condition Indexes and Variance-Decomposition Proportions: U1 Series. Two Constructed Near Dependencies (4.11c) and (4.11b): C4 = 0.03C2 − 0.6C3 + $\mathbf{e}_i$ ($i = 0, \ldots, 4$); C5 = 0.7C1 + 0.4C2 + $\mathbf{e}_3$ (unchanging).

Scaled Condition Index, $\tilde{\eta}$	Proportions of				
	var(b_1)	var(b_2)	var(b_3)	var(b_4)	var(b_5)
			U2{0, 3}		
1	.000	.000	.006	.008	.000
3	.000	.000	.056	.167	.000
4	.001	.002	.005	.003	.000
7	.000	.000	.928	.821	.000
219	.999	.998	.005	.001	1.000
			U2{1, 3}[a]		
3	.000	.000	.012	.015	.000
4	.001	.002	.000	.001	.000
21	.000	.000	.984	.982	.000
221	.999	.998	.002	.001	1.000
			U2{2, 3}[a]		
3	.000	.000	.000	.001	.000
4	.001	.002	.001	.000	.000
65	.000	.000	.997	.997	.000
222	.999	.998	.002	.002	1.000

Exhibit 4.17 Continued

Scaled Condition Index, $\tilde{\eta}$	Proportions of var(b_1)	var(b_2)	var(b_3)	var(b_4)	var(b_5)
			U2{3, 3}[a]		
4	.001	.002	.000	.000	.000
205	.061	.024	.923	.924	.059
223	.938	.974	.077	.076	.941
			U2{4, 3}[a]		
651	.001	.091	1.000	1.000	.002
221	.998	.908	.000	.000	.998

[a]Only the last several rows of the **Π** matrix are shown here; the relatively unchanging and inconsequential earlier rows have not been repeated. See text.

The following points are noteworthy:

1. The scaling problems are clearly observed, causing the role of **C2** in the near dependency associated with $\tilde{\eta}_4$ to be substantially obscured. Even by the tightest case, **U2**{4, 3}, its role is not strongly evident. Here, the dominant near dependency is that associated with $\tilde{\eta}_4 = 651$. Normally, we have found dominant dependencies to take up the largest share of the variance decomposition of its constituent variates and to mask their presence in the weaker near dependencies. Here, however, the weak involvement of **C2** in the strongest near dependency is being masked by its stronger involvement in the next weaker near dependency, that associated with $\tilde{\eta}_5 = 221$. It is clear, then, that it is a complex interplay between both the strength of a near dependency and the strength of the involvement of the variate in that near dependency that determines the corresponding variance–decomposition proportion.

2. Thus, in a case like **U2**{4, 3}, where two near dependencies seem well separated except for a bit of "bleeding through" to the stronger near dependency by one or more of the variates clearly involved in a weaker one, we might strongly suspect the simultaneous involvement of those variates even without running auxiliary regressions. Clearly, however, such auxiliary regressions would be required for sure diagnosis. The role of **C2**, for example, is clearly discernable by the $i = 3$ level in the set of supplementary correlations and regressions for the new dependency in **C4** that is given in Exhibit 4.18.

3. The preceding situation with **U1**{4, 3} is to be compared and contrasted with that, for example, for case **X1**{4} in Exhibits 4.1*a* and 4.1*b*. In many important respects these two cases look the same. Specifically, the variance

Exhibit 4.18 Supplementary Correlations and Regressions of C4 on C1, C2, and C3: U2 Series

Variate {Level}	$r(\mathbf{C4}, \widehat{\mathbf{C4}})$, $\widehat{\mathbf{C4}}=0.03\mathbf{C2}-0.6\mathbf{C3}$	C1	C2	C3	$\hat{R}^2$
C4{0}	.657	−0.170	0.112	−0.550	.529
		[−0.77]	[0.55]	[−2.94]	
C4{1}	.954	−0.054	0.056	−0.584	.917
		[−0.77]	[0.87]	[−9.86]	
C4{2}	.995	−0.017	0.038	−0.595	.991
		[−0.77]	[1.88]	[−31.80]	
C4{3}	.999	−0.005	0.033	−0.598	.999
		[−0.77]	[5.09]	[−101.00]	
C4{4}	.9999	−0.002	0.031	−0.600	.9999
		[−0.77]	[15.20]	[−320.00]	

decomposition for **C2** in case **X1**{4} is bleeding through from the low background near dependencies to the very strong near dependency associated with $\tilde{\eta}_4 = 976$, which is otherwise associated very strongly with the determination of **C3** and **C4**. Having possible simultaneity in mind, some weak involvement of **C2** in this latter near dependency might therefore be considered. But a look at the **X1**{4} supplementary regression in Exhibit 4.1*b* (as well as our prior knowledge as to the true composition of this near dependency) shows this not to be the case. Thus, the variance-decomposition proportions cannot be relied upon always to point correctly to simultaneous involvement, even in this case where a variate determined in a weaker relation seems also to be somewhat involved in a dominating near dependency. But they can be relied upon to determine the auxiliary regressions from which this information can be obtained.

Experiment 5: The V Series

This experiment consists of a single series, **V1**, also containing two truly simultaneous near dependencies but having two common variates instead of one. Moreover, a new basic data set is employed. The object is to see if any added complications arise with a slightly more involved degree of simultaneity and further to observe the stability of the diagnostics with differing data sets. The basic data set comprises the first three columns of the **V1** matrix. Two additional simultaneous near dependencies are constructed. The first is $\mathbf{C4} = 0.4\mathbf{C1} - 0.3\mathbf{C2} + 0.6\mathbf{C3} + \mathbf{e}_i$ as determined from (4.13a), and the second is $\mathbf{C5} = 0.7\mathbf{C1} + 0.4\mathbf{C2} + \mathbf{e}_j$ as determined from (4.13b). The variates **C1** and **C2** are common to both near dependencies.

Exhibit 4.19 shows the **Π** matrices for the subseries **V1**{2, j} formed by holding i fixed at 2 while varying $j = 0, \ldots, 4$. The corresponding supplementary correlations between **C5** and $\widehat{\mathbf{C5}}$ and regressions for **C5** on **C1**, **C2**, and **C3** are given in Exhibit 4.21b. Exhibit 4.20 shows the results of holding the second relation fixed at level $j = 3$ while varying the first through its range, $i = 0, \ldots, 4$. The supplementary correlations and regressions for the contrived near dependency with **C4** are given in Exhibit 4.21a.

Exhibit 4.19 Scaled Condition Indexes and Variance-Decomposition Proportions: V1 Series. Two Constructed Near Dependencies (4.13a) and (4.13b): $\mathbf{C4} = 0.4\mathbf{C1} - 0.3\mathbf{C2} + 0.6\mathbf{C3} + \mathbf{e}_2$ (unchanging); $\mathbf{C5} = 0.7\mathbf{C1} + 0.4\mathbf{C2} + \mathbf{e}_j$ ($j = 0, \ldots, 4$).

Scaled Condition Index, $\tilde{\eta}$	Proportions of				
	var(b_1)	var(b_2)	var(b_3)	var(b_4)	var(b_5)
			V1{2, 0}		
1	.000	.001	.000	.000	.009
3	.000	.003	.002	.001	.123
4	.006	.023	.002	.001	.061
66	.971	.965	.994	.998	.001
6	.023	.008	.002	.000	.806
			V1{2, 1}[a]		
3	.000	.007	.002	.002	.006
4	.008	.016	.003	.000	.003
67	.875	.884	.992	.997	.000
19	.117	.093	.003	.001	.989
			V1{2, 2}[a]		
4	.004	.009	.003	.000	.000
68	.724	.212	.859	.873	.101
60	.272	.775	.136	.125	.898
			V1{2, 3}[a]		
4	.001	.002	.003	.000	.000
67	.043	.132	.995	.998	.002
193	.956	.866	.000	.000	.998
			V1{2, 4}[a]		
4	.000	.000	.003	.000	.000
67	.005	.015	.993	.997	.000
611	.995	.985	.001	.001	1.000

[a]Only the last several rows of the **Π** matrix are shown here; the relatively unchanging and inconsequential earlier rows have not been repeated. See text.

Exhibit 4.20 Scaled Condition Indexes and Variance-Decomposition Proportions: V1 Series. Two Constructed Near Dependencies (4.13a) and (4.13b): $C4 = 0.4C1 - 0.3C2 + 0.6C3 + e_i$ $(i = 0, \ldots, 4)$; $C5 = 0.7C1 + 0.4C2 + e_3$ (unchanging).

Scaled Condition Index, $\tilde{\eta}$	Proportions of var(b_1)	var(b_2)	var(b_3)	var(b_4)	var(b_5)
			V1{0, 3}		
1	.000	.000	.006	.008	.000
3	.000	.000	.054	.171	.000
4	.001	.003	.084	.033	.000
7	.000	.003	.840	.787	.000
190	.999	.994	.016	.001	1.000
			V1{1, 3}[a]		
3	.000	.001	.018	.016	.000
4	.001	.002	.021	.004	.000
21	.004	.014	.955	.978	.000
193	.995	.983	.005	.001	1.000
			V1{2, 3}[a]		
3	.000	.001	.002	.002	.000
4	.001	.001	.003	.000	.000
67	.043	.132	.995	.998	.002
193	.956	.866	.000	.000	.998
			V1{3, 3}[a]		
4	.000	.001	.000	.000	.000
217	.764	.199	.852	.856	.118
189	.236	.800	.147	.144	.882
			V1{4, 3}[a]		
677	.899	.905	.998	.999	.000
191	.101	.094	.002	.001	1.000

[a]Only the last several rows of the **Π** matrix are shown here; the relatively unchanging and inconsequential earlier rows have not been repeated. See text.

Exhibit 4.21*a* Supplementary Correlations and Regressions of C4 on C1, C2, and C3: V1 Series

Variate {Level}	$r(\text{C4}, \widehat{\text{C4}})$, $\widehat{\text{C4}} = 0.4\text{C1} - 0.3\text{C2} + 0.6\text{C3}$	C1	C2	C3	$\hat{R}^2$
C4{0}	.715	−0.008 [−0.04]	−0.095 [−0.40]	0.817 [3.48]	.518
C4{1}	.953	0.271 [3.90]	−0.235 [−3.11]	0.668 [9.02]	.914
C4{2}	.995	0.359 [16.30]	−0.280 [−11.70]	0.622 [26.50]	.991
C4{3}	.999	0.387 [55.60]	−0.294 [−38.80]	0.607 [81.90]	.999
C4{4}	.9999	0.396 [180.00]	−0.298 [−125.00]	0.602 [257.00]	.9999

Exhibit 4.21*b* Supplementary Correlations and Regressions of C5 on C1, C2, and C3: V1 Series

Variate {Level}	$r(\text{C5}, \widehat{\text{C5}})$, $\widehat{\text{C5}} = 0.7\text{C1} + 0.4\text{C2}$	C1	C2	C3	$\hat{R}^2$
C5{0}	.652	0.625 [1.74]	0.419 [1.07]	−0.292 [−0.76]	.372
C5{1}	.950	0.676 [5.95]	0.406 [3.29]	−0.092 [−0.76]	.897
C5{2}	.995	0.692 [19.30]	0.402 [10.30]	−0.029 [−0.76]	.990
C5{3}	.999	0.698 [61.30]	0.401 [32.40]	−0.009 [−0.76]	.999
C5{4}	.9999	0.699 [194.00]	0.400 [102.00]	−0.003 [−0.76]	.9999

The following points are noteworthy:

1. No new difficulties are encountered. The presence of both relations is readily diagnosed, and the determination of variate involvement encounters the expected difficulties attendant on competing and dominating dependencies. The simultaneous involvement of **C1** and **C2** is discernable under many of the different circumstances represented in the various **Π** matrices of Exhibits 4.19 and 4.20. It is of interest to compare the behavior of **C1** and **C2** with that of the other variates as the changing near dependency gets tighter. The same pattern is seen in either exhibit: the variance–decomposition proportions of **C1** and **C2** start small and become systematically larger as the varying near dependency becomes stronger, while the variance–decomposition proportions of the other variates remain essentially unchanged.

2. The exception to the preceding occurs when there are competing dependencies. The distribution of the variance decomposition of the simultaneously involved variates across the two near dependencies is mimicked by the confounding that tends always to take place under these circumstances. Thus, in cases like **V1**{2, 2} of Exhibit 4.19 or **V1**{3, 3} of Exhibit 4.20, it is impossible to distinguish the behavior of a variate like **C3**, which belongs only to the second near dependency, from that of a variate like **C2**, which belongs to both. Auxiliary regressions are needed to discern these separate effects, as is clear from Exhibit 4.21.

3. We are again, then, in the happy position of being able to report that there is really little new to report. The diagnostics are quite able to determine the number of near dependencies, even in truly simultaneous cases, and the information contained in the **Π** matrices can directly show variate involvement for some situations and can correctly lead to auxiliary regressions to show variate involvement in all instances.

Experiment 6: The W Series

I trust the reader will indulge one more brief experiment. Recall from Section 2.3, where other collinearity diagnostics are discussed, that VIFs are criticized for being unable to diagnose the number of near dependencies correctly when there is more than one. Specifically, it is clear that if one had a five-column data matrix with two near dependencies, the first involving all five variates and the second involving, say, two of them, the VIFs would all be large. But this would also be true in the absence of the second near dependency, so this information can convey nothing about the number of near dependencies beyond 1. It is important, then, to see if the diagnostic procedure being advanced here is capable of correctly assessing such a situation, and this one-series experiment is designed to do exactly this.

In the **W1** series, two contrived near dependencies are added to the three basic columns to produce a five-column data matrix. The first near dependency is $\mathbf{C4} = \mathbf{C1} + \mathbf{e}_i$, determined in (4.15a), while the second involves all five columns,

C5 $=0.7$**C1** $+0.4$**C2** -0.3**C3** $+0.6$**C4** $+$ $\mathbf{e}_j$, determined in (4.15b). Of the 25 possible experiments, we again report only two representative subseries. Exhibit 4.22 shows the behavior of **W1**$\{2, j\}$, when i is fixed at 2 and $j=0,\ldots,4$. Exhibit 4.23 shows **W1**$\{i, 3\}$, where j is fixed at 3 and $i=0,\ldots,4$.

Exhibit 4.22 Scaled Condition Indexes and Variance-Decomposition Proportions: W1 Series. Two Constructed Near Dependencies (4.15a) and (4.15b): C4 = C1 + $\mathbf{e}_2$ (unchanging); C5 = 0.7C1 + 0.4C2 − 0.3C3 + 0.6C4 + $\mathbf{e}_j$ (j = 0, ..., 4).

Scaled Condition Index, $\tilde{\eta}$	Proportions of var(b_1)	var(b_2)	var(b_3)	var(b_4)	var(b_5)
		W1$\{2, 0\}$			
1	.000	.015	.016	.000	.007
3	.000	.096	.385	.000	.046
4	.000	.881	.487	.000	.000
66	.996	.002	.000	.996	.056
6	.004	.006	.112	.004	.891
		W1$\{2, 1\}$[a]			
3	.000	.037	.255	.000	.003
4	.000	.396	.228	.000	.000
69	.987	.051	.043	.962	.134
20	.013	.509	.466	.038	.862
		W1$\{2, 2\}$[a]			
3	.000	.004	.043	.000	.000
4	.000	.046	.035	.000	.000
47	.007	.304	.298	.521	.303
94	.993	.645	.622	.479	.696
		W1$\{2, 3\}$[a]			
3	.000	.000	.004	.000	.000
4	.000	.004	.004	.000	.000
56	.018	.020	.020	.902	.018
251	.982	.976	.972	.098	.982
		W1$\{2, 4\}$[a]			
4	.000	.000	.001	.000	.000
57	.003	.002	.002	.951	.002
775	.997	.998	.997	.049	.998

[a]Only the last several rows of the Π matrix are shown here; the relatively unchanging and inconsequential earlier rows have not been repeated. See text.

Exhibit 4.23 Scaled Condition Indexes and Variance-Decomposition Proportions: W1 Series. Two constructed near dependencies (4.15a) and (4.15b): $C4 = C1 + e_i$ ($i = 0, \ldots, 4$): $C5 = 0.7C1 + 0.4C2 - 0.3C3 + 0.6C4 + e_3$ (unchanging).

Scaled Condition Index, $\tilde{\eta}$	Proportions of				
	$\text{var}(b_1)$	$\text{var}(b_2)$	$\text{var}(b_3)$	$\text{var}(b_4)$	$\text{var}(b_5)$
			W1{0, 3}		
1	.000	.000	.000	.006	.000
3	.000	.000	.004	.031	.000
4	.000	.004	.004	.054	.000
7	.001	.002	.001	.871	.000
241	.999	.994	.991	.038	1.000
			W1{1, 3}[a]		
3	.000	.000	.004	.003	.000
4	.000	.004	.004	.004	.000
19	.003	.002	.002	.943	.002
244	.997	.994	.990	.049	.998
			W1{2, 3}[a]		
4	.000	.004	.004	.001	.000
56	.018	.020	.020	.902	.018
251	.982	.976	.972	.097	.982
			W1{3, 3}[a]		
147	.008	.295	.296	.533	.292
296	.992	.701	.696	.467	.708
			W1{4, 3}[a]		
682	.988	.133	.131	.962	.137
202	.012	.863	.861	.038	.863

[a]Only the last several rows of the Π matrix are shown here; the relatively unchanging and inconsequential earlier rows have not been repeated. See text.

The following points are noteworthy:

1. It is clear that, unlike VIFs, this diagnostic procedure has no trouble discerning the presence of the two simultaneous near dependencies, despite the fact that one of them involves all of the variates.

2. The determination of variate involvement has its usual complications, but the variance-decomposition proportions always provide the information

needed to determine appropriate auxiliary regressions, from which variate involvement can be found. In each case, at least one important variate involved in each near dependency is clearly indicated. These can be chosen as "dependent variates" in auxiliary regressions run on the remaining variates in a process that will be described more fully in Section 5.3. Moreover, it is clear in most instances that a great deal more can be learned about variate involvement, or the potential for variate involvement, directly from the variance–decomposition proportions.

3. One interesting anomaly arises here. A glance at the previous exhibits depicting the results for two coexisting or simultaneous near dependencies shows that, as the varying near dependency gets stronger, the scaled condition index for the fixed dependency remains virtually unchanged. In this experiment, however, as the near dependency with varying strength sweeps across the magnitude of the fixed near dependency, there is some noticeable, if not substantive, movement in the magnitude of the scaled condition index of the fixed near dependency. This is observable in both Exhibits 4.22 and 4.23. The meaning of this is not clear, but it certainly has no deleterious impact on the effectiveness of the diagnostic procedure.

CHAPTER 5

Summarizing and Interpreting the Collinearity Diagnostics

In Chapter 3, a collinearity diagnostic is developed that indicates the presence of collinear relations in a data matrix and determines the variates involved in each. When used in a regression context, this information also assesses the degree to which the collinear relations degrade the least-squares estimates. Chapter 4, recognizing the empirical content of this diagnostic procedure, conducts and reports a set of experiments designed to provide experience in its use and interpretation. In this chapter, we begin with a summary of this experimental evidence and then turn to more detailed aids for digesting and interpreting the information contained in the scaled condition indexes and $\mathbf{\Pi}$ matrices. Finally, several examples are given using the collinearity diagnostics with actual data.

5.1 A SUMMARY OF THE EXPERIMENTAL EVIDENCE

Before proceeding, it is worth noting that the experiments in the previous chapter are necessarily limited in scope and cannot hope to illuminate all that is to be known of the behavior of the proposed diagnostic procedure in all applications. Indeed, since the introduction of this diagnostic procedure in Belsley et al. (1980), experience with it has continued to grow and knowledge of its properties and behavior has continued to expand and be refined. It is therefore gratifying to report that all of the main considerations relative to its meaning and use have held stable over time.

This summary begins with a presentation of the experience gained from the experiments involving a single contrived near dependency. We then summarize the modifications and extensions that arise when analyzing data in which two or more near dependencies coexist or exist simultaneously.

Experience with a Single Near Dependency

1. *The Diagnostic Procedure Works.* The foremost conclusion is that the procedure for diagnosing collinearity suggested in Chapter 3 works. It works well and in accord with expectations for a wide variety of data matrices with contrived near dependencies. It is possible to determine not only the presence of a near dependency but also, subject to some qualifications that follow, the variates involved in it.

2. *The Progression of Tightness.* The tighter the underlying near dependency—as measured either by its correlation or relevant multiple correlation—, the higher the scaled condition index. Indeed, as the underlying correlations or R^2s increase along the "progression of nines," that is, $<.9$, .9, .99, .999, .9999, and so on, the scaled condition indexes increase roughly along the "progression of 10/30," that is, 1, 3, 10, 30, 100, 300, 1000, 3000, and so on. The correspondence between these two progressions, however, is not the same in every case and depends upon the type of data. A given correlation, for example, among rates-of-change data appears to correspond to a lower scaled condition index than that for levels data. Some rough generalizations do, however, seem warranted, and these are given next.

3. *Interpreting the Magnitude of the Condition Index.* Most of the experimental evidence shows that weak near dependencies (correlations of less than .9) begin to exhibit themselves with scaled condition indexes around 10 and in some cases as low as 5. A scaled condition index in the neighborhood of 15–30 tends to result from an underlying near dependency with an associated correlation of .9, usually considered to be the borderline of "tightness" in informal econometric practice, although different applied statistical disciplines have grown to adopt different conventions here. Scaled condition indexes of 100 or more appear to be large indeed in every case, causing substantial variance inflation and great potential harm to regression estimates.

If pressed to provide a value for a scaled condition index that divides large from small, 30 seems quite reasonable for many purposes. I am, however, always reluctant to give such figures because they are sometimes taken too seriously. Good statistical analysis rarely admits of mechanically applied rules of thumb. This figure clearly must be tempered somewhat by the considerations given in point 6 that follows, and we shall see in the next section that everything is indeed relative. When a value of 30 coexists with several values in excess of 3000, the 30 clearly palls into practical insignificance.

4. *Variance–Decomposition Proportions.* The rule of thumb proposed in Chapter 4—that estimates shall be deemed degraded when more than 50% of the variance of two or more coefficients is associated with a single high scaled condition index—still seems quite good, at least for cases when there is only one near dependency. Variance–decomposition proportions in excess of .5 allow the involved variates to be identified in almost all instances, even when the underlying near dependency is reasonably weak (associated correlations of .4 to

.7). Indeed, most evidence indicates that the variance–decomposition proportions of involved variates attain .8 quite early.

Like any rule of thumb, however, the .5 rule must be tempered with mercy and not be applied with unbending insistence. In particular, we shall see that the .5 rule of thumb must be refined when there are several coexisting or simultaneous near dependencies.

5. *Scaling Problems.* Essential scaling imbalance occurs when a near dependency involves several variates, some of whose contributions are orders of magnitude smaller than others. This causes the involvement of the dominated variates to be masked and more difficult to detect. Involved variates, for example, whose contribution is less than 1% of that of others in the near dependency, are dominated, and their involvement can be completely overlooked by this procedure until the scaled condition index rises to 30 or more.[1] Very strongly dominated variates (<.01%) can be masked even with scaled condition indexes in excess of 300.

6. *Data Type Matters.* As previously noted in point 2, near dependencies among rates-of-change data seem to behave slightly differently from those involving levels-type data. Exactly how data structures and the collinearity diagnostics interact is an open question, but none of the observed differences in behavior appears to be important enough to affect the efficacy of the diagnostic procedure adversely. In my own research, this has never proved to be a concern.

Experience with Coexisting and Simultaneous Near Dependencies

Coexisting near dependencies are two or more near dependencies existing together but having no variates in common. Simultaneous near dependencies have variates in common.

7. *Retention of Individuality.* While some new problems of diagnosis and interpretation are introduced, in general, it can be concluded that coexisting and simultaneous near dependencies cause the diagnostic procedure no critical problems. Subject to the modifications to be given in points 11–13, the several underlying near dependencies behave together much as they do separately. In particular, they remain *countable* (point 8) and to a great extent *separable* (point 9). Somewhat greater problems are encountered here with truly simultaneous near dependencies than with coexisting ones.

8. *Countability.* The number of near dependencies, coexisting or simultaneous, is correctly assessed in all cases by the number of high scaled condition indexes. The presence of a strong near dependency, for example, does not obscure the detection of a much weaker coexisting or simultaneous near dependency. This attribute of the diagnostic procedure presented here is to be contrasted with virtually all other suggested collinearity diagnostics.

[1]We recall from Section 2.3, however, that diagnosis based on eigenvector elements alone would always overlook this situation.

9. *Separability.* The near dependencies remain separable in the following two senses. First, near dependencies that, when existing alone, have a given scaled condition index, retain roughly the same scaled condition index when made to exist with other near dependencies regardless of their relative degrees of tightness. This is strictly true for coexisting near dependencies and substantively true for simultaneous near dependencies. Second, subject to the qualifications that follow, the individual involvement of specific variates in specific near dependencies remains observable.

10. *Isolation through Near Orthogonality.* As the theory of Section 3.2 would have it, the regression estimates of one subset of the variates can be buffered from the deleterious effects of near dependencies among another subset that is nearly orthogonal to it. Thus, under these conditions, it is possible for some regression estimates to be salvaged even in the presence of very strong collinearity (provided, of course, one has a regression algorithm that is capable of dealing correctly with such a situation[2]).

11. *Confounding of Effects with Competing Near Dependencies.* Two or more near dependencies are competing when they have scaled condition indexes of roughly equal orders of magnitude, as judged along the progression of 10/30 given in point 2. When this is the case, the variance–decomposition proportions of the variates involved in at least one of the competing near dependencies can be arbitrarily distributed among them. This can be true even if the variate is involved in only one of the competing near dependencies, thus confounding the analysis of true variate involvement. The number of coexisting near dependencies is, however, not obscured by this situation, nor is the identification of the variates that are involved in at least one of the competing near dependencies. Thus, it remains possible to diagnose correctly how many near dependencies are present and which variates are being degraded by the joint presence of those dependencies. Only information on the separate involvement of specific variates in specific competing near dependencies may be lost.

In the case of competing near dependencies, the proposed diagnostic procedure is trivially modified to allow all available information to be obtained. Specifically, the .5 rule of thumb is now no longer applicable to the variance–decomposition proportion of a specific variate but rather to the sum of the variance–decomposition proportions for that variate across the competing dependencies. That is, two or more variates are involved in at least one competing near dependency if the following conditions hold:

I* The competing near dependencies are deemed to have high scaled condition indexes and

[2]This typically means using a computational algorithm that is stable in the face of ill conditioning, such as ones based on the QR (Householder) or singular-value decompositions. Alternatively, one could use a Cholesky decomposition along with very substantial computational precision (such as the IEEE 80-bit standard). These computational issues are discussed in Belsley et al. (1980, Appendix 2B). Studies of the relative behavior of various algorithms with ill-conditioned data are given in Lesage and Simon (1985) and Simon and Lesage (1988a, b).

II* The sum of the variance–decomposition proportions of these variates across the competing near dependencies exceeds .5—even if no individual variance–decomposition proportion is greater than .5.

This is examplified in Exhibit 5.1. Here, we determine that there are two near dependencies, those with scaled condition indexes 35 and 41, and that they are both of the same order of magnitude, roughly 30. It is clear that over 90% of the variance of **C1**, **C3**, and **C4** is being determined by these two competing near dependencies, and hence all three must be involved in at least one. Note that this is true for **C1** despite the fact that none of its variance–decomposition proportions is individually greater than .5.

12. *Dominating Near Dependencies.* A dominating near dependency is one whose scaled condition index is of a higher order of magnitude along the progression of 10/30 than other near dependencies that exist along with it. Such a near dependency can become the prime determinant of the variance of the coefficient of a given variate and thus obscure information about the possible simultaneous involvement of that variate in weaker (dominated) dependencies. This is illustrated in Exhibit 5.2.

Here there are two near dependencies with high scaled condition indexes, 37 and 432, and the latter dominates. The involvement of **C3** and **C4** in this dominant near dependency is clear; however, equally clearly, we cannot rule out the potential involvement of both of these variates along with **C1** in the near dependency associated with $\tilde{\eta}_3 = 37$. Indeed, we know that at least one of these must be so involved, for we have seen in Chapter 3 that there cannot be a high scaled condition index like 37 without there also being associated nonorthogonality among the columns of the data matrix. The column **C1** must therefore be involved in a near dependency with some other columns. In this case **C2** appears to be uninvolved in any near dependency since virtually all its variance is associated only with small scaled condition indexes. So this leaves only **C3** and **C4**.

Thus, when a near dependency is dominated, as in the $\tilde{\eta}_3 = 37$ case above, the

Exhibit 5.1 Illustrative Scaled Condition Indexes and Variance–Decomposition Proportions: Competing Near Dependencies

Scaled Condition Index, $\tilde{\eta}$	Proportions of			
	var(b_1)	var(b_2)	var(b_3)	var(b_4)
1	.007	.012	.001	.003
5	.043	.799	.004	.032
35	.476	.002	.141	.438
41	.474	.187	.854	.527

Exhibit 5.2 Illustrative Variance-Decomposition Proportions and Scaled Condition Indexes: Dominating Near Dependencies

Scaled Condition Index, $\tilde{\eta}$	Proportions of $\mathrm{var}(b_1)$	$\mathrm{var}(b_2)$	$\mathrm{var}(b_3)$	$\mathrm{var}(b_4)$
1	.005	.012	.001	.003
5	.021	.988	.006	.012
37	.974	.000	.039	.058
432	.000	.000	.954	.927

high scaled condition index can still indicate the presence of a degrading collinear relation even though there is only one high variance–decomposition proportion associated with it—the involvement of other variates being obscured by their simultaneous roles in the dominating near dependency.

The case of dominant near dependencies, then, again forces us to refine our diagnostic procedure slightly:

I** A high scaled condition index is sufficient to demonstrate the presence of a degrading collinear relation.

II** But when there are dominating near dependencies, there may not be two or more high variance–decomposition proportions associated with the scaled condition index of the dominated near dependency; the sum of the variance–decomposition proportions of all involved variates across the set of high scaled condition indexes will, however, remain high ($>.5$).

If, under these circumstances, more-detailed information regarding variate involvement is needed, it will require further investigations, typically in the form of auxiliary regressions among the potentially involved variates. We will shortly examine the formation of these auxiliary regressions in some detail, but we note in the case of Exhibit 5.2 that **C3** is clearly a member of the dominant near dependency, as is **C1** of the dominated. Thus, regressions of **C1** and **C3** on **C2** and **C4** would seem in order. The auxiliary regressions, however, are not needed to determine which coefficient estimates are degraded, for we can tell directly from Exhibit 5.2 that $\mathrm{var}(b_1)$, $\mathrm{var}(b_3)$, and $\mathrm{var}(b_4)$ are involved in at least one near dependency and, hence, are all degraded.

13. *Nondegraded Estimates and Noninvolved Variates.* On occasion it is also possible to identify those variates that are not involved in any near dependency and, therefore, those regression coefficients that are not degraded by the presence of collinear relations. This is illustrated in Exhibit 5.2 where $\mathrm{var}(b_2)$ has virtually all of its variance determined in association with the relatively small

scaled condition indexes 5 and 1 and is not adversely affected by the two tighter near dependencies with scaled condition indexes of 37 and 432. The same situation arises in several of the experiments of Chapter 4 as, for example, in the case of the **X1** and **X2** series given in Exhibits 4.1*a* and 4.2*a*.

Experiences Special to Simultaneous Near Dependencies

The preceding points 7–13 apply equally well to both coexisting and simultaneous near dependencies. The following points further summarize the comparative behavior of the diagnostics between these two cases.

14. *No Troubles Arise in Determining the Presence of Truly Simultaneous Near Dependencies.* The diagnostic procedure has no trouble in determining either the presence or the number of coexisting or truly simultaneous near dependencies.

15. *Indistinguishability of Patterns for Simultaneous and Coexisting Near Dependencies.* There seems to be no systematic difference in the types of patterns for variance–decomposition proportions that can arise from coexisting and truly simultaneous near dependencies. This is certainly the case when there are competing near dependencies, where we know that the variance–decomposition proportions of a given variate can be arbitrarily distributed across the competing dependencies, whether or not the given variate is in fact involved simultaneously in each competing near dependency. And this is also true when there are dominant dependencies, where low variance–decomposition proportions can be associated with a dominated dependency regardless of possible simultaneous involvement. Thus, the variance–decomposition proportions can be used to give at best a presumption of simultaneity, which presumption, however, must be investigated further through the use of auxiliary regressions if greater detail is required.

5.2 EMPLOYING THE DIAGNOSTIC PROCEDURE

Diagnosing any given data set for the presence and composition of near dependencies and assessing the potential harm that their presence may cause least-squares regression estimates is effected by a rather straightforward series of steps, the only problems of interpretation arising when there are competing and dominating near dependencies. Here we provide the steps to be followed, giving various hints to make their enactment simpler and more transparent. The content of many of these steps will be further refined in subsequent chapters. The main purpose here is to provide enough information to get under way.

The Steps

STEP 1. Determine **X**.
STEP 2. Column-equilibrate **X**.

STEP 3. Obtain scaled condition indexes and variance-decomposition proportions.
STEP 4. Determine number of near dependencies.
STEP 5. Determine variate involvement.
STEP 6. Determine auxiliary regressions.
STEP 7. Determine unaffected variates.

1. *Determine* **X**. Any **X** matrix can, in principle, be analyzed. However, if the data are being analyzed relative to estimating a linear regression model with least squares, it is assumed that the user has a specific parameterization $\boldsymbol{\beta}^*$ in mind and has transformed the raw data (if need be) to conform, so that the resulting **X** matrix is that relevant to the model in the form $\mathbf{y} = \mathbf{X}\boldsymbol{\beta}^* + \boldsymbol{\varepsilon}$. If an intercept is appropriate to the model, it should be made explicit, so that **X** has a column of ones. On a related matter, the data should not be centered. We shall see that, contrary to much popular opinion, centering the data does not get rid of "nonessential collinearity" and will rather generally mask the role of the constant in any underlying near dependencies and produce misleading diagnostic results. We shall also see that the diagnostic results are most meaningful when the **X** data are structurally interpretable, a term whose definition is best delayed until the next chapter. When this is the case, the corresponding model is said to be in basic form, a notion discussed in Section 6.6.

2. *Column-Equilibrate* **X**. Once a specific **X** matrix has been selected for analysis, it should be column equilibrated; that is, each column should be scaled for equal Euclidean length (1.10). The reasons for this, as well as alternative scaling schemes, are discussed in the next chapter. However, unless this is done, the collinearity diagnostics can produce arbitrary results. The usual method is to scale each column $\mathbf{X}_i$ by its norm $\|\mathbf{X}_i\|$ so that the resulting $\mathbf{X}_i/\|\mathbf{X}_i\|$ has unit Euclidean length.

3. *Obtain Scaled Condition Indexes and Variance–Decomposition Proportions.* The best method for obtaining these fundamental pieces of diagnostic information is through the singular-value decomposition of the column-scaled **X** matrix (from 2 above). Then

(a) the scaled condition indexes $\tilde{\eta}_k$ are the η_k as determined as in (3.20) and
(b) the $\boldsymbol{\Pi}$ matrix of variance–decomposition proportions is determined as in Equation (3.24) and Exhibit 3.5.

In the event that software for the singular-value decomposition is not available,[3] the same formulas may instead be applied to the eigenvectors of $\mathbf{X}^T\mathbf{X}$ and the square roots of the corresponding eigenvalues. For the reasons given in Section 3.1, however, the use of the singular-value decomposition is to be preferred.

4. *Determine the Number of Near Dependencies.* Determine the number and relative strengths of the near dependencies by the scaled condition indexes

[3]See Section 3.1 for references to SVD sources.

exceeding some chosen threshold $\tilde{\eta}^*$ such as 30. The choice of this threshold is somewhat of an art form. It is not a classical significance level that must be chosen a priori and is best chosen relativistically, depending on the pattern of scaled condition indexes that arises. More will be said on this shortly. The relative strengths of the scaled condition indexes are determined by their approximate position along the progression of 10/30, that is, 1, 3, 10, 30, 100, 300, 1000, and so on.

5. *Determine Variate Involvement.* Rather generally, a variate is considered involved in, and its corresponding regression coefficient degraded by, at least one near dependency if the total proportion of its variance associated with the set of high scaled condition indexes exceeds some chosen threshold π^* such as .5. Two cases are to be considered.

Case 1: Only one near dependency is present. Here, there is only one high scaled condition index, and it is possible to determine variate involvement directly from the variance–decomposition proportions. A variate is considered involved (and its estimated coefficient degraded) if its variance–decomposition proportion associated with this single scaled condition index exceeds the threshold π^*.

Case 2: Coexisting or simultaneous near dependencies. Here there are several high scaled condition indexes. Variate involvement is now determined by aggregating the variance–decomposition proportions of each variate over the set of these several high condition indexes. Those variates whose *aggregate* proportions exceed the threshold π^* are involved in at least one of the near dependencies (and their corresponding estimated coefficients are degraded). If in addition the variance–decomposition proportion of a given variate exceeds π^* by itself in one of the several near dependencies, its involvement in that relation is typically indicated. This could be contradicted when there are competing near dependencies (several near dependencies with roughly equal condition indexes), so care should be taken here. Thus, when there are several near dependencies, the elements of the $\mathbf{\Pi}$ matrix can always be used to determine which variates are involved in at least one near dependency (and therefore which coefficients are degraded), but they cannot always be relied upon to determine which variates are involved specifically in which near dependencies.

6. *Determine Auxiliary Regressions.* Once the number of near dependencies has been determined in Step 4 and some partial knowledge of variate involvement has been determined in Step 5, auxiliary regressions among the indicated variates can be run to display the near dependencies in greater detail. As noted, these auxiliary regressions are not needed to determine either the number of near dependencies or the variates that are degraded by being in at least one of them, but they should be used if more detailed information is required to determine individual variate involvement among specific competing

or dominating near dependencies. A simple procedure for forming these auxiliary regressions is described below.

7. *Determine Unaffected Variates.* A variate is considered uninvolved in any near dependency if the total proportion of its variance associated with the set of low scaled condition indexes exceeds the threshold π^*. The set of low scaled condition indexes is, of course, the set that does not exceed the threshold $\tilde{\eta}^*$ specified in STEP 4.

5.3 SOME HINTS ON USAGE

The mass of numbers presented by the $\mathbf{\Pi}$ matrix of variance–decomposition proportions in a collinearity diagnosis can be quite overwhelming until one learns what to look for—and what to ignore. This is particularly true when p, the number of variates, gets large. Fortunately, this is an easy problem to correct, and we turn in this section to some hints for displaying, reading, interpreting, and using the collinearity diagnostics. With a little instruction and practice, the salient features of the output of a collinearity diagnosis can be almost instantly digested, literally at a glance. We begin with a discussion of the possible formats that may be chosen for the output tableaux used to present the diagnostic information.

Possible Formats

The collinearity diagnostics are comprised of two basic blocks of information, the p scaled condition indexes and the $p \times p$ $\mathbf{\Pi}$ matrix of variance–decomposition proportions. There are two natural formats for displaying this information: the row-oriented format and column-oriented format. Each has its advantages and disadvantages.

The *row-oriented format*, which has been chosen for the displays in this book, is shown in Exhibit 5.3. In this format, the scaled condition indexes and $\mathbf{\Pi}$ matrix are combined into a $p \times (p + 1)$ matrix having the scaled condition indexes displayed as the first column so that they act essentially as row headings.[4] It is clearly advantageous that the scaled condition indexes be ordered, but it makes little difference whether the order is ascending or descending. The former has been chosen in this monograph. Each row in this format, then, corresponds to a near dependency, and since it is the existence and composition of these near dependencies that is the main focus of a collinearity diagnosis, it is the structure and patterns of these rows that are the main focus of the analysis—hence the term *row-oriented* format.

[4]In Belsley et al. (1980), a column of singular values is also included in the display tableaux. The singular values, however, are quite unnecessary, and their presence just creates unneeded confusion.

Exhibit 5.3 Output Tableau for the Collinearity Diagnostics: Row-Oriented Format

Scaled Condition Index, $\tilde{\eta}$	Proportions of $\mathbf{X}_1$ $\text{var}(b_1)$	$\mathbf{X}_2$ $\text{var}(b_2)$	$\cdots$	$\mathbf{X}_p$ $\text{var}(b_p)$
$\tilde{\eta}_1$	π_{11}	π_{12}	$\cdots$	π_{1p}
$\tilde{\eta}_2$	π_{21}	π_{22}	$\cdots$	π_{2p}
$\vdots$	$\vdots$	$\vdots$		$\vdots$
$\tilde{\eta}_p$	π_{p1}	π_{p2}	$\cdots$	π_{pp}

Each column in this format can be associated equally well with a variate or a variance. We recall from Chapter 3 that, viewed strictly as a collinearity diagnostic, the π's are indirect evidence of the relevance of the variates to the various near dependencies. Within this interpretation, each column of the $\mathbf{\Pi}$ matrix is associated with a column of $\mathbf{X}$, or a variate. Viewed, however, as a regression diagnostic—where we are concerned also with the strength of the given data for estimating a regression model by least squares—, each column can be associated with the variance of the estimated coefficient of the corresponding variate. The column headings in a tableau like Exhibit 5.3, then, can either be a variance name or the name of the corresponding variate or both.

The *column-oriented format* transposes the preceding to give a tableau like Exhibit 5.4. Here, the scaled condition indexes become the column heads so that each column corresponds to a near dependency. Each row, which must sum to 1, now corresponds to a variate and/or a variance.

There are several advantages to this format. First, when p is large, since there can be more rows on a page than columns, the columns associated with each near dependency are more likely to stay together on a page rather than being broken across several pages. Visually, this makes assessment of the near dependencies easier. Second, if the scaled condition indexes are placed in descending order, only the first several columns, those corresponding to the largest scaled condition indexes, may need to be printed. Third, this format makes it somewhat easier to produce clean computer output on line printers that cannot print sideways on the page. In more flexible computer environments, however, reasonable programming can make either the row- or column-oriented formats equally effective. The main disadvantage of this format is that its presentation is less appealing to the intuition.

Keep the Numbers Simple

Regardless of the format, the numbers that are printed should be kept simple. It is, for example, completely unnecessary to print any fractional part of the scaled

Exhibit 5.4 Output Tableau for the Collinearity Diagnostics: Column-Oriented Format

Variate or Variance	Scaled Condition Index			
	$\tilde{\eta}_1$	$\tilde{\eta}_2$		$\tilde{\eta}_p$
$\mathbf{X}_1$	π_{11}	π_{21}	$\cdots$	π_{p1}
$\mathbf{X}_2$	π_{12}	π_{22}	$\cdots$	π_{p2}
$\vdots$	$\vdots$	$\vdots$		$\vdots$
$\mathbf{X}_p$	π_{1p}	π_{2p}	$\cdots$	π_{pp}

condition indexes; only the integer part is ever needed. Recall that these magnitudes increase along the progression of 10/30—1, 3, 10, 30, 100, 300, and so on. Thus, scaled condition indexes of 28 and 35 are essentially the same, as are ones of 109 and 92. It is clear that under these circumstances the value of seeing a scaled condition index of 34.859 rather than 35 is nil. Correlatively, when viewing scaled condition indexes, it is unnecessary to take the trouble mentally to digest the full number; merely an appreciation for the relative order of magnitude in the progression of 10/30 is adequate.

Similarly, the variance–decomposition proportions need not be carried out to numerous digits. The three digits to the right of the decimal shown for the π_{ij} in the displays in this monograph are wholly adequate. Leading zeros should also be suppressed; they tend only to clutter the display with unnecessary ink and distract the eye from seeing quickly the one instance in which a digit will meaningfully be to the left of the decimal, namely, in the interesting and unusual case where the variance–decomposition proportion is 1.000. Furthermore, the displays are most easily read when each π_{ij} has the same number of digits to the right of the decimal. Thus, trailing zeros are to be kept. And finally, FORTRAN-like exponential formats should be avoided at all costs. This formatting technique, so useful in some other circumstances, renders the tableaux of the collinearity diagnostics almost unreadable.

Look at the High Scaled Condition Indexes First

Given, then, the diagnostic output, the first thing to do is mentally to blot from vision everything but the scaled condition indexes and indeed to focus first only on their high end. The largest scaled condition index, which is also the scaled condition number of the data matrix, tells you immediately the worst with which you must contend. It defines for you relativistically what is "large" in the context of the given data. If large is itself absolutely small, for example, 5 or 10, then, although further analysis may be of interest, collinearity is not really a major problem besetting these data. If large is moderate, say 30–100, then there are collinearity problems and further analysis is definitely of interest. If, however, large is immense, say, 1000 or 3000, then scaled condition indexes that are several orders of magnitude smaller, even ones like 30 that might, by themselves, be of concern, are small by comparison and are not necessarily of major concern.

The next thing to look for is the possible presence of a gap in the 10/30 progression of the scaled condition indexes. This gap may occur because of a separation between a scaled condition index that is absolutely small from one that is large (a gap of the first kind), or it may occur between large scaled condition indexes that are separated by several orders of magnitude along the 10/30 progression (a gap of the second kind). Such a gap provides a natural starting place to determine the number of near dependencies.

In a progression such as 1, 3, 5, 30, the gap is of the first kind, between 30, which is high, and 5, which is absolutely low; one near dependency is indicated here. This situation is even easier to analyze in a case like 1, 3, 5, 100, where the gap is across several orders of magnitude. Likewise, in 1, 3, 5, 30, 100, there is little reason not to interpret this again as a gap between 5 and 30, indicating two near dependencies. Seemingly trickier is a progression like 1, 3, 5, 28, 32, 100. However, since 28 and 32 are really of the same order of magnitude, this is really quite similar to the preceding case except that three near dependencies are called for.

The real problems occur in highly unbalanced sequences and in very smoothly graded ones. In the case 1, 3, 5, 28, 32, 107, 1427, 3456, for example, a gap of the first kind exists between 5 and 28, indicating five near dependencies, while a gap of the second kind exists between 107 and 1427, indicating two. Clearly, this latter interpretation is taking liberties, for a scaled condition index of size 107 is indeed quite large. However, it is often advantageous to begin the analysis as simply as possible, examining only the worst cases. It is true that there may be other near dependencies that are thereby ignored, but their effects are relatively less, even if not absolutely without importance. Thus, one might begin here by assuming two near dependencies, only making refinements, if need be, at a later stage. There may, for example, be little reason to go further if it can be shown that all the variates of interest, or the relations of interest, are included in the two strongest near dependencies. On the other hand, if a variate of interest is shown not to be involved in the strongest near dependencies, its role in the next strongest set can then be investigated.

The most frustrating progressions are those that give no hint of a natural break. Consider, for example, the sequence, 1, 3, 5, 10, 22, 29, 39, 45, 72, 95, 129, 245, 373, 498. These nasty situations are more likely to arise when p, the number of variates, is large. My sympathies are often limited when confronted with cases like this with $p = 25$ or, in some cases, $p = 50$, for by and large, it has been my experience that regression equations with more than 10 variates tend to be misspecified, resulting from ad hoc specifications that should properly be modeled as part of a simultaneous system of smaller equations. But not always—and that quip does not really answer the question of how to interpret the preceding sequence of scaled condition indexes, which could arise legitimately. My inclination is to begin the analysis by picking the top three to five, not ignoring anything astronomically large but trying to keep the number of near dependencies below some reasonable proportion of p, say, 25–40%. In this case, I would probably first pick the top three.

Thus, as noted in the previous section, picking the number of near dependencies is occasionally an art form. Sometimes it can be done quite mechanically, but often some degree of judgment is required. It helps somewhat to realize that the user is not required here to state a priori what the cutoff will be, as would be the case, for example, in properly setting the test size for a classical test of hypothesis. The cutoff value is best determined relativistically according to the needs of the analysis at hand and can be changed as the analysis proceeds. The only absolute is that very small values for the scaled condition index, values of 5 or 10 and less, will rarely be of interest.

Train the Eyes to See Only the Important Variance–Decomposition Proportions

Having determined the number of near dependencies, one next looks at the variance–decomposition proportions that correspond to them, beginning with the strongest. In the displays in this monograph, this means beginning with the bottom row. In the column-oriented format of Exhibit 5.4, this would be the first column. The eye is easily trained to ignore the elements in the other rows and indeed to search first only for the large numbers in the given row, values like .8 and .9. There will almost always be some values like this in the row corresponding to the largest scaled condition index.[5] These, mentally, should stand out like they were in boldface. The columns these numbers are in indicate those variates that are definitely involved in the strongest near dependency, bearing in mind that there may be others as well, which are also being determined simultaneously in other near dependencies.

Now let the eye pick up the variance–decomposition proportions associated with the next largest scaled condition index (the penultimate row in the displays given here), and examine them for large values that indicate obvious variate involvement in this near dependency. Evidence should now also be sought as to simultaneous involvement of variates between this and the strongest near dependency. This would obviously take the form of variance–decomposition proportions distributed across the two, so that their sum is large even if no single part is. Rough sums are all that is needed here. Continuing in this way, one can build a story of the collinear structure.

The main technique, then, for absorbing the information of the variance–decomposition proportions is to train the eye first to see only the most obvious few values: the large π's in the rows corresponding to the large $\tilde{\eta}$'s. Once this

[5]An exception could occur when there is strong competition and/or simultaneity among the "strongest" near dependencies. Here the variance–decomposition proportions for the variates belonging to these near dependencies could be spread across them, producing values no higher than .4 or .5 in each cell of the $\mathbf{\Pi}$ matrix as in the case of $\mathbf{Z7}\{3, 2, 2\}$ of Exhibit 4.13. The eye also is readily trained to pick up these unmistakable patterns.

information is digested and preliminarily interpreted, allow the focus of the eye to broaden to encompass the more refined structure that occurs due to dominance and competition.

An Illustration

Consider the sequence of tableaux given in Exhibits 5.5*a*–*c*, where boldface type simulates the eye's discriminatory powers. Looking first at only the first column of Exhibit 5.5*a*, the following story begins to unfold: There are three near dependencies, one very strong, one strong, and one moderately strong. Now, picking up the large elements in the last row, the strongest near dependency involves variates 2, 5, and 8. And allowing our eyes to pick up subsequent lines as in Exhibit 5.5*b*, the story continues: The next strongest dependency involves variates 1 and 7 but could well involve 2, 5, or 8 as well, due to dominance. The third strongest near dependency could contain any of the previously involved variates but certainly involves variates 4 and 6.

At this point it is usually interesting to stop and put real names and ideas to the variates to see what all this means. For example, variate 2 might be GNP while variate 8 is lagged consumption and variate 5 is lagged investment. Then we would know that autocorrelation typical of economic time series, along with an identity in the unlagged values, is causing troubles. Such considerations can often stimulate useful ideas regarding the process that generated the given data set, ideas that can possibly give aid in carrying out appropriate corrective action. This will be discussed in Chapter 10.

Next the eye can examine what variates might be unaffected by the collinear relations. Exhibit 5.5*c* highlights the information that indicates that variate 3 seems to be associated mainly with the smaller scaled condition indexes. Its weak involvement in the stronger near dependencies is not altogether to be ignored, but these data seem well-conditioned relative to variate 3.

Exhibit 5.5*a* Interpretative Illustration: High Condition Indexes and the Strongest Near Dependency

Scaled Condition Index, $\tilde{\eta}$	Proportions of							
	var(b_1)	var(b_2)	var(b_3)	var(b_4)	var(b_5)	var(b_6)	var(b_7)	var(b_8)
1	.000	.000	.001	.000	.003	.000	.000	.000
2	.000	.000	.005	.000	.070	.000	.000	.000
5	.002	.000	.000	.003	.027	.001	.000	.000
7	.000	.000	.013	.000	.044	.001	.000	.000
11	.000	.000	.657	.001	.020	.001	.002	.000
35	.000	.000	.076	.835	.007	.852	.000	.000
153	.970	.001	.071	.083	.000	.073	.973	.000
455	.028	**.999**	.177	.078	**.829**	.072	.025	**1.000**

Exhibit 5.5*b* Interpretative Illustration: All Three Near Dependencies

Scaled Condition Index, $\tilde{\eta}$	Proportions of							
	var(b_1)	var(b_2)	var(b_3)	var(b_4)	var(b_5)	var(b_6)	var(b_7)	var(b_8)
1	.000	.000	.001	.000	.003	.000	.000	.000
2	.000	.000	.005	.000	.070	.000	.000	.000
5	.002	.000	.000	.003	.027	.001	.000	.000
7	.000	.000	.013	.000	.044	.001	.000	.000
11	.000	.000	.657	.001	.020	.001	.002	.000
35	.000	.000	.076	**.835**	.007	**.852**	.000	.000
153	**.970**	.001	.071	.083	.000	.073	**.973**	.000
455	.028	**.999**	.177	.078	**.829**	.072	.025	**1.000**

Exhibit 5.5*c* Interpretative Illustration: Noninvolvement

Scaled Condition Index, $\tilde{\eta}$	Proportions of							
	var(b_1)	var(b_2)	var(b_3)	var(b_4)	var(b_5)	var(b_6)	var(b_7)	var(b_8)
1	.000	.000	.001	.000	.003	.000	.000	.000
2	.000	.000	.005	.000	.070	.000	.000	.000
5	.002	.000	.000	.003	.027	.001	.000	.000
7	.000	.000	.013	.000	.044	.001	.000	.000
11	.000	.000	**.657**	.001	.020	.001	.002	.000
35	.000	.000	.076	.835	.007	.852	.000	.000
153	.970	.001	.071	.083	.000	.073	.973	.000
455	.028	.999	.177	.078	.829	.072	.025	1.000

An Alternative Transformation for the Scaled Condition Indexes

It has been noted that the scaled condition indexes tend to advance along the 10/30 progression, 1, 3, 10, 30, 100, 300, 1000, 3000, ..., as the corresponding near dependencies become tighter and tighter. Both the nonlinearity of this progression and the large values that occur in its more removed terms can be eliminated by using a logarithmic transformation. Thus, consider defining the transformed set of scaled condition indexes

$$\tilde{\omega}_k \equiv \log_{10} \tilde{\eta}_k, \qquad k = 1, \ldots, p. \tag{5.1}$$

The progression 0.0, 0.5, 1.0, 1.5, 2.0, 2.5, ... in the $\tilde{\omega}$'s corresponds to a progression 1, 3, 10, 32, 100, 316, 1000, 3162, ... in the $\tilde{\eta}$'s, which is, of course, essentially the same as the familiar progression of 10/30.

Thus, scaled condition indexes transformed as (5.1) are an excellent alternative means for presenting the scaled condition indexes, particularly when a linear scale is desirable, as would be the case, for example, for a graphic display of the scaled condition indexes. This has been used to advantage in the system designed by Oldford and Peters (1984, 1985) for providing guided use of these collinearity diagnostics. The graphics that result are particularly effective in helping to determine the "gaps" in the progression described above.[6] In this scale, values of (transformed) scaled condition indexes $\tilde{\omega}$ below 1.0 are of little concern, while those above 1.5 are typically worthy of notice. It is unnecessary to present these values with more than one place beyond the decimal point.

Forming the Auxiliary Regressions

When there are several near dependencies, we have seen that it is not always possible to determine from the variance–decomposition proportions alone exactly which variates are involved in which near dependencies. Of course, it is always possible to determine which variates are involved in at least one near dependency, but the presence of dominating and competing dependencies can obscure individual variate involvement. When this happens, a simple procedure can be used to form a set of auxiliary regressions that typically will display the structure of the near dependencies in greater detail. The basic idea is to use the variance–decomposition proportions to identify one variate known to be involved in each near dependency and then to regress each variate in this set on the remainder. This is exemplified in Exhibit 5.6, a variance–decomposition $\mathbf{\Pi}$ matrix for a seven-column data matrix $\mathbf{X}$.

Here we see that there are three near dependencies among the seven variates, associated with the scaled condition indexes 32, 78, and 118. Hence, in a sort of reduced form, we can express three of the seven variates in terms of the remaining four; that is, we can select three variates to regress on the remaining four to produce three descriptive auxiliary regressions displaying the three near dependencies. The t-statistics that accompany these regressions can then be used descriptively (not inferentially, because we are making no claim that these relations in fact are relevant to the process that actually generates the $\mathbf{X}$ data) to signal variate involvement.

To choose the three variates to act as the "dependent" variates for these auxiliary regressions, we need a process that will guarantee that we pick for each near dependency a variate known to be involved in it. Beginning, then, with the strongest near dependency ($\tilde{\eta} = 118$), look along its row to find large variance–decomposition proportions to signal variates known to be in this near dependency. In this case, both **C4** ($\pi = .903$) and **C7** ($\pi = .902$) appear initially to be equally good candidates. **C7** is chosen, however, because it has the remainder of its variance determined in more removed near dependencies, thereby minimizing the possibility that the values for the variance-decomposition

[6]Indeed, cluster analysis could be used to mechanize this procedure.

Exhibit 5.6 Forming the Auxiliary Regressions

Scaled Condition Index, $\tilde{\eta}$	Proportions of						
	var(b_1)	var(b_2)	var(b_3)	var(b_4)	var(b_5)	var(b_6)	var(b_7)
1	.000	.002	.000	.000	.000	.000	.000
2	.000	.000	.001	.000	.000	.120	.000
6	.000	.024	.010	.000	.002	.002	.000
8	.000	.000	.051	.005	.000	.616	.008
32	.000	.103	**.602**	.000	.097	.056	.088
78	.438	**.856**	.312	.092	.105	.141	.002
118	.562	.015	.024	.903	.796	.065	**.902**

proportions are distorted through competing near dependencies. The π value for **C7** in this last row is made bold to indicate that **C7** has been picked to be the "dependent" variate in the auxiliary regression corresponding to $\tilde{\eta} = 118$. For the next strongest near dependency ($\tilde{\eta} = 78$), there is a clear winner in **C2** ($\pi = .856$). And **C3** ($\pi = .602$) looks good for the weakest near dependency ($\tilde{\eta} = 32$). Hence, we pick **C7**, **C2**, and **C3** as the "pivots" to regress separately on the remaining variates **C1**, **C4**, **C5**, and **C6**.

The procedure just described works extremely well in a wide variety of cases but often has some troubles when there are many near dependencies or when there are several dominating near dependencies that cover up the involvement of all the variates in the weaker near dependencies. Consider, for example, the case in Exhibit 5.7.

Here again there are three near dependencies ($\tilde{\eta}$'s of 1035, 367, and 32), and there is no question of **C7**'s involvement in the strongest of the three. But there is

Exhibit 5.7 Forming the Auxiliary Regressions: Problems with Dominant Dependencies

Scaled Condition Index, $\tilde{\eta}$	Proportions of						
	var(b_1)	var(b_2)	var(b_3)	var(b_4)	var(b_5)	var(b_6)	var(b_7)
1	.000	.002	.000	.000	.000	.000	.000
2	.000	.019	.000	.000	.000	.120	.000
6	.000	.853	.000	.000	.002	.002	.000
8	.000	.000	.001	.005	.000	.616	.008
32	.000	.103	.000	.000	.097	.056	.088
367	.347	.008	.175	.092	.105	.141	.002
1035	.653	.015	.824	.903	.796	.065	.902

little information to be learned about which variates are unmistakably involved in the weaker two near dependencies. One might make a good guess about **C1** in the second strongest near dependency, but what about the third? Variates **C2** and **C6** show themselves to be effectively uninvolved in any of the near dependencies, and the strengths of the top two obscure all information relative to the third.

An extension of the preceding procedure that works well in a case like this is the following: First, make a good guess, based on what information is available, to pick a set of variates equal in number to the number of near dependencies, r. Here, for example, we might pick the three variates **C7**, **C1**, and **C5**. This latter variate is picked because there is a strong possibility that **C5** is involved in the third strongest near dependency but is being masked by its involvement in the stronger two. This might also be true of **C4**, but **C4**'s π of .000 in this third strongest near dependency is not encouraging. Second, find the condition number of the matrix composed of the remaining $p - r$ variates—in this case, **C2**, **C3**, **C4**, and **C6**. If the condition number of this matrix is of the same order of magnitude as the next smallest condition index (the $(p - r)$th, whose value is 8 in this example), then one has found a subset of the variates that possesses no near dependencies. This subset is therefore associated only with the "background" conditioning of the original data matrix and can be used as the set of regressors for the auxiliary regressions. If, on the other hand, the condition number of this matrix is of a larger order of magnitude, then try a different subset of $p - r$ variates until a matrix with a condition number with the appropriate order of magnitude is found. The variates comprising this matrix can be used as the auxiliary regressors, and the complementary r variates become the auxiliary regressands. It has not yet formally been proved that a submatrix can always be found in this way whose condition number is of the same order of magnitude as the next smallest condition index, but this is an excellent conjecture; I have never found a counterexample.

There are, of course, many other ways in which a set of auxiliary regressions could be chosen. Since these auxiliary regressions are in no way intended to specify and estimate a model describing the way the given **X** data were actually generated, any reasonable choice can serve this purely descriptive purpose of illuminating the nature of the linear near dependencies that just happen to have occurred in the given data matrix. Indeed, the specific context of any given data set may suggest a natural set of pivots, and the "automatic outcome" of the preceding procedure may not be appropriate. Interest may center, for example, in the possibly simultaneous role a specific variate plays in several of the near dependencies, and so it would be important to choose that variate as a regressor entering in all the auxiliary regressions and not as a regressand appearing in only one. This occurs in one of the case studies given in Chapter 10.

In the absence of any other considerations, however, the procedure just described for constructing auxiliary regressions has the advantages that (1) it is simple to employ, (2) it picks as a "dependent" variate for each auxiliary regression one that is known to be strongly involved in the underlying near dependency, and (3) the set of regressors (the right-hand-side variates) will necessarily be relatively well conditioned, so the auxiliary regressions should not themselves be subject to the problems of ill conditioning.

5.4 APPLICATIONS WITH ACTUAL DATA

With the very interesting exception of the Bauer data introduced in Section 3.2, we have until now employed the proposed diagnostic procedure only on data matrices with contrived near dependencies. We turn now to four matrices of actual data to see how the diagnostic procedure fares when dealing with examples of naturally occurring, uncontrived near dependencies. The first example again utilizes the Bauer matrix. The second example examines data familiar to all econometricians, those relevant to an annual, aggregate consumption function. The third example provides diagnostics of the conditioning of the data that are used in the analysis of a monetary equation in Friedman (1977). The final example uses data from an equation from the IBM econometric model.

The Modified Bauer Matrix

The modified Bauer matrix, we recall from Section 3.2, is a 6×5 matrix of integers that has an exact contrived dependency (**C4** $= 0.5$**C5**) between its last two columns. These last two columns are, in turn, orthogonal to the first three. Its purpose in Section 3.2 was to exemplify the isolation from collinearity that is afforded those variates orthogonal to, or nearly so, the variates involved in the offending near dependencies. In examining the $\boldsymbol{\Pi}$ matrix of the unscaled Bauer data given in Exhibit 3.7, the involvement of var(b_4) and var(b_5) in the exact contrived dependency is clearly observed, as is also the isolation of the first three variates from it. But, in addition, there appears an unexpected occurrence: over 97% of var(b_1), var(b_2), and var(b_3) is associated with the unscaled condition index 4785. We were not prepared to pursue this naturally arising phenomenon in Section 3.2, but we are now.

First, we note that the variance–decomposition proportions given in Exhibit 3.7 and the corresponding condition indexes cannot be given a wholly adequate interpretation since they are based on data that have not been column equilibrated, as is required in Step 2 and justified more fully in Section 6.2. Hence, in Exhibit 5.8 we present the scaled condition indexes and $\boldsymbol{\Pi}$ matrix that result from the column-equilibrated Bauer matrix.

As we analyze Exhibit 5.8, it proves instructive to feign ignorance of any prior knowledge we may have of the properties of the Bauer matrix to see how well the diagnostic procedure discovers what there is to know. The first and most obvious fact is that there is one near dependency with an astronomically large scaled condition index, essentially infinite relative to machine zero, indicating a virtually exact dependency, and that this dependency involves **C4** and **C5** alone.

Exhibit 5.8 Scaled Condition Indexes and Variance-Decomposition Proportions: Scaled Bauer Matrix

Scaled Condition Index, $\tilde{\eta}$	Proportions of var(b_1)	var(b_2)	var(b_3)	var(b_4)	var(b_5)
1	.000	.000	.000	.000	.000
1	.005	.005	.000	.000	.000
1	.001	.001	.047	.000	.000
16	.994	.994	.953	.000	.000
8×10^{19}	.000	.000	.000	1.000	1.000

It is quite safe to conclude here that the involvement of **C1**, **C2**, and **C3** in this dependency is minimal, if any.

The second most notable fact is the possible presence of a modest-strength near dependency with a scaled condition index of 16. The importance of scaling to stabilize the meaning of the condition indexes should be obvious from this example. Recall that this second largest condition index was 4785 when unscaled, a value that seems large indeed. However, we now see that the appropriately scaled condition index lies in the "between zone." Scaled condition indexes of 30 and above tend always to be consequential while those of 10 or below are usually of little consequence. A scaled condition index of 16 could be of interest, and we might want to examine auxiliary regressions to get more information. This value is particularly enticing in this case because it is removed by two orders of magnitude from the next largest scaled condition index of 1. Clearly, if we adopt the attitude that there is a second near dependency associated with this scaled condition index, it involves at least the first three columns, **C1**, **C2**, and **C3**, and one cannot rule out the possibility that it also involves **C4** and/or **C5**, their roles possibly being masked by their involvement in the overwhelmingly dominant dependency.

To display these two near dependencies through auxiliary regressions, we must first choose the two variates to act as regressands, the three remaining being the regressors. For the strongest dependency, either **C4** or **C5** is an obvious choice. There is nothing to differentiate them, so we arbitrarily pick the first, **C4**. For the weaker near dependency, either **C1** or **C2** is appropriate. Again, there is nothing to distinguish them, so we pick **C1**. Exhibit 5.9 presents the two auxiliary regressions that result when **C1** and **C4** are regressed on **C2**, **C3**, and **C5**. The regressions are based on the unscaled data, so the dependencies are displayed in terms of the original data relationships.

Both near dependencies are clearly displayed. In the first, we see the dominant, essentially perfect relation true of the modified Bauer data given in Section 3.2 in which **C4** = 0.5**C5**, exactly. The noninvolvement of **C2** and **C5** in this relation is also verified. In the second, we see a moderate relation ($\hat{R}^2 = .98$)

Exhibit 5.9 Auxiliary Regressions: Bauer Data (Unscaled)

Dependent Variate	Coefficient of C2	C3	C5	$\hat{R}^2$	$\tilde{\eta}$
C4	0.000 [0.0]	0.000 [0.0]	0.500 [∞]	1.000	∞
C1	−0.701 [−14.4]	−1.269 [−7.5]	0.000 [0.0]	.982	16

involving **C1**, **C2**, and **C3**. We can also now verify that **C5** is not involved in this weaker relation. This information could not be inferred from the $\mathbf{\Pi}$ matrix alone, since **C5**'s involvement could have been masked by the dominant dependency. The auxiliary regressions, therefore, add to our diagnostic knowledge. This weaker near dependency is the naturally occurring dependency whose presence was first suggested in Section 3.2 and whose existence is now verified. One can now conclude that all five regression estimates based on this matrix are degraded to varying degrees by the presence of two collinear relations. The variances for the coefficients of **C4** and **C5** are obviously very seriously degraded, while those for **C1**, **C2**, and **C3** are considerably less so.

It is fair to conclude that the diagnostic procedure, when applied to the modified Bauer matrix, has been very successful in uncovering all relevant properties of the collinear relations contained in it.

The Consumption Function

A relation of great historical and theoretical importance in economics is the annual, aggregate consumption function, and so we analyze the following matrix of U.S. consumption function data:

$$\mathbf{X} \equiv [\iota, \mathbf{C}(T-1), \mathbf{DPI}(T), \mathbf{r}(T), \Delta\mathbf{DPI}(T)],$$

where

$$\begin{aligned} \iota &= \text{a column of ones (the constant term)}, \\ \mathbf{C} &= \text{total consumption, 1958 dollars}, \\ \mathbf{DPI} &= \text{disposable income, 1958 dollars}, \\ \Delta\mathbf{DPI} &= \text{annual change in disposable income}, \\ \mathbf{r} &= \text{the interest rate (Moody's Aaa)}, \end{aligned}$$

and all series are annual, 1948–1974. These data are given in Exhibit 5.10.

It must be emphasized that no attempt is being made here to analyze the consumption function itself. There are many well-known, sophisticated alterations to the basic consumption data, involving, for example, the use of per-capita weightings, disaggregations by durability, the inclusion of wealth effects, and the recognition of simultaneity. Our interest here necessarily centers on the analysis of one fundamental variant without regard to additional econometric niceties; namely,

$$\mathbf{C}(T) = \beta_1 \iota + \beta_2 \mathbf{C}(T-1) + \beta_3 \mathbf{DPI}(T) + \beta_4 \mathbf{r}(T) + \beta_5 \Delta\mathbf{DPI}(T) + \varepsilon(\mathrm{T}). \tag{5.2}$$

Exhibit 5.10 Data for the Consumption Function

Year	C	r	DPI	ΔDPI
1947	206.275	2.61083	218.075	−9.375
1948	210.775	2.81667	229.7	11.625
1949	216.5	2.66	230.925	1.225
1950	230.5	2.6225	249.65	18.725
1951	232.825	2.86	255.675	6.025
1952	239.425	2.95583	263.25	7.575
1953	250.775	3.19917	275.475	12.225
1954	255.725	2.90083	278.4	2.925
1955	274.2	3.0525	296.625	18.225
1956	281.4	3.36417	309.35	12.725
1957	288.15	3.885	316.075	6.725
1958	290.05	3.7875	318.8	2.725
1959	307.3	4.38167	333.05	14.25
1960	316.075	4.41	340.325	7.275
1961	322.5	4.35	350.475	10.15
1962	338.425	4.325	367.25	16.775
1963	353.3	4.25917	381.225	13.975
1964	373.725	4.40417	408.1	26.875
1965	397.7	4.49333	434.825	26.725
1966	418.1	5.13	458.875	24.05
1967	430.1	5.50667	477.55	18.675
1968	452.725	6.175	499.05	21.5
1969	469.125	7.02917	513.5	14.45
1970	477.55	8.04	534.75	21.25
1971	496.425	7.38667	555.425	20.675
1972	527.35	7.21333	580.45	25.025
1973	552.075	7.44083	619.5	39.05
1974	539.45	8.56583	602.875	−16.625

Estimation of (5.2) with least squares results in (standard errors in parentheses)

$$\begin{aligned} \mathbf{C}(T) = {} & \underset{(3.8271)}{6.7242\boldsymbol{\iota}} + \underset{(0.2375)}{0.2454\mathbf{C}(T-1)} + \underset{(0.2076)}{0.6984\mathbf{DPI}(T)} \\ & - \underset{(1.8384)}{2.2097\mathbf{r}(T)} + \underset{(0.1834)}{0.1608\,\Delta\mathbf{DPI}(T)} \end{aligned} \tag{5.3}$$

$$R^2 = .9991, \qquad \text{SER} = 3.557, \qquad \text{DW} = 1.89, \qquad \tilde{\kappa}(\mathbf{X}) = 376.$$

Only one of these parameter estimates, that of **DPI**, is significant by a standard t-test; however, few econometricians would willingly accept the hypothesis that all the other β's are insignificantly different from zero—at least not without considerably better evidence that there are not other problems, such as data weaknesses—because strongly based a priori considerations strongly suggest the relevance of several of these variates. Furthermore, few econometricians would be happy with the confidence intervals that would result from such a regression. This dissatisfaction stems in part from the widely held belief that the consumption function data are highly ill conditioned and that the estimates based on them are too noisy to prove conclusive or be useful for many purposes. Indeed, few functions have received greater attention than the consumption function in efforts to overcome ill conditioning and to refine estimation. And a mere glance at the simple correlation matrix for these data given in Exhibit 5.11 (which we know paints a very incomplete picture) partially confirms this belief.

But how ill conditioned are these data? how many near dependencies exist among them? and how strong are they? Which variates are involved in them, giving evidence of degradation? Which estimates might benefit most from obtaining better-conditioned data, if possible, or from the introduction of appropriate prior information through Bayesian or other techniques? Answers to these questions, of course, cannot be obtained from Exhibit 5.11 alone but can be obtained from an analysis of the scaled condition indexes and variance-decomposition proportions for the consumption function data. STEPS 1–3 of the diagnostic procedure applied to the consumption function data result in the information given in Exhibit 5.12.

Exhibit 5.11 Correlation Matrix: Consumption Function Data

	C$(T-1)$	**DPI**(T)	**r**(T)	Δ**DPI**(T)
C$(T-1)$	1.000			
DPI(T)	.997	1.000		
r(T)	.975	.967	1.000	
Δ**DPI**(T)	.314	.377	.229	1.000

Exhibit 5.12 Scaled Condition Indexes and Variance-Decomposition Proportions: Consumption Function Data

Scaled Condition Index, $\tilde{\eta}$	Proportions of ι var(b_1)	C($T-1$) var(b_2)	**DPI**(T) var(b_3)	**r**(T) var(b_4)	**ΔDPI**(T) var(b_5)
1	.001	.000	.000	.000	.001
4	.004	.000	.000	.002	.136
8	.310	.000	.000	.013	.001
39	.264	.005	.005	.984	.048
376	.421	.995	.995	.001	.814

Exhibit 5.12 shows the existence of two near dependencies, one dominant with a large scaled condition index of 376 and one moderate with a scaled condition index of 39. Scanning across the bottom row, we see that the dominant relation involves **C**($T-1$), **DPI**(T), and **ΔDPI**(T). The variable **r**(T) does not seem to be involved in this stronger near dependency, but it is virtually certain that the constant term ι is. Indeed, it appears that the constant term is being shared by both near dependencies. This sort of pattern for the constant term, spread across several coexisting near dependencies, is quite common. The weaker near dependency definitely includes **r**(T), and all other variates are potentially involved, their effects clearly being dominated by their involvement in the stronger near dependency with $\tilde{\eta} = 376$.

Auxiliary regressions are required in this case to determine those variates involved in the weaker of the two near dependencies. Using the procedure described above for constructing auxiliary regressions suggests picking either **C**($T-1$) or **DPI**(T) along with **r**(T) as dependent variates to regress on the remaining three variates. Exhibit 5.13 reports the results using **DPI**(T) and **r**(T). An examination of these auxiliary regressions verifies that the dominant relation does involve ι, **C**($T-1$), **DPI**(T), and **ΔDPI**(T) and that the weaker involves ι, **DPI**(T), and **r**(T) but not **ΔDPI**(T).

Quite generally, then, we may conclude that the data on which the least-squares estimation of the consumption function (5.3) is based possess two substantive near dependencies, one very strong. Furthermore, each variate is involved in one or both of these near dependencies, and each estimate is degraded to some degree by their presence. It would appear that the estimates of the coefficients of **C**($T-1$) and **DPI**(T) are most seriously affected, followed closely by that for **ΔDPI**(T), these variates being strongly involved in either the tighter of the two near dependencies or both. The estimate of the coefficient of **r**(T) is adversely affected by its strong presence in the weaker of the two near dependencies, but in the experimental experience of Chapter 4, we found $\tilde{\eta}$'s of 39 to be large, and the $\hat{R}^2$ in Exhibit 5.13 confirms this here.

Thus, we see that all parameter estimates in (5.3) and their estimated standard errors show great potential for refinement through better conditioning of the estimation problem, either from more appropriate modeling, the introduction of better-conditioned data, or the introduction of appropriate prior information (as we shall exemplify when we continue this example in Section 10.2). One would, for example, be loath to reject the role of interest rates in the aggregate consumption function, as many have done, on the basis of the estimates like those in (5.3), and one would feel even more helpless in predicting the effects of a change in **r**(T) on aggregate consumption from such results. Thus, the econometrician's intuitive dissatisfaction with estimates of the aggregate consumption function, and his seemingly never-ending efforts to refine them, seem fully justified.

Several additional points of interest arise from this example, some of which suggest future directions for research. First, it is not surprising that the estimated coefficient of **DPI**(T) demonstrates statistical significance even in the presence of the extreme ill conditioning of the consumption function data, for **C**(T) and **DPI**(T) are phenomenally highly correlated (.9999). Indeed, it is in light of this high correlation that the seriousness of the degradation of the estimate of this parameter can be seen, for its standard error is nevertheless quite large, resulting in the very broad 95% confidence interval of [0.28, 1.11].

Second, as seen from Exhibit 5.12, no one near dependency dominates the determination of the variance of the estimate of the constant term, a fact that is verified by the auxiliary regressions in Exhibit 5.13. This estimate is nevertheless degraded since nearly 70% of its variance is associated with the two near dependencies. This lack of dominance is to be contrasted with the estimates of the coefficients of **C**($T-1$) and **DPI**(T), which also clearly enter both near dependencies but are greatly dominated by the stronger of the two. This situation suggests, in accord with intuition, that it is possible for a variate that is simultaneously weakly involved in a strong near dependency and strongly involved in a weaker near dependency to have its variance–decomposition proportions confounded across the relations. Similar results occur in the experiments of Chapter 4, but not sufficiently systematically to allow definite

Exhibit 5.13 Auxiliary Regressions: Consumption Function Data

Dependent Variate	Coefficient of				
	ι	**C**($T-1$)	Δ**DPI**(T)	$\hat{R}^2$	$\tilde{\eta}$
DPI(T)	−11.547	1.138	0.804	.9999	376
	[−4.9]	[164.9]	[11.9]		
r(T)	−1.024	0.017	−0.014	.9945	39
	[−3.9]	[22.3]	[−1.9]		

inferences to be made. Further experimentation will be required to determine whether any systematic behavior of the diagnostic procedure of this sort can be discovered.

Third, within a given near dependency, there appears to be a strong rank correlation between the relative size of the variance–decomposition proportions of the variates involved and their t-statistics in the corresponding auxiliary regressions. Comparing the variance–decomposition proportions for the near dependency with $\tilde{\eta} = 376$ in Exhibit 5.12 and the corresponding t's for the **DPI**(T) regression in Exhibit 5.13 exemplifies the point. Of course, allowance must be made for relations that are dominated (such as the one with $\tilde{\eta} = 39$) or are competing, but again there is considerable support for such a conjecture from the experiments of Chapter 4, and further experiments aimed directly to this point are suggested.

Fourth, even with this "real-world" data, the correspondence between the progression of nines of the correlations and the progression of 10/30 of the scaled condition indexes continues to hold. The near dependencies of the consumption function data are of orders of magnitudes 30 and 300, two steps apart along the progression of 10/30. Similarly, the $\hat{R}^2$s of the auxiliary regressions reported in Exhibit 5.13 are .99 and .9999, two steps apart along the progression of nines.

Fifth, we once again note the ability of these diagnostic tools to uncover complex relations among three or more variates that are overlooked by simple correlations, a problem first raised in the introduction. The simple correlation matrix of Exhibit 5.11 surely tells us that **DPI**(T) and **C**($T-1$) are closely related, but the role of **ΔDPI**(T), or equivalently, the role of **DPI**($T-1$), is not at all observable from this information. The largest simple correlation with **ΔDPI**(T) is under .4. The role of **ΔDPI**(T) in a near dependency along with **C**($T-1$) and **DPI**(T), however, is readily apparent from the variance–decomposition proportions matrix of Exhibit 5.12.

The Friedman Monetary Data

We turn now to an analysis of the conditioning of a body of monetary data relevant to an equation for the household demand for corporate bonds given in Friedman (1977), who kindly made the data available for analysis. The theoretical foundation for this analysis is described in the following excerpt[7]:

> In a world in which transactions costs are nontrivial, it is useful to represent investors' portfolio behavior by a model which determines the desired long-run equilibrium portfolio allocation together with a model which determines the short-run adjustment toward the equilibrium allocation.

[7]Reprinted with the kind permission of Benjamin Friedman and the *Journal of Political Economy*. (Copyright 1977 by the University of Chicago. All rights reserved.)

A familiar model of the selection of desired portfolio allocation, for the given investor or group of investors, is the linear homogeneous form

$$\frac{A_{it}^*}{W_t} = \sum_k^N \beta_{ik} r_{kt} + \sum_h^M \gamma_{ih} X_{ht} + \pi_i, \qquad i = 1, \ldots, N, \tag{5.4}$$

where

A_{it}^*, $i = 1, \ldots, N$ = the investor's desired equilibrium holding of the i-th asset at time period t ($\Sigma_i A_{it}^* = W_t$);

W_t = the investor's total portfolio size (wealth) at time period t;

r_{kt}, $k = 1, \ldots, N$ = the expected holding-period yield on the k-th asset at time period t;

X_{ht}, $h = 1, \ldots, M$ = the values at time period t of additional variables which influence the portfolio allocation;

and the β_{ik}, γ_{ih}, and π_i are fixed coefficients which satisfy $\Sigma_i \beta_{ik} = 0$ for all k, $\Sigma_i \gamma_{ih} = 0$ for all h, and $\Sigma_i \pi_i = 1$. The role of the wealth homogeneity constraint is to require that any shift in an asset's share in the desired equilibrium portfolio be due to movements either of relevant yields (r_k) or of other variables (X_h), rather than to overall growth of the total portfolio itself; particularly for the case of equations representing the behavior of categories of investors, this assumption seems appropriate....

Given the desired equilibrium portfolio allocation indicated by model (5.4), the usual description of investor behavior involves a shift of asset holdings which eliminates some, but not all, of the discrepancy between holdings $A_{i,t-1}$ at the end of the previous period and new desired holdings A_{it}^*. One familiar representation of the resulting portfolio adjustment process is the stock adjustment model

$$\Delta A_{it} = \sum_k^N \theta_{ik}(A_{kt}^* - A_{k,t-1}), \qquad i = 1, \ldots, N, \tag{5.5}$$

where A_{it} = the investor's actual holding of the i-th asset at time period t ($\Sigma_i A_{it} = W_t$), and the θ_{ik} are fixed coefficients of adjustment such that $0 \leqslant \theta_{ik} \leqslant 1$, $k = i$, and $\Sigma_i \theta_{ik} = 1$ for all k.

The empirical implementation by Friedman of this hypothesis is the following linear regression:

$$\begin{aligned} \mathbf{HCB}(T) = {} & \beta_1(\mathbf{HAFA}(T)\cdot\mathbf{MR}(T)) + \beta_2(\mathbf{HFA}(T-1)\cdot\mathbf{MR}(T)) \\ & + \beta_3(\mathbf{PER}(T)\cdot\mathbf{HAFA}(T)) + \beta_4((\mathbf{CPR}(T)\cdot\mathbf{HAFA}(T)) \\ & + \beta_5\mathbf{HCB}(T-1) + \beta_6\mathbf{HLA}(T-1) + \beta_7\mathbf{HE}(T-1) + \varepsilon(T), \end{aligned} \tag{5.6}$$

where

CPR = commercial paper rate,

HAFA = household net acquisition of financial assets,

HCB = household stock of corporate bonds,
HE = household stock of equities,
HFA = household stock of financial assets,
HLA = household stock of liquid assets,
MR = Moody AA utility bond rate,
PER = Standard and Poor's price-earnings ratio.

The least-squares estimates for (5.6) for the 56 quarterly periods 1960:I–1973:IV are

$$\begin{aligned}\mathbf{HCB}(T) = {} & \underset{(0.00648)}{0.0322}\mathbf{HAFA}(T)\cdot\mathbf{MR}(T) + \underset{(0.000091)}{0.000316}\mathbf{HFA}(T-1)\cdot\mathbf{MR}(T) \\ & -\underset{(0.0125)}{0.0242}\mathbf{PER}(T)\cdot\mathbf{HAFA}(T) - \underset{(0.00454)}{0.01549}\mathbf{CPR}(T)\cdot\mathbf{HAFA}(T) \qquad (5.7) \\ & +\underset{(0.0161)}{0.8847}\mathbf{HCB}(T-1) + \underset{(0.00369)}{0.00894}\mathbf{HLA}(T-1) - \underset{(0.00157)}{0.00619}\mathbf{HE}(T-1),\end{aligned}$$

$$R^2 = .99, \qquad \text{SER} = 490.6, \qquad \text{DW} = 2.25, \qquad \tilde{\kappa}(\mathbf{X}) = 112.$$

The multiple-correlation coefficient is large, as is to be expected with a smoothly growing dependent variate such as the stock of corporate bonds in an equation that also includes a lagged dependent variate. The signs of the coefficients are correct, the t-statistics are near 2 or greater in absolute value, and the Durbin–Watson statistic (biased toward serial independence in the presence of a lagged dependent variate) does not indicate autocorrelation problems. The scaled condition number $\tilde{\kappa}(\mathbf{X})$—the condition number of the column-equilibrated data matrix—, however, is large and suggests the need for a conditioning analysis and possible remedial action should the investigator deem the standard errors to be too large for the analysis at hand.

Exhibit 5.14 shows the scaled condition indexes and variance–decomposition proportions for these monetary data. Looking first at the scaled condition indexes, we see evidence of at least two near dependencies, one with a scaled condition index of 112 and one with a scaled condition index of 48. A question arises whether or not to include the next value down, $\tilde{\eta} = 21$. Doing so, however, is unlikely to add too much to our knowledge since all the variates seem to be involved in the two stronger near dependencies anyway. If variates of interest were strongly associated primarily with this scaled condition index 21 and not with the stronger ones, then the investigation of three near dependencies would be of considerably greater interest.

Thus, we consider two near dependencies, one dominant. Clearly, at least **C1**, **C2**, **C3**, **C6**, and **C7** are involved in the stronger near dependency, while **C4**, **C5**, and quite likely **C7** are involved in the weaker near dependency. Further, **C1**, **C2**, **C3**, and **C6** could conceivably be involved in this weaker near dependency, their effects possibly being masked by the dominant dependency associated with

Exhibit 5.14 Scaled Condition Indexes and Variance–Decomposition Proportions: Friedman Monetary Data

Scaled Condition Index, $\tilde{\eta}$	**HAFA**(T)·**MR**(T) var(b_1)	**HFA**($T-1$)·**MR**(T) var(b_2)	**PER**(T)·**HAFA**(T) var(b_3)	**CPR**(T)·**HAFA**(T) var(b_4)	**HCB**($T-1$) var(b_5)	**HLA**($T-1$) var(b_6)	**HE**($T-1$) var(b_7)
1	.000	.000	.000	.000	.000	.000	.000
6	.004	.002	.000	.023	.000	.001	.008
9	.001	.004	.003	.014	.097	.000	.007
15	.024	.029	.008	.201	.022	.000	.005
21	.092	.046	.019	.024	.114	.007	.009
48	.001	.224	.151	.596	.532	.029	.469
112	.878	.695	.819	.142	.235	.963	.502

$\tilde{\eta} = 112$. Auxiliary regressions are required to obtain more detailed information on the exact makeup of these two near dependencies. From among the many ways two of these variates could be chosen to be written in a linear relation with the remaining five, the procedure described above for forming auxiliary regressions would choose **C6** and **C4**. Variate **C6** is obviously strongly involved in the strongest near dependency and is a clear choice. Variates **C4** or **C5** might seem good picks for the weaker near dependency, but **C4** seems preferable since it shows less association with both the next higher and next lower scaled condition indexes. Thus, we regress **C6** and **C4** on the remaining columns **C1**, **C2**, **C3**, **C5**, and **C7** to obtain Exhibit 5.15.

From Exhibit 5.15 we see that all the variates are strongly involved in the dominant near dependency along with **C6**, a fact that for the most part could have been learned directly from the variance–decomposition proportions without recourse to the auxiliary regressions. Only **C5**'s involvement in this near dependency is subject to some possible question, although patterns like this usually turn out to show involvement. The auxiliary regressions do, however, help to make clear the roles of **C1** and **C3** in the weaker near dependency. Their roles are not sharply defined from the variance–decomposition proportions alone—although, were I pressed to make a guess based on them, I would certainly discount the role of **C1** because of its very low variance–decomposition proportion. The auxiliary regressions verify this as well as showing **C3**'s decided involvement.

Although we shall not do so here, this information is of use in determining appropriate corrective action if it seems necessary. We know, for example, that **C2**, **C3**, **C5**, and **C7** are involved in both near dependencies, **C3** and **C7** quite strongly so. Thus, new information on any combination of these variates, either in the form of new data—preferably providing points in the sampling space that differ from those extant—or prior information on their corresponding regression parameters, would help to counteract the weaknesses in this data set caused by collinearity. Such information obtained for **C1** would help to "break up" the ill conditioning relative to the stronger near dependency, which is certainly of use, but would not help with the second near dependency. We shall exemplify such corrective procedures in Chapter 10.

It is clear that the information from the auxiliary regressions is quite

Exhibit 5.15 Auxiliary Regressions: Friedman Monetary Data

Dependent Variate	Coefficient of C1	C2	C3	C5	C7	$\hat{R}^2$	$\tilde{\eta}$
C6	−1.818	0.016	2.351	3.603	0.361	.9989	112
	[−14.0]	[9.6]	[13.2]	[7.5]	[16.8]		
C4	0.015	0.010	1.336	−2.186	−0.146	.9835	48
	[0.1]	[7.5]	[9.0]	[−5.5]	[−8.1]		

complementary with that obtained from the scaled condition indexes and the variance–decomposition proportions; the two together provide a powerful and efficient tool for uncovering and analyzing the presence, degree, and content of linear near dependencies among data series.

One final point should be mentioned relative to this example. The model (5.6) is linear in the parameters but not in the variates. We have proceeded to analyze these data, then, for the presence of linear near dependencies among the nonlinear variates that comprise the columns of the data matrix **X** relevant to this model and not for near dependencies among the basic data series that make up these nonlinear transforms. The effects that such nonlinearities have on the collinearity diagnostics and more general ways of treating them are discussed and exemplified in Chapters 9 and 11.

An Equation from the IBM Econometric Model

The next example of the collinearity diagnostics makes use of a data set brought to my attention by Harry Eisenpress of IBM. It serves well both to expand our understanding of the means by which this diagnostic technique can distinguish between two seemingly intertwined linear relations within a data set and to provide an excellent example of the practical distinction between degrading and harmful collinearity made in Section 3.4 and dealt with more fully in Chapter 7.

The equation of interest is that determining the demand for consumption of nondurables and services. The basic regression model employed in this analysis is of the form

$$\mathbf{NONDUR}(T) = \beta_1\iota + \beta_2\mathbf{RATINC}(T) + \beta_3\mathbf{NONDUR}(T-1) + \varepsilon(T), \tag{5.8}$$

where

NONDUR = ratio of nondurables-and-services consumption to deflated discretionary income,

RATING = ratio of current deflated discretionary income to its lagged value.

These quarterly data series from 1955:I to 1973:IV are given in Exhibit 5.16. Equation (5.8) is estimated by least squares as

$$\underset{}{\mathbf{NONDUR}(T)} = \underset{(0.0658)}{0.6906\iota} - \underset{(0.0613)}{0.6653\mathbf{RATINC}(T)} + \underset{(0.0342)}{0.9794\mathbf{NONDUR}(T-1)}, \tag{5.9}$$

$$R^2 = .9226, \qquad \text{SER} = 0.0043, \qquad \text{DW} = 2.19, \qquad \tilde{\kappa}(\mathbf{X}) = 305.$$

The estimated standard errors are given in the parentheses, and on the basis of the t-statistics, all of which are in excess of 10, the three coefficients individually

Exhibit 5.16 Data for the IBM Equation

Date	NONDUR	RATINC	Date	NONDUR	RATINC
1954 IV	0.938	—			
1955 I	0.943	1.00589	1964 III	0.964	1.01029
II	0.936	1.01854	IV	0.962	1.00951
III	0.930	1.01470	1965 I	0.963	1.00842
IV	0.936	1.01312	II	0.961	1.01253
1956 I	0.939	1.00676	III	0.947	1.02656
II	0.937	1.00484	IV	0.955	1.01546
III	0.938	1.00486	1966 I	0.955	1.00796
IV	0.935	1.01324	II	0.961	1.00288
1957 I	0.940	0.998161	III	0.956	1.01258
II	0.938	1.00684	IV	0.951	1.00929
III	0.944	1.00366	1967 I	0.950	1.01267
IV	0.950	0.995531	II	0.949	1.00905
1958 I	0.954	0.990267	III	0.951	1.00504
II	0.956	1.00970	IV	0.950	1.00720
III	0.947	1.02483	1968 I	0.950	1.01308
IV	0.940	1.01538	II	0.944	1.01695
1959 I	0.948	1.00649	III	0.963	0.996680
II	0.944	1.01474	IV	0.963	1.00530
III	0.961	0.989521	1969 I	0.971	0.999092
IV	0.964	1.00575	II	0.970	1.00719
1960 I	0.961	1.00918	III	0.961	1.01769
II	0.970	1.00253	IV	0.963	1.00641
III	0.969	0.998267	1970 I	0.968	1.00450
IV	0.978	0.995091	II	0.947	1.02568
1961 I	0.978	1.00945	III	0.946	1.00958
II	0.977	1.01408	IV	0.952	0.995156
III	0.966	1.01200	1971 I	0.936	1.02554
IV	0.964	1.01812	II	0.933	1.01119
1962 I	0.962	1.00911	III	0.936	0.999805
II	0.963	1.00876	IV	0.938	1.00369
III	0.970	1.00128	1972 I	0.942	1.01025
IV	0.976	1.00292	II	0.953	1.00810
1963 I	0.975	1.00462	III	0.947	1.01699
II	0.976	1.00663	IV	0.934	1.02907
III	0.982	1.00757	1973 I	0.925	1.02178
IV	0.974	1.01380	II	0.918	1.00920
1964 I	0.971	1.01843	III	0.922	1.00420
II	0.960	1.02811	IV	0.914	1.00677

Exhibit 5.17 Scaled Condition Indexes and Variance–Decomposition Proportions: IBM Equation

Scaled Condition Index, $\tilde{\eta}$	Proportions of ι var(b_1)	RATINC(T) var(b_2)	NONDUR($T-1$) var(b_3)
1	.000	.000	.000
138	.046	.085	.976
305	.954	.915	.024

differ significantly from zero. At the same time, the scaled condition number is 305, indicating, according to our previous experience, at least one very strong near dependency among the three variates comprising the data matrix **X**. Thus, while the individual t-statistics are good, there is evidence that a more detailed analysis of the possible sources of collinearity within **X** is nevertheless warranted. Exhibit 5.17 presents the scaled condition indexes and variance–decomposition proportions for this data set.

Exhibit 5.17 reveals not one, but two, strong near dependencies among the columns of **X**, a dominant relation with a scaled condition index of 305, involving ι and **RATINC**(T), and a dominated, but nonetheless strong, near dependency with a scaled condition index of 138, involving **NONDUR**($T-1$) and possibly ι and/or **RATINC**(T), the effects of these latter two variates possibly being masked by their involvement in the dominant near dependency. The dominant near dependency between ι (the constant term) and **RATINC**(T) is not surprising, for **RATINC**(T), being a ratio of a relatively smooth time series to its lagged value, is clearly going to take on values around unity, as is clear from Exhibit 5.16. Auxiliary regressions are required to ascertain the nature and extent of the second, dominated near dependency. Special care, however, is required in this case in forming and interpreting the auxiliary regressions.

Since there are two near dependencies among three variates, we can consider forming auxiliary regressions by regressing any two of the variates on the third. The procedure suggested in Section 5.3 would pick ι and **NONDUR**($T-1$) as the two regressands and **RATINC**(T) as the regressor, and this would indeed be a proper way to proceed. However, some regression packages (and some regression users) become upset when the response variate is constant, and some regression packages refuse to allow the constant term to be suppressed as an explanatory variate—or when they do, they do not handle such regression statistics as F and R^2 correctly. Neither of these circumstances should be allowed to happen, but they often do. In this instance, however, we can avoid such issues altogether, for a look at Exhibit 5.17 shows we can just as well pick **RATINC**(T) and **NONDUR**($T-1$) as the regressands and ι as the regressor. Exhibit 5.18 presents the two auxiliary regressions.

Indeed, we note that both near dependencies are very strong but that the one associated with the dominant condition index 305 is the stronger, and the usual progression of nines prevails. The diagnostics are therefore quite correct in indicating two strong near dependencies in the data set, and we have identified them. There does, however, remain one very interesting question: if collinearity is so bad among the columns of **X**, how did the regression estimates in (5.9) seemingly turn out so well? Each estimated coefficient, we recall, has a t in excess of 10. But we shall see that this is only part of the story.

It is appropriate at this point to reiterate the distinction made in Section 3.4 between degrading and harmful collinearity. The presence of collinear dependencies renders tests based on least-squares estimates for a given sample size less powerful than could otherwise be the case or, in the terminology of Andrews (1989), produces a uselessly broad inner inverse power function; that is, collinearity degrades least-squares regression estimates. The degradation need not, however, be great enough actually to cause trouble for some purposes; that is, it may not actually become harmful. This is because the notion of degradation is based on the **X** data alone. The variances, however, depend not only on the elements of $(\mathbf{X}^{\mathrm{T}}\mathbf{X})^{-1}$ but also on the error variance σ^2. This is readily seen with reference to the n-dimensional geometric depiction of collinearity given in Exhibit 2.3. Clearly, even the extreme problems pictured in Exhibit 2.3c could be removed if the radius of the circle, which is directly related to the error variance, were made sufficiently small. The degradation would remain unchanged, but the harm could be arbitrarily reduced. The degradation nevertheless exists, and one would clearly be better off without it.

Thus, with regard to the estimates in Equation (5.9), we can well assume that all of the regression coefficients are seriously degraded by their involvement in the two strong near dependencies and that our knowledge of all the estimates could be made even more precise if better-conditioned data were employed. But the degradation clearly has not been terribly harmful if our interest in (5.9) centers only on tests, for example, that the coefficients individually differ significantly from zero, for each coefficient passes this test with flying colors.

If, however, our interest were in other tests of hypothesis, we would not be so fortunate. For example, it may well be of interest in a dynamic model of this sort with a lagged dependent variate to test a null hypothesis of instability,

Exhibit 5.18 Auxiliary Regressions: IBM Equation

Dependent Variate	Coefficient of ι	$\hat{R}^2$	$\tilde{\eta}$
RATINC(T)	1.009 [1067.0]	.99993	305
NONDUR($T-1$)	0.953 [561.9]	.99976	138

namely H_0: $\beta_3 = 1$, with an alternative hypothesis H_1: $\beta_3 < 1$. The calculated t for such a test here is 0.6023, and on the basis of these data we may not reject H_0. These data, therefore, which are adequate for tests of significance, are not very strong for this test of stability. However, because we know that the estimate of β_3 on which this test is based is being degraded by the involvement of its corresponding variate in a strong near dependency, we are less willing actually to accept H_0 rather than to feel that the test consequently lacks power and is inconclusive. This test of hypothesis, therefore, is actually being harmed by the presence of degrading collinearity in the sense that there is reason to believe that the introduction of better-conditioned data would result in a more refined estimate of β_3 and with it a more conclusive test of H_0: $\beta_3 = 1$.

If the test of this stability hypothesis were truly important to the investigator, the present data set would not be optimal for his needs. He would clearly be better off with a data set in which the effects of **NONDUR**($T-1$) are not so confounded with those of ι or **RATINC**(T). It may very well be the case that, even if such data were available, they would lead to the same outcome, that is, not to reject H_0: $\beta_3 = 1$. However, under such circumstances, the investigator would have increased confidence in the conclusiveness of the test of hypothesis, knowing that the acceptance region had not been enlarged by ill-conditioned data.

Collinearity, then, is harmful only if it is first degrading and then if, in addition, important tests *based on the degraded estimates* are considered inconclusive. A test of hypothesis would be inconclusive if it is unable to reject the null hypothesis, and so a test of significance would be inconclusive if it fails. It is reasonable to consider such collinearity harmful since, under these circumstances, the tests could be refined and made more useful and trustworthy (even if the outcome is the same) when based on better-conditioned data. In defining the notion of harmful collinearity, the stipulation that the collinearity first be deemed degrading is important, for the tests could not otherwise be deemed inconclusive just because they fail to reject; rather this failure should typically lend justification to the acceptance of the null hypothesis. It is the presence of the degrading collinearity that gives an alternative possibility for the weakness of the test value and so makes it reasonable to consider the test results inconclusive. We shall examine a more rigorous means for measuring the degree of harm in Chapter 7.

It is important to note that the preceding discussion should not be interpreted to mean that an investigator should continue to search out new data (should such riches be available) until a given test of hypothesis achieves a desirable outcome. Rather it is to say that, regardless of the desired outcome, tests of hypothesis on individual parameters that are based on degraded estimates tend to lack power, the confidence intervals of the estimators being enlarged by the ill conditioning, and as such, the investigator is quite justified in viewing an outcome that lies in the inflated acceptance region as being inconclusive. Of course, no similar assessment is warranted if the outcome falls in the rejection region, for one cannot be upset when an unpowerful test is nevertheless powerful enough successfully to reject a hypothesis.

CHAPTER 6

Data and Model Considerations

In this chapter we examine several data- and model-related considerations that give important detail and content to the diagnostic procedure described in the previous chapter. This includes such topics as linear transformations, choice of origin, and the form the data should take to obtain the most meaningful collinearity diagnostics. We recall from Section 5.2, for example, that STEP 1 requires that the data be transformed so that the data matrix $\mathbf{X}$ corresponds to a basic model $\mathbf{y} = \mathbf{X}\boldsymbol{\beta}^* + \boldsymbol{\varepsilon}$ whose parameters $\boldsymbol{\beta}^*$ are those of interest to the investigation at hand, and STEP 2 requires that $\mathbf{X}$ be column equilibrated—scaled so that all its columns are the same, usually unit, length. The notion of a basic model and the rationale leading to these transformational requirements have been deferred until now.

We begin, then, by examining the effects that linear transformations of the data or, equivalently, reparameterizations of the model can have on the collinearity diagnostics. We then turn to the effects of scaling and demonstrate the importance of column equilibration for obtaining meaningful diagnostics. Next, we examine issues that relate to the choice of origin. This discussion begins with examples that show that mean centering, a frequently adopted implicit choice of origin, is almost always inappropriate. These examples also set the scene for the introduction of an extremely important operative concept: *structural interpretability*, the ability to assess whether a given perturbation in the data is of consequence to the analysis or not. This in turn allows us to define a model in *basic form*, one whose data are structurally interpretable. We can then see that the collinearity diagnostics provide the most meaningful assessment of data conditioning when applied to structurally interpretable data, and hence they provide the most meaningful regression diagnostics when applied to a regression model in basic form. Finally, related data and modeling issues are examined, including a further analysis of the role of the intercept.[1]

[1]Sections 6.3–6.9 adapt freely from Belsley (1984b, 1986a).

6.1 PARAMETERIZATION AND LINEAR TRANSFORMATIONS

The procedure for diagnosing collinearity that we have examined analyzes the presence and composition of near dependencies among the columns of an $n \times p$ data matrix $\mathbf{X}$ and assesses the suitability of these data for estimating the linear regression model $\mathbf{y} = \mathbf{X}\boldsymbol{\beta} + \boldsymbol{\varepsilon}$ by least squares. What happens, however, if instead of analyzing $\mathbf{X}$ one examines a linear transformation $\mathbf{Z} \equiv \mathbf{X}\mathbf{G}^{-1}$ of these data, or equivalently, a reparameterized version of the model in the form

$$\mathbf{y} = (\mathbf{X}\mathbf{G}^{-1})(\mathbf{G}\boldsymbol{\beta}) + \boldsymbol{\varepsilon} \equiv \mathbf{Z}\boldsymbol{\delta} + \boldsymbol{\varepsilon}, \tag{6.1}$$

where $\mathbf{G}$ is a $p \times p$ nonsingular matrix? Since it is clear from (6.1) that the two notions of a *linear transformation of the data* $\mathbf{Z} \equiv \mathbf{X}\mathbf{G}^{-1}$ and a *reparameterization of the model* $\boldsymbol{\delta} \equiv \mathbf{G}\boldsymbol{\beta}$ are intimately, indeed uniquely, related in the context of a linear model, the two terms will be used essentially interchangeably, and whenever one notion is mentioned, the relevance of the other may be presumed as well.

In general, neither the condition indexes nor the variance–decomposition proportions of $\mathbf{Z} \equiv \mathbf{X}\mathbf{G}^{-1}$ are the same as those of $\mathbf{X}$. Questions, therefore, arise as to the differences to be expected when the collinearity diagnostics are applied to $\mathbf{X}$ and to $\mathbf{Z}$, and whether there might even be linear transformations that can turn an ill-conditioned $\mathbf{X}$ into a well-conditioned $\mathbf{Z}$—and if so, whether it would not then be better to use the reparameterized model in $\mathbf{Z}$ and $\boldsymbol{\delta}$. While no complete answers can be given to these questions,[2] we are able to show that (1) the dependency of the collinearity diagnostics on the choice of the data transformation in no way reduces the validity or meaningfulness of these diagnostics for assessing the conditioning of the data for any particular application, in terms of either the strengths and numbers of near dependencies or variate involvement; (2) in practice, many reparameterizations $\mathbf{G}$ will result in very similar diagnostics relative to the determination of the strengths and numbers of near dependencies (though not relative to variate involvement); and (3) a linear transformation $\mathbf{G}^{-1}$ will be able to "undo" ill conditioning in $\mathbf{X}$ only if the matrix $\mathbf{G}$ is itself ill conditioned in a manner dependent upon the nature of the ill conditioning of $\mathbf{X}$. This last proposition means (a) that such seemingly benignant transformations cannot be presumed to arise in practice, for the parameterization $\mathbf{G}$ is virtually always chosen on the basis of a priori modeling considerations, whereas the ill conditioning of $\mathbf{X}$ results from completely unrelated chance outcomes in the data, and (b) that even if such a $\mathbf{G}$ were chosen, the problems of ill conditioning would still be present since $\mathbf{G}$ is now ill conditioned and would, as a matter of practice, provide unstable computation of $\mathbf{G}^{-1}$ and $\mathbf{Z} \equiv \mathbf{X}\mathbf{G}^{-1}$.

[2]The solution to this problem clearly depends on knowledge of the relation of the singular-value decomposition of a given matrix $\mathbf{X}$ to those of the linear transform $\mathbf{X}\mathbf{G}^{-1}$, and the general solution to this problem, so far as I know, stands as an interesting but unsolved problem of numerical analysis.

Each Parameterization Is a Different Problem

It is well known that some linear combinations of regression parameters can be estimated with precision even if ill conditioning prevents precise knowledge of the individual parameters estimated.[3] Therefore, should the investigator be interested in such linear combinations, reparameterization is a benefit to his cause.[4] If, however, the investigator is not interested in such linear combinations but rather in estimates of the original parameters, then the fact that such linear combinations exist does him little good indeed. And in practice, it is unlikely that the investigator will be interested in these linear combinations, for they depend on the eigenvectors of $\mathbf{X}^T\mathbf{X}$ and thus, for the most part, have little directly to do with the investigator's model or any relevant prior considerations.

Each parameterization, then, with its corresponding data matrix, poses a separate diagnostic problem. What is clearly required is a diagnostic procedure that allows the user to assess the suitability of the data for estimating the model relevant to his choice of parameterization, and the diagnostic procedure proposed here does exactly that. If the parameters of interest are $\boldsymbol{\beta}$ so that the relevant model becomes $\mathbf{y} = \mathbf{X}\boldsymbol{\beta} + \boldsymbol{\varepsilon}$, then the diagnostics are to be applied to $\mathbf{X}$, whereas if the parameters of interest are $\boldsymbol{\delta}$ in $\mathbf{y} = \mathbf{Z}\boldsymbol{\delta} + \boldsymbol{\varepsilon} = (\mathbf{X}\mathbf{G}^{-1})(\mathbf{G}\boldsymbol{\beta}) + \boldsymbol{\varepsilon}$, then the diagnostics are to be applied to $\mathbf{Z} \equiv \mathbf{X}\mathbf{G}^{-1}$.

A simple example serves well here. Consider a Cobb–Douglas production function $\mathbf{Q} = A\mathbf{K}^{\alpha}\mathbf{L}^{\beta}\boldsymbol{\xi}$ relating aggregate output $\mathbf{Q}$ to the capital stock $\mathbf{K}$ and labor force $\mathbf{L}$ and a multiplicative lognormal error term $\boldsymbol{\xi}$. This is typically estimated as $\log\mathbf{Q} = \log A\,\boldsymbol{\iota} + \alpha\log\mathbf{K} + \beta\log\mathbf{L} + \boldsymbol{\varepsilon}$, where $\boldsymbol{\varepsilon} \equiv \log\boldsymbol{\xi}$. Assume investigator 1 is interested in knowing the individual coefficients, α and β, while investigator 2 is interested only in returns to scale, as measured by $\gamma \equiv \alpha + \beta$. Investigator 1 takes the basic data matrix $\mathbf{X} \equiv [\boldsymbol{\iota}\ \ \log\mathbf{K}\ \ \log\mathbf{L}]$, while investigator 2 formulates the model as $\log\mathbf{Q} = \log A\,\boldsymbol{\iota} + \gamma\log\mathbf{L} + \phi(\log\mathbf{K} - \log\mathbf{L}) + \boldsymbol{\varepsilon}$, where reparameterization occurs as

$$\begin{bmatrix} \log A \\ \gamma \\ \phi \end{bmatrix} = \begin{bmatrix} 1 & 0 & 0 \\ 0 & 1 & 1 \\ 0 & 1 & 0 \end{bmatrix} \begin{bmatrix} \log A \\ \alpha \\ \beta \end{bmatrix}, \tag{6.2}$$

that is, $\gamma = \alpha + \beta$, $\phi = \alpha$. For illustrative purposes, we assume that investigator 1 finds $\mathbf{X}$ ill conditioned on account of a single strong near dependency between $\log\mathbf{K}$ and $\log\mathbf{L}$ and, as a result, is unable to obtain precise estimates of α and β, although he gets a good estimate of the intercept $\log A$. Investigator 2, however, in estimating his model $\mathbf{y} = \mathbf{Z}\boldsymbol{\delta} + \boldsymbol{\varepsilon}$, where $\mathbf{Z} \equiv [\boldsymbol{\iota}\ \ \log\mathbf{L}(\log\mathbf{K} - \log\mathbf{L})]$ and $\delta \equiv [\log A, \gamma, \phi]^T$, finds that the estimate of $\gamma \equiv \alpha + \beta$ is quite well determined, and he is satisfied. This stems from the fact that the single near dependency in $\mathbf{X}$

[3]See Theil (1971, pp. 153–154); Malinvaud (1970, pp. 216–221); Silvey (1969).

[4]Of course, testing hypotheses on linear combinations $\boldsymbol{\delta} = \mathbf{G}\boldsymbol{\beta}$ of the parameters of a given model $\mathbf{y} = \mathbf{X}\boldsymbol{\beta} + \boldsymbol{\varepsilon}$ is equivalent to tests of hypotheses on the explicit parameters of an appropriately reparameterized model, $\mathbf{y} = \mathbf{Z}\boldsymbol{\delta} + \boldsymbol{\varepsilon}$, $\mathbf{Z} = \mathbf{X}\mathbf{G}^{-1}$.

between $\log \mathbf{K}$ and $\log \mathbf{L}$, which has wholly foiled investigator 1, has been transformed into a single near dependency in $\mathbf{Z}$ between $\boldsymbol{\iota}$ and the now relatively constant variate $\log \mathbf{K} - \log \mathbf{L}$. Investigator 2, therefore, finds the $\mathbf{Z}$ matrix unsuitable for estimation of $\log A$ and ϕ but quite useful for his purpose of estimating $\gamma \equiv \alpha + \beta$.

Of course, the satisfaction felt by investigator 2 in no way diminishes the dissatisfaction felt by investigator 1; investigator 1's problem is real despite the fact that another parameterization need not suffer the same fate. One should not be surprised, then, by the fact that data that are harmfully ill conditioned for one parameterization need not be so for another. What is important to realize is that the diagnostics will correctly assess the suitability of the data for each investigator's needs. Investigator 1, in applying the collinearity diagnostics to $\mathbf{X}$, will correctly discover the near dependency adversely affecting α and β and will be correctly apprised of the unsuitability of these data for his needs. Likewise, investigator 2, in analyzing $\mathbf{Z}$, will correctly discover the near dependency adversely affecting the estimates of $\log A$ and ϕ but will also correctly be apprised of the suitability of his data for estimating $\gamma \equiv \alpha + \beta$.

To summarize the foregoing, we see that each parameterization of a model presents an inherently different problem, reflecting different interests of the investigator and requiring different characteristics of the data. In general, once a parameterization $\boldsymbol{\beta}^*$ has been decided on, the data should be transformed (if need be) to conform, so that the model becomes $\mathbf{y} = \mathbf{X}\boldsymbol{\beta}^* + \boldsymbol{\varepsilon}$. Application of the diagnostics to $\mathbf{X}$ then assesses the suitability of $\mathbf{X}$ for estimating the specific parameters $\boldsymbol{\beta}^*$. If the parameterization is to be changed, the data too should be appropriately transformed and subjected to a new collinearity analysis to determine their suitability for estimating the new parameterization.

A More General Analysis

As noted, no fully general analysis of the effect of linear transformations on the collinearity diagnostics is possible, since there is no known relation between the singular-value decomposition of $\mathbf{X}$ and that of $\mathbf{X}\mathbf{G}^{-1}$ for arbitrary nonsingular $\mathbf{G}^{-1}$. The following points, however, serve to characterize many important effects that such linear transformations (reparameterizations) can have on the collinearity diagnostics.

1. In the event that there is an exact linear dependency among the columns of $\mathbf{X}$, it is clear that no linear transformation can undo it and so change the basic nature of the diagnostics. That is, for any $\mathbf{c} \neq \mathbf{0}$ such that $\mathbf{X}\mathbf{c} = \mathbf{0}$ and any nonsingular $\mathbf{G}$ (and $\mathbf{Z} \equiv \mathbf{X}\mathbf{G}^{-1}$), there exists a $\mathbf{d} \equiv \mathbf{G}\mathbf{c} \neq \mathbf{0}$ such that $\mathbf{Z}\mathbf{d} = \mathbf{0}$. Hence, $\mathbf{X}$ and $\mathbf{Z}$ have the same number of zero singular values, and the diagnostics would detect all such exact dependencies whether one analyzes $\mathbf{X}$ or $\mathbf{Z}$.

2. When the near dependencies are not exact, however, it becomes possible for an ill-conditioned data matrix to be transformed into a better-conditioned

one, but only if the linear transformation is itself ill conditioned. We see this first with a simple example and then show a more general result.

Consider the matrices

$$\mathbf{X} = \begin{bmatrix} 1 & 1-\alpha \\ 1-\alpha & 1 \end{bmatrix} \quad \text{and} \quad \mathbf{Z} = \begin{bmatrix} \alpha & 0 \\ 0 & \alpha \end{bmatrix}. \tag{6.3}$$

As α goes to zero, **X** become ill conditioned and **Z** does not. However, note that $\mathbf{Z} = \mathbf{X}\mathbf{G}^{-1}$, where

$$\mathbf{G}^{-1} = \frac{1}{2-\alpha} \begin{bmatrix} 1 & -(1-\alpha) \\ -(1-\alpha) & 1 \end{bmatrix}. \tag{6.4}$$

Hence, there is a linear transformation $\mathbf{G}^{-1}$ that takes the ill-conditioned matrix **X** (for small α) into the well-conditioned matrix **Z**. But this linear transformation itself becomes ill conditioned (as is obvious) as α goes to zero. Thus, unless the parameterization $\boldsymbol{\delta}$ associated with **Z**, that is, $\mathbf{y} = \mathbf{Z}\boldsymbol{\delta} + \boldsymbol{\varepsilon}$, is the one desired for estimation, the transformation $\mathbf{G}^{-1}\boldsymbol{\delta}$ back to the $\boldsymbol{\beta}$ associated with **X**, that is, $\mathbf{y} = \mathbf{X}\boldsymbol{\beta} + \boldsymbol{\varepsilon} = (\mathbf{Z}\mathbf{G})(\mathbf{G}^{-1}\boldsymbol{\delta}) + \boldsymbol{\varepsilon}$, reintroduces the ill conditioning—prompting one to quote the adage "you can't get something for nothing."

This result is seen more generally as follows. Consider any data matrix **X**. The condition number $\kappa(\mathbf{X}^T\mathbf{X})$ of $\mathbf{X}^T\mathbf{X}$ is $\mu^2_{\mathbf{X},\max}/\mu^2_{\mathbf{X},\min}$, and as we know from Section 3.1,

$$\mu^2_{\mathbf{X},\min} = \min_{\mathbf{c}^T\mathbf{c}=1} |\mathbf{c}^T\mathbf{X}^T\mathbf{X}\mathbf{c}| \quad \text{and} \quad \mu^2_{\mathbf{X},\max} = \max_{\mathbf{c}^T\mathbf{c}=1} |\mathbf{c}^T\mathbf{X}^T\mathbf{X}\mathbf{c}|. \tag{6.5}$$

Let $\mathbf{c}^*$ be a solution to the minimum problem and $\mathbf{c}^\circ$ be a solution to the maximum problem. Now for any nonsingular matrix **G**, consider the linear transformation $\mathbf{Z} = \mathbf{X}\mathbf{G}^{-1}$, and let

$$\mathbf{d}^* \equiv \mathbf{G}\mathbf{c}^* \quad \text{and} \quad \mathbf{d}^\circ \equiv \mathbf{G}\mathbf{c}^\circ. \tag{6.6}$$

Further, normalize these vectors to have unit length as $\tilde{\mathbf{d}}^* \equiv \mathbf{d}^*/\|\mathbf{d}^*\|$ and $\tilde{\mathbf{d}}^\circ \equiv \mathbf{d}^\circ/\|\mathbf{d}^\circ\|$. Then we have $|\tilde{\mathbf{d}}^{*T}\mathbf{Z}^T\mathbf{Z}\tilde{\mathbf{d}}^*| = \mu^2_{\mathbf{X},\min}\|\mathbf{d}^*\|^{-2} \geqslant \mu^2_{\mathbf{Z},\min}$ and $|\tilde{\mathbf{d}}^{\circ T}\mathbf{Z}^T\mathbf{Z}\tilde{\mathbf{d}}^\circ| = \mu^2_{\mathbf{X},\max}\|\mathbf{d}^\circ\|^{-2} \leqslant \mu^2_{\mathbf{Z},\max}$, where the $\mu_{\mathbf{Z},i}$ are singular values of **Z**. Hence,

$$\kappa(\mathbf{Z}) \equiv \frac{\mu_{\mathbf{Z},\max}}{\mu_{\mathbf{Z},\min}} \geqslant \frac{\mu_{\mathbf{X},\max}\|\mathbf{d}^*\|}{\mu_{\mathbf{X},\min}\|\mathbf{d}^\circ\|} = \kappa(\mathbf{X})\,\frac{\|\mathbf{d}^*\|}{\|\mathbf{d}^\circ\|} \tag{6.7}$$

We see from (6.7) that the condition number of **Z** exceeds that of **X** multiplied by a nonnegative factor $\|\mathbf{d}^*\|/\|\mathbf{d}^\circ\|$. In the case that **G** is orthogonal, $\|\mathbf{d}^*\| = \|\mathbf{d}^\circ\|$ (since $\mathbf{c}^*$ and $\mathbf{c}^\circ$ in (6.7) are unit length by definition and an orthogonal transformation does not alter length), and $\mu_{\mathbf{Z},\max} = \mu_{\mathbf{X},\max}$ and $\mu_{\mathbf{Z},\min} = \mu_{\mathbf{X},\min}$ (since $\mathbf{Z} = \mathbf{X}\mathbf{G}^{-1}$, and an orthogonal transform does not alter singular values), and hence the equality in (6.7) holds. In this case, the ill conditioning in $\mathbf{X}$ is directly reflected in the ill conditioning in $\mathbf{Z}$.

For nonorthogonal $\mathbf{G}$, however, (6.7) shows that linear transformations can possibly make things better; that is, they can transform $\mathbf{X}$ into a $\mathbf{Z}$ with a lower condition number—but only if $\|\mathbf{d}^*\|/\|\mathbf{d}^\circ\| < 1$. And indeed, if $\mathbf{X}$ is ill conditioned, so $\kappa(\mathbf{X})$ is large, $\mathbf{Z}$ can become well conditioned with a small $\kappa(\mathbf{Z})$ only if $\|\mathbf{d}^*\|/\|\mathbf{d}^\circ\|$ is very small. But this implies that some transform of a unit vector under $\mathbf{G}$, namely, $\mathbf{d}^* \equiv \mathbf{G}\mathbf{c}^*$, is small in length relative to that of another, namely, $\mathbf{d}^\circ \equiv \mathbf{G}\mathbf{c}^\circ$. And this, of course, can only occur if some eigenvalue of $\mathbf{G}$ is small relative to another, or equivalently, only if $\mathbf{G}$ is ill conditioned.

Furthermore, the more ill conditioned is $\mathbf{X}$, the more ill conditioned must be $\mathbf{G}$ if $\mathbf{Z}$ is possibly to be well conditioned. This intuitively plausible result can be proved as follows: Let $\mathbf{d} \equiv \mathbf{G}\mathbf{c}$ for any $\|\mathbf{c}\| = 1$, and let the singular-value decomposition of $\mathbf{G}$ be $\mathbf{U}\mathbf{D}\mathbf{V}^{\mathrm{T}}$. Then using the orthogonality of $\mathbf{U}$, we see that $\|\mathbf{d}\| = \|\mathbf{U}\mathbf{D}\mathbf{V}^{\mathrm{T}}\mathbf{c}\| = \|\mathbf{D}\mathbf{h}\|$, where $\mathbf{h} \equiv \mathbf{V}^{\mathrm{T}}\mathbf{c}$, and using the orthogonality of $\mathbf{V}$, we note that $\|\mathbf{h}\| = 1$. Hence, $\|\mathbf{d}\| = (\Sigma_k \mu_{\mathbf{G},k}^2 h_k^2)^{1/2}$, from which we get $\mu_{\mathbf{G},\min} \leqslant \|\mathbf{d}\| \leqslant \mu_{\mathbf{G},\max}$, where the $\mu_{\mathbf{G},k}$ are singular values of $\mathbf{G}$. Now, since this must be true for any $\mathbf{d} \equiv \mathbf{G}\mathbf{c}$ with $\|\mathbf{c}\| = 1$, it must be true for the $\mathbf{d}^*$ and $\mathbf{d}^\circ$ of (6.6). But we know from above that ill conditioning in $\mathbf{X}$ can be offset by the linear transformation $\mathbf{G}^{-1}$ only if $\|\mathbf{d}^*\| < \|\mathbf{d}^\circ\|$, and hence we have $\mu_{\mathbf{G},\min} \leqslant \|\mathbf{d}^*\| < \|\mathbf{d}^\circ\| \leqslant \mu_{\mathbf{G},\max}$. Multiplying the upper inequality of this chain by the inverse of the lower gives $\kappa(\mathbf{G}) \equiv \mu_{\mathbf{G},\max}/\mu_{\mathbf{G},\min} \geqslant \|\mathbf{d}^\circ\|/\|\mathbf{d}^*\|$. Combining this with (6.7), we obtain

$$\kappa(\mathbf{Z}) \geqslant \kappa(\mathbf{X})\kappa^{-1}(\mathbf{G}), \tag{6.8}$$

and we see that the larger $\kappa(\mathbf{X})$, the larger must be $\kappa(\mathbf{G})$ if $\kappa(\mathbf{Z})$ is to remain small.

3. The preceding result shows that transforming the data (reparameterizing the model) could improve conditioning, but only in a way that depends on the singular values of $\mathbf{X}$ (or the eigenvalues of $\mathbf{X}^{\mathrm{T}}\mathbf{X}$); the transformation defining the reparameterization must itself be ill conditioned in a way that just offsets the ill conditioning of the $\mathbf{X}$ matrix. Thus, any improvement in conditioning brought about by reparameterization depends on aspects of the matrix $\mathbf{X}$ that are almost always outside the control of the investigator, namely, the singular values of $\mathbf{X}$. It can only be fortuitous, then, that a particular parameterization chosen on a priori grounds would just undo the ill conditioning of any given data matrix. Therefore, despite its frequent recurrence as a point of interest in textbooks, little practical significance attaches to reparameterization as a means for correcting collinearity.

4. The preceding result also shows reparameterization to have little practical value for improving data conditioning in another sense. Suppose the desired parameterization is $\mathbf{y} = \mathbf{X}\boldsymbol{\beta} + \boldsymbol{\varepsilon}$ but $\mathbf{X}$ is very ill conditioned. Suppose further that it is determined that the reparameterization $\boldsymbol{\delta} = \mathbf{G}\boldsymbol{\beta}$ provides a data matrix

$\mathbf{Z} = \mathbf{X}\mathbf{G}^{-1}$ that is well conditioned. At first blush, it would seem reasonable to obtain an estimate $\mathbf{d}$ of $\boldsymbol{\delta}$ by a least-squares regression of $\mathbf{y}$ on the well-conditioned data matrix $\mathbf{Z}$ and then to estimate $\boldsymbol{\beta}$ by $\mathbf{b} = \mathbf{G}^{-1}\mathbf{d}$. This maneuver, however, cannot rid the problem of its ill conditioning, for we have seen that the ill conditioning of $\mathbf{X}$ necessarily entails the ill conditioning of $\mathbf{G}$ (and hence of $\mathbf{G}^{-1}$). Thus, the ill conditioning of $\mathbf{X}$ is reintroduced into the solution of $\mathbf{b} = \mathbf{G}^{-1}\mathbf{d}$ through the ill conditioning of $\mathbf{G}$. It is true that one can often find a reparameterization (usually without a priori interpretation) that takes us from darkness into light, but should we desire to return, the lights must again be dimmed.

5. Finally, as noted, orthogonal reparameterizations will not change the conditioning of the data matrix, since the singular values of $\mathbf{X}$ remain invariant to such transformations. Likewise, well-conditioned reparameterizations ($\mathbf{G}$ with nearly equal singular values) will, as a matter of practice, not alter the conditioning much and certainly cannot improve it greatly. This is because for $\mathbf{G}$ well conditioned, $\|\mathbf{d}^*\| \approx \|\mathbf{d}^\circ\|$ in (6.7), so that in the worst instance $\|\mathbf{d}^*\|/\|\mathbf{d}^\circ\|$ cannot be very small and the conditioning of $\mathbf{Z}$ is bounded away from being much improved over that of $\mathbf{X}$. Hence, a well-conditioned reparameterization will leave the condition indexes of the data matrix relatively unchanged. As a matter of practice, then, collinearity diagnostics applied to the data matrix $\mathbf{X}$ will also tell much about the number and relative strengths of linear dependencies that exist in any $\mathbf{Z} = \mathbf{X}\mathbf{G}^{-1}$ when the transformation matrix $\mathbf{G}$ is well conditioned.

6. Even though the numbers and strengths of collinear dependencies besetting a given data matrix may be equally well diagnosed from a well-conditioned linear transformation of that data matrix, these different data matrices can lead to very different patterns of variate involvement. Thus, suppose the best of cases, where $\mathbf{G}$ is orthogonal, and let the singular-value decomposition of $\mathbf{X}$ be $\mathbf{U}\mathbf{D}\mathbf{V}^T$. Then the singular-value decomposition of $\mathbf{Z} = \mathbf{X}\mathbf{G}^{-1}$ is simply $\mathbf{U}\mathbf{D}\tilde{\mathbf{V}}^T$, where $\tilde{\mathbf{V}} \equiv \mathbf{G}\mathbf{V}$. It is clear that the singular values, and hence the condition indexes, are unchanged but that the elements of the eigenvectors, and hence the variance–decomposition proportions, are altered. Thus, a full diagnosis of a given situation (not just the condition indexes) requires the data be put in a form appropriate to the parameterization of interest, even with orthogonal reparameterizations.

We may summarize the foregoing by noting that linear transformations of the data (reparameterizations of the model) neither cause problems for the collinearity diagnostics nor, in general, solve them. On the one hand, one cannot, except by happy accident, make use of linear transformations to relieve ill conditioning in the data. On the other hand, although the collinearity diagnostics are seen to depend upon the parameterization chosen, it is also seen that such a dependency will not, in the case of reasonably well-conditioned linear transformations, greatly alter the basic story told by the condition indexes and will not, in any event, invalidate or reduce the usefulness of the collinearity diagnostics for correctly assessing the suitability of the corresponding transformed data for estimating the parameters of the particular parameterization chosen.

6.2 COLUMN SCALING

In Section 3.3, an intuitive justification is given for the need to column-equilibrate the data matrix **X** prior to subjecting it to the collinearity diagnostics. We are now able to provide a more rigorous justification for this scaling. In addition, some refinements are suggested for other scalings that may make more sense when adequate prior information is known about the relative errors with which the data may be measured.

Column Equilibration

Column equilibration is simply column scaling for equal length. Thus, column scaling is a special case of the more general linear transformations that we have just considered, namely, the case of linear transformations of the form $\mathbf{Z} \equiv \mathbf{XB}$, where **B** is diagonal with positive elements. Unlike the general case, however, column scaling does not result in an inherently new parameterization of the regression model; rather, as we have already noted in Section 3.3, it merely changes the units in which the variates that comprise the columns of **X** are measured and hence results in the same data just called by different names. However, it is still true that different column scalings of the same **X** matrix will typically result in different singular values and, hence, cause the collinearity diagnostics to tell different stories about the conditioning of what are, from a practical point of view, essentially equivalent data sets. The question arises, therefore, as to the existence of some canonical scaling that can rid the diagnostics of this debilitating ambiguity. And indeed, we shall now see that (1) although there is an optimal scaling that does remove this ambiguity, in general, it cannot be simply determined, but (2) scaling for unit length (or more generally, for equal column length) affords a simple and effective expedient for approximating this optimal scaling.

The optimal scaling for the purpose of the collinearity diagnostics is readily determined from an examination of Theorem 3.3 of Section 3.5. There we found the inequality (3.48), reproduced here,

$$|\xi_{tk}^{ii}| \leqslant 2\kappa(\mathbf{X}), \tag{6.9}$$

that relates two measures of the conditioning of **X**: (1) the condition number of **X**, $\kappa(\mathbf{X})$, and (2) the sensitivity of the diagonal elements of $(\mathbf{X}^{\mathrm{T}}\mathbf{X})^{-1}$ to small (relative) changes in the elements of **X**, as measured by the elasticities

$$\xi_{tk}^{ii} \equiv \frac{\partial a^{ii}}{\partial x_{tk}} \frac{x_{tk}}{a^{ii}}$$

defined in (3.36). It is a simple matter to show that, whereas $\kappa(\mathbf{X})$ can be altered by changing the length of the columns of $\mathbf{X}$, that is, through column scaling, the ξ_{tk}^{ii} are invariant to such column scaling. Hence, the inequality (6.9) must be identically true for all such column scalings. That is, for the $n \times p$ data matrix $\mathbf{X}$,

$$|\xi_{tkl}^{ii}| \leq 2\kappa(\mathbf{XB}) \quad \text{for all } \mathbf{B} \in \Xi_p, \tag{6.10}$$

where Ξ_p is the set of all nonsingular diagonal matrices of size p. The bound (6.10) is obviously tightest when a scale $\mathbf{B}^*$ is chosen such that

$$\kappa(\mathbf{XB}^*) = \min_{\mathbf{B} \in \Xi_p} \kappa(\mathbf{XB}), \tag{6.11}$$

and such a scaling thereby becomes most meaningful for an analysis of the extent to which ill conditioning of the data matrix $\mathbf{X}$ can adversely affect linear regression.

Unfortunately, the general problem of optimal scaling—column scaling that results in a matrix with minimal condition number—remains unsolved. However, scaling for equal column lengths (which our unit column length is but a simple means for effecting) has known "near-optimal" properties in this regard. To wit, van der Sluis (1969) has shown that any positive-definite and symmetric $p \times p$ matrix $\mathbf{P}$ with all its diagonal elements equal has condition number

$$\kappa(\mathbf{P}) \leq p \min_{\mathbf{B} \in \Xi_p} \kappa(\mathbf{B}^T\mathbf{PB}). \tag{6.12}$$

Of course, scaling the data matrix $\mathbf{X}$ so that all its columns are of equal length results in a $p \times p$ positive-definite, real symmetric matrix $\mathbf{P} = (\mathbf{X}^T\mathbf{X})$ that has all its diagonal elements equal. Remembering that $\kappa(\mathbf{X}^T\mathbf{X}) = \kappa^2(\mathbf{X})$, we see from the van der Sluis result that column equilibration must result in a data matrix $\mathbf{X}$ with condition number

$$\kappa(\mathbf{X}) \leq \sqrt{p} \min_{\mathbf{B} \in \Xi_p} \kappa(\mathbf{XB}). \tag{6.13}$$

Hence, even though column equilibration does not necessarily result in a data matrix with optimal $\kappa(\mathbf{X})$, it cannot be off by more than a factor of $\sqrt{p}$.

In a regression context, p, the number of variates, is typically not very large. The factor of 2 or 3 that could result here in practice is very small relative to the potential variation in $\kappa(\mathbf{X})$ that could result under different scalings, a variation that can readily extend to several orders of magnitude. Thus, column equilibration is a highly desirable canonical scaling in the regression context.

A related result is of some limited additional usefulness. Forsythe and Straus (1955) show that

$$\kappa(\mathbf{P}) = \min_{\mathbf{B} \in \Xi_p} \kappa(\mathbf{B}^T\mathbf{PB}) \tag{6.14}$$

when **P** has "Young's property A," that is, when **P** takes the form

$$\begin{bmatrix} \mathbf{I}_r & \mathbf{C} \\ \mathbf{C}^T & \mathbf{I}_q \end{bmatrix}. \tag{6.15}$$

This result is of limited use in statistics and econometrics, for it will not often naturally occur that the data matrix **X** will produce a $\mathbf{P} = (\mathbf{X}^T\mathbf{X})$ that possesses Young's property A. It is of interest to note, however, that this Forsythe–Straus result will clearly be true for column equilibration in the case where $p = 2$ $(r = q = 1)$.[5]

In summary, then, Forsythe and Straus show us that column equilibration is optimal for the $p = 2$ regression model, and van der Sluis shows us that column equilibration is near optimal in general. We are therefore quite justified in adopting column scaling for unit column lengths as a canonical scaling for the data to be subjected to the collinearity diagnostics, a scaling that, as a practical matter, we have seen to work extremely well.

Alternative Scalings and Errors in Observations

Despite the general value just discussed that attaches to column equilibration, there are situations in which one might wish to adopt a different scaling scheme. This occurs when one has strong prior knowledge about the relative errors with which the different variates (columns of **X**) are measured, and a way for proceeding is suggested in Golub et al. (1976).

We recall from Section 3.4 that the condition number of a data matrix **X** gives a multiplication factor by which perturbations in the data can get "blown up" to produce a magnified change in the calculated least-squares solution **b**. So, if $\kappa(\mathbf{X}) = 10^r$, then shifts of the order of magnitude of 10^d in the data can become shifts in the order of magnitude of 10^{d+r} in **b**. Thus, for any given scaling, including column equilibration, the resulting condition number tells what could happen when any variate is shifted by a given order of magnitude. But it may well be the case that we know a priori that a shift in, say, the tenths digit (the first digit to the right of the decimal) of the variate in the first column of **X** does not possess the same significance to the analysis at hand as a shift in the tenths digit of the variate in the second column.

The first column of **X**, for example, could be a series on GNP, with entries like 3,423.233, which are trusted to the first digit to the right of the decimal. The second column could be an investment series for some industry known to be cagy about its reporting practices, having entries like 423.49, which are trusted only to the "tens" digit, that is, the second digit to the left of the decimal. Under these circumstances, changes (perturbations) in the GNP series in the second place to the right of the decimal and changes in the investment series in the first

[5]The $p = 2$ case is often called the "bivariate-regression" case, but this widely used term is really ill conceived. Typically, the term *bivariate-regression* is intended to refer to the $p = 2$ model $\mathbf{y} = \alpha + \beta\mathbf{x} + \boldsymbol{\varepsilon}$. But if that model is the bivariate model, then what is the $p = 1$ model $\mathbf{y} = \beta\mathbf{x} + \boldsymbol{\varepsilon}$ and what is the $p = 2$ homogeneous model $\mathbf{y} = \alpha\mathbf{z} + \beta\mathbf{x} + \boldsymbol{\varepsilon}$?

place to the left must result in data series that are considered observationally equivalent to the original series. If, then, we wished collinearity diagnostics that reflect perturbations matching the confidence we have in the various data series, we should scale them so that a given additive perturbation has the same significance for each variate. If, for example, we multiplied GNP by 10 (giving figures like 34,232.33) and divided investment by 10 (giving figures like 42.349), then additive changes in the first digit to the right of the decimal, such as ± 0.1, would be considered observationally inconsequential for each variate and would have the same significance relative to the error of measurement for each variate. In doing so, we have rescaled to obtain an equilibration relative to the errors of observation rather than relative to the data; that is, rather than scaling $\mathbf{X}$ so each of its columns has equal length, $\mathbf{X}$ is scaled so that the errors in each column have roughly equal norm.

More generally, the following procedure will effect such a scaling. Consider the data matrix $\mathbf{X} = [\mathbf{X}_1 \cdots \mathbf{X}_p]$, and let d_i be the place relative to the units digit (the first digit to the left of the decimal point) to which the data series $\mathbf{X}_i$ is trusted (this would be -1 for the preceding GNP series and $+1$ for the investment series). Shifts in the $(d_i - 1)$th place of $\mathbf{X}_i$ are therefore considered inconsequential. Define $k_i \equiv 10^{-d_i}$ and $\mathbf{K} \equiv \text{diag}(k_1, \ldots, k_p)$. Then, applying the collinearity diagnostics to the scaled matrix $\mathbf{Z} \equiv \mathbf{XK}$ would produce condition indexes that reflect the differing degrees of precision with which the data are known.

Furthermore, since the condition number obtained in this way can be interpreted as a bound on the potential relative sensitivity of the least-squares estimates $\mathbf{b}$ to relative changes in the data that correspond to the errors of observation, this condition number provides an overall bound on the effects of errors in observation. Care must be taken in this interpretation, however, since (a) the bound cannot be applied to each coefficient estimate individually, but only to the overall relative norm $\|\delta\mathbf{b}\|/\|\mathbf{b}\|$—if $\mathbf{b}$ is dominated by a large element, the bound will apply to it more meaningfully than to the smaller elements—and (b) the bound is based on the matrix $\mathbf{Z}$ that already contains the errors.[6]

On a correlative note, we recall from Section 3.1 that the minimal singular value of a data matrix $\mathbf{Z}$ provides an absolute measure of the distance of $\mathbf{Z}$ from exact collinearity. Thus, with the scaling $\mathbf{K}$, if the spectral norm of the matrix of errors is of the same magnitude as the minimal singular value of $\mathbf{Z}$, we have the possibility that the observed data derive from an inherently singular generation process; that is, it becomes possible (but not necessary) that it is only because of the errors in observation that the data are not exactly collinear. While such considerations are interesting, they carry us far beyond the intent of this work, which deals with diagnostics for the conditioning of a given data set as observed and not of the process that generated it.

[6]See Stewart (1987) for an interesting discussion of the effects of errors in observation on the VIF-based collinearity indices and a procedure that does provide sensitivity bounds on individual estimates, but under the rather limiting conditions that the linear model $\mathbf{y} = \mathbf{Xb}$ is exact (that is, $\mathbf{e} \equiv \mathbf{0}$), and that only one column of $\mathbf{X}$ is measured with error.

Returning, then, to the alternate scaling **K**, we note that there are several shortcomings with this scaling procedure. First, it requires good a priori information about the number of correct digits published for each given data series. This information often is not available or even knowable. Second, the procedure assumes that the "significance" to the analysis is the same as the significance to the data. That is, while this procedure balances the effects of the variates with respect to their errors of measurement, this weighting may have nothing to do with the relative importance of the various series to the analysis at hand. Third, when a data series has a very wide range, such as would a historic, annual GNP series, a comparable additive shift in different elements of the series may not be a very meaningful exercise. A shift in the units digit of GNP for 1890 is quite a different thing from a shift in the units digit of GNP for 1990. The simple scaling undertaken here is unable to account for such cases where the measurement error may be proportionate rather than additive. We shall, however, provide a more general means for proceeding in Section 11.1 when the conditioning of nonlinear models is examined. Thus, unless there are well-defined a priori reasons for adopting a different weighting scheme, it is strongly suggested that the rather agnostic column equilibration be used.

6.3 MEAN CENTERING: EXAMPLES OF ITS INEFFICACY

The topic of the choice of origin must necessarily be dominated by a discussion of mean centering. This is so because mean centering is considered by some to be an appropriate choice of origin for assessing collinearity. Marquardt (1980), for example, states that mean centering removes "nonessential ill conditioning," and Weisberg (1980), Montgomery and Peck (1982), Gunst (1983), and Stewart (1987) also advocate mean centering in this context. Others adopt the choice of mean centering implicitly through an automatic tendency to "standardize" their data, or equivalently, to analyze the correlation matrix **R** corresponding to the data matrix **X**.

Yet it is readily shown that mean centering produces collinearity diagnostics that are often, indeed typically, highly distorted and misleading. In Belsley et al. (1980), Belsley (1984b, 1986a), and the examples that follow, it is demonstrated that the ill conditioning due to the intercept is not "nonessential" and that the collinearity diagnostics that result from using mean-centered data not only ignore the effects of this important source of ill conditioning but also give diagnostic information that is almost always relevant to the wrong problem. Further evidence supporting this from a computational perspective is offered in Lesage and Simon (1985) and Simon and Lesage (1988a, b).

It is not difficult to find reasons why the desire to center is so strong in this context; two important ones come immediately to mind. First, in the linear regression context, statisticians are weaned on the fact that the constant term can be removed from the problem through mean centering without affecting the slope estimates and related tests of hypotheses. This, combined with the fact that in many models the intercept seems to be without inherent interest and merely goes along for the ride, appears to produce a general feeling that the constant term is a bit of a nuisance that can be excised with impunity. While such ideas may be true for some aspects of estimating linear models with linear unbiased estimators, they are not true in general. The same sort of invariance to centering, for example, need not pertain to biased estimators or to nonlinear estimators or to estimators of nonlinear models.[7] And it is similarly faulty reasoning to think that invariance properties that apply to linear statistical estimators should also apply to collinearity diagnostics. Indeed, quite to the contrary, we shall see that appropriate collinearity diagnostics are not, and should not be, invariant to centering.

Second, as was noted in Chapter 1, there is a general tendency to confuse the two notions of collinearity and correlation, many practitioners thinking them to be the same. Here the feeling seems to be that, since the constant cannot be correlated with anything, it cannot be collinear with anything. Removing it, therefore, cannot affect a collinearity analysis. However, we have seen in Sections 1.4 and 2.1 that this simply is not so. Two variates can indeed be collinear without being correlated (see again Exhibit 2.1), and hence, the constant term must be retained to have a meaningful collinearity analysis. The following two examples should demonstrate the issues involved.

Example 1

Consider the data in Exhibit 6.1 in which $\mathbf{y}$ has been generated according to the $p = 3$ model

$$\mathbf{y} = \beta_1 \mathbf{X}_1 + \beta_2 \mathbf{X}_2 + \beta_3 \mathbf{X}_3 + \boldsymbol{\varepsilon} \tag{6.16}$$

with the three variates $\mathbf{X}_1$, $\mathbf{X}_2$, and $\mathbf{X}_3$, an error term that is iid normal with mean zero, and parameter values $\beta_1 = 3.0$, $\beta_2 = 0.6$, and $\beta_3 = 0.9$. Here, $\mathbf{X}_1 \equiv \boldsymbol{\iota}$, the vector of ones, but we continue to call it $\mathbf{X}_1$ to highlight that it is part of the basic data matrix $\mathbf{X} \equiv [\mathbf{X}_1 \ \mathbf{X}_2 \ \mathbf{X}_3]$.

It is readily verified that these data are highly ill conditioned. If this is not obvious from examination, we note that the uncentered R^2 of $\mathbf{X}_3$ regressed on $\mathbf{X}_1$ and $\mathbf{X}_2$ is .99999, implying an uncentered VIF exceeding 10^5, and the scaled condition number $\tilde{\kappa}(\mathbf{X}) = 1342$, which we know to be astronomical. It is also readily verified that the two mean-centered variates

$$\tilde{\mathbf{X}}_i \equiv \mathbf{M}\mathbf{X}_i \equiv \mathbf{X}_i - \bar{X}_i \boldsymbol{\iota}, \qquad i = 2, 3, \tag{6.17}$$

are orthogonal and, in contrast to the $\mathbf{X}_i$'s, are perfectly conditioned with a

[7]The inappropriateness of mean centering for CUSUM testing is demonstrated in Ploberger and Krämer (1987).

Exhibit 6.1 Data for Mean-Centering Example 1

Observation	$\mathbf{y}$	$\mathbf{X}_1 \equiv \iota$	$\mathbf{X}_2$	$\mathbf{X}_3$
1	2.69385	1.0	0.996926	1.000060
2	2.69402	1.0	0.997091	0.998779
3	2.70052	1.0	0.997300	1.000680
4	2.68559	1.0	0.997813	1.002420
5	2.70720	1.0	0.997898	1.000650
6	2.69550	1.0	0.998140	1.000500
7	2.70417	1.0	0.998556	0.999596
8	2.69699	1.0	0.998737	1.002620
9	2.69327	1.0	0.999414	1.003210
10	2.68999	1.0	0.999678	1.001300
11	2.70003	1.0	0.999926	0.997579
12	2.70200	1.0	0.999995	0.998597
13	2.70938	1.0	1.000630	0.995316
14	2.70094	1.0	1.000950	0.995966
15	2.70536	1.0	1.001180	0.997125
16	2.70754	1.0	1.001770	0.998951
17	2.69519	1.0	1.002310	1.001020
18	2.70170	1.0	1.003060	1.001860
19	2.70451	1.0	1.003940	1.003530
20	2.69532	1.0	1.004690	1.000210

scaled condition number $\tilde{\kappa}(\tilde{\mathbf{X}}) = 1.0$. In (6.17) $\bar{X}_i$ is simply the mean of the elements of $\mathbf{X}_i$, and $\mathbf{M}$ is the mean-centering operator defined in (1.17).

Since there is no collinearity among the mean-centered data, they must be considered to be trouble free by those who would advocate centering prior to diagnosing for collinearity. This is not so, however, for although the constant term seems to have been removed by mean centering ($\tilde{\mathbf{X}}_1 = \mathbf{0}$) along with its consequent conditioning problems, its effects have only been disguised. As we shall see—here as a practical matter and in Section 6.8 somewhat more rigorously—, the ill conditioning survives in the centering transformation itself, and the constant is effectively redistributed among the transformed variates. As a result, all of the conditioning problems affecting the estimation of the clearly ill-conditioned "basic model" (6.16) also beset the estimation of the equivalent, seemingly perfectly conditioned "centered model"[8]

$$\tilde{\mathbf{y}} = \beta_2\tilde{\mathbf{X}}_2 + \beta_3\tilde{\mathbf{X}}_3 + \tilde{\boldsymbol{\varepsilon}}, \tag{6.18a}$$

$$\beta_1 = \bar{y} - \beta_2\bar{X}_2 - \beta_3\bar{X}_3 - \bar{\varepsilon}. \tag{6.18b}$$

[8]It is immaterial in Equation (6.18a), of course, whether $\mathbf{y}$ is mean centered or not. If, for example, (6.18a) were replaced by $\mathbf{y} = \tilde{\beta}\iota + \beta_2\tilde{\mathbf{X}}_2 + \beta_3\tilde{\mathbf{X}}_3 + \tilde{\boldsymbol{\varepsilon}}$, no change in the estimates or conditioning would occur since both $\tilde{\mathbf{X}}_2$ and $\tilde{\mathbf{X}}_3$ are orthogonal to ι. The estimate of $\tilde{\beta}$ would simply be $\bar{y}$, and β_1 continues to be related by and estimated with reference to (6.18b).

We recall from Section 3.4 that the consequences of these conditioning problems are twofold: computational and statistical. Computationally, ill conditioning means that small relative changes in the **X** and **y** data can produce large relative changes in the least-squares estimates **b**, and statistically, ill conditioning means inflated variances for these estimates. We now demonstrate that mean centering removes neither of these problems.

Consider first the basic regression of **y** on $\mathbf{X}_1$, $\mathbf{X}_2$, and $\mathbf{X}_3$, giving

$$\begin{aligned} \mathbf{y} = \underset{(0.784)}{3.191}&\mathbf{X}_1 + \underset{(0.555)}{0.810}\mathbf{X}_2 - \underset{(0.555)}{1.302}\mathbf{X}_3, \\ R^2 = .310, \quad & F = 3.82, \quad \tilde{\kappa}(\mathbf{X}) = 1342, \end{aligned} \tag{6.19}$$

and next the centered regression of $\tilde{\mathbf{y}}$ on $\tilde{\mathbf{X}}_2$ and $\tilde{\mathbf{X}}_3$, giving

$$\begin{aligned} \tilde{\mathbf{y}} &= \underset{(0.555)}{0.810}\tilde{\mathbf{X}}_2 - \underset{(0.555)}{1.302}\tilde{\mathbf{X}}_3, \\ \text{with implied intercept} &= \bar{y} - 0.810\bar{X}_2 + 1.302\bar{X}_3 \\ &= \underset{(0.784)}{3.191} \\ R^2 = .310, \quad & \tilde{\kappa}(\tilde{\mathbf{X}}) = 1.0. \end{aligned} \tag{6.20}$$

Regarding the problem of inflated variances, we note that there is ample evidence that collinearity has inflated variances in the estimation of the basic model (6.19). Despite the fact that $\mathbf{X}_2$ and $\mathbf{X}_3$ are known to generate **y** and are jointly significant (from the F-test) at the 5% level, only b_2 is individually so. But we also note from (6.20) that the centered estimates *and their standard errors* are unchanged, so mean centering has certainly not reduced this variance inflation. This should come as no surprise; it is a well-known fact that this will occur, but its significance in this context is apparently easily overlooked. Clearly mean centering can in no way change, much less reduce, any variance inflation due to collinearity.

Regarding the problem of the sensitivity of the regression estimates to small changes in the data, let us perturb $\mathbf{X}_2$ and $\mathbf{X}_3$ by a relatively small error to give

$$\mathbf{X}_i^\circ \equiv \mathbf{X}_i + \boldsymbol{\delta}\mathbf{X}_i, \qquad i = 2, 3, \tag{6.21}$$

where the random $\boldsymbol{\delta}\mathbf{X}$'s are scaled so that $\|\boldsymbol{\delta}\mathbf{X}_i\| = 0.01\,\|\mathbf{X}_i\|$. The regression of **y** on the slightly perturbed variates $\mathbf{X}_1$ ($\equiv \boldsymbol{\iota}$), $\mathbf{X}_2^\circ$, and $\mathbf{X}_3^\circ$ gives

$$\begin{aligned} \mathbf{y} = \underset{(0.173)}{2.978}&\mathbf{X}_1 - \underset{(0.139)}{0.111}\mathbf{X}_2^\circ - \underset{(0.134)}{0.167}\mathbf{X}_3^\circ, \\ R^2 = .137, \quad & \tilde{\kappa}(\mathbf{X}^\circ) = 265. \end{aligned} \tag{6.22}$$

Comparing (6.22) with (6.19) shows that the small relative change in the **X**'s has indeed resulted in a large relative change in the estimates. Letting **b** denote the original estimates and $\mathbf{b}^\circ$ the perturbed, $\|\mathbf{b} - \mathbf{b}^\circ\|/\|\mathbf{b}\| = 0.4172$; that is, a 1% relative change in the **X**'s results in over a 40% relative change in the estimators. That the change is not even larger or nearer the maximal change possible as indicated by (3.19) is due only to the fact that we have chosen not to perturb $\mathbf{X}_1 \equiv \iota$. Although we shall see shortly that it is not unreasonable to perturb the "constant" in a sensitivity analysis, we do not do so here in order to retain compatibility with the centered results that follow. The sensitivity of **b** is, however, quite dramatic enough without perturbing $\mathbf{X}_1$, so there can be no doubt here about the presence of ill conditioning.

Now examine the corresponding perturbed, mean-centered regression of $\tilde{\mathbf{y}}$ on $\tilde{\mathbf{X}}_2^\circ$ and $\tilde{\mathbf{X}}_3^\circ$, where $\tilde{\mathbf{X}}_i^\circ \equiv \mathbf{X}_i^\circ - \bar{X}_i^\circ \iota$, $i = 2, 3$, giving

$$\begin{aligned}
\tilde{\mathbf{y}} &= \underset{(0.139)}{-0.111\tilde{\mathbf{X}}_2^\circ} - \underset{(0.134)}{0.167\tilde{\mathbf{X}}_3^\circ}, \\
&\text{with implied intercept} = \underset{(0.173)}{2.978}, \\
R^2 &= .137, \qquad \tilde{\kappa}(\tilde{\mathbf{X}}^\circ) = 1.2.
\end{aligned} \tag{6.23}$$

Again, as should be obvious but somehow goes unappreciated in this context, the perturbation of the **X**'s has altered the estimates of the mean-centered equation (6.23) in exactly the same way as those of the basic equation (6.22), demonstrating that the centered-data estimates retain the same sensitivity as well as the same variance inflation as the uncentered-data estimates despite the perfect conditioning of the centered data matrix $\tilde{\mathbf{X}} \equiv [\tilde{\mathbf{X}}_2 \ \tilde{\mathbf{X}}_3]$.

From this example, we must conclude that (a) the ill conditioning of the basic data has caused true problems for least-squares estimation, giving inflated variances and extremely sensitive parameter estimates; (b) mean centering has altered neither of these problems; and (c) diagnosing the conditioning of the mean-centered data $[\tilde{\mathbf{X}}_2 \ \tilde{\mathbf{X}}_3]$ (which are perfectly conditioned) overlooks this situation, whereas diagnosing the basic data $[\mathbf{X}_1 \ \mathbf{X}_2 \ \mathbf{X}_3]$ does not.

Conclusion (a), of course, is specific to this example, but conclusion (b) clearly results from the mechanics of the least-squares procedure and must hold generally. Expressed somewhat differently, it indicates that, at least for least-squares estimation of a truly linear model, there is no such thing as "nonessential" ill conditioning relative to centering. Conclusion (c) is sufficient to demonstrate that, in general, mean centering can remove from the data the information that is needed to assess conditioning correctly. Taken together, conclusions (b) and (c) show that, whereas the ill conditioning as measured by $\tilde{\kappa}(\mathbf{X})$ cannot be considered nonessential, the fine conditioning as measured by $\tilde{\kappa}(\tilde{\mathbf{X}})$ can certainly be considered nonsensical.

Example 2

The previous example is intended to provide a crystal-clear counterexample to the efficacy of mean centering as a means for removing ill conditioning. It uses data that, when mean centered, are perfectly conditioned, so if mean centering has any value in this regard at all, it should be seen here—and it is not. Interestingly, however, several of those who commented on its first publication—see the comments accompanying Belsley (1984b)—became embroiled in the specifics of the data set to the detriment of the inevitability and universality of its conclusions. Extraneous issues such as low R^2s, implied domains, the importance of the constant term to the analysis, and the nature of the example appear to have distracted attention from the central result. These issues are answered in the reply to Belsley (1984b) and, with one exception, need not be reexamined here.

The exception is the nature of the example, for some object to using contrived rather than actual data. But of course, the objection should be exactly the reverse. We can only truly assess the value of a diagnostic (or statistical) procedure when its behavior is tested in a situation in which the truth is known. If problems with a statistical procedure can be shown to occur when it is used in a situation in which we know what is going on, they certainly will not disappear just because we decide to use it in a situation in which we do not. I therefore still hold that the preceding example provides the sharpest means for demonstrating the inefficacy of mean centering. But the many concerns voiced do suggest that a less idiosyncratic example may be beneficial as well.

So let us consider a second example constructed to emulate a more "standard" looking data set, and indeed we shall see several of its more important elements in some of the real-life examples given in later chapters. We need not repeat the detail of the previous experiment, for the idea will be clear with a simple sensitivity analysis. The data in Exhibit 6.2 are relevant to the model

$$\mathbf{y} = \beta_1 \boldsymbol{\iota} + \beta_2 \mathbf{X}_2 + \beta_3 \mathbf{X}_3 + \beta_4 \mathbf{X}_4 + \boldsymbol{\varepsilon}, \tag{6.24}$$

with $\beta_1 = 5.0$, $\beta_2 = 0.3$, $\beta_3 = 0.9333$, $\beta_4 = 0.4$, and $\boldsymbol{\varepsilon}$ iid normal with mean zero and variance 0.1.

Exhibit 6.3 shows the scaled condition indexes and variance–decomposition proportions of the basic data matrix $\mathbf{X} \equiv [\boldsymbol{\iota}\ \mathbf{X}_2\ \mathbf{X}_3\ \mathbf{X}_4]$. From this we see the presence of two strong near dependencies, the stronger with $\tilde{\eta} = 308$ involving $\mathbf{X}_2$ and $\mathbf{X}_3$ and the slightly weaker one with $\tilde{\eta} = 252$ likely involving all four variates but certainly involving the constant variate $\boldsymbol{\iota}$ and $\mathbf{X}_4$. In comparison, Exhibit 6.4 shows the scaled condition indexes and variance–decomposition proportions for the mean-centered data corresponding to the centered model

$$\tilde{\mathbf{y}} = \beta_2 \tilde{\mathbf{X}}_2 + \beta_3 \tilde{\mathbf{X}}_3 + \beta_4 \tilde{\mathbf{X}}_4 + \tilde{\boldsymbol{\varepsilon}}, \tag{6.25}$$

Exhibit 6.2 Data for Mean-Centering Example 2

Observation	y	X_2	X_3	X_4
1	5.40551	9.43474	7.01697	10.6733
2	6.41417	9.00556	6.67755	12.4143
3	4.14120	10.4510	7.85542	9.25088
4	4.54715	10.3146	7.74277	9.35618
5	5.61870	9.36443	7.05775	11.1377
6	4.26429	10.7270	8.02466	8.76072
7	3.18105	11.5927	8.68041	6.53162
8	4.70980	10.1615	7.64178	9.46036
9	6.54516	8.86543	6.73278	12.3553
10	5.73221	9.97447	7.37513	10.3976
11	6.05554	8.67871	6.53295	12.3337
12	5.66113	9.14190	6.94166	11.3969
13	7.18856	8.86809	6.61631	12.2169
14	4.97143	9.80247	7.46327	9.80596
15	3.24592	10.8986	8.31965	8.06527
16	5.53840	9.02447	6.81657	12.2756
17	4.12843	10.3523	7.84951	8.83951
18	6.21633	8.91739	6.71936	12.5908
19	7.73236	8.02686	5.92669	13.8066
20	7.07766	8.32836	6.31519	13.4595

where, as before, $\tilde{\mathbf{X}}_i \equiv \mathbf{MX}_i$, $i = 2, 3, 4$. Here there is one, considerably more moderate near dependency with $\tilde{\eta} = 26$ involving $\tilde{\mathbf{X}}_2$ and $\tilde{\mathbf{X}}_3$; some might also look to the possibility of a second, weak near dependency with $\tilde{\eta} = 14$ possibly involving all three centered variates.

Exhibit 6.3 Scaled Condition Indexes and Variance-Decomposition Proportions: Mean-Centering Example 2. Basic, Uncentered Data

Scaled Condition Index, $\tilde{\eta}$	Proportions of ι var(b_1)	X_2 var(b_2)	X_3 var(b_3)	X_4 var(b_4)
1	.000	.000	.000	.000
9	.000	.000	.001	.009
252	.998	.078	.116	.990
308	.002	.922	.883	.001

Exhibit 6.4 Scaled Condition Indexes and Variance-Decomposition Proportions: Mean-centering Example 2. Mean-Centered Data

Scaled Condition Index, $\tilde{\eta}$	Proportions of $\tilde{\mathbf{X}}_2$ var(b_1)	$\tilde{\mathbf{X}}_3$ var(b_2)	$\tilde{\mathbf{X}}_4$ var(b_3)
1	.001	.001	.002
14	.097	.094	.998
26	.902	.905	.000

Certainly the mean centering has produced a friendlier set of diagnostics. And if it is correct, as claimed by some, that it is these mean-centered diagnostics that provide the information relevant to the conditioning of this regression problem, then small relative changes in the data should cause relative changes in the regression estimates that correspond to these figures and not those for the basic, uncentered data in Exhibit 6.3. So let us see if this is the case.

Equation (6.26) shows the least-squares estimates of (6.25) using the mean-centered original data $\tilde{\mathbf{X}} \equiv [\tilde{\mathbf{X}}_2\ \tilde{\mathbf{X}}_3\ \tilde{\mathbf{X}}_4]$, and Equation (6.27) shows this regression using mean-centered perturbations of the original data $\tilde{\mathbf{X}}^\circ \equiv [\tilde{\mathbf{X}}_2^\circ\ \tilde{\mathbf{X}}_3^\circ\ \tilde{\mathbf{X}}_4^\circ]$. The perturbations $\boldsymbol{\delta}\mathbf{X}_i \equiv \mathbf{X}_i^\circ - \mathbf{X}_i$, $i = 2, 3, 4$, are constructed so that their spectral (and Frobenius) norms are less than 1% of that of the original data; that is, $\|\boldsymbol{\delta}\hat{\mathbf{X}}\| < 0.01\|\hat{\mathbf{X}}\|$, where $\boldsymbol{\delta}\hat{\mathbf{X}} \equiv [\boldsymbol{\delta}\mathbf{X}_2\ \boldsymbol{\delta}\mathbf{X}_3\ \boldsymbol{\delta}\mathbf{X}_4]$ and $\hat{\mathbf{X}} \equiv [\mathbf{X}_2\ \mathbf{X}_3\ \mathbf{X}_4]$.[9]

$$
\begin{aligned}
\tilde{\mathbf{y}} &= \underset{(0.862)}{1.8091\tilde{\mathbf{X}}_2} - \underset{(1.130)}{3.5925\tilde{\mathbf{X}}_3} + \underset{(0.257)}{0.1759\tilde{\mathbf{X}}_4}, \\
R^2 &= .947, \qquad \hat{R}^2 = .997, \qquad \tilde{\kappa}(\tilde{\mathbf{X}}) = 26,
\end{aligned} \tag{6.26}
$$

$$
\begin{aligned}
\tilde{\mathbf{y}} &= -\underset{(0.698)}{0.6670\tilde{\mathbf{X}}_2^\circ} + \underset{(0.922)}{0.3279\tilde{\mathbf{X}}_3^\circ} + \underset{(0.281)}{0.4039\tilde{\mathbf{X}}_4^\circ}, \\
R^2 &= .915, \qquad \tilde{\kappa}(\tilde{\mathbf{X}}^\circ) = 16.
\end{aligned} \tag{6.27}
$$

It is clear that these small relative shifts in the basic data have caused a rather dramatic relative change in the estimated slope parameters of the mean-centered regression. Again letting $\mathbf{b}$ denote the estimates in (6.26) and $\mathbf{b}^\circ$ those in (6.27), we find that $\|\mathbf{b} - \mathbf{b}^\circ\|/\|\mathbf{b}\| = 1.153$, indicating a 115.3% relative change in the least-squares slope estimates as a result of less than a 1% relative

[9] In this example, the percentage change $\|\boldsymbol{\delta}\hat{\mathbf{X}}\|/\|\hat{\mathbf{X}}\|$ is 0.93% using the Frobenius norm and 0.62% using the spectral norm.

change in the data! But if the scaled condition number $\tilde{\kappa}(\tilde{\mathbf{X}}) = 26$ of the mean-centered data given in Exhibit 6.4 were relevant to this situation, the greatest relative shift we could expect to get, as seen from (3.19), is approximately $2\tilde{\kappa}(\tilde{\mathbf{X}}) = 52\%$ since the uncentered $\hat{R}^2$ here is close to 1.

Once again, then, we see that the conditioning of the mean-centered data seriously understates the potential for problems due to collinearity and that, rather generally, the conditioning diagnostics applied to mean-centered data provide misleading information about the true conditioning of a set of data relevant to a regression model having a constant term. (When a model does not have a constant term, of course, mean centering is completely inappropriate.)

We further see that the misinformation does not pertain just to the estimates of the intercept term. Mean centering also removes the information needed to assess correctly the problems caused to the estimates of those slope parameters that are also involved in the collinear relations with the constant term. In this example, all the slope parameters are so involved, and their sensitivity is very clearly not correctly assessed by the conditioning of the mean-centered data. This conclusion is worthy of notice because some have argued that one should examine the conditioning of the original, uncentered data only if the estimate of the constant term is of theoretical interest to the analysis, and otherwise the data should be mean centered.[10] Clearly this is not so, for we see that the conditioning of the mean-centered data can fail to diagnose the potential sensitivity (corruption) of the estimates of the slope parameters *whether or not the intercept is of any theoretical importance to the analysis at hand.*

6.4 STRUCTURAL INTERPRETABILITY

The preceding examples, in addition to discrediting the value of mean centering in analyzing collinearity, rather more generally demonstrate that it matters very much in what form the data are put in order to assess their conditioning meaningfully and that of the corresponding least-squares regression model. What, then, is the proper form to pick in practice? To answer this question, we first review what a condition number indicates in general and then determine the special conditions under which it also indicates something meaningful about the data.

Giving Meaning to the Condition Number

We know from the analysis of Section 3.1 that the condition number $\kappa(\mathbf{X})$ tells us roughly the potential relative change in the least-squares solution $\mathbf{b}$ that can result from a small relative shift in the data $\mathbf{X}$. Examination of (3.19) shows that there could be a factor of 2 or more in this relation, depending upon the

[10] See, e.g., Gunst (1983) and Stewart (1987).

tightness of the regression. But for reasonably tight regressions, $\kappa(\mathbf{X})$ provides the proper order of magnitude for this potential relative sensitivity. Thus, it is productive to view the condition number $\kappa(\mathbf{X})$ as a rough magnification factor by which a 1% relative change in the data $\mathbf{X}$ could effect a relative shift in the least-squares estimates $\mathbf{b}$.

Now, taken by itself, this condition number is just that—a number. Its values, large or small, cannot be given meaning unless the situation it measures can be given meaning. To see this, let us examine several cases. First, suppose that the data relevant to a given regression model are in a form that allows us to interpret, from our understanding of the underlying real-life situation they measure, that a 1% relative change in these data is marginally inconsequential or unimportant to the analysis. Larger relative shifts begin to produce "new situations"; smaller ones do not. Under these circumstances it is possible to put meaning to the condition number, for we know that, if the scaled condition number were large, an unimportant relative change in the data could be magnified into a substantially greater relative shift in the parameter estimates, and we can assess whether this situation is disturbing to the statistical analysis at hand.[11] Similarly, if under the same circumstances the scaled condition number were small, we could rest content knowing that inconsequential relative shifts in the data could produce at most small, presumably inconsequential relative shifts in the least-squares estimates. The fact that our knowledge of the underlying real-life situation being described by the regression model enables us to interpret that a 1% relative change in the data is of marginal consequence to the analysis allows us to give direct meaning to the condition number of the data.

As a second case, let us suppose that the data relevant to a regression model are again interpretable relative to the underlying real-life situation and that their scaled condition number is 300. This means that a 1% change in the data could produce roughly a 300% relative shift in the regression estimates, and on the face of it, this looks bad. However, let us suppose that further investigation reveals that a 1% relative shift in the data really corresponds to a very substantive change in the conditions of the underlying situation. Since we can hardly be upset to discover that substantive relative changes in the data can produce large relative changes in the regression estimates, the value of 300 for the condition number seems to overstate the true conditioning of this problem. Rather, let us suppose it is determined that a 0.1% shift in the data is more nearly of marginal consequence. Relative changes of this magnitude could produce relative shifts in $\mathbf{b}$ only of the order of magnitude of 30, one-tenth of 300, and this value of 30 certainly provides a more meaningful measure of the conditioning of this problem than the original 300. It tells us the degree to which marginally inconsequential shifts in the data could affect the parameter estimates, and if this number is too high, we have good reason to be upset. So our ability to assess the meaning of a relative shift in the data has allowed us to temper, refine, and reinterpret the meaning that attaches to the raw condition number of 300. But

[11] A discussion of a complete conditioning analysis that also takes into account assessing the "consequence" of the shift in the least-squares estimates is given in Section 11.1.

the important thing to note from this example is that, despite any refinement that might be indicated, the condition number is meaningfully interpretable only because we are able to assess the significance to the analysis of relative changes in the data in the first place.

As a third case, let us suppose we have a regression model with a data set whose condition number is numerically small, say 1.6, indicating that a 1% relative shift in the data might produce something of the order of magnitude of a 1.6% relative change in the least-squares estimates. At first blush, this seems excellent. But suppose that further investigation reveals these data to be in a form in which we cannot meaningfully assess whether a given relative change in them corresponds to an important alteration in the underlying real-life situation—as, we shall shortly see, can readily occur if the data have been, for example, mean centered. In this case, the condition number 1.6 is without interpretive value since it does us little good to know $\mathbf{b}$ will shift by at most 1.6% in response to an uninterpretable 1% relative change in the $\mathbf{X}$'s. This is a classic example of GIGO (garbage in, garbage out).

From the preceding discussion, we can see that the collinearity diagnostics will provide the most meaningful information about the sensitivity of a least-squares solution and therefore the most meaningful collinearity diagnostics when they are applied to data whose form allows a given numerical relative change to be meaningfully assessed as being consequential or inconsequential relative to the real-life situation being modeled by the regression equation. Such data will be called *structurally interpretable.*

Structurally Interpretable Data

Structurally interpretable variates are ones whose numeric values and (relative) differences derive meaning and interpretation from the context they measure. The use of the term *structural* here relates to its meaning in econometrics. Loosely, an econometric model is said to be in structural form when its equations directly pair up with the modeler's vision of the separately identifiable, autonomous real-life activities whose interactions comprise the phenomenon being investigated.[12] Clearly, most linear transformations of structural models, even if mathematically valid, are no longer structural, since arbitrary linear combinations of structural equations typically no longer pair up with anything that "actually exists."

Since the elements of the structural equations pair up with their real-life counterparts, they are thereby afforded direct interpretability; that is, the elements of the model are given "meaning" in terms of the observable meaning possessed by their real-life counterparts, and hence they become structurally interpretable. Aggregate consumption C, for example, in the Keynesian consumption function is structurally interpretable since it is a measure that pairs up

[12]For further reading on these topics, see Belsley (1986b, 1988b) and the bibliographies contained there. A particularly important reference in econometrics is Haavelmo (1944).

with the real-life, economically definable, observable, and measurable notion of the aggregate result of the behavior of consumers. Most economists could tell quite a bit about C. First, they would note that $C = \$0$ implies the absence of consumption; that is, there is a natural origin for the measurements of C. Second, measured relative to this origin, recent consumption in current dollars in the United States is approximately \$3.227 trillion annually. Third, a shift in C by 0.1% is relatively small, both because its overall economic impact is not great and because such changes are likely within its error of measurement. Fourth, scale changes (changes in units of measurement, such as trillions of dollars to billions of dollars) do not affect this interpretability.

But if we were arbitrarily to change this origin of measurement, the resulting variate need no longer be structurally interpretable. If, for example, we were to mean-center the consumption data, little interpretation could be given to the values of the centered consumption variate $\tilde{C} \equiv C - \bar{C}$. To see this, ask the nearest economist what $\tilde{C} = \$0$ means,[13] or $\tilde{C} = \$1000$, or if $\tilde{C}$ were changed by 10% would the change be large or small? consequential or inconsequential to the analysis at hand? It is clear that our knowledge about the economic concept of consumption lends meaning to the consumption variate C but not, taken by itself, to the mean-centered consumption variate $\tilde{C}$. Thus, in order to answer the preceding questions, the economist would have to return to the original, interpretable C variate from which the $\tilde{C}$ values came. Any interpretability of $\tilde{C}$, then, is derived from C. So we must conclude that, taken by themselves, mean-centered variates are not structurally interpretable. And rather more generally, we see that similar considerations show that arbitrary shifts of origin render structurally interpretable variates uninterpretable.

Now we can answer the question posed previously about which data form is best suited for obtaining meaningful collinearity diagnostics. The collinearity diagnostics are most meaningful when applied to data that are structurally interpretable, that is, data placed in a form in which it is possible, from knowledge of the underlying real-life situation they measure, to stipulate what specific changes in their values constitute consequential or inconsequential alterations in that situation. We also see that, to obtain this structural interpretability, it is important to pick an appropriate origin of measurement.

6.5 ORIGINS OF MEASUREMENT

Not all variates encountered in regression analyses have natural origins as does the consumption variate C above. A temperature of 0° Celsius, for example, does not imply the "absence of temperature." And we would be hard put to know in general if a 10% change in Celsius temperature were large or small. Near the freezing point of water, it could well be small, but not near the boiling point. To make such an assessment, it is clear that one would typically have to know more about the structure of the real situation in which the measure is being used. One

[13] And since everybody is his own best economist, you can ask yourself.

can conceive of physical experiments, such as those dealing with Boyle's law or entropy, in which the structural interpretation of temperature would be with respect to absolute zero (0° Kelvin = −273.16° Celsius). In other cases, the structural interpretation might in fact center upon the freezing point of water, such as in a model of water viscosity. In psychological experiments dealing with human comfort, some origin relating to 20°C might be more appropriate. In general, it is a priori knowledge of the structure of the underlying situation being modeled that indicates the appropriate origin of measurement against which relative changes can be assessed.

Even variates like price, weight, profits, or acceleration, which seem to have natural origins of zero, might be structurally interpretable with respect to different origins in other situations. The Dow–Jones average, for example, for a long time had a psychological plateau of 800, then 1200, and now something more like 3000. If one were modeling such aspects of stockmarket behavior, it is quite possible that consequential and inconsequential relative changes should be assessed with respect to some (possibly moving) origin of measurement determined by these plateaux and not necessarily with respect to zero.

In assessing the conditioning of a least-squares problem and collinearity among the data, then, we wish the data to be structurally interpretable. This means that (1) the investigator must be able to pick an origin against which consequential and inconsequential relative changes in the data can be appropriately assessed and (2) it is the data measured relative to this origin that should be subjected to the diagnostic analysis.

Point 2 does not imply that the data may not perhaps be advantageously transformed for estimation or other purposes. It does mean, however, that such transformed data are not generally relevant to assessing conditioning and collinearity and that, for this purpose, the data should be transformed back into the structurally interpretable form. With regard to point 1, if the investigator is unable to provide the appropriate origin of measurement, there is an indication of an incomplete understanding of the real-life situation being modeled, and the meaning of the diagnostics (and quite possibly the estimates and other aspects of the statistical analysis[14]) must necessarily reflect this arbitrariness.

6.6 THE BASIC MODEL AND THE BASIC DATA

With the preceding in mind, we define the linear model

$$\mathbf{y} = \mathbf{X}\boldsymbol{\beta} + \boldsymbol{\varepsilon} = \beta_1\mathbf{X}_1 + \cdots + \beta_p\mathbf{X}_p + \boldsymbol{\varepsilon} \tag{6.28}$$

to be in *basic form* when the $[\mathbf{X}_1 \cdots \mathbf{X}_p]$, after perhaps being transformed and measured relative to an appropriate origin, are structurally interpretable. Such a

[14]Of course, arbitrariness with respect to the origin will not typically affect the interpretability of all aspects of a statistical analysis, at least not when dealing with linear models. It will, however, make impossible a full assessment of model validity, since various of the model's elements are admittedly not understood.

set of X's will be called the *basic data*. It is this form that is referred to in the discussion of STEP 1 of Section 5.2. Three remarks apply to these notions.

First, while the origin of measurement is integral to the basic data, rescaling the X's clearly has no effect on the relative changes, and hence the units in which the X's are measured are irrelevant to this basic formulation.

Second, it is assumed that the basic data are linear. While there is nothing to stop one from applying the collinearity diagnostics to data that are nonlinear transformations of structurally interpretable variates, and doing so can often provide interesting and useful information, it is clear that the discussion given above does not extend directly or easily to such transformed data sets. These issues are taken up in Chapter 9, where we will be able to provide appropriate collinearity diagnostics for models that contain logged variates, and in Chapter 11, where we give more general techniques for assessing the conditioning of nonlinear models.

Third, it is quite acceptable for the basic model to be homogeneous (no intercept) if a priori considerations indicate this to be appropriate. If, however, there are no such a priori reasons for so restricting the basic model, there must be an intercept term in the basic model and a vector of ones should be included in the basic data matrix **X**. Please note that this is not a matter of whether the "situation at the origin is meaningful" or whether the origin is within or without the range of the observed X's. For even if the situation at the origin is not meaningful (e.g., blood pressure when the heart rate is zero), the sensitivity of the regression estimates to changes in the data can only be meaningfully assessed using basic data. And even if the data have been transformed from the basic data for other purposes (such as mean centering for the purpose of obtaining an intercept estimate "at the center of the data"), they should be returned to their basic form for conducting the most meaningful collinearity diagnostics.

6.7 GENERAL CONSIDERATIONS ON CENTERING

We are now in a position to examine more closely the effects of centering on the collinearity diagnostics. At first, centering will again refer to mean centering, but thereafter more general forms of centering will be considered.

A Formal Analysis of Mean Centering

In Section 6.3 we gave two examples that showed that mean centering is ineffective in removing ill conditioning from a given basic data set. We can now see why this is so. Suppose we have formulated a regression model in basic form as

$$\mathbf{y} = \mathbf{X}\boldsymbol{\beta} + \boldsymbol{\varepsilon} = \beta_1 \boldsymbol{\iota} + \beta_2 \mathbf{X}_2 + \cdots + \beta_p \mathbf{X}_p + \boldsymbol{\varepsilon} \qquad (6.29)$$

with the basic data matrix $\mathbf{X} \equiv [\boldsymbol{\iota} \ \mathbf{X}_2 \cdots \mathbf{X}_p]$. Corresponding to this, of course, is the centered model

$$\tilde{\mathbf{y}} = \tilde{\mathbf{X}}\tilde{\boldsymbol{\beta}} + \tilde{\boldsymbol{\varepsilon}}, \tag{6.30a}$$

$$\beta_1 = \bar{y} - \beta_2\bar{X}_2 - \cdots - \beta_p\bar{X}_p - \bar{\varepsilon}, \tag{6.30b}$$

where $\tilde{\boldsymbol{\beta}}$ is the set of slope parameters from the basic model, that is, $\boldsymbol{\beta} \equiv [\beta_1, \tilde{\boldsymbol{\beta}}^{\mathrm{T}}]^{\mathrm{T}}$, and $\tilde{\mathbf{X}} \equiv [\tilde{\mathbf{X}}_2 \cdots \tilde{\mathbf{X}}_p]$ is the matrix of mean-centered data. Let $\mathbf{b}$ be the least-squares estimates of $\boldsymbol{\beta}$ using the basic model and $\tilde{\mathbf{b}}$ be those of $\tilde{\boldsymbol{\beta}}$ using the mean-centered model. It is well known, as the two examples in Section 6.3 illustrate, that the least-squares estimates $\tilde{\mathbf{b}}$ of $\tilde{\boldsymbol{\beta}}$ obtained by estimating the centered model are identical to the least-squares estimates of the slope parameters of $\boldsymbol{\beta}$ obtained by estimating the basic model, that is, $\mathbf{b} \equiv [b_1, \tilde{\mathbf{b}}^{\mathrm{T}}]^{\mathrm{T}}$, and that the same estimate of the intercept (but not its standard error) may also be obtained from the obvious reworking of (6.30b). And so, it appears we have a choice how to estimate the $\boldsymbol{\beta}$'s, either with the basic or the mean-centered model, and how to assess the data conditioning, either with the basic or the mean-centered data. Common sense should suggest that the two are equivalent (barring truncation error problems, as discussed below), and hence there really is not a choice.

But the following logic seems seductive. It will be noticed that mean centering seems to improve data conditioning. We show, for example, in Section 6.9 that $\tilde{\kappa}(\tilde{\mathbf{X}}) \leqslant \tilde{\kappa}(\mathbf{X})$, and indeed, in practice $\tilde{\kappa}(\tilde{\mathbf{X}}) \ll \tilde{\kappa}(\mathbf{X})$! Thus, mean centering seems to rid us of "nonessential" ill conditioning, and it appears we are better off assessing the conditioning $\tilde{\kappa}(\tilde{\mathbf{X}})$ of the mean-centered data than the conditioning $\tilde{\kappa}(\mathbf{X})$ of the basic data. The flaw in this logic, as we are now able to see, is that the conditioning of the mean-centered data $\tilde{\kappa}(\tilde{\mathbf{X}})$ gives us information about the wrong problem; namely, it tells us about the sensitivity of the least-squares solution $\tilde{\mathbf{b}}$ to numerically small relative changes in the mean-centered data $\tilde{\mathbf{X}}$. But we have seen that changes in the mean-centered data are almost always without structural interpretability, and so therefore is the magnitude of $\tilde{\kappa}(\tilde{\mathbf{X}})$.[15] The proper diagnostic is still the one showing the sensitivity of the $\mathbf{b}$'s to interpretably small changes in the basic data, which, of course, is given by the condition number $\tilde{\kappa}(\mathbf{X})$ of the basic data $\mathbf{X}$. Mean centering, then, typically removes from the data the interpretability that makes conditioning diagnostics meaningful, and hence a conditioning diagnostic based on mean-centered data does not relate meaningfully to the real situation being diagnosed. The removal of "nonessential" ill conditioning through mean centering is thus a will-o'-the-wisp that is as effective in improving the actual conditioning of a given data set as breathing hot breath on a thermometer is in heating up a room.

Some aspects of this may be appreciated heuristically by noting that mean

[15] When $\tilde{\kappa}(\tilde{\mathbf{X}})$ is small, this leads to situations like that described in case 3 of Section 6.4. However, $\tilde{\kappa}(\tilde{\mathbf{X}})$ need not be wholly devoid of meaning in situations where it is large. Since $\tilde{\kappa}(\tilde{\mathbf{X}}) \leqslant \tilde{\kappa}(\mathbf{X})$, a large $\tilde{\kappa}(\tilde{\mathbf{X}})$ still indicates a large $\tilde{\kappa}(\mathbf{X})$ and hence ill conditioning with respect to $\mathbf{X}$. The problem is that a small $\tilde{\kappa}(\tilde{\mathbf{X}})$ need not indicate the absence of ill conditioning with respect to $\mathbf{X}$.

centering throws away information: one can always go from the basic data to the mean-centered data, but given only the mean-centered data, one cannot retrieve the basic data. Contained in this lost information is that needed to diagnose some aspects of the conditioning of the least-squares estimator, and hence the diagnostics applied to the mean-centered data cannot, in general, tell the whole story.

The preceding result is equally relevant to the case in which the least-squares estimates of the basic model, the **b**'s, may have been calculated by a regression algorithm that uses mean-centered data, that is, when **b** is calculated as $[b_1, \tilde{\mathbf{b}}^T]^T$. In this case, the conditioning of the **b**'s with respect to the basic data and the relation between $\tilde{\kappa}(\tilde{\mathbf{X}})$ and $\tilde{\kappa}(\mathbf{X})$ may be better understood by realizing that **b** now really results from a sequence of two transformations: (i) the transformation of the basic data into the mean-centered data and (ii) the regression with the centered data, or

$$\begin{aligned} &\text{(i)}\ \tilde{\mathbf{X}} \equiv \mathbf{M}[\mathbf{X}_2 \cdots \mathbf{X}_p], \\ &\text{(ii)}\ \tilde{\mathbf{b}} = (\tilde{\mathbf{X}}^T\tilde{\mathbf{X}})^{-1}\tilde{\mathbf{X}}^T\tilde{\mathbf{y}}. \end{aligned} \tag{6.31}$$

The desired conditioning of $\mathbf{b} \equiv [b_1, \tilde{\mathbf{b}}^T]^T$ with respect to the basic **X**'s is now seen loosely to be the product of the conditioning of these two transformations. The conditioning of (ii) is $\tilde{\kappa}(\tilde{\mathbf{X}})$, while that of (i) can be indefinitely large, since it is singular in the direction of the "constant" vector $\boldsymbol{\iota}$, that is, $\mathbf{M}\boldsymbol{\iota} = \mathbf{0}$.

We can see the ill conditioning of the **M** transformation geometrically, since this translates into the fact that it is possible for small relative changes in the length of a vector somewhat collinear with $\boldsymbol{\iota}$ to produce massively larger relative changes in the length of the corresponding mean-centered vector. Using the geometry of mean centering given in Exhibit 1.5, this situation is depicted in Exhibit 6.5. Here, the $\boldsymbol{\iota}$ vector is oriented vertically, and its orthogonal complement is represented by the horizontal axis. The vector **x** is collinear with (has a small angle with) $\boldsymbol{\iota}$. Now we recall that $\tilde{\mathbf{x}}$, the mean-centered transform of **x**, is simply the orthogonal projection of **x** into the orthogonal complement of $\boldsymbol{\iota}$. So in Exhibit 6.5 we see that a perturbation $\boldsymbol{\delta}\mathbf{x}$ whose length is less than a third that of **x** can nevertheless produce a perturbed vector $\mathbf{x}^\circ$ whose centered length $\|\tilde{\mathbf{x}}^\circ\|$ is nearly 400% that of the length $\|\tilde{\mathbf{x}}\|$ of centered **x**. Clearly, the greater the collinearity between **x** and $\boldsymbol{\iota}$ (the smaller the angle between them), the more exaggerated this phenomenon can become.

It is therefore quite possible for the centered regression (ii) to be very well conditioned with respect to the mean-centered data $\tilde{\mathbf{X}}$ ($\tilde{\kappa}(\tilde{\mathbf{X}})$ small) yet the true conditioning to be very poor with respect to the basic data **X** because some of the basic data are involved in near dependencies with the constant vector $\boldsymbol{\iota}$ so that the mean-centering transformation (i) is very ill conditioned. This, of course, is exactly the situation depicted in the two examples given in Section 6.3, and it is the reason that we find that mean centering can mask the role of the constant term in any near dependencies among the columns of the basic data matrix. It is

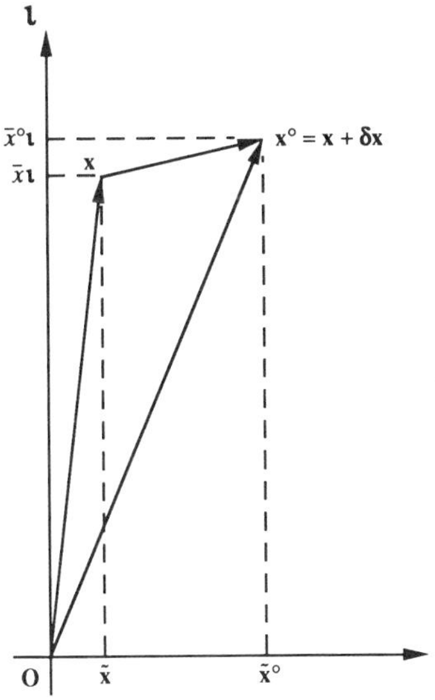

Exhibit 6.5 Ill conditioning of the mean-centering transformation when **x** is collinear with ι.

also interesting to note that the analysis of Section 3.5 can be used to show that it is precisely the circumstances in which the basic data **X** are poorly conditioned but the centered data $\tilde{\mathbf{X}}$ are well conditioned that transformation (i) will necessarily be poorly conditioned. That is, the seeming benefits of better conditioning in (ii) that result from mean centering are in fact necessarily compensated for by the concomitant poor conditioning of the centering transformation (i).

Model- versus Data-Dependent Centering

Until now centering has denoted deviations from the mean. Other centers are, however, possible. Rather more generally, we can define centering to be the removal of a constant, thought in some sense to be central to the problem, from each term in a data series. Then, a variate **x** will be said to be centered with respect to α, denoted $\mathbf{x}^{(\alpha)}$, if

$$\mathbf{x}^{(\alpha)} \equiv \mathbf{x} - \alpha\iota. \tag{6.32}$$

It will prove useful to distinguish between model- and data-dependent centerings. A *data-dependent* centering is one for which $\alpha = \alpha(\mathbf{x})$, that is, α is a function of the variate being centered. Mean centering, for example, is a data-dependent centering with $\alpha(\mathbf{x}) = \bar{x}$. So also are centerings based on any other descriptive "central values" for the data, such as mode, median, maximum, or minimum. *Model-dependent* centerings occur when α is determined from a priori model considerations, independent of the actual data being centered. We encountered such a centering in Section 6.5 when the freezing point of water (e.g., $\alpha = 32°F$) was chosen for the origin.

The Inappropriateness of Data-Dependent Centerings

Data-dependent centerings typically result in data series that are inappropriate for diagnosing data conditioning. The shift in origin implicit in such centerings is usually arbitrary since it is derived from the specific data series at hand and would change for each different realization of the given data series to which the model would apply. Such data-dependent centerings are therefore not associated with any structural element of the model, and the resulting centered variate must typically lack structural interpretability. We have seen, for example, how mean centering destroys the structural interpretability of the consumption variate C. Clearly the meaning that attaches to the same value, say $\tilde{C} = \$1000$, for mean-centered consumption would vary wildly depending upon the period and range of values over which the consumption series was being centered. Yet, presumably, the same consumption function (model) would apply to numerous such data series. Thus, there can be no stable structural meaning to $\alpha = \bar{C}$.

Further insight into the inappropriateness of data-dependent centerings for conditioning diagnostics arises from recognizing that all such centerings that employ a "representative" data element behave essentially like mean centering when acting on variates collinear with $\boldsymbol{\iota}$ and hence become ill conditioned exactly as the mean-centering transformation $\mathbf{M}$. To see this, center the variate $\mathbf{x} = (x_1 \cdots x_n)^T$ by any representative value α, which, being representative, must surely obey $x_{\min} \leqslant \alpha \leqslant x_{\max}$, where $x_{\min}$ and $x_{\max}$ are, respectively, the smallest and largest elements of $\mathbf{x}$. Define $z \equiv \max|x_i - \bar{x}|$. Then, necessarily, $\bar{x} - z \leqslant x_{\min} \leqslant \alpha \leqslant x_{\max} \leqslant \bar{x} + z$, and hence

$$0 \leqslant |\alpha - \bar{x}| \leqslant z. \tag{6.33}$$

Now, as $\mathbf{x} \to \boldsymbol{\iota}$ (i.e., as we consider $\mathbf{x}$ whose elements become more nearly equal, or as the angle between $\mathbf{x}$ and $\boldsymbol{\iota}$ decreases), $\Sigma_i (x_i - \bar{x})^2 \to 0$, and hence $(x_i - \bar{x}) \to 0$ for all i. Thus, $z = \max|x_i - \bar{x}| \to 0$, and from (6.33), $\alpha \to \bar{x}$. For vectors approaching $\boldsymbol{\iota}$, then, all such data-dependent centerings are equivalent to mean centering and must consequently suffer the same ill conditioning as $\mathbf{M}$ in transforming vectors collinear with $\boldsymbol{\iota}$.

The Appropriateness of Some Model-Dependent Centerings

Unlike data-dependent centerings, model-dependent centerings are based on model elements that may possess structural meaning and hence could lead to structurally interpretable variates. We have already seen in Section 6.5 that some model-dependent centerings are properly employed to produce data appropriate to conditioning diagnostics. When such centerings produce structurally interpretable data, they are not only appropriate, they are essential for meaningful collinearity diagnostics. All other centers, however, whether data or model dependent, are improper for this diagnostic purpose, although they may possess some other computational or statistical virtue.

There is one case in which mean centering could estimate an appropriate model-dependent centering and hence become a proper centering for assessing conditioning. If it were felt from prior considerations that an observed variate **x** should be centered relative to its model-interpretable but unknown *population* mean in order to be structurally interpretable *and* if it were also felt that the mean of the observed sample in **x** is suitable to provide a good estimator of this population mean (if, for example, the elements of **x** constitute a random sample from a fixed distribution), then the mean-centered variate $\tilde{\mathbf{x}}$ would *estimate* a structurally interpretable variate. While such conditions could possibly arise in experimental situations, they are certainly not very likely to occur with nonexperimental data.

Centering When Approximating Nonlinearities

It is sometimes thought good to center when estimating a linear equation known to be a local approximation to a nonlinear model. Suppose in Exhibit 6.6 that the true nonlinear model is $\mathbf{y} = f(\mathbf{x}) + \boldsymbol{\varepsilon}$, written in terms of the structurally interpretable, basic data **y** and **x**, and that the indicated data points have been generated from it.

Suppose also these data are used to estimate a linear local approximation to f by least squares giving

$$\mathbf{y} = a\boldsymbol{\iota} + b\mathbf{x} + \mathbf{e}. \tag{6.34}$$

Now, in this instance, it is often noted that the constant term a has no structural meaning since $x = 0$ lies outside the range of the approximation. Hence, one

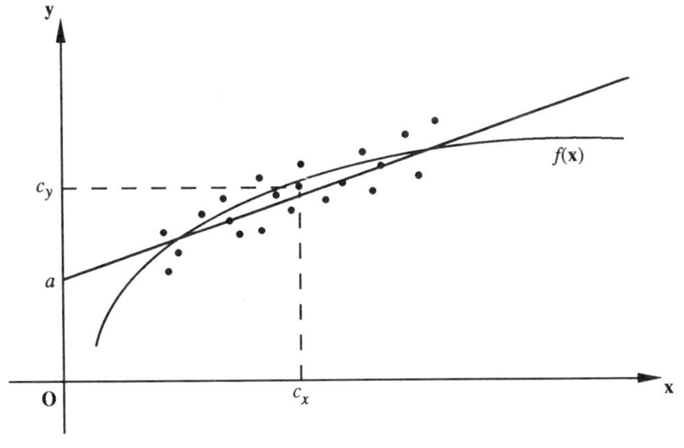

Exhibit 6.6 Centering in linear approximations to nonlinear models.

might obtain a more useful "constant" if the data were centered about locally representative values (not necessarily means) such as c_x and c_y to give $\tilde{\mathbf{y}} \equiv \mathbf{y} - c_y\boldsymbol{\iota}$ and $\tilde{\mathbf{x}} \equiv \mathbf{x} - c_x\boldsymbol{\iota}$, and if $\tilde{\mathbf{y}}$ were regressed on $\boldsymbol{\iota}$ and $\tilde{\mathbf{x}}$ giving

$$\tilde{\mathbf{y}} = a^{\circ}\boldsymbol{\iota} + b^{\circ}\tilde{\mathbf{x}} + \mathbf{e}. \tag{6.35}$$

Since we have not necessarily used mean centering, there will, in general, be a constant term here, and its estimate will be interpretable as a centralized value of $\mathbf{y}$ when $\mathbf{x}$ takes on a value representative for the approximation. Thus, such a transformation may indeed allow the constant, *as printed in the regression output*, to be more readily interpreted. But it is to be emphasized that this is a convenience only; nothing really substantive has been gained here, for clearly (6.35) and (6.34) are really the same regression, related by

$$\begin{aligned} a^{\circ} &= a - c_y + bc_x \\ b^{\circ} &= b. \end{aligned} \tag{6.36}$$

What we can learn from one regression we can equally well learn from the other through a simple linear transformation.

But the question naturally arises, which data should be assessed for conditioning: $\mathbf{X} = [\boldsymbol{\iota}\ \ \mathbf{x}]$ or $\tilde{\mathbf{X}} = [\boldsymbol{\iota}\ \ \tilde{\mathbf{x}}]$? This is readily answered since, as always, we most meaningfully measure the potential sensitivity in the estimates with respect to relative changes in the structurally interpretable, basic data. Therefore, we seek $\tilde{\kappa}(\mathbf{X})$ and not $\tilde{\kappa}(\tilde{\mathbf{X}})$.

Once again we see that, regardless of how the data may be transformed for the purposes of executing a regression, it is the basic data, not the transformed data, that are relevant to diagnosing the conditioning of the regression problem, and it is this conditioning that applies to all such transformed regressions as well.[16]

Centering for Computational Advantage

Mean centering, by reducing the number of significant digits that must be maintained, is often employed with computers with short word lengths to reduce possible problems due to truncation errors. This practice is also useful when memory availability is limited or when one wishes to take advantage of the

[16]Similar considerations apply to transformations such as those that replace two collinear basic variates $\mathbf{x}_1$ and $\mathbf{x}_2$ by the pair $\mathbf{x}_1$ and $\mathbf{x}_2 - \mathbf{x}_1$ in a ploy that often appears to improve conditioning. In some instances this will not work because even though $\mathbf{x}_1$ and $\mathbf{x}_2$ are structurally interpretable, the difference $\mathbf{x}_2 - \mathbf{x}_1$ will not be. In other instances, however, this difference may also be structurally interpretable. This would be true, for example, if income $\mathbf{Y}$ and consumption $\mathbf{C}$ were replaced by income $\mathbf{Y}$ and savings $\mathbf{S} \equiv \mathbf{Y} - \mathbf{C}$. Care must be taken here, however, in comparing the conditioning of these two data sets because "inconsequential" relative shifts in consumption are of a very different order of magnitude from those for savings.

increased speeds that accompany single-precision, floating-point arithmetic. In any event, it is clear that such centering is motivated by reasons extraneous to assessing conditioning, and all conditioning diagnostics should still employ only the basic (typically uncentered) data.

It is also thought that regression algorithms generally behave better when digesting the "better-conditioned" centered data. However, this ignores the fact that better-conditioned centered data imply a poorly conditioned centering transform, and computationally this could be "out of the frying pan, into the fire."[17] Furthermore, with well-designed regression algorithms (see Belsley, 1974) practical conditioning problems begin to arise only after the scaled condition number of the basic data becomes astronomically large, of the order of 40,000. The conditioning of the famous Longley (1967) data, for example, is 43,167, and the computational abilities of good regression algorithms are not adversely affected by these data. Unfortunately, condition numbers that are even one-fortieth of this magnitude already indicate regression results that could be practically useless, even if all calculations were exact. Recall, for instance, the data in Example 1 of Section 6.3. These data have a scaled condition number of 1340, and we found that minute changes in them resulted in massive shifts in the estimates. So, under these circumstances, it is quite possible to adopt the philosophy "if you don't like the estimates you've got, just change the data by an insignificant amount." Thus, computational problems arise only after the data are so ill conditioned as to be effectively useless anyway. If, however, you still wish to center for computational reasons, please do, but use the basic, not the centered, data for diagnosing ill conditioning.

Centering and Assessing Conditioning When There Is No Constant

Suppose structural considerations indicate there should be no constant term in the basic regression (6.28), giving the homogeneous model $\mathbf{y} = \beta_1\mathbf{X}_1 + \cdots + \beta_p\mathbf{X}_p + \boldsymbol{\varepsilon}$, with $\mathbf{X}_1 \neq \boldsymbol{\iota}$. Of course, our interpretation of the scaled condition number holds whether or not $\boldsymbol{\iota}$ is in $\mathbf{X} \equiv [\mathbf{X}_1 \cdots \mathbf{X}_p]$, and hence the appropriate conditioning measure is still $\tilde{\kappa}(\mathbf{X})$. It is illuminating, however, to see the damage that mean centering can do if such a transform were used in this situation, say, for computational reasons. Consider the regression

$$\mathbf{y} = b_1\mathbf{X}_1 + b_2\mathbf{X}_2 + \mathbf{e} \tag{6.37}$$

with the basic data matrix

$$\mathbf{X} = \begin{bmatrix} 1 & 0 \\ 0 & 1 \\ 0 & 1 \end{bmatrix}.$$

[17]The studies of Lesage and Simon (1985) and Simon and Lesage (1988a, b) are very interesting in this and related contexts.

It is clear that $\mathbf{X}_1$ and $\mathbf{X}_2$ are well conditioned: they are orthogonal, and their scaled condition number $\tilde{\kappa}(\mathbf{X}) = 1$. If, however, we mean-center these data, we obtain

$$\tilde{\mathbf{X}} = \begin{bmatrix} \frac{2}{3} & -\frac{2}{3} \\ -\frac{1}{3} & \frac{1}{3} \\ -\frac{1}{3} & \frac{1}{3} \end{bmatrix},$$

a matrix that we note to be rank deficient.

Thus, mean centering when there is no intercept term can transform well-conditioned data into highly ill conditioned data. Exhibit 6.7 shows this geometrically. Here $\mathbf{X}_1$ and $\mathbf{X}_2$ are quite well conditioned, the angle between them being close to a right angle, while the centered variates $\mathbf{MX}_1$ and $\mathbf{MX}_2$ are exactly collinear.

Mean centering, then, is inappropriate in the context of estimating homogeneous equations. It certainly should not be employed for diagnosing collinearity but also not even for computational reasons.

6.8 THE CONSTANT

Much confusion surrounding centering arises because of some commonly held misconceptions about the "constant term." This section aims at several of these issues with the goal of showing that, for the most part, despite much practice to the contrary, the constant is most reasonably viewed as just another element in a regression analysis that plays a role no different from any other "variate."

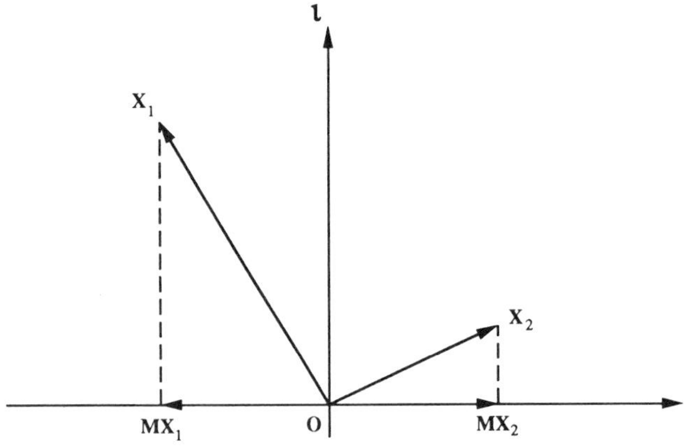

Exhibit 6.7 Ill conditioning due to mean centering.

Structural versus Algorithmic Roles of ι: Variates versus Vectors

We recall from Section 1.5 that every data series $\mathbf{X}_i$ $(i = 1, \ldots, p)$ in a regression analysis, including a "constant column" ($\mathbf{X}_1 \equiv \iota$), plays two separate roles: one structural, the other algorithmic.

The Structural Role of Variates

Structurally, a series $\mathbf{X}_i$ is a measured representation of some definable aspect of the underlying real situation being modeled. In this context, we often call $\mathbf{X}_i$ a *variate*. Sometimes this term is reserved for series that actually vary, but this is an unwarranted restriction since those structurally unchanging aspects of the underlying situation are properly measured and modeled by the "constant variate" $\mathbf{X}_1 \equiv \iota$. In a truly linear model, these constant aspects are simply the expected value of the response variate $\mathbf{y}$, conditional on all other $\mathbf{X}_i$'s being zero. If it is known a priori that there are no such constant elements, then ι should not be included. This might be true in a demand equation, for example, when theoretical considerations indicate that it should be homogeneous of degree zero in all prices and income. In other words, in proper modeling, the decision to include a constant should be made as consciously as that to include any other variate.[18] Often the constant is included to "sop up" excluded effects or because the model is assumed to be a linear approximation to a nonlinear situation. This is acceptable practice unless it is known a priori that $\beta_1 = 0$. Often, however, the constant is included simply because the regression package includes it automatically. This convenience can lead to bad practice.

The Algorithmic Role of Vectors

Algorithmically, once it has been determined structurally that $\mathbf{X}_1 \equiv \iota$ belongs in the regression, the vector ι takes on a life of its own that is in every sense the equal of any other vector $\mathbf{X}_i$ included in analysis. We can see this very easily with reference to Exhibit 1.9, which shows the n-dimensional geometry of the least-squares procedure. Here we note that a least-squares regression of $\mathbf{y}$ on $\mathbf{X} = [\mathbf{X}_1 \cdots \mathbf{X}_p]$ is a mechanical process, having nothing to do with any underlying structural considerations, and that performs an orthogonal projection of the *vector* $\mathbf{y}$ on the linear space $\mathscr{S}$ spanned by the *vectors* that comprise the columns of $\mathbf{X}$. This is true whether or not the first column of $\mathbf{X}$ happens to be the vector ι. If it is, however, then ι becomes a basis vector determining one dimension of $\mathscr{S}$ and plays a role exactly the same as every other

[18]Regression packages that automatically include an intercept term may provide a convenience but do us no favor in otherwise understanding the role this term plays in regression. Regression packages that further do not even allow the constant term to be suppressed are simply in error. Similar levels of criticism can be voiced against regression texts that isolate, either explicitly or implicitly, the intercept from the other variates. This occurs, for example, in those that refer to regressions on $p + 1$ regressors, where the $+1$, of course, is the intercept—or those that assume that the variates are standardized (scaled and centered), a transformation that is equivalent to assuming that the constant term is always present and, if nothing else, lacks generality.

$\mathbf{X}_i$. Indeed, assume in Exhibit 1.9 that $\mathbf{X}_1$ is ι. Now, in your mind's eye, rotate the figure in Exhibit 1.9 rigidly so that all vectors, $\mathbf{y}$ and $\mathbf{X}_i$'s alike, change but keep exactly the same relation to each other. Clearly the least-squares regression process (the orthogonal projection problem) remains completely unchanged—but now no vector is ι. In its role in the regression process, then, ι is no more or less a "constant" than any other vector $\mathbf{X}_i$. It is constant only in its structural interpretation.

Perturb a Constant?

The distinction between the structural and algorithmic roles of ι helps to remove the qualm, sometimes voiced, that the sensitivity aspects of a conditioning analysis cannot apply to ι (or any dummy variate) since ι is fixed in value and cannot be perturbed. But we see that, among other faults, this confuses the two interpretations of ι—for ι is constant and fixed in value only as a structural concept but can be perturbed just like any other vector for the purposes of determining the sensitivity of the output of the projection algorithm to changes in its inputs.

The Constant and Collinearity

Another misplaced use of the constant term occurs in some definitions of collinearity. Weisberg (1980), for example, defines two variates $\mathbf{X}_1$ and $\mathbf{X}_2$ as being exactly collinear if there is a linear equation such as

$$\mathbf{X}_2 = \alpha_0 \iota + \alpha_1 \mathbf{X}_1. \tag{6.38}$$

Although the inclusion of ι in this definition could perhaps prove reasonable (albeit awkward) if all regression models had constants, it becomes unduly restrictive and even misleading in the context of homogeneous models.

Consider a regression in the context of the two-variate homogeneous model

$$\mathbf{y} = \beta_1 \mathbf{X}_1 + \beta_2 \mathbf{X}_2 + \boldsymbol{\varepsilon}, \tag{6.39}$$

where $\mathbf{X}_1 \neq \iota$ and Equation (6.38) holds with $\alpha_0 \neq 0$. According to this definition, there is exact collinearity between $\mathbf{X}_1$ and $\mathbf{X}_2$, and so presumably, no regression should be possible. But in fact, the data matrix $\mathbf{X} = [\mathbf{X}_1 \ \mathbf{X}_2] = [\mathbf{X}_1 \quad \alpha_0\iota + \alpha_1\mathbf{X}_1]$ will have full rank so long as $\alpha_0 \neq 0$, and a regression may indeed properly be run.[19] Of course, it would be impossible to estimate the three-variate, nonhomogeneous model

$$\mathbf{y} = \beta_0 \iota + \beta_1 \mathbf{X}_1 + \beta_2 \mathbf{X}_2 + \boldsymbol{\varepsilon} \tag{6.40}$$

[19]It is easily shown here that $\det(\mathbf{X}^T\mathbf{X}) = \alpha_0^2 n^2 s_1^2$, where $s_1^2 = \Sigma(X_{1t} - \bar{X}_1)^2/n$, the sample variance of the elements of $\mathbf{X}_1$. Thus, $\mathbf{X}^T\mathbf{X}$ will be singular only if $\alpha_0 = 0$ or $\mathbf{X}_1 = \iota$, neither of which holds here.

with these data, since $\mathbf{X} = [\boldsymbol{\iota}\ \ \mathbf{X}_1\ \ \mathbf{X}_2]$ would have less than full rank. But this is not because there is collinearity between the two variates $\mathbf{X}_1$ and $\mathbf{X}_2$; rather it is because of collinearity among the three variates $\boldsymbol{\iota}$, $\mathbf{X}_1$, and $\mathbf{X}_2$. Appropriate definitions of collinearity, then, such as those given in Chapter 3, do not make any special inclusion of the constant term.

Centering Does Not Get Rid of the Constant Term

It is generally thought that centering "gets rid" of the constant term. But this is not the case; centering merely redistributes the constant among all the variates so that it continues to be present, but not explicitly. Begin with

$$\mathbf{y} = \beta_1\boldsymbol{\iota} + \beta_2\mathbf{X}_2 + \cdots + \beta_p\mathbf{X}_p + \boldsymbol{\varepsilon}. \tag{6.41}$$

Taking means gives

$$\bar{y} = \beta_1 + \beta_2\bar{X}_2 + \cdots + \beta_p\bar{X}_p + \bar{\varepsilon}, \tag{6.42}$$

from which we express the constant as

$$\beta_1 = \bar{y} - \beta_2\bar{X}_2 - \cdots - \beta_p\bar{X}_p - \bar{\varepsilon}. \tag{6.43}$$

Postmultiplying (6.42) by $\boldsymbol{\iota}$ and subtracting from (6.41), we get the mean-centered model

$$\mathbf{y} - \bar{y}\boldsymbol{\iota} = \beta_2(\mathbf{X}_2 - \bar{X}_2\boldsymbol{\iota}) + \cdots + \beta_p(\mathbf{X}_p - \bar{X}_p\boldsymbol{\iota}) + (\boldsymbol{\varepsilon} - \bar{\varepsilon}\boldsymbol{\iota}), \tag{6.44a}$$

or

$$\tilde{\mathbf{y}} = \beta_2\tilde{\mathbf{X}}_2 + \cdots + \beta_p\tilde{\mathbf{X}}_p + \tilde{\boldsymbol{\varepsilon}}. \tag{6.44b}$$

In the form (6.44b), there appears to be no constant term, but in the more transparent form (6.44a) we see through to the fact that the information in (6.43) is indeed present; it is just incorporated into the other variates—the constant term is still there.

In fact, $\boldsymbol{\iota}$ does not even play a unique role in such "reductive" transformations of the regression, for we may similarly transform (6.41) with respect to any of the explanatory variates $\mathbf{X}_i$. To emphasize the parallel between $\boldsymbol{\iota}$ and any other $\mathbf{X}_i$ in this role of "centering," we first (re)establish a well-known result with $\boldsymbol{\iota}$ in a way that allows for immediate reapplication with $\mathbf{X}_i$.

Let $b_1^*, \ldots, b_p^*$ be the least-squares solution to

$$\mathbf{y} = b_1\boldsymbol{\iota} + \cdots + b_p\mathbf{X}_p + \mathbf{e}, \tag{6.45}$$

and consider the corresponding transformed (centered with respect to $\boldsymbol{\iota}$) problem

$$\mathbf{My} = b_2\mathbf{MX}_2 + \cdots + b_p\mathbf{MX}_p + \mathbf{Me}, \tag{6.46}$$

recalling that $\mathbf{M}\boldsymbol{\iota} \equiv \mathbf{0}$. It is easy to show the well-known result that $b_2^*, \ldots, b_p^*$ are also the least-squares solution to the mean-centered problem (6.46)—for consider any other values $b_2, \ldots, b_p$. Put these, along with b_1^*, in (6.45) and denote the resulting residual by $\mathbf{e}^\circ \equiv \mathbf{y} - b_1^*\boldsymbol{\iota} - b_2\mathbf{X}_2 - \cdots - b_p\mathbf{X}_p$. Denote the least-squares residual by $\mathbf{e}^* \equiv \mathbf{y} - b_1^*\boldsymbol{\iota} - b_2^*\mathbf{X}_2 - \cdots - b_p^*\mathbf{X}_p$. Since $\mathbf{e}^*$ results from an orthogonal projection of $\mathbf{y}$ into $\mathscr{S}$, the linear space spanned by the $\mathbf{X}_i$'s (including $\boldsymbol{\iota}$), and $\mathbf{e}^\circ$ results from a projection (not orthogonal) of $\mathbf{y}$ into the same space, there must be an $\boldsymbol{\xi} \in \mathscr{S}$, orthogonal to $\mathbf{e}^*$, such that $\mathbf{e}^\circ = \mathbf{e}^* + \boldsymbol{\xi}$. Thus, the "centered" residual $\mathbf{Me}^*$ that results from putting $b_2^*, \ldots, b_p^*$ into (6.46) has least sum of squares since it must obey $\mathbf{e}^{\circ T}\mathbf{Me}^\circ = (\mathbf{e}^* + \boldsymbol{\xi})^T\mathbf{M}(\mathbf{e}^* + \boldsymbol{\xi}) = \mathbf{e}^{*T}\mathbf{Me}^* + \boldsymbol{\xi}^T\mathbf{M}\boldsymbol{\xi} \geqslant \mathbf{e}^{*T}\mathbf{e}^*$, recalling that $\mathbf{Me}^* = \mathbf{e}^*$ and $\boldsymbol{\xi}^T\mathbf{e}^* = 0$.

But we may establish an analogous result for $\mathbf{X}_i$. Consider the projection matrix

$$\mathbf{M}_{(i)} \equiv \mathbf{I} - \mathbf{X}_i(\mathbf{X}_i^T\mathbf{X}_i)^{-1}\mathbf{X}_i^T. \tag{6.47}$$

This matrix, like $\mathbf{M}$, is idempotent with rank $n - 1$ and has the readily established property that it transforms any vector $\mathbf{x} \in \mathfrak{R}^n$ into the residuals from the regression of that vector onto $\mathbf{X}_i$. It is clear that for $\mathbf{X}_i = \boldsymbol{\iota}$, (6.47) reduces to $\mathbf{M}$ in (1.17), so we have a proper generalization of the notion of mean centering.

If we premultiply (6.45) by $\mathbf{M}_{(i)}$, we get the model "centered" with respect to $\mathbf{X}_i$,

$$\begin{aligned}\mathbf{M}_{(i)}\mathbf{y} &= b_1\mathbf{M}_{(i)}\boldsymbol{\iota} + b_2\mathbf{M}_{(i)}\mathbf{X}_2 + \cdots + b_i\mathbf{M}_{(i)}\mathbf{X}_i + \cdots + b_p\mathbf{M}_{(i)}\mathbf{X}_p + \mathbf{M}_{(i)}\mathbf{e} \\ &= b_1\mathbf{M}_{(i)}\boldsymbol{\iota} + b_2\mathbf{M}_{(i)}\mathbf{X}_2 + \cdots + \mathbf{0} + \cdots + b_p\mathbf{M}_{(i)}\mathbf{X}_p + \mathbf{M}_{(i)}\mathbf{e}.\end{aligned} \tag{6.48}$$

We have now "removed" $\mathbf{X}_i$ from the model—actually no more so than $\boldsymbol{\iota}$ was removed from the mean-centered model above—since $\mathbf{M}_{(i)}\mathbf{X}_i = \mathbf{0}$, and the new variates $\mathbf{M}_{(i)}\mathbf{X}_j$ $(j \neq i)$ are the residuals of the regression of $\mathbf{X}_j$ on $\mathbf{X}_i$. Of course, $\mathbf{M}_{(i)}\boldsymbol{\iota}$ is now no longer zero and not even "constant." One may now make use of the previous argument to show that a regression of $\mathbf{M}_{(i)}\mathbf{y}$ on $\mathbf{M}_{(i)}\boldsymbol{\iota}$, $\mathbf{M}_{(i)}\mathbf{X}_2, \ldots,$ $\mathbf{M}_{(i)}\mathbf{X}_{i-1}, \mathbf{M}_{(i)}\mathbf{X}_{i+1}, \ldots, \mathbf{M}_{(i)}\mathbf{X}_p$ results in the least-squares solution $b_1^*, \ldots, b_{i-1}^*,$ $b_{i+1}^*, \ldots, b_p^*$ and that

$$b_i^* = \alpha_y - b_1^*\alpha_1 - \cdots - b_{i-1}^*\alpha_{i-1} - b_{i+1}^*\alpha_{i+1} - \cdots - b_p^*\alpha_p, \tag{6.49}$$

where α_j is the regression coefficient of the regression of $\mathbf{X}_j$ on $\mathbf{X}_i$ and α_y that of the regression of $\mathbf{y}$ on $\mathbf{X}_i$.[20] One need only substitute $\mathbf{M}_{(i)}$ for $\mathbf{M}$ and recognize

[20]Of course, in the event of mean centering (where $\mathbf{X}_i \equiv \boldsymbol{\iota}$), these α's are just the sample means, $\alpha_y = \bar{y}$ and $\alpha_j = \bar{X}_j$.

that $\mathbf{M}_{(i)}\mathbf{e}^* = \mathbf{e}^*$, since $\mathbf{e}^*$ is necessarily orthogonal to all the $\mathbf{X}_i$. The transformed model (6.48), then, is equivalent to the original model (6.45) in exactly the same way that the mean-centered model (6.46) is.

Thus, any variate $\mathbf{X}_i$ can be used in place of $\boldsymbol{\iota}$ for the purpose of "reducing" the model; there is nothing special in $\boldsymbol{\iota}$. For those who would center before diagnosing for collinearity, this result should be quite disturbing, for now it appears that we could equally well, and with identical motive, "center" the data by regressing them on any $\mathbf{X}_i$ rather than $\boldsymbol{\iota}$, and each resulting centered data set would have its own condition number, which is shown in the next section always to be better than that of the original data. Which, then, would they suggest to be picked? Fortunately, we already know the answer to this question: none. These all tell the wrong story—that is, they indicate only the conditioning of the corresponding transformed regression and ignore that of the preceding transformation $\mathbf{M}_{(i)}$ that produces the "centered" variates,[21] and this transformation is necessarily perfectly ill conditioned in the direction of $\mathbf{X}_i$ since $\mathbf{M}_{(i)}\mathbf{X}_i = \mathbf{0}$.

6.9 APPENDIX: CENTERING AND CONDITIONING

It has been noted several times in passing that mean centering reduces the condition number of a given data set (although not in any way that is useful for reducing collinearity). It is the purpose of this appendix to show this effect formally. We begin with a very brief review of the relevant definitions and notation.

We recall from (3.18) that the condition number $\kappa(\mathbf{A})$ of an $n \times p$ matrix $\mathbf{A}$ is $\mu_{\max}/\mu_{\min}$, the ratio of its maximal to its minimal singular values, and from (3.26) that the scaled condition number $\tilde{\kappa}(\mathbf{A})$ of a matrix $\mathbf{A}$ is defined to be the ratio of the maximal to minimal singular values of $\mathbf{A}$ after its columns have first been scaled to have unit (or, equivalently, equal) length. Thus, if $\mathbf{A} = [\mathbf{A}_1 \cdots \mathbf{A}_p]$, $s_i = (\mathbf{A}_i^{\mathrm{T}}\mathbf{A}_i)^{-1/2}$ and $\mathbf{S} \equiv \operatorname{diag}(s_1, \ldots, s_p)$,

$$\tilde{\kappa}(\mathbf{A}) = \kappa(\mathbf{AS}). \tag{6.50}$$

We have seen in Section 6.2 that this scaled condition number "tends" to be smallest among all possible column scalings $\mathbf{AB}$ of $\mathbf{A}$, where $\mathbf{B} \in \Xi_p$, the set of all $p \times p$ nonsingular, diagonal matrices, in the sense of (6.13), namely that $\tilde{\kappa}(\mathbf{A})$ cannot exceed the smallest such condition number by more than the typically

[21] Indeed, the $\mathbf{M}_{(i)}$ of (6.47) are data-dependent "centerings" more broadly conceived than those given by (6.32) in Section 6.7. Here $\mathbf{x} \to \mathbf{x} - \alpha\mathbf{X}_i$, where $\alpha = (\mathbf{X}_i^{\mathrm{T}}\mathbf{X}_i)^{-1}\mathbf{X}_i^{\mathrm{T}}\mathbf{x}$. Although the definition of centering given in (6.32) could easily be broadened to encompass such transformations, there is little practical need to do so. Model-dependent centerings, the only ones useful for conditioning diagnostics, must be of the form (6.32), since model-dependent centers must apply equally well to all data series measuring a given variate and, hence, cannot distinguish among individual elements.

small factor of $\sqrt{p}$. We shall denote this tendency, defined rigorously by (6.13), by the symbol $\circledS$, giving Result 6.1.

Result 6.1. For any $n \times p$ matrix $\mathbf{A}$, $\tilde{\kappa}(\mathbf{A}) \circledS \kappa(\mathbf{A})$.

It is this result, applied to $\mathbf{A} = \mathbf{X}$ in the least-squares context, that allows us to know that it is the scaled condition number $\tilde{\kappa}(\mathbf{X})$, and not $\kappa(\mathbf{X})$, that provides the most meaningful diagnostic information about the conditioning of the least-squares solution $\mathbf{b} = (\mathbf{X}^T\mathbf{X})^{-1}\mathbf{X}^T\mathbf{y}$.

Now we are prepared to examine the following theorem that allows us to see the effect of centering on the conditioning of a matrix. I am grateful to Dr. David Gay of Bell Laboratories for the basis of this proof.

Theorem 6.1. Let the $n \times p$ matrix $\mathbf{A}$ of full rank p be partitioned as $\mathbf{A} = [\mathbf{A}_1 \ \mathbf{A}_2]$, and obtain from it $\mathbf{B} = [\mathbf{A}_1 \ \mathbf{B}_2]$, where $\mathbf{B}_2 \equiv \mathbf{M}_1\mathbf{A}_2$ with $\mathbf{M}_1 \equiv \mathbf{I} - \mathbf{A}_1(\mathbf{A}_1^T\mathbf{A}_1)^{-1}\mathbf{A}_1^T$ (i.e., $\mathbf{B}_2$ is the orthogonal projector of $\mathbf{A}_2$ on the column space of $\mathbf{A}_1$—thus, if $\mathbf{A}_1 \equiv \iota$, $\mathbf{B}_2$ is the matrix of mean-centered variates). Then

$$\kappa(\mathbf{A}) \geqslant \kappa(\mathbf{B}) \geqslant \kappa(\mathbf{B}_2). \tag{6.51}$$

Proof. It suffices to consider the case where $\mathbf{A}$ is square and upper triangular, since we can factor $\mathbf{A} = \mathbf{QR}$, where $\mathbf{Q}^T\mathbf{Q} = \mathbf{I}_p$ and $\mathbf{R}$ is square and upper triangular, and the singular values of $\mathbf{R}$ are the same as those of $\mathbf{A}$.[22] Thus, $\mathbf{A}$ and $\mathbf{A}^{-1}$ have the block forms

$$\mathbf{A} = \begin{bmatrix} \mathbf{A}_{11} & \mathbf{A}_{21} \\ \mathbf{0} & \mathbf{A}_{22} \end{bmatrix} \quad \text{and} \quad \mathbf{A}^{-1} = \begin{bmatrix} \mathbf{A}_{11}^{-1} & \mathbf{A}_{21}^{-1} \\ \mathbf{0} & \mathbf{A}_{22}^{-1} \end{bmatrix}, \tag{6.52}$$

where $\mathbf{A}_{21}^{-1} \equiv -\mathbf{A}_{11}^{-1}\mathbf{A}_{21}\mathbf{A}_{22}^{-1}$. And $\mathbf{B}$ and $\mathbf{B}^{-1}$ have the forms

$$\mathbf{B} = \begin{bmatrix} \mathbf{A}_{11} & \mathbf{0} \\ \mathbf{0} & \mathbf{A}_{22} \end{bmatrix} \quad \text{and} \quad \mathbf{B}^{-1} = \begin{bmatrix} \mathbf{A}_{11}^{-1} & \mathbf{0} \\ \mathbf{0} & \mathbf{A}_{22}^{-1} \end{bmatrix}. \tag{6.53}$$

By definition, the spectral norms (3.13) of $\mathbf{B}$ and $\mathbf{B}^{-1}$ are $\|\mathbf{B}\| = \sup_{\|\mathbf{x}\|=1} \|\mathbf{Bx}\|$ and $\|\mathbf{B}^{-1}\| = \sup_{\|\mathbf{y}\|=1} \|\mathbf{B}^{-1}\mathbf{y}\|$, respectively, and the condition

[22]The factorization of $\mathbf{A}$ into $\mathbf{Q}$ and $\mathbf{R}$ is called, quite naturally, the QR decomposition. On this, see Golub (1969), Stewart (1973), or Belsley et al. (1980). Now, let the SVD of $\mathbf{R} = \mathbf{UDV}^T$, so the diagonal elements of $\mathbf{D}$ are the singular values of $\mathbf{R}$. Then $\mathbf{A} = \mathbf{QR} = \mathbf{QUDV}^T \equiv \tilde{\mathbf{U}}\mathbf{DV}^T$, where $\tilde{\mathbf{U}} \equiv \mathbf{QU}$. Since $\tilde{\mathbf{U}}$ is column orthogonal, this last expression is seen to be the SVD of $\mathbf{A}$, and hence $\mathbf{A}$ has the same singular values as $\mathbf{R}$.

number of $\mathbf{B}$ is $\kappa(\mathbf{B}) = \|\mathbf{B}\|\|\mathbf{B}^{-1}\|$. Let $\mathbf{x}^*$ and $\mathbf{y}^*$ be solutions to these two suprema, respectively. If we partition

$$\mathbf{x}^* = \begin{pmatrix} \mathbf{x}_1^* \\ \mathbf{x}_2^* \end{pmatrix} \quad \text{and} \quad \mathbf{y}^* = \begin{pmatrix} \mathbf{y}_1^* \\ \mathbf{y}_2^* \end{pmatrix}$$

commensurately with $\mathbf{B}$, we show Lemma 6.1.

Lemma 6.1. It is always possible to pick $\mathbf{x}^*$ so that either $\mathbf{x}_1^* = \mathbf{0}$ ($\|\mathbf{x}_2^*\| = 1$) or $\mathbf{x}_2^* = \mathbf{0}$ ($\|\mathbf{x}_1^*\| = 1$), and similarly for $\mathbf{y}^*$.

Proof. For any $\mathbf{x} \equiv [\mathbf{x}_1^T, \mathbf{x}_2^T]^T$ with $\|\mathbf{x}\| = 1$,

$$\|\mathbf{B}\mathbf{x}\|^2 = \|\mathbf{A}_{11}\mathbf{x}_1\|^2 + \|\mathbf{A}_{22}\mathbf{x}_2\|^2 \leqslant \max[\|\mathbf{A}_{11}\|^2, \|\mathbf{A}_{22}\|^2]. \tag{6.54}$$

The first equal sign in (6.54) is due to the block diagonality of $\mathbf{B}$, and the inequality is due to the fact that $\|\mathbf{x}\|^2 = \|\mathbf{x}_1\|^2 + \|\mathbf{x}_2\|^2 = 1$.

Now suppose $\|\mathbf{A}_{11}\| \geqslant \|\mathbf{A}_{22}\|$. By definition, there exists $\|\hat{\mathbf{x}}_1\| = 1$ such that $\|\mathbf{A}_{11}\hat{\mathbf{x}}_1\| = \|\mathbf{A}_{11}\|$. Letting $\hat{\mathbf{x}} \equiv [\mathbf{x}_1^T, \mathbf{0}^T]^T$, we see that $\|\mathbf{B}\hat{\mathbf{x}}\| = \|\mathbf{A}_{11}\hat{\mathbf{x}}_1\| = \|\mathbf{A}_{11}\|$. Thus, $\hat{\mathbf{x}}$ allows $\|\mathbf{B}\mathbf{x}\|$ to reach its upper bound $\|\mathbf{A}_{11}\|$ among $\|\mathbf{x}\| = 1$, and hence $\mathbf{x}^* = \hat{\mathbf{x}}$ satisfies Lemma 6.1 with $\mathbf{x}_2^* = \mathbf{0}$. A similar argument derives an $\mathbf{x}^*$ with $\mathbf{x}_1^* = \mathbf{0}$ if $\|\mathbf{A}_{11}\| \leqslant \|\mathbf{A}_{22}\|$. □

Recalling that $\|\mathbf{A}\mathbf{x}\| = (\mathbf{x}^T\mathbf{A}^T\mathbf{A}\mathbf{x})^{1/2}$, we see that, for either $\mathbf{x}_1^* = \mathbf{0}$ or $\mathbf{x}_2^* = \mathbf{0}$,

$$\begin{aligned} \|\mathbf{A}\| &\geqslant \|\mathbf{A}\mathbf{x}^*\| \\ &= [\mathbf{x}_1^{*T}\mathbf{A}_{11}^T\mathbf{A}_{11}\mathbf{x}_1^* + 2\mathbf{x}_2^{*T}\mathbf{A}_{21}^T\mathbf{A}_{11}\mathbf{x}_1^* + \mathbf{x}_2^{*T}(\mathbf{A}_{21}^T\mathbf{A}_{21} + \mathbf{A}_{22}^T\mathbf{A}_{22})\mathbf{x}_2^*]^{1/2} \\ &\geqslant (\mathbf{x}_1^{*T}\mathbf{A}_{11}^T\mathbf{A}_{11}\mathbf{x}_1^* + \mathbf{x}_2^{*T}\mathbf{A}_{22}^T\mathbf{A}_{22}\mathbf{x}_2^*)^{1/2} = \|\mathbf{B}\mathbf{x}^*\| = \|\mathbf{B}\|. \end{aligned} \tag{6.55}$$

A similar argument in $\mathbf{y}$ shows that $\|\mathbf{A}^{-1}\| \geqslant \|\mathbf{B}^{-1}\|$. Thus

$$\kappa(\mathbf{A}) = \|\mathbf{A}\|\|\mathbf{A}^{-1}\| \geqslant \|\mathbf{B}\|\|\mathbf{B}^{-1}\| = \kappa(\mathbf{B}), \tag{6.56}$$

and the left-hand inequality in (6.51) is shown.

Now, for any matrix $\mathbf{D}$, denote by $\mu_{\min}(\mathbf{D})$ the $\inf_{\|\mathbf{z}\|=1}\|\mathbf{D}\mathbf{z}\|$ and by $\mu_{\max}(\mathbf{D})$ the $\sup_{\|\mathbf{z}\|=1}\|\mathbf{D}\mathbf{z}\|$. These, of course, are just the minimal and maximal singular values of $\mathbf{D}$. Then the second inequality in (6.51) can be seen from Lemma 6.2.

Lemma 6.2. Let the $n \times p$ matrix $\mathbf{C}$ be partitioned as $\mathbf{C} = [\mathbf{C}_1 \ \ \mathbf{C}_2]$. Then $\kappa(\mathbf{C}) \geqslant \kappa(\mathbf{C}_2)$.

Proof. Pick $\|\mathbf{z}_2^*\| = 1$ such that $\|\mathbf{C}_2\mathbf{z}_2^*\| = \mu_{\min}(\mathbf{C}_2)$ and $\|\mathbf{w}_2^*\| = 1$ such that $\|\mathbf{C}_2\mathbf{w}_2^*\| = \mu_{\max}(\mathbf{C}_2)$, and define

$$\mathbf{z}^* \equiv \begin{pmatrix} \mathbf{0} \\ \mathbf{z}_2^* \end{pmatrix} \quad \text{and} \quad \mathbf{w}^* \equiv \begin{pmatrix} \mathbf{0} \\ \mathbf{w}_2^* \end{pmatrix}.$$

Now, by construction and definition,

$$\mu_{\min}(\mathbf{C}_2) \equiv \|\mathbf{C}_2\mathbf{z}_2^*\| = \|\mathbf{C}\mathbf{z}^*\| \geqslant \min_{\|\mathbf{z}\|=1} \|\mathbf{C}\mathbf{z}\| \equiv \mu_{\min}(\mathbf{C}) \tag{6.57}$$

and

$$\mu_{\max}(\mathbf{C}_2) \equiv \|\mathbf{C}_2\mathbf{w}_2^*\| = \|\mathbf{C}\mathbf{w}^*\| \leqslant \max_{\|\mathbf{w}\|=1} \|\mathbf{C}\mathbf{w}\| \equiv \mu_{\max}(\mathbf{C}). \tag{6.58}$$

Hence,

$$\kappa(\mathbf{C}) \equiv \frac{\mu_{\max}(\mathbf{C})}{\mu_{\min}(\mathbf{C})} \geqslant \frac{\mu_{\max}(\mathbf{C}_2)}{\mu_{\min}(\mathbf{C}_2)} \equiv \kappa(\mathbf{C}_2). \quad \square \tag{6.59}$$

Applying this result to **B** in (6.51) completes the proof to Theorem 6.1. □

Now, let **S** be the diagonal matrix that scales **A** to have unit column lengths as in (6.50), with diagonal blocks $\mathbf{S}_1$ and $\mathbf{S}_2$ partitioned commensurately with $\mathbf{A}_1$ and $\mathbf{A}_2$ of Theorem 6.1 so $\mathbf{AS} = [\mathbf{A}_1\mathbf{S}_1 \;\; \mathbf{A}_2\mathbf{S}_2]$. Then we have

$$\tilde{\kappa}(\mathbf{A}) = \kappa(\mathbf{AS}) = \kappa([\mathbf{A}_1\mathbf{S}_1 \;\; \mathbf{A}_2\mathbf{S}_2]) \geqslant \kappa(\mathbf{B}_2\mathbf{S}_2) \circledcirc\!\!\!\geqslant \tilde{\kappa}(\mathbf{B}_2). \tag{6.60}$$

The first equality holds by the definition of the scaled condition number, and the first inequality holds by Theorem 6.1 along with the recognition that $\mathbf{B}_2\mathbf{S}_2 \equiv \mathbf{M}_1\mathbf{A}_2\mathbf{S}_2$. The final inequality merely reflects Result 6.1, that the scaled condition number of $\mathbf{B}_2$ tends to be less than that for any other scaling. Thus, we have Corollary 6.1.

Corollary 6.1. For the matrix **A** of Theorem 6.1, $\tilde{\kappa}(\mathbf{A}) \circledcirc\!\!\!\geqslant \tilde{\kappa}(\mathbf{B}_2)$.

Since $\mathbf{B}_2$ is the matrix of the columns of $\mathbf{A}_2$ after they have been "centered" relative to those in $\mathbf{A}_1$, this translates as "the centered-data matrix (broadly interpreted) will strongly tend to have a lower scaled condition number than the basic-data matrix." In the special case that **A** is a data matrix **X** with a first column of ones and $\mathbf{A}_1$ in Theorem 6.1 is $\mathbf{X}_1 \equiv \iota$, the matrix $\mathbf{B}_2$ is simply mean-centered $\tilde{\mathbf{X}}$ and we have Corollary 6.2.

Corollary 6.2. For a data matrix **X** with constant column ι, $\tilde{\kappa}(\mathbf{X}) \circledcirc\!\!\!\geqslant \tilde{\kappa}(\tilde{\mathbf{X}})$.

That this result should *not* be interpreted to mean that data should be mean centered before being diagnosed for collinearity has been amply demonstrated in various preceding sections.

CHAPTER 7

Harmful Collinearity and Short Data

The collinearity diagnostics developed in the preceding chapters, consisting of scaled condition indexes, variance–decomposition proportions, and auxiliary regressions, allow us to determine the presence of collinear relations in a data matrix **X** and to determine the variates involved in each. Furthermore, used as a regression diagnostic, this information allows us to determine which least-squares estimates are being degraded by the presence of the collinear relations. But as noted in Section 3.4, degrading collinearity need not be harmful, and while the collinearity diagnostics can determine degrading collinearity, they cannot alone determine whether the collinearity is degrading enough also to be harmful. In this chapter, we remove this shortcoming by combining the collinearity diagnostics with a measure of *signal-to-noise* that allows us to define and determine the presence of harmful collinearity. In so doing, we also discover that collinearity has a sister problem: *short data*. Together, collinearity and short data constitute the sources of data weaknesses in the context of estimating linear models by least squares.

We begin with a discussion that leads to the introduction of a statistical test for the degree of signal-to-noise present in a given data set relative to the estimation of a given subset of parameters by least squares. A notion of *adequacy* is then defined, allowing this test to become one for adequate/inadequate signal-to-noise. This test is then combined with the previously derived collinearity diagnostics to produce a test for harmful collinearity and short data. An example is given using a commercial paper rate equation from the Michigan Quarterly Econometric Model. Tables of critical values for most test situations are given in Section 7.8.[1]

[1]The material in this chapter is based on Belsley (1982).

7.1 BACKGROUND

It proves useful to risk a mild degree of repetition by first motivating the main concepts informally relative to the estimation of a single parameter β_i. This helps to set the scene for the tests that are developed formally in subsequent sections and are quite general, dealing with arbitrary subsets of parameters.

Weak or Uninformative Data

We have seen repeatedly in the foregoing chapters how data that are beset with problems like collinearity can reduce the quality of the least-squares estimates $\mathbf{b} = (\mathbf{X}^T\mathbf{X})^{-1}\mathbf{X}^T\mathbf{y}$ of the linear model $\mathbf{y} = \mathbf{X}\boldsymbol{\beta} + \boldsymbol{\varepsilon}$ (with $E\boldsymbol{\varepsilon} = \mathbf{0}$, $E\boldsymbol{\varepsilon}\boldsymbol{\varepsilon}^T = \sigma^2\mathbf{I}$) by inflating some or all of the parameter variances $\text{var}(b_i) \equiv \sigma_{b_i}^2$, $i = 1, \ldots, p$, which are, of course, the diagonal elements of the data-dependent matrix $\sigma^2(\mathbf{X}^T\mathbf{X})^{-1}$.[2] Because of the link that clearly exists between these data problems and inflated variances, it is reasonable to define the data matrix $\mathbf{X}$ as possessing a *weak-data problem* or an *uninformative-data problem* relative to the least-squares estimation of a parameter β_i if it can be determined that the variance $\sigma_{b_i}^2$ of the b_i based on those $\mathbf{X}$ is in some sense "too large." We adopt just such a definition here, and it becomes the purpose of this chapter to develop both an appropriate measure of "too large" and a practical means for assessing when a variance is so. Since data weaknesses apply to a specific data matrix $\mathbf{X}$, we view the least-squares estimator $\mathbf{b}$ and its variances $\sigma_{b_i}^2$ conditionally (in the statistical sense) on the given $\mathbf{X}$.

The Need for a Measure Like Signal-to-Noise

Recall in Section 3.4 that we showed that the ill effects of collinearity could be counteracted by a sufficiently small error variance σ^2, so that not all collinearity need be harmful. This is readily seen with reference to Exhibit 2.3*c*, where, if σ^2, and therefore the radius of the circle, becomes sufficiently small, even in this extreme case, the potential spread of least-squares estimates can become sufficiently narrow to be acceptable. To assess harmful collinearity, then, it is clear that the diagnostic procedure must also be able to incorporate information on the size of σ^2 and therefore on the sizes of the parameter variances $\sigma_{b_i}^2$.

Standard diagnostic and statistical tests cannot assess the size of the parameter variances $\sigma_{b_i}^2$ and so cannot be used to determine the presence of harmful collinearity. The collinearity diagnostics that we have developed in the preceding chapters are based on the $\mathbf{X}$ data alone. Thus, while these diagnostics can signal the presence of collinear relations among the columns of $\mathbf{X}$ and, hence, can indicate the potential for data problems, they are devoid of information on σ^2 and, hence, cannot determine when the parameter variances

[2]Here we are dealing with the true variances, $\sigma_{b_i}^2$, which are to be distinguished from the estimated variances $s_{b_i}^2$ based on the diagonal elements of $s^2(\mathbf{X}^T\mathbf{X})^{-1}$, where $s^2 = \mathbf{e}^T\mathbf{e}/(n-p)$ and $\mathbf{e}$ is the least-squares residual vector.

$\sigma_{b_i}^2$ are large or small. Likewise, traditional statistical tests of location, t's and F's, employ null hypotheses under which the resulting test statistics are centrally distributed (having a noncentrality parameter of zero) *regardless of the parameter variances* $\sigma_{b_i}^2$. Thus, they cannot provide further tests on the sizes of these parameter variances. A test for uninformative data, then, must add to these tools the ability to test for the size of the parameter variances $\sigma_{b_i}^2$.

Of course, whether any particular $\sigma_{b_i}^2$ is large or small is necessarily relative, and hence tests on the absolute size of $\sigma_{b_i}^2$ are not suitable to this need. Rather, we must measure $\sigma_{b_i}^2$ relative to some suitable standard. Since $\sigma_{b_i}^2$ is effectively a measure of the noise inherent in the least-squares estimate b_i of β_i, an appropriate standard would be the nonnoise, or signal, inherent in this least-squares estimate, namely, β_i itself. Clearly, if $\sigma_{b_i}^2$ were large relative to β_i, the situation would be problemful, whereas if $\sigma_{b_i}^2$ were small relative to β_i, our concerns would be less. This suggests the value for our purposes of using a test based on the magnitude of the *signal-to-noise* parameter of the least-squares estimator b_i of the ith regression parameter β_i, namely,

$$\tau \equiv \frac{\beta_i}{\sigma_{b_i}}. \tag{7.1}$$

The inverse of τ is also called the *coefficient of variation.*[3]

Given from prior considerations that $\beta_i \neq 0$, a test that τ is high (low) is also a test that σ_{b_i} is relatively low (high) and, hence, is indicative of the absence (presence) of weak or uninformative data. A choice of units, or of the parameterization of the model, is not a concern here, since the signal-to-noise ratio is seen to be invariant to linear transformations of the data.

7.2 SOME PRELIMINARY REMARKS

It Is Maintained That the Relevant Parameters Are Not Zero

It is noted in the preceding discussion that, when conducting a test for signal-to-noise, it is necessary to maintain a priori that $\beta_i \neq 0$. This condition is important, for if the null hypothesis H_0: $\beta_i = 0$ is allowed to be entertained, one cannot distinguish between having a low signal-to-noise τ because of a relatively high σ_{b_i} or because of the truth of H_0, and hence one cannot assess the presence of weak data. It often happens in practice that it is known a priori that $\beta_i \neq 0$. This occurs, for example, when it can be maintained that the variable $\mathbf{X}_i$, the ith column of $\mathbf{X}$, definitely belongs in the regression model, and its presence is therefore not subject to test, no matter how poor its parameter estimates may be. An "income" term, for example, surely belongs in the consumption function, as does a "surface area" term in a model of heat dissipation.

[3]See, e.g., Wilks (1962).

The focus of this test for signal-to-noise, then, is distinctly different from that of conventional tests of hypotheses. We do not test here whether a variable belongs in the model ($H_0: \beta_i = 0$) or whether a parameter is a particular value ($H_0: \beta_i = \beta_i^*$). Rather, we test whether the variance of a given estimate, which depends on the data, is sufficiently small relative to the expected value of that estimate to rule out problemful data weaknesses such as collinearity or short data as being underlying problems in need of resolution; that is, we test $H: |\tau| > \tau^*$ for some suitably chosen threshold for signal-to-noise, τ^*.

The Relation of τ to t

The signal-to-noise parameter τ bears a superficial resemblance to the usual t-statistic,

$$t = \frac{b_i}{s_{b_i}}, \tag{7.2}$$

but the two concepts are obviously fundamentally different—τ is a parameter, while t is a random variable. And indeed, the random variable t is distributed as a noncentral Student's t with noncentrality parameter τ; that is, the signal-to-noise parameter τ is identifiable with the noncentrality parameter of the distribution of t. In subsequent sections, we shall deal more generally with F (t^2-like) statistics that are noncentrally distributed with noncentrality parameter τ^2.

These notions help to relate the test for signal-to-noise developed here to conventional tests. Both tests use the t-statistic (7.2) as a natural means for estimating τ. Under conventional tests ($H_0: \beta_i = \beta_i^*$), however, $\tau \equiv 0$ for all σ_{b_i}; that is, t is centrally distributed regardless of σ_{b_i}, since, in general, $t = (b_i - \beta_i^*)/s_{b_i}$ is distributed with a noncentrality parameter $\tau \equiv (\beta_i - \beta_i^*)/\sigma_{b_i}$, which under $H_0: \beta_i = \beta_i^*$ implies $\tau \equiv 0$. Such tests, therefore, cannot provide further information on the relative size of σ_{b_i}. By contrast, under the test for signal-to-noise proposed here, it is assumed that $\beta_i \neq 0$, so that the t-statistic becomes noncentrally distributed with $\tau \neq 0$. Thus, a test that $|\tau|$ exceeds some chosen threshold $\tau^* > 0$ does indeed provide information on the magnitude of σ_{b_i} relative to β_i and, hence, on the presence of uninformative data.

Assessing the Size of t in a Test for Signal-to-Noise

An immediate consequence of the preceding is that t values thought to be large for conventional tests of hypothesis need not be large for tests on the presence of nonzero signal-to-noise and, hence, for the absence of data problems. A t of 4 in a test of significance, for example, typically indicates its corresponding coefficient to be quite significantly different from zero (even if normality cannot be assumed). We shall see, however, that this same t of 4 may not suffice to accept a hypothesis that the signal-to-noise exceeds some reasonably chosen threshold level and, hence, to accept the hypothesis that σ_{b_i} is sufficiently small relative to β_i to rule out the presence of data weaknesses. Under these conditions, the investigator is apprised of the presence of a data weakness and may feel that, despite the significance of the estimate, an analysis based on more or better-conditioned data or one incorporating prior information of some sort is warranted before the regression results can be used for understanding a key structural parameter or for providing the basis for some important policy recommendation. We shall see an example of the value of introducing prior information in such a context in Section 10.3.

Collinearity versus Short Data

The test we develop here for adequate signal-to-noise is useful as a diagnostic for data weaknesses, but it cannot, by itself, distinguish between the two data weaknesses: harmful collinearity and short data. A rigorous definition of short data awaits Section 7.6, but loosely it occurs when a variate $\mathbf{X}_i$ has little length $\|\mathbf{X}_i\|$ so that the term $\beta_i\mathbf{X}_i$ adds little signal to the overall determination of $\mathbf{y}$. This inability to distinguish between collinearity and short data is overcome by combining the test for adequate signal-to-noise with the collinearity diagnostics developed earlier. Harmful collinearity is defined as inadequate signal-to-noise occurring simultaneously with collinearity, while short data is defined as inadequate signal-to-noise without concurrent collinearity. Since linear transformations of the data do not affect signal-to-noise but can reduce collinearity, we see that such transformations cannot remove data weakness but can only alter its form. It has been suggested, for example, that the strong collinearity that exists between income $\mathbf{Y}$ and consumption $\mathbf{C}$ in the consumption function can be reduced by transforming the equation to one using $\mathbf{Y}$ along with the more nearly orthogonal savings $\mathbf{S} \equiv \mathbf{Y} - \mathbf{C}$. In fact, all one has done here is to transform the collinearity between $\mathbf{Y}$ and $\mathbf{C}$ into the short data of $\mathbf{S}$.

Harmful collinearity and short data, then, are really just two sides of the same coin. But in correcting for their effects, we shall see that there is typically an advantage in knowing which type of problem is occurring relative to the particular parameterization being estimated, for this helps to focus corrective activity. Thus, the two diagnostic procedures, the test for adequate signal-to-noise and the previously developed collinearity diagnostics, strongly complement one another in analyzing weak-data problems, each having independent value. The test for adequate signal-to-noise can detect the presence of weak data but cannot determine its cause: collinearity or short data. The collinearity diagnostics, by contrast, cannot detect the presence of weak data but can determine whether an already detected data weakness is due to collinearity or short data. Furthermore, they can help determine the structure of a collinear relation and, hence, help to direct where new data or prior information can most advantageously be employed.

Filling a Need

Finally, it can now be seen that the test procedure provided here makes substantial headway in filling the legitimate needs implicit in the following remark by Smith and Campbell (1980):

> The essential problem with VIF and similar measures [of collinearity] is that they ignore the parameters while trying to assess the information given by the data. Clearly, an evaluation of the strength of the data depends on the scale and nature of the parameters. One cannot label a variance or confidence interval (or, even worse, a part of the variance) as large or small without knowing what the parameter is and how much precision is required in the estimate of that parameter. In particular, a seemingly large variance may be quite satisfactory if the parameter is very large.

7.3 SIGNAL-TO-NOISE AND A TEST FOR ITS MAGNITUDE

Consider then, once again, the linear regression model

$$\mathbf{y} = \mathbf{X}\boldsymbol{\beta} + \boldsymbol{\varepsilon}, \tag{7.3}$$

where, as usual, $\mathbf{y}$ is an n-vector, $\mathbf{X}$ is an $n \times p$ data matrix, and $\boldsymbol{\beta}$ is a p-vector of unknown parameters. Here, we also assume that $\boldsymbol{\varepsilon}$ is normally distributed, $\boldsymbol{\varepsilon} \sim N_n(\mathbf{0}, \sigma^2\mathbf{I})$. Since our concern is with tests on an arbitrary subset β_2 of p_2 of the elements of $\boldsymbol{\beta}$, we partition (7.3) as

$$\mathbf{y} = \mathbf{X}_1\boldsymbol{\beta}_1 + \mathbf{X}_2\boldsymbol{\beta}_2 + \boldsymbol{\varepsilon}, \tag{7.4}$$

where $\mathbf{X}_1$ is $n \times p_1$, $\mathbf{X}_2$ is $n \times p_2$, $\boldsymbol{\beta}_1$ is a p_1-vector, $\boldsymbol{\beta}_2$ is a p_2-vector, and $p_1 + p_2 = p$. We similarly partition the least-squares estimator $\mathbf{b} = (\mathbf{X}^T\mathbf{X})^{-1}\mathbf{X}^T\mathbf{y}$ of $\boldsymbol{\beta}$ to give $\mathbf{b} = [\mathbf{b}_1^T, \mathbf{b}_2^T]^T$.

Now, as is well known,[4]

$$\mathbf{b}_2 = (\mathbf{X}_2^T\mathbf{M}_1\mathbf{X}_2)^{-1}\mathbf{X}_2^T\mathbf{M}_1\mathbf{y}, \tag{7.5}$$

where $\mathbf{M}_1 \equiv \mathbf{I} - \mathbf{X}_1(\mathbf{X}_1^T\mathbf{X}_1)^{-1}\mathbf{X}_1^T$, and $(\mathbf{X}_1^T\mathbf{X}_1)^{-1}$ is assumed to exist. And hence, under the normality assumption on $\boldsymbol{\varepsilon}$, the marginal distribution of the p_2-vector $\mathbf{b}_2$ is

$$\mathbf{b}_2 \sim N_{p_2}(\boldsymbol{\beta}_2, \mathbf{V}(\mathbf{b}_2)), \tag{7.6}$$

where $\mathbf{V}(\mathbf{b}_2)$ is the variance–covariance matrix of $\mathbf{b}_2$ and, conditional on $\mathbf{X}$, is

$$\mathbf{V}(\mathbf{b}_2) = \sigma^2(\mathbf{X}_2^T\mathbf{M}_1\mathbf{X}_2)^{-1}. \tag{7.7}$$

[4] Theil (1971, Chapter 3).

Defining Signal-to-Noise

Let $\boldsymbol{\beta}_2^\circ$ be any p_2-vector. We define the *signal-to-noise* of the least-squares estimator $\mathbf{b}_2$ of $\boldsymbol{\beta}_2$ relative to $\boldsymbol{\beta}_2^\circ$, denoted τ^2, as

$$\tau^2 \equiv (\boldsymbol{\beta}_2 - \boldsymbol{\beta}_2^\circ)^{\mathrm{T}}\mathbf{V}^{-1}(\mathbf{b}_2)(\boldsymbol{\beta}_2 - \boldsymbol{\beta}_2^\circ). \tag{7.8}$$

It is readily seen that this quadratic form appropriately generalizes the concept of signal-to-noise given in (7.1), clearly reducing to that expression in the case that $p_2 = 1$, $\boldsymbol{\beta}_2^\circ = \mathbf{0}$, and $\mathbf{b}_2 = b_i$.

The generality that arises from introducing $\boldsymbol{\beta}_2^\circ$ will prove useful later. For the moment, it is simply any arbitrary point in a p_2-dimensional parameter space, but eventually it will be identified with a hypothesized value for $\boldsymbol{\beta}_2$. The special case where $\boldsymbol{\beta}_2^\circ = \mathbf{0}$ is an important, but by no means only, instance. Also, unless context requires it, we shall find it convenient to suppress the phrase "relative to $\boldsymbol{\beta}_2^\circ$" when referring to signal-to-noise, but its presence should always be kept in mind. Furthermore, while the term signal-to-noise properly refers to the magnitude τ, we typically deal with τ^2 and find it convenient to call this magnitude by the same name. No confusion arises on this account.

Testing for Signal-to-Noise

We wish to be able to test for the level of signal-to-noise. To this end, we will say that the signal-to-noise τ^2 significantly exceeds some hypothesized value τ_*^2 when we are able to reject the null hypothesis $A_0: \tau^2 = \tau_*^2$ in favor of the alternative hypothesis $A_1: \tau^2 > \tau_*^2$ at some chosen test size α. The symbols A_0 and A_1 are used here to avoid confusion with the usual tests of hypothesis on coefficient values; H_0 and H_1 will be reserved for this latter context.

In light of (7.6), we have[5]

$$(\mathbf{b}_2 - \boldsymbol{\beta}_2^\circ)^{\mathrm{T}}\mathbf{V}^{-1}(\mathbf{b}_2)(\mathbf{b}_2 - \boldsymbol{\beta}_2^\circ) \sim \chi_{p_2}^2(\lambda), \tag{7.9}$$

where $\chi_{p_2}^2(\lambda)$ denotes the noncentral chi-squared distribution with p_2 degrees of freedom and noncentrality parameter

$$\lambda \equiv (\boldsymbol{\beta}_2 - \boldsymbol{\beta}_2^\circ)^{\mathrm{T}}\mathbf{V}^{-1}(\mathbf{b}_2)(\boldsymbol{\beta}_2 - \boldsymbol{\beta}_2^\circ). \tag{7.10}$$

And letting $\mathbf{e} \equiv \mathbf{y} - \mathbf{Xb}$ denote the regression residuals, we know that their sum of squares $s^2 \equiv \mathbf{e}^{\mathrm{T}}\mathbf{e}/(n-p)$ obeys

$$(n-p)s^2/\sigma^2 \sim \chi_{n-p}^2, \tag{7.11}$$

a central chi-squared distribution with $n - p$ degrees of freedom, independent of

[5]Rao (1973, pp. 181–192) or Anderson (1958, p. 113).

(7.9).[6] Thus, the ratio of (7.9) to (7.11), each divided by its respective degrees of freedom, produces the test statistic [recalling (7.7)]

$$\phi^2 \equiv \frac{(\mathbf{b}_2 - \boldsymbol{\beta}_2^\circ)^T(\mathbf{X}_2^T\mathbf{M}_1\mathbf{X}_2)(\mathbf{b}_2 - \boldsymbol{\beta}_2^\circ)}{p_2 s^2} \sim F_{n-p}^{p_2}(\lambda), \tag{7.12}$$

which is distributed as a noncentral Fisher F with p_2 and $n - p$ degrees of freedom and noncentrality parameter λ as in (7.10).[7]

Now the desired test for signal-to-noise comes from comparing (7.10) with (7.8), for we immediately recognize that

$$\lambda \equiv \tau^2; \tag{7.13}$$

that is, the noncentrality parameter λ relevant to the distribution of ϕ^2 is identically the signal-to-noise parameter τ^2. Hence, under A_0, we have $\lambda = \tau_*^2$ and

$$\phi^2 \sim F_{n-p}^{p_2}(\tau_*^2). \tag{7.14}$$

We therefore have the following test for A_0: $\tau^2 = \tau_*^2$ against A_1: $\tau^2 > \tau_*^2$: Adopt a test size α, such as 0.05 or 0.01, and calculate

$$F_\alpha \equiv {}_{1-\alpha}F_{n-p}^{p_2}(\tau_*^2), \tag{7.15}$$

the $(1 - \alpha)$-critical value for the noncentral F with p_2 and $n - p$ degrees of freedom and noncentrality parameter τ_*^2. If $\phi^2 \leqslant F_\alpha$, accept A_0; if $\phi^2 > F_\alpha$, reject A_0 and accept A_1. In accepting A_1 we can also say that τ^2 is significantly greater than τ_*^2 at significance level α.

In passing, it is of interest to note that this test for signal-to-noise is related to that of Toro-Vizcarrendo and Wallace (1968) for determining when linear restrictions on $\boldsymbol{\beta}$ will reduce the mean-squared error of the estimates. The τ_*^2 for their test is $\frac{1}{2}$. As it relates to collinearity, their test can be used to determine when the removal of a set of collinear variates reduces variances by more than it increases bias. Our orientation here, however, is quite different, since we wish (a) to define the presence of data weaknesses, of which collinearity is but one; (b) to determine when estimates of parameters of variates known to be involved in the regression model (and hence not to be removed) are beset with such problems; and (c) to be able to set τ_*^2 meaningfully for different levels of stringency. A procedure for accomplishing this latter objective is introduced shortly.

Two aspects of the test for signal-to-noise deserve comment. First, the test statistic ϕ^2 in (7.12) used to test A_0 is readily seen to be the same as the F-statistic conventionally used to test the null hypothesis H_0: $\boldsymbol{\beta}_2 = \boldsymbol{\beta}_2^\circ$ against the

[6] Theil (1971, Chapter 3).

[7] Rao (1973, p. 216) or Anderson (1958, p. 114).

alternative $H_1: \boldsymbol{\beta}_2 \neq \boldsymbol{\beta}_2^\circ$.[8] Indeed, in the case where $p_2 = 1$ with $\boldsymbol{\beta}_2^\circ = \mathbf{0}$, ϕ^2 is recognized to be the square of the standard t-statistic (7.2). However, it is important to realize that, although the test statistic is the same (a fact that facilitates its calculation), the tests in general are not. Testing for A_0 is equivalent to testing for H_0 only for $\tau_*^2 = 0$. This is seen by noting that, in addition to the fact that ϕ^2 is the usual F-statistic, (a) $\tau^2 = 0$ if and only if $\boldsymbol{\beta}_2 = \boldsymbol{\beta}_2^\circ$ and (b) under H_0 (A_0 with $\tau_*^2 = 0$), the F-distribution in (7.14) becomes a central F.

Second, if $\boldsymbol{\beta}_2 \neq \boldsymbol{\beta}_2^\circ$, a test that τ^2 is high is also interpretable as a test that the variance–covariance structure $\mathbf{V}(\mathbf{b}_2)$ of $\mathbf{b}_2$ is low relative to $\boldsymbol{\beta}_2 - \boldsymbol{\beta}_2^\circ$. Similarly, an inability to reject A_0 for at least a modest value of τ_*^2 can be considered as evidence for large or "inflated" variances of $\mathbf{b}_2$, an interpretation we exploit in Section 7.6 in diagnosing the presence of harmful data problems.

7.4 DEFINING AND TESTING FOR ADEQUATE/INADEQUATE SIGNAL-TO-NOISE

The test for signal-to-noise given in the preceding section has one practical drawback: it requires knowledge of $\boldsymbol{\beta}_2$ and $\mathbf{V}(\mathbf{b}_2)$ to stipulate directly a value for τ_*^2. In this section we propose a practical and intuitively appealing definition for an *adequate level of signal-to-noise* that does not require knowledge of $\boldsymbol{\beta}_2$ and $\mathbf{V}(\mathbf{b}_2)$. This measure is an increasing function of a single, selectable parameter $\gamma \in [0, 1)$, and it can be made stringent by choosing γ near unity or relaxed by choosing γ small. This level of adequate signal-to-noise is combined with the previous test for signal-to-noise to produce a test for adequate/inadequate signal-to-noise.

Isodensity Ellipsoids as a Measure of Adequate Signal-to-Noise

We begin by introducing the *γ-isodensity ellipsoid* for the least-squares estimator $\mathbf{b}_2$ of $\boldsymbol{\beta}_2$. For each $0 \leqslant \gamma < 1$, this p_2-dimensional ellipsoid is the set of all $\mathbf{b}_2$ such that

$$(\mathbf{b}_2 - \boldsymbol{\beta}_2)^{\mathrm{T}} \mathbf{V}^{-1}(\mathbf{b}_2)(\mathbf{b}_2 - \boldsymbol{\beta}_2) \leqslant {}_\gamma\chi^2_{p_2}, \tag{7.16}$$

where ${}_\gamma\chi^2_{p_2}$ is the γ-critical value for the central chi-squared distribution with p_2 degrees of freedom. Exhibit 7.1 depicts one member of this family of ellipses for the case with $p_2 = 2$ in which $\mathbf{b}_2$ is comprised of b_i and b_j, the ith and jth elements of $\mathbf{b}$.

The concept of the isodensity ellipsoid, or closely related ones, occurs in the literature under a number of different names. Cramér (1946, Section 21.10) calls one element of this family the *ellipse of concentration* and, in a different context

[8]See Goldberger (1964, pp. 173–176) or Theil (1971, pp. 139–141). This F-statistic is also known as Hotelling's (1931) T^2-statistic—see Anderson (1958, Chapter 5).

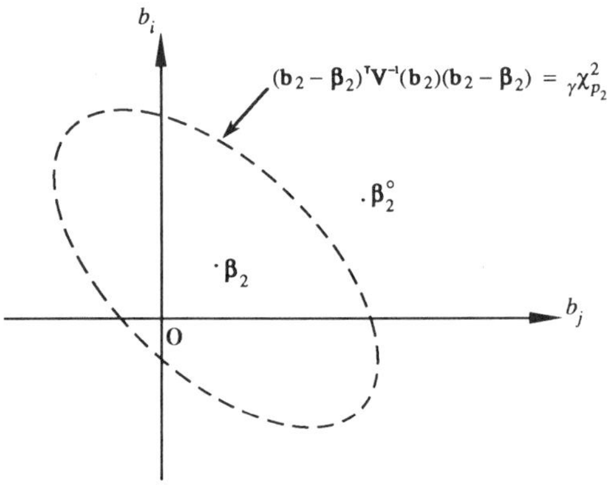

Exhibit 7.1 The γ-isodensity ellipsoid for $\mathbf{b}_2$.

(Section 21.6), calls the family of such curves *ellipses of inertia*. Malinvaud (1970, p. 154) borrows the name *concentration ellipsoid*. This is a good name that would have been used here except that it appears too easily confused with the commonly occurring but distinctly different concept of the confidence ellipse employed in describing regions of confidence for a subset of parameters. Wonnacott and Wonnacott (1979) call this identical family of ellipsoids the *isoprobability ellipses*, but this intuitively appealing term is somewhat of a misnomer, since the probability associated with every outcome $\mathbf{b}_2$ is zero. Presumably they have isodensity in mind. Anderson (1958, p. 57) in fact names this family the *ellipsoids of constant density*, a wholly appropriate, albeit slightly awkward name. The term *isodensity ellipsoid* seems a proper compromise.

Now, since the fact that $\mathbf{b}_2 \sim N_{p_2}(\boldsymbol{\beta}_2, \mathbf{V}(\mathbf{b}_2))$ implies

$$(\mathbf{b}_2 - \boldsymbol{\beta}_2)^T\mathbf{V}^{-1}(\mathbf{b}_2)(\mathbf{b}_2 - \boldsymbol{\beta}_2) \sim {}_{\gamma}\chi^2_{p_2}, \tag{7.17}$$

that is, that the quadratic form in (7.17) has a central chi-squared distribution with p_2 degrees of freedom, the γ-isodensity ellipsoid is necessarily the ellipsoid of smallest volume that contains any particular least-squares outcome from $\mathbf{b}_2 \sim N_{p_2}(\boldsymbol{\beta}_2, \mathbf{V}(\mathbf{b}_2))$ with probability γ. As noted, these isodensity ellipsoids for the random vector $\mathbf{b}_2$ should not be confused with the superficially similar but conceptually distinct notion of a confidence ellipsoid for a parameter $\boldsymbol{\beta}_2$. These latter ellipsoids, defined for a particular least-squares estimate $\mathbf{b}_2^*$, are the set of $\xi \in \mathscr{R}^{p_2}$ such that

$$(\xi - \mathbf{b}_2^*)^T\mathbf{S}^{-1}(\mathbf{b}_2)(\xi - \mathbf{b}_2^*) \leqslant p_2\, {}_{\gamma}F^{p_2}_{n-p}, \tag{7.18}$$

where $\mathbf{S}^{-1}(\mathbf{b}_2) = s^2(\mathbf{X}_2^T\mathbf{M}_1\mathbf{X}_2)^{-1}$, and ${}_{\gamma}F^{p_2}_{n-p}$ is the γ-critical level for the central F with p_2 and $n-p$ degrees of freedom. The confidence ellipsoids are centered on the least-squares estimate $\mathbf{b}_2^*$ and have the property that, conditional on $\mathbf{X}$, they will bracket the true value $\boldsymbol{\beta}_2$ with probability γ. By contrast, the γ-isodensity ellipsoids are centered on the true parameter value $\boldsymbol{\beta}_2$ and have the property that, conditional on $\mathbf{X}$, they will contain any particular least-squares outcome $\mathbf{b}_2$ with probability γ.

The γ-isodensity ellipsoids therefore define regions of likely and unlikely outcomes for the least-squares estimator $\mathbf{b}_2$ (given $\mathbf{X}$ and hence $\mathbf{V}(\mathbf{b}_2)$). In Exhibit 7.1, for example, if the ellipse that is shown corresponds to $\gamma = 0.95$, an outcome such as $\mathbf{b}_2 = \boldsymbol{\beta}_2^\circ$ that lies outside of the ellipsoid would be unlikely. This would be even more so if γ is 0.99 or 0.999. If, however, the ellipse of Exhibit 7.1 corresponds to $\gamma = 0.3$, an outcome like $\boldsymbol{\beta}_2^\circ$ would hardly be surprising. Thus, the family of γ-isodensity ellipsoids provides a natural measure of the *probabilistic distance* between any point $\boldsymbol{\beta}_2^\circ$ and the true mean $\boldsymbol{\beta}_2$.[9] This leads to the following definition:

Definition 7.1. The *probabilistic distance* between $\boldsymbol{\beta}_2^\circ$ and $\boldsymbol{\beta}_2$, relative to the least-squares estimator $\mathbf{b}_2 \sim N_{p_2}(\boldsymbol{\beta}_2, \mathbf{V}(\mathbf{b}_2))$, is the γ that determines the isodensity ellipsoid that is centered on $\boldsymbol{\beta}_2$ and includes $\boldsymbol{\beta}_2^\circ$ on its boundary, that is, the γ such that

$$(\boldsymbol{\beta}_2^\circ - \boldsymbol{\beta}_2)^{\mathrm{T}}\mathbf{V}^{-1}(\mathbf{b}_2)(\boldsymbol{\beta}_2^\circ - \boldsymbol{\beta}_2) = {}_{\gamma}\chi^2_{p_2}. \tag{7.19}$$

Clearly, $\gamma = 0$ if and only if $\boldsymbol{\beta}_2 = \boldsymbol{\beta}_2^\circ$, and there is otherwise a one-to-one monotone-increasing mapping between γ on $[0, 1)$ and ${}_{\gamma}\chi^2_{p_2}$ on $[0, \infty)$.

This measure has immediate application for our needs, for we see from (7.8) that the left-hand side of (7.19) is τ^2, the signal-to-noise of $\mathbf{b}_2$ relative to $\boldsymbol{\beta}_2^\circ$. The probabilistic distance, then, also provides a natural means for assessing the size of signal-to-noise τ^2 that does not require knowledge of $\boldsymbol{\beta}_2$ and $\mathbf{V}(\mathbf{b}_2)$. A level of signal-to-noise τ^2 can be considered large if it corresponds to a large probabilistic separation of $\boldsymbol{\beta}_2^\circ$ from $\boldsymbol{\beta}_2$, that is, if it equals a value of ${}_{\gamma}\chi^2_{p_2}$ for a γ chosen near unity, say, 0.90, or 0.95, or even 0.999. A weak level of signal-to-noise corresponds to little probabilistic separation and has a magnitude for ${}_{\gamma}\chi^2_{p_2}$ chosen small, say, 0.75, or 0.5, or even 0.0. Exhibit 7.2 shows how ${}_{\gamma}\chi^2_{p_2}$ varies with γ for $p_2 = 1$.

For any choice of $\gamma \in [0, 1)$, then, call the magnitude

$$\tau_\gamma^2 \equiv {}_{\gamma}\chi^2_{p_2} \tag{7.20}$$

the *threshold of adequacy at level* γ, where we recall ${}_{\gamma}\chi^2_{p_2}$ is the γ-critical value of the central chi-squared distribution with p_2 degrees of freedom. The signal-to-noise τ^2 of the least-squares estimator $\mathbf{b}_2$ of $\boldsymbol{\beta}_2$ relative to $\boldsymbol{\beta}_2^\circ$ will be said to be *adequate at level* γ if

$$\tau^2 > \tau_\gamma^2. \tag{7.21}$$

[9]In the $p_2 = 1$ case, where $\boldsymbol{\beta}_2$ consists of a single element β_k from the vector $\boldsymbol{\beta}$, this measure is simply the probability associated with the event that a least-squares outcome will be less than or equal to a multiple $m \equiv |\beta_k - \beta_k^\circ|/\sigma_{b_k}$ standard errors away from the mean value β_k.

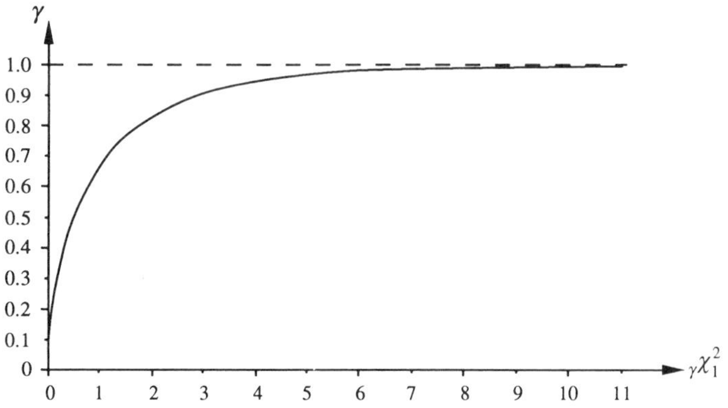

Exhibit 7.2 Relation of γ to probabilistic distance when $p_2 = 1$.

A level $\tau^2 \leqslant \tau_\gamma^2$ will be said to be *inadequate at level* γ.

The threshold of adequacy τ_γ^2 may be set by the researcher according to need. It may be set stringently by choosing γ near unity or relaxed by choosing smaller values for γ. Precisely which value is most useful for a particular application is somewhat like choosing a size for a test of hypothesis, something that experience, empirical experimentation, or convention determines. Different choices will clearly be appropriate to different applications. In the example that follows in Section 7.7, two different choices, $\gamma = 0.90$ and $\gamma = 0.999$, are compared.

Testing for Adequate/Inadequate Signal-to-Noise

The test for signal-to-noise given in Section 7.3 can now be used to test for the "adequacy condition" (7.21) to produce the following test for adequate signal-to-noise:

1. Choose a level $0 \leqslant \gamma < 1$ defining adequacy and determining $\tau_\gamma^2 \equiv {}_\gamma\chi_{p_2}^2$ and consider the null hypothesis $A_0(\gamma)$: $\tau^2 = \tau_\gamma^2$ versus the alternative hypothesis $A_1(\gamma)$: $\tau^2 > \tau_\gamma^2$, which is condition (7.21).
2. Choose a test size α (say, 0.05 or 0.01) for the test of signal-to-noise of Section 7.3 and calculate the critical value F_α from (7.15) for $\tau_*^2 = \tau_\gamma^2$ (for which, see the tables in Section 7.8).
3. Calculate ϕ^2 from (7.12) for the relevant parameters $\mathbf{b}_2$ and point of reference $\boldsymbol{\beta}_2^\circ$ (which will often be $\mathbf{0}$—this is illustrated in Section 7.7).
4. Then $\phi^2 > F_\alpha$ rejects $A_0(\gamma)$ in favor of $A_1(\gamma)$ and, hence, accepts the presence of adequate signal-to-noise, while $\phi^2 \leqslant F_\alpha$ accepts $A_0(\gamma)$, which will be interpreted to accept the presence of inadequate signal-to-noise.

Section 7.8 provides tables for the critical values F_α required in step 2 for values of $\alpha = 0.01, 0.05, 0.10, 0.25$; for values of $\gamma = 0.75, 0.90, 0.95, 0.99, 0.999, 0.9999, 0.99999, 0.999999$; and for degrees of freedom $p_2 = 1(1)6(2)10$, and $n - p = 10(1)20(2)30(5)40(10)60(20)100(50)300$ and 1000. These ranges should be relevant to most applications.

7.5 RELATION OF THE TEST FOR ADEQUATE SIGNAL-TO-NOISE TO CONVENTIONAL TESTS OF HYPOTHESIS

In this section we explore in greater detail some aspects of the relation between the test just derived for adequate signal-to-noise and the usual tests of hypothesis $H_0: \boldsymbol{\beta}_2 = \boldsymbol{\beta}_2^\circ$ against the alternative hypothesis $H_1: \boldsymbol{\beta}_2 \neq \boldsymbol{\beta}_2^\circ$, which will be interpreted to subsume the closely related concepts of tests of significance and confidence intervals.

As typically described in textbooks, in testing $H_0: \boldsymbol{\beta}_2 = \boldsymbol{\beta}_2^\circ$ against $H_1: \boldsymbol{\beta}_2 \neq \boldsymbol{\beta}_2^\circ$, a test statistic is examined. If it falls in the acceptance region, H_0 is accepted; otherwise H_0 is rejected and H_1 is accepted.

In practice, however, tests are not so straightforwardly conducted. Often the acceptance of H_0 can have deleterious repercussions as, for example, when a variate $\mathbf{X}_i$ thought to be important to the analysis at hand is found to have an insignificant least-squares estimate b_i. In such circumstances, few practitioners would unquestioningly accept $H_0: \beta_i = 0$ and toss $\mathbf{X}_i$ from the model, for the costs of such an action can be great. Structurally, it has the cost of removing a theoretically important driving force from the model. Statistically, it weighs the usually small gains that accrue to one added degree of freedom against the possibility of introducing substantial bias due to misspecification from omitted variates. Rather, the practitioner would typically realize that data problems can greatly inflate the acceptance region, rendering standard tests unpowerful and substantially broadening interval estimates.[10]

The preceding highlights an interesting practical aspect of the use of conventional tests of hypothesis and the interpretation given to their test statistics. Since data problems are known to cause inflated parameter variances and a loss of power, their suspected presence can raise suspicion about the quality of the estimated regression results and any tests based on them. But in the absence of anything better, low values for the t-statistic are often used, somewhat informally and certainly circularly, as diagnostic evidence for the presence of such data problems, particularly when an important variate has an insignificant parameter estimate.

Of course, conventional t-tests are not designed to serve such heavy double duty—providing both a test of hypothesis on coefficient values and a diagnostic for problems due to data weaknesses; whereas low t's may indicate data weaknesses, high t's need not indicate their absence.

[10]Often producing an Andrews (1989) inner inverse (non) power region that engulfs all relevant alternatives.

To serve this diagnostic need, a test for a loss of power or inflated variances is required. This task might be accomplished through the use of the power function, but this requires knowledge of $\boldsymbol{\beta}_2$ and $\mathbf{V}(\mathbf{b}_2)$. However, in testing H_0 against H_1 when H_1 is true, a test based on the least-squares estimator $\mathbf{b}_2$ of $\boldsymbol{\beta}_2$ will have greater power, ceteris paribus, the greater is τ^2, the signal-to-noise of $\mathbf{b}_2$ relative to $\boldsymbol{\beta}_2^\circ$.[11] This suggests the relevance of assessing the magnitude of signal-to-noise as a diagnostic for the presence of inflated variances and data problems, and this is precisely what the test for adequate signal-to-noise is designed to achieve.[12]

The test for adequate signal-to-noise, then, does not substitute for, nor is it intended to substitute for, conventional testing procedures. Rather, it complements these procedures, adding to them a tool for indicating when data problems may be causing inflated variances. Thus, when tests of significance (H_0: $\boldsymbol{\beta}_2 = \mathbf{0}$) or tests of hypothesis (H_0: $\boldsymbol{\beta}_2 = \boldsymbol{\beta}_2^\circ$) are called for, standard test procedures are appropriately adopted. When, however, it is known a priori that $\boldsymbol{\beta}_2 \neq \mathbf{0}$, and tests of data weakness are called for, the test for adequate signal-to-noise is relevant.

7.6 DIAGNOSING HARMFUL DATA PROBLEMS: COLLINEARITY AND SHORT DATA

We have seen that a low value of signal-to-noise τ^2 can indicate the presence of inflated variances that can reduce the power of statistical tests of significance and hypothesis and render interval estimates broader than may be desirable. We have also indicated that many practitioners, when confronted with such symptoms as "low t's," often consider them as informal evidence of the presence of data problems. Here, we show how this process can be given a proper basis through the test for adequate signal-to-noise. Specifically, given the true structural parameters $\boldsymbol{\beta}_2$ and σ^2 and the presumption that H_0: $\boldsymbol{\beta}_2 = \boldsymbol{\beta}_2^\circ$ is not true, τ^2 is small precisely when there are data problems either in the form of collinearity or of short data, a term that receives rigorous definition as we

[11] This can be seen for the case of $p_2 = 1$ in an example given in Wilks (1962, pp. 396–397).

[12] In a different context, a related use of signal-to-noise τ^2 is suggested by Anderson (1958, p. 115), who writes:

> Emma Lehmer (1944) has computed tables of ϕ [a transformation of τ^2] for given significance level and given probability of a Type II error. Her tables can be used to see what value of τ^2 is needed to make the probability of acceptance of the null hypothesis significantly low when $\boldsymbol{\mu} \neq \mathbf{0}$. For instance, if we want to be able to reject the hypothesis $\boldsymbol{\mu} = \mathbf{0}$ on the basis of a sample for a given $\boldsymbol{\mu}$ and $\boldsymbol{\Sigma}$, we may be able to choose N so that $N\boldsymbol{\mu}^{\mathrm{T}}\boldsymbol{\Sigma}^{-1}\boldsymbol{\mu} = \tau^2$ is sufficiently large. Of course, the difficulty with these considerations is that we usually do not know exactly the values of $\boldsymbol{\mu}$ and $\boldsymbol{\Sigma}$ (hence, τ^2) for which we want the probability of rejecting a certain value.

Our use of the concept of adequate signal-to-noise as measured by the probabilistic distance given in Section 7.4 is designed to circumvent this last-mentioned practical shortcoming.

proceed. Thus, under these circumstances, inadequate signal-to-noise directly reflects data weaknesses, and a test for inadequate signal-to-noise is diagnostic of these problems.

The Four Causes of Low Signal-to-Noise

We can rewrite (7.8), the expression for τ^2, the signal-to-noise of the least-squares estimator $\mathbf{b}_2$ of $\boldsymbol{\beta}_2$ relative to $\boldsymbol{\beta}_2^\circ$, in a manner that allows us to see how four separately interpretable elements come together to determine its value. We begin by repeating (7.8),

$$\tau^2 \equiv (\boldsymbol{\beta}_2 - \boldsymbol{\beta}_2^\circ)^{\mathrm{T}}\mathbf{V}^{-1}(\mathbf{b}_2)(\boldsymbol{\beta}_2 - \boldsymbol{\beta}_2^\circ),$$

and then reformulate $\mathbf{V}^{-1}(\mathbf{b}_2)$ as follows:

$$\begin{aligned}\sigma^2\mathbf{V}^{-1}(\mathbf{b}_2) &\equiv \mathbf{X}_2^{\mathrm{T}}\mathbf{M}_1\mathbf{X}_2\\ &= \mathbf{X}_2^{\mathrm{T}}\mathbf{X}_2 - \hat{\mathbf{X}}_2^{\mathrm{T}}\hat{\mathbf{X}}_2\\ &= \mathbf{X}_2^{\mathrm{T}}\mathbf{X}_2[\mathbf{I} - (\mathbf{X}_2^{\mathrm{T}}\mathbf{X}_2)^{-1}\hat{\mathbf{X}}_2^{\mathrm{T}}\hat{\mathbf{X}}_2],\\ &= \mathbf{X}_2^{\mathrm{T}}\mathbf{X}_2(\mathbf{I} - \mathbf{P}),\end{aligned} \tag{7.22}$$

where in the first step $\mathbf{M}_1 \equiv \mathbf{I} - \mathbf{X}_1(\mathbf{X}_1^{\mathrm{T}}\mathbf{X}_1)^{-1}\mathbf{X}_1^{\mathrm{T}}$, in the second step $\hat{\mathbf{X}}_2 \equiv \mathbf{X}_1(\mathbf{X}_1^{\mathrm{T}}\mathbf{X}_1)^{-1}\mathbf{X}_1^{\mathrm{T}}\mathbf{X}_2$, in the third step we assume $(\mathbf{X}_2^{\mathrm{T}}\mathbf{X}_2)^{-1}$ exists, and in the fourth step $\mathbf{P} \equiv (\mathbf{X}_2^{\mathrm{T}}\mathbf{X}_2)^{-1}\hat{\mathbf{X}}_2^{\mathrm{T}}\hat{\mathbf{X}}_2$.

This last matrix, $\mathbf{P}$, generalizes the uncentered multiple-correlation coefficient $\hat{R}^2$ of (2.4a). Indeed, if $\mathbf{X}_2$ were a single column ($p_2 = 1$), then $\mathbf{P}$ is simply $\hat{R}^2 = \hat{\mathbf{X}}_2^{\mathrm{T}}\hat{\mathbf{X}}_2/\mathbf{X}_2^{\mathrm{T}}\mathbf{X}_2$, the uncentered R^2 from a regression of $\mathbf{X}_2$ on $\mathbf{X}_1$, the remaining columns of $\mathbf{X}$. More generally, if at the one extreme the p_2 columns of $\mathbf{X}_2$ are orthogonal to those of $\mathbf{X}_1$, then it is readily seen that $\hat{\mathbf{X}}_2 = \mathbf{0}$ and $\mathbf{P} = \mathbf{0}$; while at the other extreme, if the columns of $\mathbf{X}_2$ are perfectly linearly related to those of $\mathbf{X}_1$, then $\hat{\mathbf{X}}_2 = \mathbf{X}_2$ and $\mathbf{P} = \mathbf{I}$. The greater the collinearity, then, between the columns belonging to $\mathbf{X}_2$ with those belonging to $\mathbf{X}_1$, the "closer" is $\mathbf{P}$ to $\mathbf{I}$. Several scalar measures of closeness could be used here, such as det($\mathbf{P}$) or the spectral norm $\|\mathbf{P}\|$. We shall, in fact, show this in another context in Section 11.4.

Now (7.8) becomes

$$\tau^2 \equiv (\boldsymbol{\beta}_2 - \boldsymbol{\beta}_2^\circ)^{\mathrm{T}}[\mathbf{X}_2^{\mathrm{T}}\mathbf{X}_2(\mathbf{I} - \mathbf{P})](\boldsymbol{\beta}_2 - \boldsymbol{\beta}_2^\circ)/\sigma^2. \tag{7.23}$$

Here we see how four separately interpretable elements come together to determine how a low value for τ^2 can occur. First, the closer $\boldsymbol{\beta}_2^\circ$ is to $\boldsymbol{\beta}_2$, ceteris paribus, the smaller is τ^2. Second, the greater the inherent noise σ^2, ceteris paribus, the lower is τ^2. Third, the greater the degree of collinearity, ceteris paribus, the closer $\mathbf{P}$ is to $\mathbf{I}$ and the smaller becomes τ^2. Fourth, the shorter the "length" of $\mathbf{X}_2$, that is, $\mathbf{X}_2^{\mathrm{T}}\mathbf{X}_2$, as measured, for example, by its spectral norm

$\|\mathbf{X}_2\|$, ceteris paribus, the lower τ^2 tends to be. It is of interest to note that, in a given regression situation, one cannot increase the signal-to-noise τ^2 simply by rescaling $\mathbf{X}_2$ (changing its units of measurement) to make it longer, since any such rescaling of $\mathbf{X}_2$ entails a compensating inverse scaling for $\boldsymbol{\beta}_2$, leaving τ^2 unchanged. Indeed, we note that τ^2 has the desirable property of being invariant to linear transformations on $\mathbf{X}$.

Now, the first two factors determining τ^2 deal with the unknown but unalterable structural parameters $\boldsymbol{\beta}_2$ and σ^2 along with the hypothesized value for $\boldsymbol{\beta}_2^\circ$, all of which are fixed for a given regression situation. Thus, if it is assumed true a priori that $\boldsymbol{\beta}_2 \neq \boldsymbol{\beta}_2^\circ$, τ^2 can become small precisely on account of the third and fourth factors, namely, a strong degree of collinearity or a "short" set of explanatory variates $\mathbf{X}_2$.

Determining Harmful Collinearity and Short Data

In light of the preceding discussion, it is natural to designate collinearity as being harmful when it can simultaneously be demonstrated that (a) collinearity exists and (b) inadequate signal-to-noise exists relative to some or all of the parameter estimates of the variates known to be involved in the collinear relations.

The matrix $\mathbf{P}$ introduced above might seem to provide a reasonable basis for defining the presence of collinearity, but as we have well established in Section 2.3, correlation-like measures have several serious faults in this context. From the current perspective, two practical problems stand out. First, while they are near unity when there is very strong collinearity and near zero when there is near orthogonality, no value has been determined that provides a meaningful dividing line between the presence and absence of collinearity. Second, $\mathbf{P}$ can be calculated only after the collinear relations among the columns of $\mathbf{X}$ have been discovered, and so it cannot be used to diagnose them.

But of course, the collinearity diagnostics developed in the earlier chapters will serve our needs quite well, allowing us quickly to determine the number of near dependencies among the columns of $\mathbf{X}$ as well as the variates involved and hence to partition the model as in (7.4). The basis for this partitioning is straightforward, quite similar to that used in forming the auxiliary regressions. From the number of scaled condition indexes of $\mathbf{X}$ that are deemed to be high, one determines p_2, the number of collinear relations. From the variance–decomposition proportions and/or the auxiliary regressions, one determines $\mathbf{X}_2$, the p_2 columns of $\mathbf{X}$ that can be written as linear combinations of $\mathbf{X}_1$, the remaining $p_1 = p - p_2$ columns of $\mathbf{X}$. Since this set of $\mathbf{X}_2$ contains variates that, when adjoined to the otherwise well-conditioned $\mathbf{X}_1$, cause the full set of variates in $\mathbf{X}$ to be ill conditioned, harmful collinearity can be shown if the parameter estimates $\mathbf{b}_2$ possess inadequate signal-to-noise. Of course, this partitioning is not unique since if there are p_2 relations among the p columns of $\mathbf{X}$, there are many ways p_2 of the variates could be selected to regress on the remaining p_1. This nonuniqueness is of limited importance, however, since no matter how $\mathbf{X}_2$ is picked, the essence of the ensuing argument continues to hold.

A procedure for determining harmful collinearity, then, results from the sequence of (1) applying the collinearity diagnostics of the preceding chapters to determine an $\mathbf{X}_1$ and $\mathbf{X}_2$ followed by (2) a test for adequate signal-to-noise for the $\mathbf{b}_2$ that correspond to the $\mathbf{X}_2$ indicated in 1. Harmful collinearity will be said to occur when both conditions prevail. More generally, we can examine four possible outcomes of this sequence of tests. These are designated in Exhibit 7.3.

(I) In situation I of Exhibit 7.3, everything seems right with the world; neither collinearity nor inadequate signal-to-noise is present. This situation is desirable for both structural estimation and forecasting.

(II) Situation II depicts nonharmful collinearity; that is, collinearity is present but has not resulted in inadequate signal-to-noise. This is not to say that, ceteris paribus, better-conditioned data would not be desirable (for this is always the case) but rather that the ill effects of collinearity have been mitigated by the presence of relatively small σ^2 and/or long $\mathbf{X}_2$. This situation augurs well for the use of the model for forecasting purposes, particularly, but not necessarily only, if the collinear relations are expected to continue into the forecasting period.

(III) Situation III depicts the case where there is no collinearity in $\mathbf{X}$ but inadequate signal-to-noise is nevertheless present. We have seen above that, for given $\boldsymbol{\beta}_2 \neq \boldsymbol{\beta}_2^\circ$ and σ^2, this situation can only arise when the "length" of $\mathbf{X}_2$ is short. Thus, while data problems exist, collinearity is not the culprit, and we use this situation to define the presence of *short data*. It is of interest to note that, should inadequate signal-to-noise be shown to exist for the full set $\mathbf{b}$ relative to $\boldsymbol{\beta}^\circ = \mathbf{0}$ in this situation, the quality of information derivable from least-squares estimation would be poor not only for the estimation of individual parameters but also for estimates of linear combinations of the parameters, since there are not strong near dependencies among the columns of $\mathbf{X}$. This set of circumstances, then, would be bad for either structural estimation or forecasting.

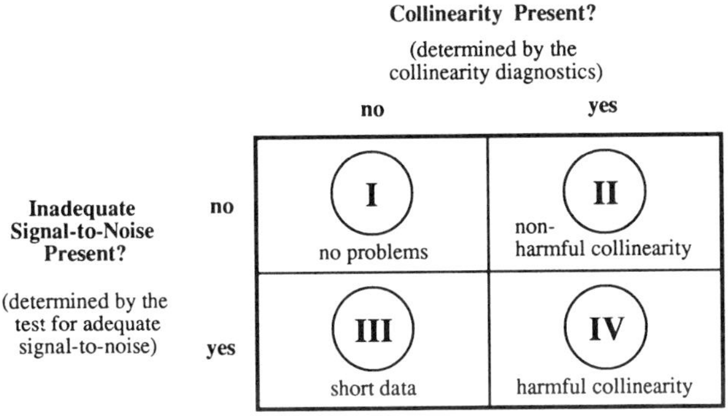

Exhibit 7.3 Interpreting the test for harmful collinearity.

(IV) Situation IV, of course, defines *harmful collinearity*: the joint occurrence of collinearity and inadequate signal-to-noise. This situation is rather generally harmful to structural estimation but not necessarily to forecasting if the collinear relations are expected to extend into the forecast period. A case study that shows how to use these diagnostics to improve forecasts when harmful collinearity does not extend into the forecast period is given in Section 10.3.

7.7 AN EXAMPLE OF DIAGNOSING HARMFUL COLLINEARITY

We will now illustrate the foregoing techniques with a commercial paper rate equation from the Michigan Quarterly Econometric Model (MQEM) for the U.S. economy. This model and the data were kindly made available by Saul Hymans and Phil Howrey. A full description of the model employed here is to be found in Gardner and Hymans (1978).

The commercial paper rate equation has the form

$$\mathbf{CPR}(T) = \beta_1\iota + \beta_2\mathbf{TB}(T) + \beta_3\mathbf{TB}(T-1) + \beta_4\mathbf{S1}(T) + \beta_5\mathbf{S2}(T) + \beta_6\mathbf{S3}(T) + \beta_7\mathbf{D}(T) + \beta_8\mathbf{INF}(T) + \beta_9\mathbf{CPR}(T-1) + \boldsymbol{\varepsilon}(T), \tag{7.24}$$

where

$\mathbf{CPR}(T)$ = the commercial paper rate, percentage per annum,

$\mathbf{TB}(T)$ = the 90-day treasury bill rate, percentage per annum,

$\mathbf{Si}(T)$ = a seasonal dummy for quarter $i = 1, 2, 3$,

$\mathbf{D}(T)$ = a dummy for the first three quarters 1974 (see Gardner and Hymans, 1978),

$\mathbf{INF}(T)$ = the rate of inflation, $(p_t - p_{t-2})/p_{t-2}$.

The data are quarterly, 1955:I–1975:IV, giving 84 observations and 75 degrees of freedom. Although $\mathbf{TB}(T)$ and $\mathbf{INF}(T)$ are endogenous, this equation is nevertheless estimated in the MQEM by ordinary least squares. We also treat all the right-hand side variates as predetermined.

Estimation of (7.24) by least squares gives

$$\begin{aligned}\mathbf{CPR}(T) = {} & \underset{(1.325)}{5.589}\iota + \underset{(0.041)}{0.985}\mathbf{TB}(T) - \underset{(0.095)}{0.395}\mathbf{TB}(T-1) - \underset{(0.037)}{0.053}\mathbf{S1}(T) + \underset{(0.037)}{0.082}\mathbf{S2}(T) \\ & + \underset{(0.038)}{0.057}\mathbf{S3}(T) + \underset{(0.153)}{1.680}\mathbf{D}(T) - \underset{(1.340)}{5.489}\mathbf{INF}(T) + \underset{(0.066)}{0.511}\mathbf{CPR}(T-1),\end{aligned} \tag{7.25}$$

$$R^2 = .993, \qquad \text{SER} = 0.189, \qquad \text{DW} = 1.28, \qquad \tilde{\kappa}(\mathbf{X}) = 206.$$

Following the diagnostic procedure given in the preceding section, we first apply the collinearity diagnostics to determine the number of near dependencies and variate involvement, and then we test for inadequate signal-to-noise.

The Collinearity Diagnostics

The large scaled condition number of 206 in (7.25) assures us that at least one strong near dependency exists among the columns of **X**. A more complete diagnostic picture is given by the $\mathbf{\Pi}$ table of scaled condition indexes and variance–decomposition proportions given in Exhibit 7.4.

Here we see that there are two near dependencies, the expected strong one, with the scaled condition index of 206, and a somewhat weaker one, with a scaled condition index of 58. The first of these is dominant and certainly contains ι and **INF**(T), indicating little variation in this latter variate since none of the other variates shows very much involvement in this dominant dependency. The second contains **TB**($T-1$), **CPR**($T-1$), and possibly $\boldsymbol{\iota}$ and **INF**(T), these latter two variates having their potential involvement masked by their presence in the dominant dependency. Greater detail on the composition of these near dependencies can be obtained from auxiliary regressions. The procedure given in Section 5.3 for forming auxiliary regressions would select either ι or **INF**(T) along with **TB**($T-1$). Exhibit 7.5 shows the auxiliary regressions with $\boldsymbol{\iota}$ and **TB**($T-1$) regressed on the remaining variates.

These results confirm a strong near dependency between $\boldsymbol{\iota}$ and **INF**(T), which also includes **CPR**($T-1$) at a moderate level and **D**(T) weakly. Certainly the dominant relation here is between $\boldsymbol{\iota}$ and **INF**(T). The second near dependency involves **TB**($T-1$) and **CPR**($T-1$) strongly and **TB**(T) moderately and has weak, indeed inconsequential, involvement of **S1**(T) and **S3**(T). Thus, the five variates ι, **TB**(T), **TB**($T-1$), **INF**(T), and **CPR**($T-1$) are substantially involved in one or both of the near dependencies, while the variates **S1**(T), **S2**(T), **S3**(T), and **D**(T) are essentially free from such involvement.

Inadequate Signal-to-Noise

Unless some specific value for $\boldsymbol{\beta}_2^\circ$ is indicated by the statistical context under investigation, signal-to-noise will most frequently be examined relative to the origin, $\boldsymbol{\beta}_2^\circ = \mathbf{0}$. We begin here by examining each coefficient individually—that is, a set of tests with $p_2 = 1$—and turn, when indicated, to joint tests of interest.

Choices must be made for the test size α and for the parameter γ determining the level of adequacy in (7.20). For purposes of illustration, we choose $\alpha = 0.05$ and examine two different choices for γ of 0.90 and 0.999. A γ of 0.9 is a modest value, illustrating a choice that might be made if the regression analysis were of less importance. A γ of 0.999 is more stringent and illustrates a choice suitable for assessing the quality of parameter estimates that are considered important. With this latter value, the test will be more sensitive to data inadequacies and, hence, will apprise the user more readily of their presence.

Exhibit 7.4 Scaled Condition Indexes and Variance–Decomposition Proportions: Commercial Paper Rate Equation (MQEM)

Scaled Condition Index, $\tilde{\eta}$	Proportions of ι var(b_1)	**TB**(T) var(b_2)	**TB**($T-1$) var(b_3)	**S1**(T) var(b_4)	**S2**(T) var(b_5)	**S3**(T) var(b_6)	**D**(T) var(b_7)	**INF**(T) var(b_8)	**CPR**($T-1$) var(b_9)
1	.000	.000	.000	.000	.000	.000	.002	.000	.000
2	.000	.000	.000	.096	.103	.096	.005	.000	.000
2	.000	.000	.000	.031	.000	.000	.723	.000	.000
3	.000	.000	.000	.000	.629	.578	.000	.000	.000
3	.000	.000	.000	.787	.210	.203	.049	.000	.000
6	.000	.009	.003	.002	.014	.002	.129	.000	.002
19	.000	.589	.008	.022	.000	.034	.002	.000	.085
58	.001	.402	.977	.062	.042	.087	.015	.001	.809
206	.999	.000	.012	.000	.002	.000	.075	.999	.104

Exhibit 7.5 Auxiliary Regressions: Commercial Paper Rate Equation (MQEM)

Dependent Variate	Coefficient of **TB**(T)	**S1**(T)	**S2**(T)	**S3**(T)	**D**(T)	**INF**(T)	**CPR**($T-1$)	$\hat{R}^2$	$\tilde{\eta}$
ι	0.0026	0.0002	−0.0008	−0.0007	−0.0338	1.0096	−0.0110	.999	206
	[1.01]	[0.07]	[−0.24]	[−0.24]	[−2.68]	[206.16]	[−5.08]		
TB($T-1$)	0.3030	0.1029	0.0777	−0.1255	−0.1748	−0.1206	0.6121	.985	58
	[8.39]	[2.36]	[1.79]	[−2.86]	[−0.99]	[−1.76]	[20.22]		

Exhibit 7.6 summarizes the results for assessing the adequacy of the signal-to-noise for each parameter estimate b_i taken individually ($p_2 = 1$) relative to the origin, $\beta_i^\circ = 0$. Column 1 indicates the particular variate whose parameter is being tested for inadequate signal-to-noise. Column 2 indicates from the previous collinearity analysis whether the particular variate is involved in a near dependency. Column 3 gives the value of the test statistic ϕ^2 for signal-to-noise calculated from (7.12).

Harmful Collinearity

Columns 4 and 5 report whether there is inadequate signal-to-noise and, if so, what sort of data problem accompanies it. Column 4 is relevant to the case when the more modest level of adequacy is chosen at $\gamma = 0.90$. In this instance the critical value for $F_{0.05}$ is 11.3. This is taken from Exhibit 7.9*b* of Section 7.8 for $p_2 = 1$ and $n - p = 75$. A rough interpolation is employed. From column 4 we see that none of the collinear variates also has a ϕ^2 exceeding this critical level. In this instance, then, we have degrading but not harmful collinearity.

Column 5 reports the same results relative to the more stringent choice for a level of adequacy of $\gamma = 0.999$. The critical level for this case, obtained from Exhibit 7.9*e* for 1 and 75 (interpolated) degrees of freedom, is roughly 26.0. At this level of adequacy, there is harmful collinearity evidenced for the estimates of three coefficients, β_1 ($\mathbf{\iota}$), β_3 (**TB**($T-1$)), and β_8 (**INF**(T)). The two other variates showing degradation, β_2 (**TB**(T)) and β_9 (**CPR**($T-1$)), show no evidence of harm due to collinearity, at least not relative to the origin. Thus, if the estimates of β_1, β_3, or β_8 are of importance to a particular analysis of this commercial paper rate equation, the researcher is apprised that collinear relations have significantly reduced the quality of these estimates and that efforts for increasing the information set upon which these estimates are based, either through better-conditioned data or through the introduction of appropriate prior information, would be worthwhile.

We shall deal with such corrective action and provide examples in Chapter 10. But for the moment, note that this analysis clearly helps to determine where prior information is most usefully applied. Since $\mathbf{\iota}$ and **INF**(T) are strongly collinear, prior information on β_1 ($\mathbf{\iota}$) or β_8 (**INF**(T)) will increase the quality of both of these estimates *even if this prior information is directed at only one of the affected variates.* Similarly, since **CPR**($T-1$) is involved in both near dependencies identified by the collinearity diagnostics, prior information on β_9 (**CPR**($T-1$)) can help the quality not only of its own estimate but also those of the estimates of all the other variates involved in near dependencies with it, namely, β_3 (**TB**($T-1$)) and β_1 ($\mathbf{\iota}$), and this will be true *even if the estimate of β_9 is not directly harmed by the presence of collinearity.*

We can also investigate whether the estimates of the parameters of variates involved in collinear relations show jointly inadequate signal-to-noise. Since there are two near dependencies, we take $p_2 = 2$. If we select $\mathbf{X}_2$ to contain $\mathbf{\iota}$ and **TB**($T-1$), the two dependent variates used in the auxiliary regressions, the joint

Exhibit 7.6 Summary of Adequacy of Signal-to-Noise (S/N) and Possible Data Problems for Each Coefficient: Commerical Paper Rate Equation (MQEM), $p_2 = 1$, $\beta_i^\circ = 0$, $\alpha = 0.05$

(1) Parameter and Corresponding Variate	(2) Degraded Due to Collinearity?	(3) ϕ^2	(4) $\gamma = 0.90$, $F_{0.05} = 11.3$		(5) $\gamma = 0.999$, $F_{0.05} = 26.0$	
			Inadequate S/N	Possible Data Problem	Inadequate S/N	Possible Data Problem
β_1, $\boldsymbol{\iota}$	Yes	17.9	No	None	Yes	Harmful collinearity
β_2, **TB**(T)	Yes	566.0	No	None	No	None
β_3, **TB**($T-1$)	Yes	17.3	No	None	Yes	Harmful collinearity
β_4, **S1**(T)	No	2.0	Yes	Short data	Yes	Short data
β_5, **S2**(T)	No	4.9	Yes	Short data	Yes	Short data
β_6, **S3**(T)	No	2.2	Yes	Short data	Yes	Short data
β_7, **D**(T)	No	119.9	No	None	No	None
β_8, **INF**(T)	Yes	16.8	No	None	Yes	Harmful collinearity
β_9, **CPR**($T-1$)	Yes	59.7	No	None	No	None

test on $\mathbf{b}_2 \equiv [b_1, b_3]^T$ relative to the origin has a ϕ^2 of 16.04. The critical values are approximately 8.32 for $\gamma = 0.90$ (from Exhibit 7.9*b*) and 16.06 for $\gamma = 0.999$ (from Exhibit 7.9*e*). At the weaker γ there is no evidence of jointly harmful collinearity, while at the stronger γ there is. In this latter case, not only are the individual estimates of β_1 and β_3 of poor quality, but so also is the estimate of their joint influence. This need not hold for all such joint tests. If one were interested in the joint adequacy of the estimates of β_1 ($\mathbf{\iota}$) and β_9 (**CRP**($T-1$)), the ϕ^2 for this joint test is 31.79, which exceeds both critical values. In this latter case, it is interesting to note that, while collinearity is not harmful at $\gamma = 0.999$ to an analysis based on the joint effects of b_1 and b_9, as seen from Exhibit 7.6, it is harmful to an analysis based on the individual effects, at least of b_1.

We can also consider tests relative to $\boldsymbol{\beta}_2^\circ$s other than the origin. In a dynamic model of this sort with a lagged dependent variate, stability requires that $|\beta_9| < 1$. Given that the estimate b_9 of β_9 is seriously degraded by collinearity, we might wish to determine whether there is harmful collinearity relative to H_0: $\beta_9 = 1$. If there is not, one would have great faith in these data for testing the stability of the equation; if there is, one would understandably hesitate to accept H_0: $\beta_9 = 1$ on the basis of these data. The ϕ^2 relevant to this test is 54.79, greatly in excess of the critical value of 26.0 for $\gamma = 0.999$. Thus, collinearity is not harmful to b_9 in this context either.

Short Data

From Exhibit 7.6, we note that there is inadequate individual signal-to-noise for the estimates of the coefficients of the three seasonal dummies **S1**(T), **S2**(T), and **S3**(T), which are, however, effectively free from involvement in collinear relations.

A researcher with no strong prior belief in this seasonal effect could, then, easily discount the value of these dummies. A test for the joint adequacy of signal-to-noise for these three dummies ($p_2 = 3$) fails for both $\gamma = 0.90$ and $\gamma = 0.999$. The ϕ^2 relevant to this test is 3.71, and the appropriate critical value is roughly 7.0 for $\gamma = 0.90$ and 12.4 for $\gamma = 0.999$. Since this researcher is willing to entertain the hypothesis that these three coefficients are jointly equal to zero, a reasonable strategy would simply be to test this hypothesis by a conventional test of significance and, if it fails, either remove the variates or possibly simplify their structure. The 0.05-critical value for this test (from standard F tables) is 2.73. The joint effect is therefore significant but clearly not well determined and not suitable for assessing important analytical questions that may hinge on seasonality.

By way of contrast, a researcher with strong prior beliefs in the relevance of this seasonal effect would be disappointed indeed. Not only are the estimates of low quality, but also the data cannot be held responsible. Here, one might question the seasonal specification in (7.24), but if the linear structure is considered appropriate, only a much larger sample could further reduce these seasonal parameter variances. Short of such riches, the user must depend upon strongly formulated a priori information if better quality estimates for β_4, β_5, and β_6 are required. Since these variates are not involved in collinear relations with other variates, the prior information must be directed specifically at the seasonal pattern and not at any coinvolved terms.

Finally, we note that the dummy $\mathbf{D}(T)$ is not involved in a collinear relation and that its parameter estimate possesses adequate signal-to-noise even for the choice of $\gamma = 0.999$. It must be concluded that the data are thoroughly suitable to providing a high-quality estimate of β_7.

7.8 APPENDIX: CRITICAL VALUES FOR ${}_{1-\alpha}F^{p_2}_{n-p}({}_{\gamma}\chi^2_{p_2})$

In order to apply the test for adequate signal-to-noise given above, critical values are required for

$$ {}_{1-\alpha}F^{p_2}_{n-p}({}_{\gamma}\chi^2_{p_2}), \tag{7.26} $$

that is, the $(1-\alpha)$-critical value for the noncentral F with p_2 and $n-p$ degrees of freedom and noncentrality parameter $\tau^2_\gamma \equiv {}_{\gamma}\chi^2_{p_2}$. Here, we recall

α = test size for the test of signal-to-noise, e.g., 0.05 or 0.10,

n = number of observations,

p = number of explanatory (independent) variates, including the constant term if present,

p_2 = number of explanatory variates under test.

Exhibits 7.7–7.10, grouped according to increasing values of γ within each α, provide values for (7.26) for $\gamma = 0.75$, 0.90, 0.95, 0.99, 0.999, 0.9999, 0.99999, 0.999999; $\alpha = 0.01$, 0.05, 0.10, 0.25; $p_2 = 1(1)6(2)10$; and $n - p = 10(1)20(2)30(5)40(10)60(20)100(50)300$ and 1000. These values have been computed using a modified version of the noncentral distribution algorithms given in Bargmann and Ghosh (1964). The exhibits are numbered according to the following scheme:

α	Exhibits	and	γ	Subletter
0.01	7.7		0.75	a
0.05	7.8		0.90	b
0.10	7.9		0.95	c
0.25	7.10		0.99	d
			0.999	e
			0.9999	f
			0.99999	g
			0.999999	h

Exhibit 7.7*a*   Critical Values for Adequate Signal-to-Noise: $\alpha = 0.25$, $\gamma = 0.75$

Degrees of Freedom, Denominator $(n - p)$	Degrees of Freedom, Numerator (p_2) 1	2	3	4	5	6	8	10
10	3.771	3.892	3.808	3.726	3.660	3.607	3.526	3.468
11	3.729	3.838	3.750	3.666	3.598	3.543	3.460	3.402
12	3.694	3.794	3.702	3.616	3.546	3.490	3.407	3.347
13	3.666	3.757	3.662	3.574	3.503	3.447	3.362	3.301
14	3.641	3.726	3.628	3.539	3.467	3.409	3.323	3.262
15	3.620	3.670	3.599	3.508	3.435	3.377	3.290	3.228
16	3.602	3.676	3.574	3.482	3.408	3.349	3.261	3.198
17	3.586	3.656	3.552	3.458	3.384	3.325	3.236	3.172
18	3.572	3.638	3.532	3.438	3.363	3.303	3.213	3.149
19	3.560	3.622	3.515	3.420	3.344	3.284	3.193	3.129
20	3.549	3.608	3.499	3.404	3.327	3.266	3.175	3.110
22	3.530	3.584	3.473	3.375	3.298	3.236	3.144	3.078
24	3.514	3.564	3.450	3.352	3.274	3.212	3.119	3.052
26	3.500	3.547	3.432	3.333	3.254	3.191	3.097	3.030
28	3.489	3.532	3.416	3.316	3.237	3.173	3.078	3.011
30	3.479	3.520	3.402	3.302	3.222	3.158	3.062	2.994
35	3.460	3.495	3.375	3.273	3.192	3.127	3.030	2.961
40	3.445	3.476	3.355	3.252	3.170	3.104	3.006	2.936
50	3.425	3.451	3.327	3.222	3.139	3.072	2.973	2.902
60	3.412	3.434	3.308	3.202	3.118	3.051	2.951	2.878
80	3.396	3.413	3.285	3.178	3.093	3.025	2.923	2.849
100	3.386	3.400	3.272	3.163	3.078	3.009	2.906	2.832
150	3.373	3.384	3.254	3.145	3.058	2.989	2.885	2.809
300	3.359	3.367	3.236	3.125	3.038	2.968	2.862	2.786
1000	3.350	3.356	3.223	3.112	3.024	2.953	2.846	2.769

Exhibit 7.7*b*   Critical Values for Adequate Signal-to-Noise: $\alpha = 0.25$, $\gamma = 0.90$

Degrees of Freedom, Denominator $(n - p)$	Degrees of Freedom, Numerator (p_2) 1	2	3	4	5	6	8	10
10	6.140	5.358	4.923	4.649	4.459	4.318	4.120	3.986
11	6.065	5.279	4.844	4.571	4.381	4.240	4.042	3.908
12	6.003	5.215	4.780	4.506	4.316	4.176	3.978	3.845
13	5.952	5.161	4.726	4.452	4.262	4.122	3.924	3.791
14	5.908	5.115	4.680	4.406	4.217	4.076	3.879	3.745
15	5.871	5.076	4.640	4.367	4.177	4.036	3.839	3.706
16	5.838	5.042	4.606	4.333	4.143	4.002	3.805	3.671
17	5.810	5.013	4.577	4.303	4.113	3.972	3.774	3.641
18	5.785	4.986	4.550	4.276	4.086	3.945	3.748	3.614
19	5.762	4.963	4.526	4.253	4.062	3.921	3.724	3.590
20	5.742	4.942	4.506	4.231	4.041	3.900	3.702	3.568
22	5.708	4.906	4.469	4.195	4.005	3.863	3.665	3.531
24	5.679	4.877	4.439	4.165	3.974	3.833	3.634	3.500
26	5.656	4.852	4.414	4.139	3.949	3.807	3.608	3.474
28	5.635	4.830	4.392	4.118	3.927	3.785	3.586	3.452
30	5.618	4.812	4.374	4.099	3.908	3.766	3.567	3.432
35	5.583	4.776	4.337	4.062	3.870	3.728	3.529	3.393
40	5.557	4.748	4.310	4.034	3.842	3.700	3.500	3.364
50	5.521	4.711	4.272	3.995	3.803	3.661	3.460	3.323
60	5.497	4.686	4.246	3.970	3.778	3.634	3.433	3.296
80	5.467	4.655	4.215	3.938	3.745	3.602	3.400	3.262
100	5.450	4.637	4.196	3.919	3.726	3.582	3.380	3.241
150	5.427	4.613	4.172	3.894	3.701	3.557	3.353	3.214
300	5.403	4.588	4.147	3.869	3.675	3.530	3.326	3.186
1000	5.386	4.572	4.130	3.851	3.657	3.512	3.307	3.167

Exhibit 7.7c Critical Values for Adequate Signal-to-Noise: $\alpha = 0.25$, $\gamma = 0.95$

Degrees of Freedom, Denominator $(n-p)$	Degrees of Freedom, Numerator (p_2)							
	1	2	3	4	5	6	8	10
10	8.002	6.444	5.726	5.302	5.016	4.809	4.524	4.334
11	7.897	6.346	5.632	5.210	4.927	4.721	4.438	4.250
12	7.810	6.265	5.555	5.135	4.853	4.648	4.367	4.180
13	7.739	6.198	5.490	5.072	4.791	4.587	4.307	4.121
14	7.678	6.141	5.436	5.019	4.739	4.536	4.256	4.070
15	7.625	6.092	5.388	4.973	4.694	4.491	4.212	4.027
16	7.580	6.050	5.348	4.933	4.654	4.452	4.174	3.989
17	7.540	6.012	5.312	4.898	4.620	4.418	4.140	3.956
18	7.505	5.980	5.280	4.867	4.589	4.388	4.110	3.926
19	7.474	5.950	5.252	4.839	4.562	4.361	4.084	3.900
20	7.446	5.924	5.227	4.815	4.538	4.336	4.060	3.876
22	7.398	5.879	5.183	4.772	4.496	4.295	4.019	3.835
24	7.358	5.842	5.147	4.737	4.461	4.260	3.984	3.801
26	7.325	5.811	5.117	4.707	4.431	4.231	3.956	3.772
28	7.297	5.784	5.091	4.682	4.406	4.206	3.931	3.748
30	7.272	5.761	5.069	4.660	4.385	4.184	3.909	3.726
35	7.223	5.715	5.024	4.616	4.341	4.142	3.866	3.683
40	7.187	5.681	4.992	4.584	4.309	4.109	3.834	3.651
50	7.137	5.634	4.946	4.538	4.264	4.064	3.790	3.606
60	7.104	5.603	4.915	4.508	4.234	4.035	3.760	3.576
80	7.062	5.564	4.877	4.471	4.197	3.998	3.722	3.538
100	7.038	5.541	4.855	4.449	4.175	3.975	3.700	3.516
150	7.007	5.511	4.826	4.420	4.146	3.946	3.670	3.486
300	6.973	5.480	4.796	4.390	4.116	3.916	3.640	3.455
1000	6.950	5.459	4.775	4.369	4.095	3.895	3.619	3.433

Exhibit 7.7d Critical Values for Adequate Signal-to-Noise: $\alpha = 0.25$, $\gamma = 0.99$

Degrees of Freedom, Denominator $(n-p)$	Degrees of Freedom, Numerator (p_2)							
	1	2	3	4	5	6	8	10
10	12.416	8.924	7.521	6.739	6.230	5.870	5.385	5.069
11	12.233	8.779	7.392	6.618	6.116	5.759	5.280	4.969
12	12.081	8.660	7.285	6.519	6.021	5.668	5.194	4.885
13	11.958	8.560	7.196	6.435	5.942	5.591	5.121	4.815
14	11.852	8.475	7.120	6.365	5.874	5.526	5.059	4.756
15	11.760	8.402	7.055	6.304	5.816	5.470	5.006	4.704
16	11.682	8.339	6.998	6.251	5.765	5.421	4.959	4.659
17	11.612	8.284	6.948	6.204	5.721	5.378	4.918	4.619
18	11.551	8.235	6.904	6.163	5.682	5.340	4.882	4.584
19	11.497	8.191	6.865	6.126	5.646	5.306	4.849	4.552
20	11.448	8.152	6.830	6.093	5.615	5.276	4.820	4.524
22	11.364	8.085	6.770	6.036	5.561	5.223	4.770	4.475
24	11.295	8.029	6.720	5.990	5.516	5.179	4.728	4.434
26	11.237	7.982	6.677	5.950	5.478	5.142	4.692	4.400
28	11.187	7.942	6.641	5.916	5.445	5.111	4.662	4.370
30	11.144	7.908	6.610	5.887	5.417	5.083	4.636	4.344
35	11.059	7.839	6.548	5.828	5.361	5.029	4.583	4.293
40	10.996	7.788	6.502	5.785	5.319	4.988	4.544	4.254
50	10.908	7.718	6.438	5.724	5.260	4.931	4.488	4.200
60	10.850	7.670	6.395	5.684	5.222	4.893	4.451	4.164
80	10.778	7.612	6.342	5.634	5.173	4.846	4.405	4.118
100	10.735	7.577	6.311	5.604	5.144	4.817	4.377	4.090
150	10.681	7.532	6.269	5.565	5.106	4.780	4.341	4.054
300	10.623	7.486	6.227	5.524	5.066	4.741	4.303	4.017
1000	10.583	7.453	6.197	5.496	5.039	4.715	4.277	3.991

Exhibit 7.7e Critical Values for Adequate Signal-to-Noise: $\alpha = 0.25$, $\gamma = 0.999$

Degrees of Freedom, Denominator	Degrees of Freedom, Numerator (p_2)							
$(n-p)$	1	2	3	4	5	6	8	10
10	18.854	12.421	10.000	8.693	7.882	7.281	6.514	6.023
11	18.547	12.205	9.818	8.531	7.712	7.141	6.385	5.902
12	18.295	12.027	9.669	8.397	7.589	7.024	6.278	5.801
13	18.084	11.878	9.544	8.286	7.486	6.927	6.188	5.716
14	17.906	11.752	9.438	8.190	7.397	6.843	6.112	5.644
15	17.752	11.643	9.347	8.108	7.321	6.771	6.046	5.581
16	17.619	11.549	9.267	8.036	7.255	6.709	5.988	5.526
17	17.502	11.466	9.197	7.974	7.196	6.654	5.937	5.478
18	17.399	11.392	9.135	7.918	7.145	6.605	5.892	5.436
19	17.307	11.327	9.080	7.868	7.099	6.561	5.851	5.397
20	17.224	11.268	9.030	7.824	7.057	6.522	5.815	5.363
22	17.083	11.168	8.945	7.747	6.986	6.454	5.752	5.303
24	16.965	11.084	8.874	7.683	6.926	6.398	5.700	5.254
26	16.867	11.014	8.815	7.629	6.876	6.350	5.656	5.211
28	16.782	10.954	8.764	7.583	6.833	6.309	5.618	5.175
30	16.710	10.902	8.720	7.544	6.796	6.274	5.585	5.144
35	16.565	10.798	8.632	7.464	6.722	6.204	5.519	5.081
40	16.458	10.721	8.566	7.404	6.666	6.150	5.469	5.034
50	16.308	10.614	8.475	7.322	6.589	6.076	5.400	4.967
60	16.209	10.543	8.414	7.266	6.537	6.027	5.353	4.922
80	16.086	10.454	8.338	7.197	6.472	5.965	5.295	4.866
100	16.013	10.401	8.293	7.156	6.433	5.928	5.260	4.832
150	15.917	10.332	8.234	7.102	6.382	5.879	5.214	4.787
300	15.820	10.261	8.173	7.046	6.329	5.828	5.165	4.741
1000	15.752	10.212	8.130	7.008	6.293	5.794	5.132	4.708

Exhibit 7.7f Critical Values for Adequate Signal-to-Noise: $\alpha = 0.25$, $\gamma = 0.9999$

Degrees of Freedom, Denominator	Degrees of Freedom, Numerator (p_2)							
$(n-p)$	1	2	3	4	5	6	8	10
10	25.376	15.887	12.422	10.583	9.428	8.627	7.579	6.915
11	24.936	15.599	12.190	10.381	9.244	8.457	7.426	6.774
12	24.574	15.361	11.998	10.214	9.093	8.316	7.300	6.657
13	24.272	15.162	11.837	10.073	8.965	8.198	7.194	6.558
14	24.015	14.992	11.700	9.954	8.857	8.097	7.103	6.474
15	23.793	14.846	11.582	9.850	8.763	8.010	7.025	6.401
16	23.601	14.719	11.479	9.760	8.681	7.934	6.956	6.337
17	23.432	14.607	11.388	9.681	8.609	7.867	6.896	6.281
18	23.283	14.508	11.308	9.611	8.546	7.807	6.842	6.231
19	23.150	14.420	11.237	9.549	8.488	7.754	6.794	6.186
20	23.031	14.341	11.172	9.492	8.437	7.706	6.751	6.146
22	22.826	14.205	11.062	9.395	8.349	7.624	6.676	6.076
24	22.656	14.092	10.970	9.315	8.275	7.555	6.614	6.018
26	22.513	13.997	10.893	9.247	8.213	7.497	6.561	5.969
28	22.391	13.916	10.826	9.188	8.160	7.448	6.516	5.926
30	22.285	13.845	10.769	9.138	8.114	7.404	6.477	5.890
35	22.075	13.705	10.654	9.037	8.021	7.318	6.398	5.816
40	21.918	13.600	10.568	8.961	7.952	7.253	6.339	5.760
50	21.700	13.454	10.449	8.855	7.855	7.162	6.255	5.681
60	21.555	13.357	10.369	8.785	7.790	7.101	6.199	5.629
80	21.375	13.236	10.269	8.697	7.709	7.025	6.129	5.562
100	21.268	13.163	10.210	8.644	7.660	6.978	6.087	5.522
150	21.128	13.069	10.131	8.574	7.596	6.918	6.031	5.468
300	20.982	12.970	10.051	8.502	7.529	6.855	5.972	5.413
1000	20.882	12.903	9.995	8.453	7.484	6.812	5.932	5.375

Exhibit 7.7g Critical Values for Adequate Signal-to-Noise: $\alpha = 0.25$, $\gamma = 0.99999$

Degrees of Freedom, Denominator $(n-p)$	Degrees of Freedom, Numerator (p_2)							
	1	2	3	4	5	6	8	10
10	31.951	19.338	14.814	12.437	10.955	9.934	8.605	7.770
11	31.373	18.976	14.529	12.194	10.738	9.735	8.430	7.609
12	30.897	18.676	14.294	11.994	10.558	9.570	8.285	7.476
13	30.498	18.425	14.097	11.825	10.408	9.432	8.162	7.364
14	30.158	18.212	13.929	11.681	10.279	9.313	8.058	7.269
15	29.866	18.027	13.784	11.557	10.168	9.211	7.968	7.186
16	29.612	17.867	13.658	11.448	10.070	9.122	7.888	7.113
17	29.388	17.726	13.546	11.353	9.985	9.043	7.819	7.049
18	29.191	17.601	13.448	11.268	9.909	8.973	7.757	6.992
19	29.014	17.489	13.360	11.192	9.841	8.911	7.702	6.941
20	28.856	17.389	13.281	11.124	9.780	8.855	7.652	6.895
22	28.584	17.216	13.144	11.007	9.675	8.758	7.565	6.816
24	28.358	17.073	13.031	10.909	9.587	8.677	7.493	6.749
26	28.168	16.952	12.935	10.827	9.513	8.608	7.432	6.693
28	28.005	16.849	12.853	10.756	9.449	8.549	7.380	6.645
30	27.864	16.759	12.782	10.695	9.394	8.498	7.334	6.602
35	27.584	16.580	12.640	10.572	9.283	8.396	7.243	6.518
40	27.375	16.446	12.533	10.480	9.200	8.319	7.174	6.454
50	27.083	16.260	12.384	10.351	9.084	8.211	7.077	6.364
60	26.888	16.135	12.285	10.265	9.006	8.139	7.012	6.303
80	26.646	15.980	12.161	10.157	8.908	8.048	6.930	6.226
100	26.502	15.886	12.086	10.092	8.849	7.993	6.880	6.180
150	26.312	15.765	11.988	10.007	8.771	7.920	6.814	6.118
300	26.117	15.638	11.887	9.918	8.691	7.845	6.746	6.054
1000	25.982	15.552	11.817	9.857	8.635	7.793	6.698	6.009

Exhibit 7.7h Critical Values for Adequate Signal-to-Noise: $\alpha = 0.25$, $\gamma = 0.999999$

Degrees of Freedom, Denominator $(n-p)$	Degrees of Freedom, Numerator (p_2)							
	1	2	3	4	5	6	8	10
10	38.536	22.776	17.184	14.267	12.456	11.214	9.605	8.599
11	37.816	22.339	16.848	13.983	12.206	10.986	9.408	8.420
12	37.223	21.978	16.570	13.749	11.999	10.798	9.244	8.272
13	36.725	21.674	16.336	13.552	11.824	10.640	9.106	8.147
14	36.301	21.416	16.137	13.384	11.676	10.504	8.988	8.040
15	35.936	21.192	15.965	13.238	11.547	10.387	8.886	7.947
16	35.617	20.998	15.815	13.112	11.435	10.285	8.797	7.866
17	35.338	20.827	15.682	13.000	11.336	10.194	8.718	7.794
18	35.090	20.675	15.565	12.901	11.248	10.114	8.648	7.730
19	34.869	20.540	15.460	12.812	11.169	10.043	8.585	7.673
20	34.670	20.418	15.366	12.732	11.098	9.978	8.528	7.622
22	34.328	20.208	15.204	12.594	10.976	9.866	8.431	7.533
24	34.045	20.034	15.068	12.480	10.874	9.773	8.349	7.458
26	33.805	19.886	14.954	12.383	10.788	9.694	8.280	7.395
28	33.600	19.760	14.856	12.299	10.714	9.627	8.220	7.341
30	33.423	19.651	14.771	12.227	10.650	9.568	8.169	7.293
35	33.068	19.432	14.601	12.082	10.521	9.450	8.065	7.198
40	32.804	19.268	14.473	11.974	10.424	9.361	7.986	7.126
50	32.434	19.039	14.294	11.821	10.287	9.236	7.876	7.025
60	32.188	18.886	14.175	11.719	10.196	9.152	7.801	6.956
80	31.880	18.695	14.025	11.591	10.081	9.046	7.707	6.869
100	31.696	18.580	13.935	11.514	10.012	8.982	7.650	6.817
150	31.454	18.429	13.816	11.412	9.921	8.898	7.574	6.746
300	31.205	18.273	13.693	11.306	9.826	8.810	7.496	6.674
1000	31.034	18.166	13.609	11.233	9.760	8.749	7.441	6.623

Exhibit 7.8a Critical Values for Adequate Signal-to-Noise: $\alpha = 0.10$, $\gamma = 0.75$

Degrees of Freedom, Denominator $(n-p)$	Degrees of Freedom, Numerator (p_2) 1	2	3	4	5	6	8	10
10	7.340	6.570	6.100	5.800	5.591	5.436	5.221	5.076
11	7.193	6.408	5.934	5.631	5.421	5.266	5.049	4.904
12	7.073	6.276	5.799	5.495	5.284	5.127	4.910	4.764
13	6.974	6.167	5.687	5.382	5.169	5.012	4.794	4.648
14	6.891	6.076	5.593	5.286	5.073	4.916	4.696	4.550
15	6.820	5.998	5.513	5.205	4.991	4.833	4.613	4.466
16	6.758	5.930	5.444	5.135	4.921	4.762	4.541	4.393
17	6.704	5.872	5.384	5.074	4.859	4.700	4.478	4.329
18	6.657	5.820	5.331	5.020	4.805	4.645	4.422	4.273
19	6.615	5.774	5.284	4.973	4.756	4.596	4.373	4.224
20	6.578	5.733	5.242	4.930	4.713	4.553	4.329	4.179
22	6.514	5.664	5.170	4.857	4.640	4.479	4.254	4.103
24	6.461	5.606	5.111	4.798	4.579	4.417	4.192	4.040
26	6.417	5.558	5.062	4.747	4.528	4.366	4.139	3.987
28	6.380	5.517	5.020	4.705	4.485	4.322	4.095	3.942
30	6.347	5.482	4.984	4.668	4.448	4.285	4.057	3.903
35	6.284	5.413	4.913	4.596	4.375	4.210	3.981	3.826
40	6.236	5.361	4.860	4.542	4.320	4.155	3.924	3.768
50	6.171	5.290	4.787	4.468	4.245	4.079	3.846	3.688
60	6.128	5.244	4.739	4.418	4.195	4.028	3.794	3.635
80	6.075	5.186	4.680	4.358	4.133	3.966	3.730	3.569
100	6.043	5.152	4.644	4.322	4.096	3.928	3.691	3.530
150	6.002	5.108	4.600	4.275	4.049	3.880	3.641	3.478
300	5.958	5.062	4.552	4.228	4.000	3.830	3.590	3.425
1000	5.928	5.032	4.521	4.195	3.967	3.796	3.554	3.389

Exhibit 7.8b Critical Values for Adequate Signal-to-Noise: $\alpha = 0.10$, $\gamma = 0.90$

Degrees of Freedom, Denominator $(n-p)$	Degrees of Freedom, Numerator (p_2) 1	2	3	4	5	6	8	10
10	10.933	8.716	7.710	7.121	6.729	6.445	6.060	5.805
11	10.687	8.486	7.490	6.907	6.519	6.238	5.857	5.606
12	10.487	8.300	7.311	6.733	6.348	6.070	5.692	5.443
13	10.321	8.146	7.163	6.589	6.206	5.931	5.555	5.308
14	10.182	8.016	7.039	6.468	6.087	5.813	5.440	5.194
15	10.063	7.905	6.932	6.364	5.985	5.712	5.341	5.096
16	9.960	7.810	6.841	6.274	5.898	5.626	5.255	5.012
17	9.871	7.726	6.761	6.196	5.821	5.550	5.181	4.938
18	9.792	7.653	6.690	6.128	5.753	5.483	5.115	4.873
19	9.722	7.588	6.628	6.067	5.693	5.424	5.056	4.815
20	9.660	7.530	6.572	6.013	5.640	5.371	5.004	4.763
22	9.553	7.431	6.477	5.920	5.548	5.280	4.915	4.674
24	9.466	7.350	6.399	5.843	5.473	5.205	4.841	4.600
26	9.392	7.282	6.333	5.779	5.410	5.143	4.779	4.539
28	9.330	7.224	6.278	5.725	5.356	5.089	4.726	4.486
30	9.276	7.174	6.230	5.678	5.309	5.043	4.680	4.441
35	9.170	7.075	6.135	5.585	5.218	4.952	4.590	4.350
40	9.092	7.002	6.064	5.516	5.150	4.884	4.522	4.283
50	8.983	6.901	5.967	5.420	5.055	4.791	4.429	4.190
60	8.911	6.835	5.903	5.358	4.993	4.729	4.367	4.127
80	8.823	6.753	5.824	5.280	4.916	4.652	4.290	4.050
100	8.770	6.704	5.777	5.233	4.870	4.606	4.244	4.004
150	8.704	6.642	5.716	5.173	4.810	4.546	4.184	3.943
300	8.631	6.577	5.654	5.112	4.748	4.485	4.122	3.881
1000	8.584	6.533	5.611	5.069	4.706	4.443	4.080	3.838

Exhibit 7.8c Critical Values for Adequate Signal-to-Noise: $\alpha = 0.10$, $\gamma = 0.95$

Degrees of Freedom, Denominator $(n-p)$	Degrees of Freedom, Numerator (p_2)							
	1	2	3	4	5	6	8	10
10	13.665	10.282	8.858	8.049	7.518	7.139	6.628	6.295
11	13.337	10.000	8.598	7.801	7.279	6.907	6.404	6.076
12	13.070	9.772	8.387	7.600	7.085	6.718	6.222	5.898
13	12.849	9.583	8.212	7.434	6.924	6.560	6.070	5.750
14	12.663	9.423	8.065	7.294	6.789	6.428	5.942	5.626
15	12.504	9.288	7.939	7.174	6.673	6.315	5.833	5.519
16	12.368	9.170	7.830	7.070	6.572	6.217	5.738	5.426
17	12.248	9.068	7.736	6.980	6.485	6.132	5.656	5.345
18	12.143	8.978	7.653	6.901	6.408	6.057	5.583	5.274
19	12.050	8.899	7.579	6.830	6.340	5.990	5.518	5.210
20	11.967	8.828	7.513	6.767	6.279	5.930	5.460	5.154
22	11.825	8.706	7.400	6.660	6.174	5.828	5.361	5.056
24	11.708	8.606	7.308	6.571	6.088	5.744	5.279	4.976
26	11.611	8.522	7.230	6.497	6.016	5.673	5.210	4.908
28	11.528	8.451	7.164	6.434	5.955	5.613	5.151	4.850
30	11.456	8.390	7.107	6.379	5.902	5.561	5.101	4.800
35	11.315	8.269	6.995	6.272	5.797	5.458	5.000	4.701
40	11.210	8.179	6.911	6.192	5.719	5.381	4.925	4.627
50	11.066	8.055	6.796	6.081	5.611	5.275	4.821	4.524
60	10.970	7.973	6.720	6.008	5.540	5.205	4.752	4.456
80	10.852	7.872	6.626	5.917	5.452	5.118	4.666	4.370
100	10.782	7.812	6.570	5.863	5.399	5.066	4.615	4.319
150	10.695	7.735	6.497	5.793	5.330	4.998	4.548	4.252
300	10.601	7.655	6.423	5.721	5.260	4.929	4.479	4.183
1000	10.536	7.600	6.372	5.672	5.211	4.881	4.432	4.136

Exhibit 7.8d Critical Values for Adequate Signal-to-Noise: $\alpha = 0.10$, $\gamma = 0.99$

Degrees of Freedom, Denominator $(n-p)$	Degrees of Freedom, Numerator (p_2)							
	1	2	3	4	5	6	8	10
10	20.024	13.817	11.402	10.078	9.228	8.630	7.835	7.323
11	19.489	13.413	11.051	9.756	8.926	8.342	7.565	7.065
12	19.055	13.085	10.766	9.495	8.680	8.107	7.345	6.855
13	18.695	12.813	10.529	9.278	8.476	7.912	7.162	6.680
14	18.392	12.584	10.330	9.095	8.304	7.747	7.008	6.532
15	18.134	12.389	10.160	8.939	8.156	7.606	6.876	6.406
16	17.911	12.220	10.012	8.804	8.029	7.484	6.761	6.296
17	17.716	12.073	9.884	8.686	7.918	7.378	6.661	6.200
18	17.546	11.943	9.771	8.582	7.820	7.284	6.573	6.115
19	17.394	11.828	9.671	8.490	7.733	7.201	6.494	6.040
20	17.259	11.726	9.582	8.408	7.656	7.127	6.424	5.973
22	17.028	11.551	9.429	8.267	7.522	6.999	6.304	5.857
24	16.838	11.407	9.303	8.151	7.413	6.894	6.204	5.761
26	16.678	11.286	9.197	8.054	7.321	6.805	6.121	5.680
28	16.543	11.183	9.107	7.971	7.242	6.730	6.049	5.612
30	16.426	11.095	9.030	7.900	7.175	6.665	5.988	5.552
35	16.196	10.920	8.876	7.758	7.040	6.536	5.865	5.434
40	16.026	10.790	8.763	7.653	6.941	6.440	5.774	5.346
50	15.790	10.610	8.605	7.507	6.802	6.307	5.647	5.223
60	15.634	10.492	8.501	7.411	6.711	6.218	5.563	5.141
80	15.442	10.346	8.372	7.291	6.597	6.108	5.458	5.038
100	15.328	10.258	8.296	7.220	6.529	6.043	5.395	4.977
150	15.183	10.145	8.195	7.127	6.440	5.957	5.312	4.896
300	15.031	10.030	8.094	7.032	6.350	5.869	5.228	4.813
1000	14.928	9.950	8.022	6.967	6.287	5.809	5.170	4.756

Exhibit 7.8e Critical Values for Adequate Signal-to-Noise: $\alpha = 0.10$, $\gamma = 0.999$

Degrees of Freedom, Denominator $(n-p)$	Degrees of Freedom, Numerator (p_2) 1	2	3	4	5	6	8	10
10	29.157	18.749	14.887	12.820	11.515	10.607	9.414	8.655
11	28.302	18.166	14.407	12.396	11.126	10.242	9.083	8.345
12	27.609	17.692	14.016	12.050	10.809	9.946	8.813	8.092
13	27.034	17.298	13.691	11.763	10.545	9.699	8.588	7.882
14	26.550	16.967	13.418	11.521	10.323	9.491	8.398	7.704
15	26.137	16.684	13.184	11.314	10.133	9.313	8.236	7.552
16	25.780	16.440	12.982	11.135	9.969	9.158	8.095	7.420
17	25.469	16.226	12.806	10.978	9.825	9.024	7.972	7.304
18	25.195	16.039	12.651	10.841	9.698	8.905	7.863	7.202
19	24.952	15.872	12.513	10.718	9.586	8.799	7.766	7.111
20	24.736	15.723	12.390	10.609	9.485	8.704	7.680	7.029
22	24.365	15.469	12.179	10.421	9.312	8.542	7.531	6.890
24	24.060	15.259	12.005	10.267	9.170	8.408	7.408	6.773
26	23.804	15.083	11.859	10.137	9.050	8.295	7.305	6.676
28	23.587	14.933	11.734	10.026	8.948	8.199	7.217	6.593
30	23.400	14.804	11.627	9.931	8.860	8.116	7.140	6.521
35	23.030	14.549	11.415	9.741	8.685	7.952	6.989	6.377
40	22.755	14.359	11.257	9.600	8.555	7.829	6.875	6.270
50	22.375	14.097	11.038	9.405	8.374	7.658	6.717	6.120
60	22.125	13.923	10.893	9.275	8.254	7.544	6.612	6.020
80	21.815	13.708	10.714	9.114	8.105	7.403	6.481	5.895
100	21.630	13.580	10.607	9.018	8.015	7.318	6.402	5.820
150	21.390	13.414	10.466	8.893	7.898	7.207	6.298	5.720
300	21.145	13.242	10.324	8.764	7.779	7.093	6.192	5.618
1000	20.976	13.125	10.225	8.676	7.697	7.015	6.119	5.548

Exhibit 7.8f Critical Values for Adequate Signal-to-Noise: $\alpha = 0.10$, $\gamma = 0.9999$

Degrees of Freedom, Denominator $(n-p)$	Degrees of Freedom, Numerator (p_2) 1	2	3	4	5	6	8	10
10	38.327	23.609	18.278	15.464	13.702	12.485	10.899	9.898
11	37.137	22.843	17.669	14.938	13.228	12.048	10.510	9.539
12	36.170	22.220	17.173	14.509	12.842	11.692	10.192	9.246
13	35.368	21.704	16.761	14.153	12.522	11.395	9.928	9.003
14	34.692	21.268	16.413	13.852	12.251	11.145	9.704	8.796
15	34.115	20.895	16.116	13.595	12.019	10.931	9.513	8.620
16	33.616	20.573	15.859	13.373	11.818	10.745	9.347	8.466
17	33.181	20.292	15.634	13.178	11.642	10.582	9.202	8.332
18	32.798	20.044	15.436	13.006	11.487	10.439	9.073	8.213
19	32.458	19.825	15.260	12.854	11.350	10.311	8.959	8.108
20	32.155	19.628	15.103	12.718	11.226	10.197	8.857	8.013
22	31.635	19.292	14.834	12.484	11.015	10.001	8.681	7.850
24	31.207	19.014	14.611	12.291	10.840	9.839	8.536	7.715
26	30.848	18.781	14.424	12.128	10.693	9.703	8.414	7.602
28	30.543	18.583	14.265	11.990	10.568	9.587	8.309	7.504
30	30.280	18.412	14.128	11.871	10.460	9.486	8.219	7.420
35	29.759	18.073	13.856	11.634	10.245	9.286	8.039	7.253
40	29.372	17.821	13.653	11.457	10.084	9.137	7.904	7.127
50	28.836	17.471	13.371	11.211	9.861	8.929	7.716	6.952
60	28.482	17.240	13.185	11.048	9.712	8.791	7.590	6.834
80	28.043	16.953	12.953	10.845	9.528	8.618	7.434	6.687
100	27.782	16.781	12.814	10.724	9.417	8.515	7.340	6.599
150	27.440	16.558	12.633	10.565	9.271	8.378	7.215	6.481
300	27.094	16.329	12.447	10.401	9.122	8.239	7.088	6.361
1000	26.852	16.172	12.320	10.290	9.020	8.143	6.999	6.277

Exhibit 7.8g Critical Values for Adequate Signal-to-Noise: $\alpha = 0.10$, $\gamma = 0.99999$

Degrees of Freedom, Denominator	Degrees of Freedom, Numerator (p_2)							
$(n - p)$	1	2	3	4	5	6	8	10
10	47.530	28.430	21.616	18.049	15.830	14.306	12.328	11.087
11	45.993	27.480	20.877	17.422	15.274	13.797	11.882	10.682
12	44.744	26.707	20.276	16.912	14.820	13.382	11.519	10.350
13	43.706	26.064	19.776	16.487	14.442	13.037	11.216	10.074
14	42.832	25.523	19.354	16.129	14.123	12.746	10.960	9.841
15	42.085	25.059	18.994	15.822	13.850	12.496	10.740	9.640
16	41.439	24.658	18.681	15.556	13.613	12.279	10.550	9.466
17	40.875	24.308	18.408	15.323	13.406	12.090	10.383	9.314
18	40.378	23.999	18.167	15.118	13.223	11.922	10.236	9.180
19	39.937	23.725	17.953	14.936	13.061	11.773	10.105	9.060
20	39.543	23.480	17.761	14.772	12.915	11.640	9.987	8.952
22	38.868	23.059	17.433	14.492	12.665	11.411	9.785	8.767
24	38.311	22.712	17.162	14.261	12.459	11.221	9.618	8.614
26	37.844	22.421	16.934	14.066	12.285	11.062	9.477	8.485
28	37.446	22.173	16.739	13.900	12.136	10.925	9.356	8.374
30	37.103	21.959	16.571	13.757	12.008	10.808	9.252	8.278
35	36.423	21.533	16.238	13.471	11.753	10.573	9.044	8.087
40	35.917	21.217	15.989	13.259	11.562	10.398	8.889	7.944
50	35.216	20.777	15.644	12.962	11.296	10.153	8.671	7.744
60	34.752	20.486	15.414	12.765	11.120	9.990	8.526	7.609
80	34.177	20.124	15.129	12.520	10.899	9.786	8.344	7.441
100	33.834	19.907	14.958	12.372	10.766	9.664	8.235	7.340
150	33.383	19.624	14.734	12.178	10.592	9.502	8.089	7.204
300	32.926	19.334	14.504	11.981	10.413	9.336	7.940	7.066
1000	32.611	19.136	14.346	11.844	10.290	9.222	7.837	6.969

Exhibit 7.8h Critical Values for Adequate Signal-to-Noise: $\alpha = 0.10$, $\gamma = 0.999999$

Degrees of Freedom, Denominator	Degrees of Freedom, Numerator (p_2)							
$(n - p)$	1	2	3	4	5	6	8	10
10	56.720	33.223	24.918	20.596	17.920	16.087	13.719	12.240
11	54.831	32.087	24.052	19.870	17.281	15.508	13.218	11.789
12	53.294	31.162	23.345	19.278	16.760	15.036	12.810	11.420
13	52.018	30.394	22.757	18.785	16.326	14.643	12.469	11.113
14	50.941	29.745	22.261	18.369	15.960	14.311	12.181	10.853
15	50.020	29.189	21.836	18.012	15.646	14.026	11.934	10.629
16	49.223	28.708	21.468	17.703	15.373	13.778	11.720	10.435
17	48.527	28.288	21.146	17.433	15.135	13.562	11.532	10.266
18	47.914	27.917	20.862	17.194	14.924	13.371	11.366	10.115
19	47.368	27.588	20.609	16.982	14.737	13.201	11.218	9.982
20	46.881	27.293	20.383	16.792	14.569	13.048	11.086	9.862
22	46.046	26.788	19.996	16.465	14.281	12.787	10.858	9.655
24	45.357	26.370	19.675	16.195	14.043	12.570	10.669	9.484
26	44.778	26.019	19.405	15.968	13.842	12.387	10.510	9.339
28	44.285	25.720	19.175	15.774	13.670	12.231	10.374	9.216
30	43.860	25.462	18.976	15.606	13.522	12.096	10.256	9.108
35	43.016	24.949	18.581	15.273	13.227	11.827	10.020	8.894
40	42.388	24.566	18.286	15.024	13.006	11.625	9.844	8.734
50	41.514	24.034	17.875	14.676	12.697	11.344	9.597	8.508
60	40.936	23.680	17.602	14.444	12.492	11.156	9.432	8.358
80	40.218	23.241	17.261	14.156	12.235	10.921	9.225	8.168
100	39.788	22.978	17.057	13.982	12.081	10.780	9.101	8.054
150	39.226	22.632	16.789	13.753	11.877	10.591	8.934	7.901
300	38.650	22.278	16.513	13.519	11.668	10.400	8.765	7.744
1000	38.257	22.036	16.325	13.357	11.523	10.267	8.646	7.634

Exhibit 7.9a Critical Values for Adequate Signal-to-Noise: $\alpha = 0.05$, $\gamma = 0.75$

Degrees of Freedom, Denominator	Degrees of Freedom, Numerator (p_2)							
$(n - p)$	1	2	3	4	5	6	8	10
10	10.477	8.872	8.056	7.563	7.230	6.988	6.656	6.438
11	10.191	8.579	7.766	7.274	6.942	6.700	6.370	6.152
12	9.960	8.344	7.531	7.041	6.710	6.469	6.140	5.923
13	9.770	8.150	7.339	6.850	6.519	6.279	5.950	5.734
14	9.610	7.989	7.178	6.690	6.360	6.120	5.792	5.575
15	9.475	7.852	7.042	6.554	6.224	5.985	5.657	5.441
16	9.358	7.734	6.925	6.438	6.108	5.869	5.541	5.325
17	9.257	7.631	6.824	6.337	6.008	5.768	5.441	5.224
18	9.168	7.542	6.735	6.248	5.919	5.680	5.352	5.136
19	9.090	7.462	6.656	6.170	5.841	5.602	5.274	5.058
20	9.020	7.392	6.586	6.100	5.772	5.533	5.205	4.988
22	8.901	7.272	6.467	5.982	5.653	5.414	5.086	4.870
24	8.803	7.174	6.370	5.885	5.556	5.318	4.989	4.772
26	8.722	7.092	6.288	5.804	5.476	5.237	4.908	4.691
28	8.653	7.022	6.219	5.735	5.407	5.168	4.839	4.622
30	8.593	6.963	6.160	5.676	5.348	5.109	4.780	4.562
35	8.476	6.846	6.044	5.560	5.232	4.993	4.663	4.445
40	8.389	6.759	5.958	5.475	5.147	4.907	4.577	4.358
50	8.271	6.640	5.840	5.357	5.029	4.789	4.458	4.237
60	8.193	6.562	5.762	5.280	4.951	4.711	4.380	4.158
80	8.096	6.466	5.667	5.184	4.856	4.616	4.283	4.060
100	8.039	6.409	5.610	5.128	4.800	4.559	4.225	4.002
150	7.965	6.336	5.538	5.054	4.726	4.485	4.150	3.925
300	7.884	6.260	5.465	4.981	4.652	4.409	4.074	3.848
1000	7.832	6.211	5.414	4.931	4.600	4.358	4.022	3.795

Exhibit 7.9b Critical Values for Adequate Signal-to-Noise: $\alpha = 0.05$, $\gamma = 0.90$

Degrees of Freedom, Denominator	Degrees of Freedom, Numerator (p_2)							
$(n - p)$	1	2	3	4	5	6	8	10
10	15.066	11.583	10.080	9.219	8.653	8.248	7.701	7.344
11	14.606	11.177	9.700	8.854	8.298	7.900	7.364	7.015
12	14.236	10.850	9.393	8.560	8.012	7.621	7.093	6.749
13	13.932	10.581	9.142	8.319	7.778	7.391	6.870	6.530
14	13.677	10.356	8.931	8.117	7.581	7.199	6.683	6.347
15	13.461	10.166	8.753	7.946	7.415	7.035	6.524	6.191
16	13.275	10.002	8.600	7.798	7.272	6.895	6.388	6.057
17	13.114	9.860	8.467	7.671	7.147	6.773	6.269	5.940
18	12.972	9.735	8.350	7.559	7.038	6.666	6.165	5.838
19	12.847	9.625	8.247	7.460	6.942	6.572	6.073	5.747
20	12.736	9.527	8.156	7.372	6.856	6.488	5.991	5.666
22	12.546	9.361	8.000	7.222	6.710	6.344	5.851	5.529
24	12.390	9.224	7.872	7.099	6.590	6.227	5.736	5.416
26	12.260	9.110	7.765	6.996	6.490	6.129	5.640	5.321
28	12.150	9.014	7.675	6.909	6.406	6.045	5.558	5.240
30	12.056	8.931	7.598	6.835	6.333	5.974	5.488	5.171
35	11.870	8.768	7.445	6.688	6.189	5.833	5.350	5.035
40	11.732	8.648	7.332	6.579	6.083	5.728	5.248	4.933
50	11.544	8.482	7.177	6.430	5.937	5.584	5.106	4.793
60	11.419	8.374	7.075	6.332	5.841	5.490	5.013	4.700
80	11.266	8.240	6.950	6.211	5.723	5.373	4.898	4.586
100	11.176	8.161	6.876	6.139	5.653	5.304	4.829	4.518
150	11.063	8.059	6.779	6.046	5.561	5.213	4.740	4.428
300	10.942	7.955	6.681	5.951	5.469	5.122	4.649	4.337
1000	10.856	7.883	6.616	5.886	5.404	5.059	4.587	4.275

Exhibit 7.9*c* Critical Values for Adequate Signal-to-Noise: $\alpha = 0.05$, $\gamma = 0.95$

Degrees of Freedom, Denominator $(n - p)$	Degrees of Freedom, Numerator (p_2)							
	1	2	3	4	5	6	8	10
10	18.519	13.550	11.517	10.378	9.637	9.112	8.408	7.952
11	17.919	13.057	11.070	9.958	9.235	8.723	8.037	7.592
12	17.436	12.660	10.711	9.620	8.912	8.410	7.737	7.302
13	17.040	12.334	10.416	9.343	8.646	8.152	7.491	7.063
14	16.708	12.062	10.170	9.111	8.424	7.936	7.285	6.863
15	16.426	11.830	9.960	8.914	8.235	7.754	7.109	6.692
16	16.183	11.632	9.780	8.745	8.072	7.596	6.958	6.546
17	15.973	11.460	9.624	8.598	7.932	7.459	6.827	6.418
18	15.789	11.309	9.487	8.469	7.808	7.339	6.712	6.307
19	15.626	11.175	9.366	8.355	7.699	7.233	6.611	6.208
20	15.481	11.056	9.259	8.254	7.601	7.139	6.520	6.119
22	15.234	10.854	9.076	8.081	7.436	6.978	6.365	5.969
24	15.032	10.689	8.926	7.940	7.300	6.846	6.238	5.845
26	14.863	10.551	8.800	7.822	7.186	6.736	6.132	5.741
28	14.720	10.434	8.694	7.722	7.090	6.642	6.042	5.653
30	14.597	10.333	8.603	7.636	7.007	6.562	5.964	5.577
35	14.355	10.135	8.424	7.466	6.844	6.403	5.811	5.427
40	14.177	9.990	8.292	7.341	6.724	6.286	5.698	5.316
50	13.932	9.789	8.109	7.169	6.558	6.124	5.541	5.162
60	13.770	9.657	7.989	7.056	6.448	6.017	5.438	5.061
80	13.572	9.495	7.842	6.916	6.314	5.885	5.310	4.935
100	13.454	9.398	7.754	6.833	6.234	5.807	5.234	4.860
150	13.304	9.274	7.641	6.726	6.129	5.705	5.134	4.761
300	13.152	9.149	7.526	6.616	6.024	5.602	5.033	4.661
1000	13.046	9.061	7.447	6.541	5.950	5.531	4.964	4.593

Exhibit 7.9*d* Critical Values for Adequate Signal-to-Noise: $\alpha = 0.05$, $\gamma = 0.99$

Degrees of Freedom, Denominator $(n - p)$	Degrees of Freedom, Numerator (p_2)							
	1	2	3	4	5	6	8	10
10	26.500	17.969	14.690	12.905	11.765	10.966	9.907	9.229
11	25.555	17.273	14.094	12.364	11.260	10.486	9.461	8.805
12	24.794	16.714	13.615	11.929	10.853	10.099	9.101	8.463
13	24.170	16.254	13.221	11.571	10.518	9.781	8.805	8.181
14	23.647	15.870	12.891	11.272	10.238	9.515	8.557	7.945
15	23.204	15.544	12.611	11.017	10.000	9.288	8.347	7.744
16	22.823	15.264	12.371	10.799	9.796	9.094	8.165	7.571
17	22.492	15.021	12.162	10.609	9.618	8.925	8.007	7.421
18	22.202	14.808	11.979	10.442	9.462	8.776	7.869	7.288
19	21.946	14.619	11.817	10.295	9.324	8.645	7.746	7.171
20	21.719	14.452	11.673	10.164	9.202	8.528	7.637	7.067
22	21.331	14.166	11.428	9.941	8.992	8.328	7.450	6.889
24	21.013	13.932	11.227	9.758	8.821	8.165	7.297	6.742
26	20.748	13.737	11.059	9.605	8.677	8.028	7.169	6.619
28	20.523	13.572	10.917	9.475	8.555	7.912	7.060	6.515
30	20.331	13.430	10.795	9.364	8.451	7.812	6.966	6.425
35	19.951	13.150	10.554	9.144	8.244	7.614	6.780	6.247
40	19.671	12.943	10.376	8.981	8.091	7.468	6.643	6.115
50	19.285	12.659	10.130	8.757	7.880	7.266	6.453	5.932
60	19.032	12.472	9.969	8.609	7.741	7.133	6.328	5.811
80	18.720	12.242	9.770	8.427	7.570	6.969	6.172	5.661
100	18.535	12.106	9.652	8.319	7.468	6.871	6.080	5.572
150	18.298	11.929	9.499	8.179	7.334	6.743	5.957	5.454
300	18.055	11.748	9.344	8.036	7.200	6.613	5.834	5.334
1000	17.891	11.625	9.235	7.937	7.107	6.525	5.750	5.251

Exhibit 7.9e Critical Values for Adequate Signal-to-Noise: $\alpha = 0.05$, $\gamma = 0.999$

Degrees of Freedom, Denominator	Degrees of Freedom, Numerator (p_2)							
$(n - p)$	1	2	3	4	5	6	8	10
10	37.891	24.107	19.023	16.312	14.604	13.418	11.865	10.879
11	36.422	23.120	18.217	15.604	13.958	12.816	11.320	10.372
12	35.239	22.325	17.568	15.033	13.438	12.331	10.881	9.962
13	34.268	21.671	17.034	14.564	13.009	11.931	10.519	9.624
14	33.455	21.124	16.587	14.171	12.651	11.596	10.216	9.341
15	32.766	20.660	16.208	13.838	12.346	11.312	9.958	9.100
16	32.173	20.262	15.882	13.551	12.084	11.067	9.736	8.893
17	31.659	19.915	15.598	13.301	11.856	10.854	9.542	8.712
18	31.208	19.612	15.350	13.082	11.656	10.667	9.372	8.553
19	30.809	19.343	15.130	12.889	11.479	10.501	9.222	8.413
20	30.455	19.104	14.935	12.716	11.321	10.354	9.088	8.287
22	29.851	18.697	14.601	12.422	11.052	10.102	8.859	8.073
24	29.356	18.363	14.327	12.181	10.831	9.895	8.671	7.896
26	28.943	18.084	14.099	11.979	10.646	9.722	8.513	7.748
28	28.593	17.848	13.905	11.808	10.489	9.575	8.379	7.622
30	28.292	17.645	13.738	11.661	10.354	9.449	8.264	7.514
35	27.700	17.244	13.409	11.370	10.087	9.199	8.036	7.299
40	27.263	16.948	13.166	11.155	9.890	9.013	7.866	7.140
50	26.660	16.541	12.830	10.858	9.617	8.757	7.631	6.918
60	26.265	16.272	12.609	10.662	9.437	8.587	7.476	6.771
80	25.777	15.942	12.337	10.420	9.214	8.378	7.282	6.589
100	25.488	15.745	12.174	10.276	9.081	8.253	7.167	6.480
150	25.111	15.491	11.962	10.088	8.908	8.088	7.016	6.335
300	24.732	15.231	11.748	9.896	8.731	7.922	6.862	6.188
1000	24.467	15.052	11.601	9.767	8.611	7.808	6.756	6.088

Exhibit 7.9f Critical Values for Adequate Signal-to-Noise: $\alpha = 0.05$, $\gamma = 0.9999$

Degrees of Freedom, Denominator	Degrees of Freedom, Numerator (p_2)							
$(n - p)$	1	2	3	4	5	6	8	10
10	49.290	30.140	23.230	19.590	17.315	15.746	13.704	12.418
11	47.273	28.857	22.215	18.718	16.533	15.026	13.066	11.832
12	45.650	27.824	21.398	18.015	15.902	14.446	12.551	11.359
13	44.316	26.974	20.725	17.436	15.383	13.968	12.127	10.969
14	43.200	26.263	20.162	16.952	14.948	13.567	11.770	10.641
15	42.253	25.659	19.684	16.540	14.578	13.226	11.468	10.363
16	41.438	25.139	19.272	16.186	14.260	12.933	11.207	10.123
17	40.731	24.688	18.914	15.878	13.983	12.678	10.980	9.913
18	40.111	24.292	18.600	15.608	13.740	12.454	10.781	9.730
19	39.563	23.942	18.323	15.368	13.525	12.255	10.604	9.567
20	39.075	23.630	18.075	15.155	13.333	12.078	10.446	9.421
22	38.243	23.099	17.654	14.791	13.006	11.776	10.177	9.173
24	37.561	22.663	17.307	14.492	12.736	11.527	9.955	8.968
26	36.992	22.298	17.017	14.242	12.511	11.319	9.770	8.796
28	36.509	21.989	16.772	14.030	12.320	11.142	9.612	8.650
30	36.094	21.724	16.560	13.847	12.155	10.990	9.476	8.524
35	35.276	21.199	16.143	13.486	11.829	10.688	9.206	8.274
40	34.672	20.811	15.834	13.219	11.588	10.465	9.005	8.088
50	33.838	20.275	15.406	12.848	11.253	10.155	8.727	7.830
60	33.290	19.923	15.125	12.604	11.032	9.950	8.543	7.658
80	32.615	19.487	14.776	12.302	10.758	9.696	8.314	7.445
100	32.213	19.228	14.569	12.122	10.595	9.544	8.177	7.317
150	31.689	18.890	14.298	11.885	10.380	9.344	7.995	7.147
300	31.160	18.549	14.024	11.646	10.163	9.142	7.811	6.975
1000	30.794	18.315	13.836	11.483	10.015	9.002	7.685	6.856

Exhibit 7.9g Critical Values for Adequate Signal-to-Noise: $\alpha = 0.05$, $\gamma = 0.99999$

Degrees of Freedom, Denominator $(n-p)$	Degrees of Freedom, Numerator (p_2) 1	2	3	4	5	6	8	10
10	60.706	36.117	27.365	22.792	19.951	18.000	15.473	13.890
11	58.128	34.536	26.143	21.758	19.035	17.166	14.744	13.228
12	56.053	33.262	25.158	20.925	18.296	16.492	14.156	12.694
13	54.346	32.214	24.346	20.238	17.688	15.938	13.671	12.253
14	52.918	31.336	23.667	19.664	17.178	15.472	13.265	11.884
15	51.704	30.591	23.090	19.174	16.744	15.077	12.919	11.568
16	50.661	29.949	22.593	18.753	16.370	14.736	12.620	11.297
17	49.755	29.392	22.160	18.387	16.045	14.439	12.360	11.060
18	48.960	28.902	21.781	18.066	15.760	14.178	12.132	10.852
19	48.256	28.469	21.445	17.781	15.507	13.947	11.930	10.668
20	47.630	28.084	21.146	17.527	15.282	13.741	11.749	10.503
22	46.563	27.426	20.635	17.094	14.896	13.390	11.440	10.222
24	45.687	26.885	20.216	16.737	14.580	13.100	11.186	9.990
26	44.955	26.434	19.865	16.439	14.314	12.857	10.973	9.795
28	44.334	26.050	19.566	16.186	14.089	12.651	10.791	9.629
30	43.800	25.721	19.310	15.968	13.895	12.473	10.635	9.486
35	42.747	25.069	18.803	15.536	13.510	12.121	10.325	9.202
40	41.968	24.587	18.427	15.216	13.225	11.860	10.094	8.991
50	40.892	23.920	17.907	14.773	12.829	11.497	9.774	8.697
60	40.184	23.481	17.564	14.480	12.568	11.257	9.561	8.501
80	39.310	22.937	17.139	14.117	12.243	10.958	9.296	8.257
100	38.790	22.614	16.885	13.900	12.049	10.780	9.138	8.111
150	38.110	22.192	16.553	13.615	11.794	10.543	8.927	7.917
300	37.423	21.760	16.216	13.327	11.534	10.304	8.714	7.719
1000	36.952	21.468	15.986	13.128	11.356	10.140	8.567	7.582

Exhibit 7.9h Critical Values for Adequate Signal-to-Noise: $\alpha = 0.05$, $\gamma = 0.999999$

Degrees of Freedom, Denominator $(n-p)$	Degrees of Freedom, Numerator (p_2) 1	2	3	4	5	6	8	10
10	72.093	42.053	31.454	25.946	22.537	20.205	17.194	15.317
11	68.947	40.172	30.025	24.751	21.489	19.257	16.377	14.582
12	66.412	38.657	28.872	23.788	20.644	18.493	15.718	13.988
13	64.327	37.409	27.922	22.994	19.947	17.862	15.174	13.498
14	62.581	36.364	27.127	22.329	19.363	17.334	14.717	13.086
15	61.098	35.476	26.450	21.763	18.866	16.884	14.328	12.736
16	59.822	34.711	25.868	21.276	18.437	16.496	13.993	12.434
17	59.712	34.046	25.361	20.852	18.064	16.159	13.701	12.170
18	57.739	33.463	24.916	20.479	17.737	15.862	13.445	11.939
19	56.878	32.946	24.522	20.149	17.447	15.599	13.217	11.734
20	56.110	32.486	24.170	19.855	17.188	15.364	13.014	11.550
22	54.802	31.700	23.571	19.352	16.745	14.964	12.666	11.236
24	53.727	31.054	23.077	18.939	16.381	14.633	12.380	10.977
26	52.829	30.514	22.665	18.592	16.076	14.356	12.140	10.760
28	52.066	30.055	22.314	18.298	15.817	14.121	11.936	10.575
30	51.411	29.661	22.012	18.044	15.594	13.918	11.760	10.416
35	50.115	28.880	21.414	17.542	15.151	13.516	11.410	10.099
40	49.156	28.301	20.971	17.170	14.822	13.217	11.149	9.862
50	47.831	27.501	20.357	16.652	14.365	12.801	10.787	9.533
60	46.957	26.972	19.951	16.310	14.062	12.526	10.546	9.314
80	45.877	26.318	19.448	15.886	13.686	12.183	10.246	9.040
100	45.234	25.927	19.147	15.632	13.461	11.978	10.066	8.876
150	44.394	25.415	18.752	15.296	13.165	11.706	9.827	8.656
300	43.537	24.896	18.351	14.958	12.863	11.430	9.584	8.434
1000	42.957	24.541	18.076	14.723	12.655	11.239	9.415	8.279

Exhibit 7.10a Critical Values for Adequate Signal-to-Noise: $\alpha = 0.01$, $\gamma = 0.75$

Degrees of Freedom, Denominator	Degrees of Freedom, Numerator (p_2)							
$(n-p)$	1	2	3	4	5	6	8	10
10	19.588	15.480	13.652	12.598	11.903	11.408	10.742	10.311
11	18.702	14.667	12.879	11.848	11.170	10.686	10.036	9.616
12	18.002	14.026	12.270	11.259	10.594	10.120	9.482	9.670
13	17.434	13.509	11.780	10.785	10.130	9.663	9.036	8.630
14	16.965	13.083	11.376	10.394	9.749	9.288	8.669	8.268
15	16.571	12.726	11.038	10.068	9.430	8.974	8.362	7.966
16	16.236	12.422	10.751	9.791	9.159	8.708	8.102	7.709
17	15.947	12.161	10.505	9.553	8.926	8.479	7.878	7.489
18	15.695	11.935	10.291	9.346	8.725	8.281	7.684	7.297
19	15.474	11.736	10.103	9.165	8.548	8.107	7.514	7.129
20	15.278	11.560	9.938	9.005	8.392	7.954	7.364	6.981
22	14.948	11.264	9.658	8.736	8.129	7.695	7.110	6.731
24	14.679	11.024	9.432	8.518	7.915	7.485	6.905	6.529
26	14.456	10.825	9.245	8.337	7.739	7.312	6.736	6.361
28	14.269	10.658	9.087	8.186	7.591	7.166	6.593	6.220
30	14.109	10.515	8.953	8.056	7.465	7.042	6.471	6.100
35	13.796	10.236	8.692	7.804	7.217	6.800	6.234	5.865
40	13.566	10.033	8.500	7.620	7.039	6.623	6.060	5.693
50	13.253	9.756	8.240	7.369	6.794	6.382	5.824	5.459
60	13.049	9.576	8.072	7.207	6.635	6.226	5.670	5.306
80	12.800	9.356	7.866	7.008	6.442	6.035	5.482	5.120
100	12.654	9.227	7.745	6.892	6.328	5.922	5.372	5.010
150	12.464	9.059	7.588	6.741	6.179	5.776	5.228	4.867
300	12.279	8.896	7.432	6.589	6.033	5.632	5.085	4.725
1000	12.116	8.779	7.329	6.487	5.931	5.534	4.989	4.630

Exhibit 7.10b Critical Values for Adequate Signal-to-Noise: $\alpha = 0.01$, $\gamma = 0.90$

Degrees of Freedom, Denominator	Degrees of Freedom, Numerator (p_2)							
$(n-p)$	1	2	3	4	5	6	8	10
10	26.951	19.783	16.844	15.200	14.133	13.379	12.372	11.722
11	25.617	18.686	15.852	14.268	13.241	12.515	11.546	10.922
12	24.564	17.823	15.072	13.535	12.540	11.836	10.898	10.294
13	23.712	17.126	14.443	12.945	11.975	11.290	10.376	9.788
14	23.009	16.552	13.926	12.460	11.511	10.841	9.947	9.371
15	22.419	16.072	13.493	12.054	11.123	10.465	9.588	9.023
16	21.918	15.664	13.126	11.710	10.793	10.146	9.283	8.728
17	21.486	15.314	12.810	11.414	10.510	9.872	9.021	8.474
18	21.111	15.010	12.536	11.157	10.265	9.634	8.794	8.253
19	20.781	14.743	12.296	10.932	10.050	9.426	8.595	8.060
20	20.490	14.507	12.084	10.734	9.860	9.242	8.419	7.889
22	19.998	14.109	11.726	10.399	9.539	8.932	8.122	7.601
24	19.599	13.787	11.437	10.127	9.280	8.681	7.882	7.367
26	19.268	13.520	11.197	9.903	9.065	8.473	7.683	7.174
28	18.990	13.296	10.996	9.715	8.885	8.298	7.516	7.011
30	18.753	13.105	10.824	9.554	8.731	8.150	7.373	6.872
35	18.289	12.732	10.489	9.240	8.431	7.859	7.094	6.601
40	17.949	12.459	10.245	9.011	8.212	7.646	6.891	6.402
50	17.487	12.088	9.912	8.699	7.914	7.357	6.613	6.132
60	17.186	11.847	9.696	8.497	7.720	7.169	6.432	5.955
80	16.819	11.553	9.433	8.250	7.483	6.940	6.212	5.740
100	16.603	11.380	9.278	8.105	7.344	6.805	6.082	5.612
150	16.322	11.156	9.076	7.916	7.163	6.629	5.912	5.446
300	16.050	10.933	8.877	7.730	6.984	6.456	5.744	5.281
1000	15.873	10.789	8.738	7.600	6.860	6.337	5.632	5.170

Exhibit 7.10c Critical Values for Adequate Signal-to-Noise: $\alpha = 0.01$, $\gamma = 0.95$

Degrees of Freedom, Denominator $(n - p)$	Degrees of Freedom, Numerator (p_2) 1	2	3	4	5	6	8	10
10	32.429	22.880	19.098	17.012	15.670	14.726	13.472	12.666
11	30.742	21.572	17.947	15.950	14.667	13.764	12.565	11.796
12	29.412	20.542	17.043	15.116	13.878	13.008	11.853	11.112
13	28.337	19.712	16.314	14.444	13.243	12.399	11.279	10.561
14	27.451	19.029	15.714	13.891	12.721	11.898	10.807	10.108
15	26.708	18.458	15.213	13.429	12.284	11.480	10.412	9.728
16	26.077	17.972	14.787	13.037	11.914	11.124	10.077	9.407
17	25.534	17.555	14.422	12.700	11.595	10.819	9.790	9.130
18	25.062	17.193	14.104	12.408	11.319	10.554	9.540	8.890
19	24.648	16.875	13.826	12.152	11.077	10.322	9.321	8.679
20	24.282	16.595	13.581	11.925	10.863	10.117	9.127	8.493
22	23.665	16.122	13.167	11.544	10.503	9.771	8.801	8.179
24	23.164	15.739	12.831	11.235	10.211	9.491	8.536	7.925
26	22.749	15.422	12.554	10.979	9.969	9.259	8.318	7.714
28	22.400	15.155	12.321	10.765	9.766	9.064	8.134	7.537
30	22.103	14.928	12.122	10.582	9.593	8.898	7.976	7.385
35	21.521	14.484	11.734	10.224	9.255	8.574	7.670	7.089
40	21.096	14.160	11.451	9.963	9.008	8.337	7.445	6.873
50	20.517	13.719	11.065	9.608	8.672	8.014	7.139	6.577
60	20.141	13.433	10.815	9.377	8.454	7.804	6.940	6.385
80	19.682	13.084	10.509	9.095	8.187	7.547	6.696	6.149
100	19.412	12.878	10.330	8.930	8.030	7.396	6.553	6.010
150	19.060	12.613	10.096	8.714	7.826	7.200	6.366	5.827
300	18.718	12.345	9.865	8.501	7.623	7.005	6.180	5.647
1000	18.490	12.167	9.704	8.356	7.485	6.874	6.056	5.525

Exhibit 7.10d Critical Values for Adequate Signal-to-Noise: $\alpha = 0.01$, $\gamma = 0.99$

Degrees of Freedom, Denominator $(n - p)$	Degrees of Freedom, Numerator (p_2) 1	2	3	4	5	6	8	10
10	44.983	29.798	24.053	20.952	18.982	17.609	15.799	14.646
11	42.449	28.004	22.547	19.603	17.735	16.433	14.718	13.626
12	40.455	26.593	21.363	18.544	16.756	15.509	13.869	12.825
13	38.846	25.456	20.409	17.690	15.966	14.765	13.185	12.179
14	37.521	24.521	19.625	16.988	15.317	14.153	12.622	11.648
15	36.412	23.739	18.969	16.402	14.774	13.641	12.151	11.204
16	35.470	23.074	18.412	15.904	14.314	13.207	11.751	10.826
17	34.661	22.504	17.934	15.476	13.918	12.834	11.408	10.502
18	33.957	22.009	17.519	15.104	13.575	12.510	11.110	10.220
19	33.341	21.575	17.156	14.779	13.274	12.226	10.848	9.973
20	32.796	21.191	16.834	14.492	13.008	11.975	10.617	9.755
22	31.877	20.545	16.293	14.007	12.559	11.552	10.228	9.386
24	31.132	20.021	15.854	13.614	12.196	11.209	9.911	9.087
26	30.516	19.588	15.491	13.289	11.895	10.925	9.650	8.840
28	29.998	19.223	15.186	13.016	11.642	10.686	9.430	8.631
30	29.557	18.913	14.926	12.783	11.427	10.483	9.242	8.453
35	28.694	18.307	14.418	12.328	11.006	10.085	8.874	8.105
40	28.064	17.864	14.047	11.996	10.698	9.794	8.606	7.850
50	27.206	17.262	13.542	11.544	10.278	9.397	8.239	7.502
60	26.648	16.870	13.214	11.250	10.006	9.139	8.000	7.275
80	25.968	16.393	12.813	10.890	9.672	8.824	7.707	6.996
100	25.569	16.112	12.578	10.679	9.476	8.638	7.534	6.832
150	25.048	15.747	12.271	10.404	9.220	8.395	7.308	6.616
300	24.535	15.387	11.967	10.130	8.966	8.155	7.085	6.401
1000	24.177	15.135	11.759	9.946	8.795	7.993	6.933	6.255

Exhibit 7.10e Critical Values for Adequate Signal-to-Noise: $\alpha = 0.01$, $\gamma = 0.999$

Degrees of Freedom, Denominator $(n - p)$	Degrees of Freedom, Numerator (p_2)							
	1	2	3	4	5	6	8	10
10	62.768	39.356	30.789	26.242	23.388	21.412	18.832	17.201
11	58.978	36.869	28.787	24.501	21.812	19.951	17.521	15.987
12	55.998	34.915	27.215	23.134	20.573	18.803	16.492	15.033
13	53.596	33.340	25.948	22.032	19.576	17.878	15.662	14.264
14	51.620	32.046	24.907	21.126	18.755	17.117	14.979	13.631
15	49.966	30.962	24.035	20.368	18.069	16.480	14.408	13.102
16	48.562	30.043	23.296	19.724	17.486	15.940	13.923	12.652
17	47.356	29.253	22.660	19.172	16.986	15.475	13.506	12.265
18	46.309	28.568	22.109	18.692	16.551	15.072	13.144	11.929
19	45.392	27.967	21.626	18.271	16.170	14.719	12.827	11.635
20	44.581	27.436	21.199	17.900	15.833	14.406	12.546	11.374
22	43.214	26.541	20.479	17.273	15.266	13.879	12.072	10.934
24	42.106	25.816	19.895	16.765	14.805	13.452	11.688	10.577
26	41.190	25.216	19.412	16.345	14.424	13.098	11.370	10.282
28	40.420	24.712	19.007	15.991	14.104	12.800	11.102	10.032
30	39.764	24.282	18.661	15.690	13.830	12.546	10.873	9.820
35	38.481	23.442	17.984	15.101	13.296	12.049	10.426	9.403
40	37.545	22.829	17.490	14.670	12.905	11.686	10.098	9.098
50	36.270	21.994	16.817	14.083	12.371	11.190	9.650	8.680
60	35.442	21.451	16.379	13.701	12.024	10.866	9.358	8.407
80	34.430	20.788	15.844	13.233	11.599	10.471	9.000	8.072
100	33.836	20.398	15.529	12.958	11.349	10.237	8.788	7.874
150	33.061	19.890	15.121	12.599	11.021	9.932	8.510	7.614
300	32.297	19.389	14.711	12.243	10.697	9.629	8.234	7.355
1000	31.767	19.048	14.436	12.004	10.479	9.420	8.046	7.178

Exhibit 7.10f Critical Values for Adequate Signal-to-Noise: $\alpha = 0.01$, $\gamma = 0.9999$

Degrees of Freedom, Denominator $(n - p)$	Degrees of Freedom, Numerator (p_2)							
	1	2	3	4	5	6	8	10
10	80.483	48.718	37.310	31.320	27.585	25.015	21.677	19.581
11	75.401	45.537	34.821	29.197	25.692	23.280	20.150	18.185
12	71.408	43.038	32.866	27.529	24.204	21.918	18.950	17.088
13	68.190	41.025	31.291	26.185	23.006	20.819	17.982	16.203
14	65.543	39.370	29.995	25.080	22.020	19.916	17.186	15.475
15	63.329	37.985	28.911	24.155	21.195	19.160	16.520	14.866
16	61.450	36.809	27.991	23.370	20.494	18.518	15.954	14.348
17	59.835	35.799	27.201	22.696	19.892	17.966	15.468	13.903
18	58.434	34.923	26.514	22.110	19.370	17.486	15.045	13.516
19	57.206	34.154	25.913	21.597	18.911	17.066	14.674	13.176
20	56.121	33.476	25.382	21.143	18.506	16.695	14.347	12.876
22	54.291	32.331	24.485	20.378	17.823	16.068	13.794	12.370
24	52.808	31.403	23.758	19.757	17.268	15.559	13.344	11.958
26	51.581	30.635	23.157	19.243	16.809	15.138	12.972	11.616
28	50.550	29.990	22.651	18.811	16.423	14.783	12.659	11.329
30	49.672	29.440	22.220	18.442	16.094	14.481	12.392	11.084
35	47.954	28.364	21.376	17.721	15.449	13.889	11.867	10.602
40	46.700	27.578	20.759	17.193	14.977	13.455	11.483	10.249
50	44.990	26.506	19.918	16.473	14.332	12.862	10.958	9.766
60	43.880	25.809	19.371	16.004	13.912	12.476	10.615	9.450
80	42.523	24.957	18.701	15.430	13.398	12.002	10.194	9.061
100	41.725	24.456	18.307	15.092	13.094	11.722	9.945	8.831
150	40.684	23.800	17.792	14.649	12.698	11.355	9.617	8.528
300	39.656	23.154	17.281	14.210	12.303	10.990	9.292	8.227
1000	38.950	22.719	16.940	13.908	12.032	10.742	9.069	8.020

Exhibit 7.10*g* Critical Values for Adequate Signal-to-Noise: $\alpha = 0.01$, $\gamma = 0.99999$

Degrees of Freedom, Denominator $(n - p)$	Degrees of Freedom, Numerator (p_2)							
	1	2	3	4	5	6	8	10
10	98.178	57.971	43.709	36.273	31.661	28.500	24.411	21.855
11	91.782	54.096	40.737	33.774	29.458	26.499	22.674	20.285
12	86.756	51.052	38.402	31.811	27.726	24.927	21.309	19.050
13	82.707	48.598	36.520	30.229	26.331	23.660	20.209	18.054
14	79.377	46.581	34.972	28.928	25.183	22.617	19.303	17.235
15	76.591	44.893	33.677	27.839	24.222	21.744	18.545	16.548
16	74.226	43.461	32.578	26.914	23.406	21.004	17.901	15.966
17	72.195	42.230	31.633	26.119	22.705	20.366	17.347	15.464
18	70.432	41.161	30.813	25.429	22.095	19.813	16.866	15.028
19	68.886	40.224	30.094	24.824	21.561	19.328	16.444	14.646
20	67.521	39.396	29.458	24.289	21.089	18.898	16.071	14.308
22	65.218	38.000	28.386	23.387	20.292	18.174	15.440	13.736
24	63.350	36.868	27.516	22.654	19.645	17.586	14.928	13.272
26	61.806	35.931	26.796	22.048	19.110	17.099	14.504	12.887
28	60.507	35.143	26.191	21.538	18.659	16.689	14.147	12.563
30	59.400	34.471	25.674	21.102	18.274	16.338	13.842	12.285
35	57.235	33.156	24.663	20.250	17.520	15.652	13.243	11.742
40	55.653	32.195	23.924	19.626	16.968	15.150	12.804	11.343
50	53.496	30.883	22.914	18.774	16.214	14.462	12.203	10.796
60	52.094	30.030	22.256	18.219	15.722	14.014	11.810	10.438
80	50.380	28.986	21.451	17.538	15.118	13.463	11.328	9.997
100	49.371	28.371	20.976	17.136	14.762	13.138	11.042	9.736
150	48.059	27.566	20.356	16.609	14.294	12.709	10.666	9.391
300	46.752	26.771	19.737	16.086	13.830	12.285	10.292	9.049
1000	45.874	26.228	19.315	15.728	13.513	11.994	10.035	8.812

Exhibit 7.10*h* Critical Values for Adequate Signal-to-Noise: $\alpha = 0.01$, $\gamma = 0.999999$

Degrees of Freedom, Denominator $(n - p)$	Degrees of Freedom, Numerator (p_2)							
	1	2	3	4	5	6	8	10
10	115.797	67.150	50.030	41.147	35.658	31.906	27.070	24.059
11	108.076	62.579	46.577	38.276	33.148	29.644	25.129	22.318
12	102.010	58.988	43.863	36.020	31.176	27.866	23.603	20.950
13	97.122	56.095	41.676	34.202	29.586	26.433	22.372	19.846
14	93.102	53.715	39.878	32.706	28.278	25.254	21.360	18.938
15	89.739	51.723	38.372	31.454	27.183	24.267	20.512	18.177
16	86.884	50.033	37.094	30.391	26.254	23.428	19.791	17.530
17	84.431	48.580	35.996	29.477	25.454	22.707	19.172	16.974
18	82.302	47.318	35.042	28.683	24.760	22.081	18.633	16.490
19	80.435	46.212	34.205	27.987	24.151	21.532	18.161	16.066
20	78.785	45.235	33.466	27.372	23.612	21.046	17.743	15.691
22	76.003	43.585	32.218	26.333	22.703	20.225	17.037	15.057
24	73.746	42.247	31.205	25.489	21.964	19.559	16.464	14.541
26	71.879	41.140	30.367	24.791	21.353	19.006	15.988	14.114
28	70.309	40.208	29.661	24.203	20.838	18.541	15.588	13.753
30	68.969	39.413	29.059	23.701	20.398	18.144	15.245	13.445
35	66.350	37.858	27.880	22.719	19.537	17.366	14.574	12.841
40	64.435	36.720	27.017	21.999	18.905	16.795	14.082	12.397
50	61.821	35.166	25.838	21.014	18.042	16.014	13.406	11.787
60	60.121	34.154	25.069	20.372	17.478	15.503	12.964	11.388
80	58.040	32.914	24.127	19.584	16.785	14.875	12.420	10.896
100	56.814	32.183	23.570	19.119	16.375	14.504	12.099	10.604
150	55.212	31.224	22.842	18.507	15.838	14.016	11.674	10.219
300	53.625	30.277	22.119	17.901	15.301	13.530	11.252	9.836
1000	52.554	29.622	21.623	17.486	14.939	13.196	10.960	9.569

CHAPTER 8

Collinearity-Influential Observations

As we have seen, collinearity occurs when there are near dependencies among the columns of the data matrix **X**. It is possible, however, for a small subset of one or more observations—that is, rows of the data matrix—to be greatly responsible for the presence or absence of a near dependency. We can see from Exhibit 8.1*a*, for example, how a single data point can ameliorate collinearity between two variates (or mask it, depending upon one's point of view), and we can see from Exhibit 8.1*b* how a single point can create collinearity. The situation in Exhibit 8.1*a* is similar to that shown in Exhibit 2.2*f*. For reasons to be made clear, such data points have been dubbed *collinearity-influential observations*, and we examine them in this chapter.

Collinearity-influential observations were noted in Belsley et al. (1980) and Cook and Weisberg (1982) and have prompted much recent work. Mason and Gunst (1985) formalized the possibility of collinearity-influential observations by showing that the conditioning of a data set can be made arbitrarily poor by systematically extending the distance between an outlying observation and the main body of data and, further, that it is possible for r separate such outliers similarly to induce up to $r - 1$ separate near dependencies. Numerous interesting efforts have since been made to provide appropriate diagnostics for these and related situations.

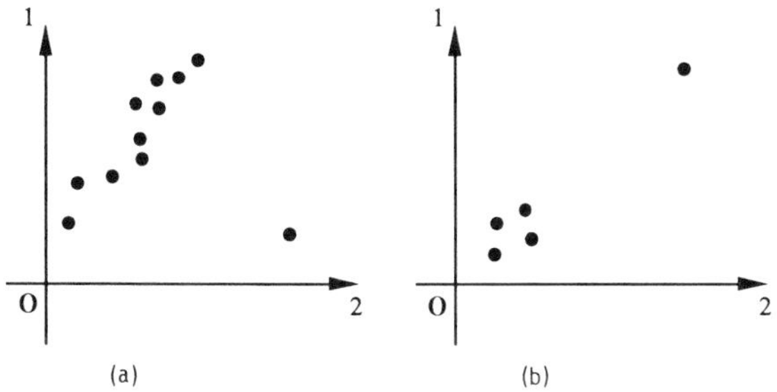

Exhibit 8.1 (*a*) Collinearity except for an outlier. (*b*) Collinearity due to an outlier.

In the next section we provide a brief review of some necessary notions from the field of influential-data diagnostics. In Section 8.2 some of the diagnostics that have been proposed for identifying collinearity-influential observations are described and examples of their use are given in Section 8.3. A final section evaluates these procedures and provides some cautionary remarks.

8.1 INFLUENCE AND LEVERAGE

Influential Observations

The term *influence* is used rather generally to denote a situation in which a perturbation in the conditions under which a model is estimated produces substantive changes in some of its important estimated values. Thus, one might perturb a model assumption, such as linearity or homoskedasticity, and assess what happens to the parameter estimates. Or one might perturb some element of the data, perhaps by deleting an observation or a small group of observations. Indeed, in the regression setting, an observation has traditionally been termed influential if its removal (the deletion of its row $\mathbf{x}_i^T$ from $\mathbf{X}$) has a disproportionate effect on, say, one or more of the estimated coefficients, the predicted (or fitted) values, the residuals, or the estimated variance–covariance structure of the coefficients. This is readily illustrated in Exhibit 8.1*a* if we were to view the axes as y and x, respectively. It is clear that regression lines fitted to these data with and without the presence of the outlier would have radically different estimated coefficients and residuals, and this outlier is therefore an influential observation. Diagnostic methods that help to identify such observations are *influential-data diagnostics*. These topics are treated in detail in Belsley et al. (1980), Cook and Weisberg (1982), and Chatterjee and Hadi (1988).

We have also noted that the deletion of a single observation can alter patterns of collinearity among the columns of $\mathbf{X}$. But the use of the term influence for this purpose is an extension beyond the traditions described above in the regression setting, since these patterns are not part of the estimates of the model but rather are part of the conditions under which the model is estimated. The use of the term *collinearity influential* is therefore useful in this context to help keep this distinction in mind.[1] The effects of the deletion of collinearity-influential observations on the various regression estimates, then, are indirect: the observation affects the data conditioning, which in turn can affect the estimation results. Thus, the ultimate assessment of such influence is less in terms of its direct effects on collinearity than its indirect effects on important estimation results.

[1]The term influence is appropriately applied, however, when used in a closely related context: assessing the effects of a perturbation on a principal-component analysis. Whereas in the regression context, the estimation results are obtained conditionally upon $\mathbf{X}$ and hence the eigensystem of $\mathbf{X}$ is simply a characteristic of the input data, in the principal-component context, the eigensystem of the multivariate data is part of the estimated result and hence can be influenced by a perturbation in the traditional sense. For an examination of influence in this context, see Critchley (1985) and Wang and Nyquist (1989).

One can similarly consider small groups of observations that together, but not necessarily individually, are influential. If in our mind's eye we were to picture a second outlier in Exhibit 8.1*a* that was very close to the existing one, the deletion of either one alone would not have much effect on the regression line, whereas their joint removal would. Thus, the detection of groups of influential data presents a significantly more difficult diagnostic task than detecting single influential observations. The simple expedient of dropping observations becomes a combinatorial nightmare when applied to multiple deletions. Ingenious attempts to circumvent this problem are described in the previous references, and we shall examine some below in the collinearity context.

Leverage

Influential observations are necessarily either y_i values that are outliers, $\mathbf{x}_i^{\mathrm{T}}$ values with *leverage*, or both—although the converse is not true. The term leverage applies only to the explanatory variates and is intended to convey something close to what intuition would suggest: an outlying $\mathbf{x}_i^{\mathrm{T}}$ that, by its distance from the rest of the data, exerts a great force in determining various estimated (fitted) properties of the least-squares regression plane. This is illustrated in Exhibit 8.1*b* (again viewing the axes as y and x) and quite strikingly in Exhibit 2.2*f*. In fitting a regression line to these data by least squares, not only will the outlier have a great effect in determining the estimated slope, but also it will do so in a way that guarantees a small residual for this observation, or equivalently, a fitted value $\hat{y}_i$ very close to the actual value y_i. Thus, if this outlier were shifted slightly upward or downward, the slope of the regression line would change accordingly, directly pivoting about the other points so that the fitted value that corresponds to the outlier will also change nearly in tandem. It is clear that the same would not be true for modest movements in any of the other data points. This outlier, then, has high leverage. In the context of collinearity, interest centers on influence due to elements of the **X** matrix with high leverage.

The Hat Matrix

More formally, we note that the fitted values $\hat{\mathbf{y}}$ are related to the actual values $\mathbf{y}$ by

$$\hat{\mathbf{y}} \equiv \mathbf{X}\mathbf{b} = \mathbf{X}(\mathbf{X}^{\mathrm{T}}\mathbf{X})^{-1}\mathbf{X}^{\mathrm{T}}\mathbf{y} \equiv \mathbf{H}\mathbf{y}, \tag{8.1}$$

where

$$\mathbf{H} \equiv \mathbf{X}(\mathbf{X}^{\mathrm{T}}\mathbf{X})^{-1}\mathbf{X}^{\mathrm{T}}. \tag{8.2}$$

The matrix $\mathbf{H} = (h_{ij})$, then, is the projection matrix that carries any vector $\mathbf{y} \in \mathscr{R}^n$ into its orthogonal projection onto the space spanned by the columns of **X**. Since in the context of the regression of **y** on **X** these projections (the fitted values) are typically denoted by a hat as in (8.1), Hoaglin and Welsch (1978) call this the *hat matrix*. Like any projection matrix, $\mathbf{H}$ has a number of very strong properties. Specifically, it is readily shown that

$$\operatorname{rank}(\mathbf{H}) = \operatorname{rank}(\mathbf{X}) = \operatorname{tr}(\mathbf{H}) = p, \tag{8.3a}$$

there are p eigenvalues of $\mathbf{H}$ that are equal to 1 and $n - p$ equal to 0, and (8.3b)

$$0 \leqslant h_{ii} \leqslant 1, \qquad i = 1, \dots, n. \tag{8.3c}$$

From (8.1) we see that $\partial \hat{y}_i / \partial y_i = h_{ii}$, and hence the diagonal elements of the hat matrix,

$$h_{ii} = \mathbf{x}_i^{\mathrm{T}}(\mathbf{X}^{\mathrm{T}}\mathbf{X})^{-1}\mathbf{x}_i, \tag{8.4}$$

directly indicate the sensitivity of the ith fitted value $\hat{y}_i$ to changes in the ith actual value y_i. It can also be shown that the h_{ii} figure prominently in many other influential-data diagnostics measuring the sensitivity of the coefficients, the residuals, and the error covariance structure to deletions of the corresponding $\mathbf{x}_i^{\mathrm{T}}$. The h_{ii}, then, provide a formal measure of leverage. The double subscript is often dropped in this usage in favor of h_i ($\equiv h_{ii}$). From (8.3c) we see that this measure has a maximum value of unity. Rows of $\mathbf{X}$ corresponding to large values of h_{ii} are said to have high leverage. Since from (8.3a) we have $\Sigma_{i=1}^{n} h_{ii} = p$, h_{ii} has an average value of p/n. This leads Belsley et al. (1980) to suggest a value of $2p/n$ for determining high leverage.

The hat matrix is very easily computed from the singular-value decomposition since substituting $\mathbf{X} = \mathbf{U}\mathbf{D}\mathbf{V}^{\mathrm{T}}$ into (8.2) gives

$$\mathbf{H} = \mathbf{U}\mathbf{U}^{\mathrm{T}}, \tag{8.5}$$

and hence, the leverage of $\mathbf{x}_i^{\mathrm{T}}$, the ith row of $\mathbf{X}$, can be expressed as

$$h_i = \mathbf{u}_i^{\mathrm{T}}\mathbf{u}_i = \sum_j u_{ij}^2. \tag{8.6}$$

The expression (8.5) can also be used to calculate a hat matrix when $\mathbf{X}$ is of less than full rank or is very ill conditioned.

8.2 DIAGNOSING COLLINEARITY-INFLUENTIAL OBSERVATIONS

We first consider diagnostics for the presence of a single collinearity-influential observation, beginning with some theoretical underpinnings and then defining three diagnostics. We then turn briefly to diagnostics for sets of collinearity-influential observations.

Assessing Collinearity-Influential Observations

A natural means for examining the role played by a given row $\mathbf{x}_i^T$ in effecting the patterns of collinearity among the columns of $\mathbf{X}$ is to determine the effect of that row on the singular values, and therefore the condition indexes, of $\mathbf{X}$.[2] That is, we can compare the singular values of $\mathbf{X}$ with those of $\mathbf{X}_{(i)}$, where this latter notation rather generally denotes a matrix with its ith row deleted. Various elements of this problem have been examined in Dorsett (1982), Kempthorne (1985), Chatterjee and Hadi (1988), Hadi (1988), Nyquist (1989), Wang and Nyquist (1989), and Walker (1989). The current state of affairs relative to our needs is summed up in the following result proven in the Hadi references:

Theorem 8.1 (Hadi). Let the SVD of $\mathbf{X} = \mathbf{UDV}^T$, giving singular values $\mathbf{D} = \text{diag}(\mu_1, \ldots, \mu_p)$, and define $\mathbf{Z} \equiv \mathbf{XV}$. Then the singular values of $\mathbf{X}_{(i)}$ and $\mathbf{Z}_{(i)}$ are the same and are the (positive) solutions γ to

$$\left[1 - \sum_{j=1}^{p} \frac{z_{ij}^2}{\mu_j^2 - \gamma^2}\right] \prod_{j=1}^{p} (\mu_j^2 - \gamma^2). \tag{8.7}$$

Unfortunately, a general closed-form solution to this problem has proved intractable, but solutions do exist for two special cases, both of which are capable of motivating interesting collinearity-influential diagnostics.

The first special case was initially shown by Kempthorne (1985), giving a solution to (8.7) for $p = 2$. Specifically:

Theorem 8.2 (Kempthorne). For $p = 2$, the condition number of $\mathbf{X}_{(i)}$ is

$$\kappa_{(i)} = \sqrt{\frac{1 + (1 - 4/C_i)^{1/2}}{1 - (1 - 4/C_i)^{1/2}}}, \tag{8.8}$$

where

$$C_i = \frac{[\text{tr}(\mathbf{X}^T\mathbf{X}) - \mathbf{x}_i^T\mathbf{x}_i]^2}{|\mathbf{X}^T\mathbf{X}|(1 - h_{ii})}. \tag{8.9}$$

Kempthorne's original statement of this result assumes that $\mathbf{X}$ is scaled for unit column lengths, which we know to be desirable, but Chatterjee and Hadi (1988) show this requirement is unnecessary for (8.9) to hold.

The second special case was initially shown by Dorsett (1982) and provides a solution to (8.7) in the case where $\mathbf{x}_i = \alpha\mathbf{v}_j$, that is, when the ith row of $\mathbf{X}$ is exactly in the direction of the jth eigenvector $\mathbf{v}_j$ of $\mathbf{X}^T\mathbf{X}$. Specifically:

[2]One could also base diagnostics for collinearity-influential observations on other measures of collinearity. Hadi and Velleman (1987) and Chatterjee and Hadi (1988), for example, develop such diagnostics using Stewart's (1987) VIF-based collinearity indices. My concerns with such measures are discussed in Section 2.3.

Theorem 8.3 (Dorsett). Let the SVD of $\mathbf{X} = \mathbf{UDV}^T$. Then, if $\mathbf{x}_i = \alpha \mathbf{v}_j$, the singular values of $\mathbf{X}_{(i)}$ are

$$\begin{cases} \mu_k & \text{for } k \neq j, \\ \sqrt{\mu_k^2 - \alpha^2} & \text{for } k = j. \end{cases} \tag{8.10}$$

Thus, if $\mathbf{x}_i$ is in the direction of the jth eigenvector of $\mathbf{X}^T\mathbf{X}$, only the jth singular value of $\mathbf{X}$ (jth eigenvalue of $\mathbf{X}^T\mathbf{X}$) will be affected, and it will be reduced by an amount determined by α. This result is easily understood at an intuitive level. We recall from Section 3.1 that the eigenvectors of $\mathbf{X}^T\mathbf{X}$ lie in the directions of the principal axes of the ellipse $\xi^T(\mathbf{X}^T\mathbf{X})^{-1}\xi = 1$ and that the lengths of these axes are the eigenvalues of $\mathbf{X}^T\mathbf{X}$ (the squares of the singular values of $\mathbf{X}$). Thus, the eigenvector corresponding to the maximal eigenvalue lies along the major axis of the ellipse, while that corresponding to the next largest eigenvalue lies along the next longest axis, and so on. The preceding result, then, states that removing an observation lying in the direction of a principal axis merely reduces the length of that axis, leaving all others unchanged.

The implications of this for the effect on the condition indexes, η_i, is straightforward. If $\mathbf{x}_i$ lies exactly in the direction of the eigenvector corresponding to the largest singular value $\mu_{\max}$, then its deletion will cause all the condition indexes to be reduced in the same proportion. If $\mathbf{x}_i$ lies exactly in the direction of any other eigenvector, its deletion will cause the corresponding condition index to be increased, leaving the others unchanged. In particular, the deletion of an observation lying in the direction of the eigenvector corresponding to the smallest singular value, $\mu_{\min}$, will directly affect the largest condition index of $\mathbf{X}$, that is, its condition number.

Diagnostics for Individual Collinearity-Influential Observations

A General Diagnostic

Since no general closed-form solution exists for (8.7), a truly general diagnostic must rely on a more direct, less elegant method. Chatterjee and Hadi (1988) suggest basing such a diagnostic upon

$$H_i \equiv \frac{|\kappa_{(i)} - \kappa|}{\kappa}, \tag{8.11}$$

where $\kappa = \kappa(\mathbf{X})$, the condition number of $\mathbf{X}$, and $\kappa_{(i)} = \kappa(\mathbf{X}_{(i)})$. That is, the collinearity influence of each row of $\mathbf{X}$ would be measured by the relative change in the condition number that results from its deletion. A plot of the H_i against i would readily identify those observations having a substantive effect on the condition number of $\mathbf{X}$. This is a good start, but taken as it stands, this idea has several important weaknesses.

First, we have already seen that the condition number is most effective in measuring the degree of ill conditioning only when it is properly scaled. From this perspective, it would be better to base H_i on the scaled condition numbers $\tilde{\kappa} = \tilde{\kappa}(\mathbf{X})$ and $\tilde{\kappa}_{(i)} = \tilde{\kappa}(\mathbf{X}_{(i)})$, giving the diagnostic

$$\tilde{H}_i \equiv \frac{|\tilde{\kappa}_{(i)} - \tilde{\kappa}|}{\tilde{\kappa}}. \tag{8.12}$$

Second, while a relative-change measure such as (8.11) or (8.12) has some intuitive appeal—and will no doubt do the job of highlighting influential observations—, the actual condition numbers have magnitudes that convey useful information that is lost in this transformation. For example, H_i or $\tilde{H}_i$ can be meaninglessly large. A situation with $\tilde{\kappa} = 6$ and $\tilde{\kappa}_{(i)} = 3$ would produce a large $\tilde{H}_i$ but would not be particularly interesting since a scaled condition index of 6 is most unexciting. I would prefer simply to see $\tilde{\kappa}$ and $\tilde{\kappa}_{(i)}$ together so that I can assess whether the change reflects a meaningful difference.

Third, the diagnostic suggested here deals only with the condition number, the largest of the condition indexes, and ignores (indeed, overlooks) information on the remaining condition indexes. We have seen from Theorem 8.3 that it is quite possible for an observation to be strongly influential relative to, say, the second strongest near dependency without affecting the strongest. If this second largest condition index were itself large, this observation would certainly have to be classified as collinearity influential despite its being ignored by diagnostics like (8.11) and (8.12). Thus, I would prefer to see the effects of the row deletions on the full set of scaled condition indexes of $\mathbf{X}_{(i)}$, that is, in comparison with the scaled condition indexes $\tilde{\boldsymbol{\eta}} = (\tilde{\eta}_1, \ldots, \tilde{\eta}_p)^{\mathrm{T}}$ of $\mathbf{X}$, one should examine

$$\tilde{\boldsymbol{\eta}}_{(i)} = (\tilde{\eta}_{1(i)}, \ldots, \tilde{\eta}_{p(i)})^{\mathrm{T}}. \tag{8.13}$$

We shall call these the *row-deleted scaled condition indexes*, or *deletion condition indexes* for short.

Finally, this is a computationally rather intensive diagnostic. It requires $n + 1$ full runs of the singular-value decomposition. Chatterjee and Hadi (1988) are well aware of this problem and offer a computationally simpler means for estimating $\kappa_{(i)}$ in (8.11) based on a power method for approximating the maximal and minimal eigenvalues of a matrix.[3] This method seems to work very well, but it cannot be used to determine the full set of deletion condition indexes in (8.13). Fortunately, computationally intensive diagnostics no longer present the obstacles they once did. My experiences with calculating the full set of $\tilde{\boldsymbol{\eta}}_{(i)}$ for

[3]This computational problem can also be finessed using the influence function techniques developed in Nyquist (1989) and Wang and Nyquist (1989). There it is shown that one of the sample influence functions measuring the influence of each observation on the condition number is equivalent to the H_i diagnostic given in (8.11), while another empirical influence function, approximating the first, is "easy to compute and admits fairly simple interpretations of collinearity-influential points." However, the other problems mentioned in relation to H_i remain.

modest-sized problems ($n \leqslant 200$, $p \leqslant 10$) on a Macintosh II shows this to be a very acceptable computational task.

As a general diagnostic, then, a table, or set of graphs, presenting the $\tilde{\boldsymbol{\eta}}$ for a base of comparison followed by the n values $\tilde{\boldsymbol{\eta}}_{(i)}$ of the deletion condition indexes from (8.13) is always an available option.

The Trace-to-Det Ratio C_i

A second diagnostic for individual collinearity-influential observations makes use of the C_i from (8.9). It can be seen from (8.8) that, at least for the case when $p=2$, $\kappa_{(i)}$ increases with C_i, and C_i depends, for given trace and determinant, on the length of the observation $\mathbf{x}_i^{\mathrm{T}}$ and its leverage h_{ii}. Relatively high values for C_i point to collinearity-influential points, and a list or plot of C_i for $i = 1, \ldots, n$ becomes a simple diagnostic.

This diagnostic too has problems. First, its theoretical relevance has been shown only for $p = 2$. Chatterjee and Hadi (1988), however, feel this diagnostic has value more generally, and my own experience agrees with this assessment. Second, the C_i are not scale invariant, and like any method employing condition indexes, this diagnostic is best applied to an $\mathbf{X}$ matrix scaled for unit column lengths. Even if this is done, however, the C_i's determined by (8.9) and the condition number determined by (8.8) are relevant to an $\mathbf{X}_{(i)}$ that has not been rescaled for unit column lengths, and this is a weakness. Third, C_i can give diagnostic information only about the condition number; it necessarily ignores the effects of $\mathbf{x}_i^{\mathrm{T}}$ on the other condition indexes. But as a quick, crude diagnostic for single collinearity-influential observations, it seems reasonably effective.

Leverage Components

Walker (1989) exploits the Dorsett result of (8.10) to derive another diagnostic for collinearity-influential observations based on combining information on leverage and elements of the $\mathbf{U}$ matrix from the SVD of $\mathbf{X}$. Specifically, consider the SVD of $\mathbf{X} = \mathbf{UDV}^{\mathrm{T}}$, so $\mathbf{x}_i = \Sigma_{k=1}^{p} u_{ik}\mu_k \mathbf{v}_k$, and assume $\mathbf{x}_i = \alpha \mathbf{v}_j$. Then $u_{ik} = 0$ for $k \neq j$ and $u_{ij} = \alpha/\mu_j$. Putting this result into (8.6) gives

$$h_{ii} = u_{ij}^2 = \alpha^2/\mu_j^2. \tag{8.14}$$

Applying this to Theorem 8.3 gives the following corollary:

Corollary 8.1 (Walker). Let $\mathbf{x}_i = \alpha \mathbf{v}_j$. Then:

(a) If j is the index corresponding to the maximal singular value,

$$\eta_{k(i)} = \eta_k \sqrt{1 - h_{ii}} \quad \text{for } k \neq j. \tag{8.15a}$$

(b) If j is otherwise,

$$\begin{cases} \eta_{k(i)} = \eta_k \dfrac{1}{\sqrt{1 - h_{ii}}} & \text{for } k = j, \\ \eta_k & \text{for } k \neq j. \end{cases} \tag{8.15b}$$

Proof. If j is the index of the maximal singular value, then $\eta_{k(i)} \equiv \sqrt{\mu_k^2 - \alpha^2}/\mu_k$. Using (8.14) leads directly to (8.15a). If j is other than the index of the maximal singular value, $\eta_{k(i)}$ is unaffected unless $k = j$, and then $\eta_{k(i)} \equiv \mu_{\max}/\sqrt{\mu_k^2 - \alpha^2}$. □

It is seen from this corollary that when $\mathbf{x}_i = \alpha \mathbf{v}_j$, the effect on the condition indexes of deleting $\mathbf{x}_i^{\mathrm{T}}$ is directly related to that observation's leverage h_{ii}, which, from the first equality in (8.14), is u_{ij}^2. Thus, the square of the (ij)th element of $\mathbf{U}$ can be viewed as a measure of the component of $\mathbf{x}_i$'s leverage in the direction of the jth eigenvector. Combining this with (8.6), which states that in general $h_{ii} = \Sigma_j u_{ij}^2$, Walker (1989) concludes that "a large value of any of the u_{ij}s, say u_{ik}, suggests that the deletion of $\mathbf{x}_i^{\mathrm{T}}$ would affect primarily the kth condition index." Since the cutoff for h_{ii} is usually chosen at $2p/n$, Walker suggests considering a u_{ik} to be large when it accounts for at least half of this, that is, when u_{ik}^2 is larger than p/n.

This diagnostic has several advantages. It is very easy to compute, and it provides information on all the condition indexes, not just the maximal one. Its main disadvantage is that, even if the $\mathbf{U}$ matrix is initially obtained from a scaled $\mathbf{X}$ matrix, the condition indexes on the right-hand sides of (8.15a) and (8.15b) that underlie this method are unscaled, for Theorem 8.3 does not apply when $\mathbf{X}_{(i)}$ is rescaled to have unit column lengths. We shall see that this is somewhat of a limitation to the value of this diagnostic.

Diagnosing Sets of Collinearity-Influential Observations

As already noted, problems literally multiply when one attempts to diagnose the presence of several coexisting collinearity-influential observations. Methods effective in discerning a single collinearity-influential observation can fail altogether when two high-leverage outliers are near one another, each masking the deletion of the other. The obvious stratagem of deleting all possible subsets of observations, even if only small subsets, leads quickly by combinatorics to fantastically extensive computations. Efforts to hold this task to manageable proportions have been made by Hadi and Wells (1990), Kempthorne (1989), and Walker (1989).

Hadi and Wells (1990) provides a generalization of Theorem 8.1 for multiple-row deletions along with a corresponding means for approximating the condition number of the row-deleted matrix. The diagnostic that results from this technique is analogous to the related single-observation method (8.11) and has the same limitations. Namely, it deals only with the condition number (and not the full set of condition indexes), and it does not allow for the row-deleted matrix to be rescaled. Furthermore, this method provides no means for identifying influential subgroups other than by the brute-force method of all possible subsets (at least up to a given size).

In an altogether different vein, Kempthorne (1989) uses the following theorem to reduce the search for collinearity-influential sets to one of finding vectors in the column space of **X** that have zeros in appropriate places.

Theorem 8.4 (Kempthorne). Let **X** be of full rank. Then a group of cases I is rank (collinearity) influential if and only if there exists a $\boldsymbol{\zeta} \neq \mathbf{0}$ in the column space of **X** satisfying $\zeta_i = 0$ for all $i \notin I$.

Kempthorne provides two methods for conducting this search, one based on Kaiser's (1958) varimax/quartimax rotation and one based on projection pursuit (Huber, 1985). Even though these methods are greatly simpler than multiple deletions, they are still relatively arduous tasks that use techniques not generally found in the practitioner's everyday bag of tools.

A much simpler means for identifying sets of collinearity-influential observations that seems to have considerable promise is a direct extension of Walker's (1989) method described above. Walker extends Theorem 8.3 to show:

Theorem 8.5 (Walker). Let $\mathbf{X} = \mathbf{U}\mathbf{D}\mathbf{V}^{\mathrm{T}}$ with $\mathbf{D} = \mathrm{diag}(\mu_1, \ldots, \mu_p)$ and suppose two rows of **X** have the form $\mathbf{x}_1^{\mathrm{T}} = \alpha\mathbf{v}_i^{\mathrm{T}}$ and $\mathbf{x}_2^{\mathrm{T}} = \beta\mathbf{v}_j^{\mathrm{T}}$. If these two rows are deleted from **X**, the singular values of the resulting matrix are

$$\mu_1, \ldots, \mu_{i-1}, \sqrt{\mu_i^2 - \alpha^2}, \mu_{i+1}, \ldots, \mu_{j-1}, \sqrt{\mu_j^2 - \beta^2}, \mu_{j+1}, \ldots, \mu_p, \qquad (8.16a)$$

and if the two points lie in the same direction, so that $\mathbf{x}_1^{\mathrm{T}} = \alpha\mathbf{v}_i^{\mathrm{T}}$ and $\mathbf{x}_2^{\mathrm{T}} = \beta\mathbf{v}_i^{\mathrm{T}}$, the resulting singular values are

$$\mu_1, \ldots, \mu_{i-1}, \sqrt{\mu_i^2 - \alpha^2 - \beta^2}, \mu_{i+1}, \ldots, \mu_p. \qquad (8.16b)$$

Result (8.16b) shows the effects of masking.

In light of this theorem, Walker suggests as a diagnostic for sets of collinearity-influential observations a simple extension of the diagnostic given before for single collinearity-influential points. That is, choose for further examination only that subset of observations $\mathbf{x}_i^{\mathrm{T}}$ having some large u_{ij}^2, large again being defined as greater than a size-adjusted cutoff p/n.

8.3 ILLUSTRATIONS OF DIAGNOSING COLLINEARITY-INFLUENTIAL OBSERVATIONS

The Cement Data

The pleasantly short data set in Exhibit 8.2 gives 14 observations on the curing temperature **y** associated with cement compounds whose component weights (in percentages of the total) comprise the **X** data. These data have a venerable history extending back to Woods et al. (1932). They have also been analyzed in this context in Chatterjee and Hadi (1988), Stewart (1987), Nyquist (1989), and Wang and Nyquist (1989). They are being reanalyzed here because they provide a good testing ground for comparing the various one-at-a-time diagnostics discussed above, that is, those suitable to discovering a single collinearity-influential observation.

Because the rows of **X** add up roughly to 100%, no constant term is usually included. In this and other respects, I should add for my own protection that, although these data are usefully illustrative for our current needs, I am making no attempt whatsoever at justifying their relevance to any meaningful real-life process.

To begin the analysis, we note from the scaled condition indexes and variance–decomposition proportions given in Exhibit 8.3 that these data are quite well conditioned. The largest scaled condition index is only 12, and perhaps further analysis is unnecessary.

Let us, however, apply the three one-at-a-time diagnostics to see if there are any collinearity-influential rows in **X**. The first diagnostic is the sequence of

Exhibit 8.2 Cement Data

Case	y	X_1	X_2	X_3	X_4	X_5
1	85.5	6	7	26	60	2.5
2	76.0	15	1	29	52	2.3
3	110.4	8	11	56	20	5.0
4	90.6	8	11	31	47	2.4
5	103.5	6	7	52	33	2.4
6	109.8	9	11	55	22	2.4
7	108.0	17	3	71	6	2.1
8	71.6	22	1	31	44	2.2
9	97.0	18	2	54	22	2.3
10	122.7	4	21	47	26	2.5
11	83.1	23	1	40	34	2.2
12	115.4	9	11	66	12	2.6
13	116.3	8	10	68	12	2.4
14	62.6	18	1	17	61	2.1

Exhibit 8.3 Scaled Condition Indexes and Variance–Decomposition Proportions: Complete Cement Data

Scaled Condition Index, $\tilde{\eta}$	Proportions of				
	$\mathbf{X}_1$ var(b_1)	$\mathbf{X}_2$ var(b_2)	$\mathbf{X}_3$ var(b_3)	$\mathbf{X}_4$ var(b_4)	$\mathbf{X}_5$ var(b_5)
1	.004	.006	.003	.007	.004
3	.044	.114	.004	.038	.001
4	.042	.038	.057	.264	.000
8	.297	.459	.000	.003	.704
12	.613	.383	.936	.688	.291

Exhibit 8.4 Scaled Condition Indexes and Row-deleted Scaled Condition Indexes: Cement Data

	Scaled Condition Indexes of $\mathbf{X}$			
	$\tilde{\eta}_5$	$\tilde{\eta}_4$	$\tilde{\eta}_3$	$\tilde{\eta}_2$
	11.9	8.4	3.8	2.5
	Scaled Condition Indexes of $\mathbf{X}_{(i)}$			
Case	$\tilde{\eta}_{(i)5}$	$\tilde{\eta}_{(i)4}$	$\tilde{\eta}_{(i)3}$	$\tilde{\eta}_{(i)2}$
1	12.8	8.4	4.1	2.5
2	11.8	8.3	3.7	2.6
3	70.9	10.1	3.7	2.5
4	11.5	8.2	3.9	2.4
5	13.6	8.2	3.7	2.5
6	11.4	8.2	3.7	2.5
7	11.7	8.3	4.4	2.5
8	12.0	8.2	3.6	2.6
9	11.4	8.1	3.9	2.5
10	13.9	9.4	3.9	2.6
11	12.0	8.3	3.7	2.6
12	11.4	8.1	3.8	2.5
13	11.9	8.1	3.8	2.5
14	11.6	8.2	3.7	2.7

deletion condition indexes $\tilde{\boldsymbol{\eta}}_{(i)}$ relevant to the scaled row-deleted matrices $\mathbf{X}_{(i)}$, $i = 1, \ldots, n$. These, along with the base scaled condition indexes of $\mathbf{X}$ are given in Exhibit 8.4. Needless to say, only the four largest condition indexes are given (in decreasing order) since the smallest is always 1.

It is clear that observation 3 is having a substantial effect, increasing the condition number from a very modest magnitude to a large one in excess of 70 and slightly elevating the second largest condition index as well, although this latter change is of little concern since scaled condition indexes of 10 are usually not too important. It is important to note here that it is only because we are examining the actual scaled condition indexes in this diagnostic and not the relative changes as in (8.11) or (8.12) that the preceding interpretative considerations beyond the opening phrase are possible.

Observation 3, then, is an example of what is often called a collinearity-masking observation. With observation 3 included, the data conditioning is quite good; without it, the conditioning is substantively poorer. Of course, everything is relative, and whereas some might call this a collinearity-masking observation, others might call it a collinearity-breaking observation. We shall return to this point in the next section.

In larger data sets, a graphical presentation of the columns of Exhibit 8.4 would probably allow the information to be digested more easily.

Next, we examine the trace-to-det-ratio diagnostic C_i. The values for C_i are not normalized and can be quite large. However, the absolute values of C_i are without interpretation, so it is only their relative size that matters. Thus, in Exhibit 8.5, we present $C_i/\min(C_i)$. Once again observation 3 stands out as being

Exhibit 8.5 Relative Trace-to-Det Ratio: Cement Data

Case	Relative Trace-to-Det Ratio
1	1.52
2	1.12
3	66.96
4	1.09
5	1.38
6	1.00
7	1.40
8	1.21
9	1.07
10	2.74
11	1.26
12	1.05
13	1.15
14	1.22

collinearity influential. This diagnostic, then, does seem to have more general value than indicated by the assumption of $p = 2$ that underlies its derivation.

Finally, the u_{ij}^2 relevant to the Walker diagnostic are presented in Exhibit 8.6. The last five columns are ordered according to decreasing values in their associated singular values. Also displayed here in the second column are the leverage values h_{ii} for each observation. Two observations, 3 and 10, are seen to possess high leverage, but of these, only observation 3 has a substantial part of that leverage associated with a single singular value. So observation 3 is seen by this diagnostic also to be collinearity influential.

It should be noted that it is the component of observation 3 associated with the second smallest singular value that is large. According to Corollary 8.1, then, the main effect of deleting this observation should be seen on the second largest condition index. A glance at Exhibit 8.4, however, shows that it is the largest condition index that is most affected, although some effect on the second largest is also seen. The reasons for this are severalfold. First, Corollary 8.1 applies accurately only if all the weight is associated with a single singular value. Here we see that observation 3 also has some weight associated with the smallest singular value, and this would affect the largest condition index. Also, Corollary 8.1 does not apply to a deleted matrix $\mathbf{X}_{(i)}$ that has been rescaled for unit column lengths, whereas the condition indexes in Exhibit 8.4 are so scaled. At first it would seem that, if $\mathbf{X}$ were scaled to begin with, $\mathbf{X}_{(i)}$ would be nearly so. But recall that we are deleting high-leverage observations, which can have large

Exhibit 8.6 Leverage and Leverage Components u_{ij}^2: Cement Data

		Leverage Component Associated with				
Case	Leverage	$\tilde{\mu}_5$	$\tilde{\mu}_4$	$\tilde{\mu}_3$	$\tilde{\mu}_2$	$\tilde{\mu}_1$
1	0.435	0.066	0.004	0.215	0.033	0.117
2	0.220	0.060	0.096	0.024	0.015	0.025
3	0.988[a]	0.107	0.054	0.006	0.600[b]	0.222
4	0.213	0.075	0.005	0.117	0.015	0.001
5	0.355	0.060	0.005	0.003	0.028	0.259
6	0.132	0.072	0.034	0.002	0.020	0.003
7	0.383	0.059	0.003	0.273	0.008	0.039
8	0.301	0.069	0.139	0.001	0.023	0.069
9	0.186	0.062	0.032	0.091	0.000	0.001
10	0.716[a]	0.094	0.244	0.064	0.215	0.099
11	0.326	0.069	0.114	0.035	0.033	0.075
12	0.188	0.074	0.057	0.044	0.004	0.008
13	0.248	0.067	0.051	0.049	0.002	0.078
14	0.308	0.064	0.161	0.074	0.003	0.006

[a]Value greater than cutoff of $2p/n$.
[b]Value greater than cutoff of p/n.

differential effects on the relative lengths of the columns of **X**. Thus, rescaling can redistribute the effects across the condition indexes in a more complicated fashion than indicated by Corollary 8.1. And as we shall see in the next section, it is the rescaled results that are the more informative as to what is really happening.

In this rather simple case, then, with a single collinearity-influential observation, all three diagnostic procedures seem to be working well. The full set of row-deleted condition indexes $\tilde{\boldsymbol{\eta}}_{(i)}$, however, tells the most complete story.

The Modified Consumption Function Data

In this example, in order to test the efficacy of the Walker suggestion for identifying sets of collinearity-influential observations, I have modified the consumption function data introduced in Exhibit 5.10 to include two collinearity-influential outliers lying in the direction of the eigenvector corresponding to the largest singular value. This is done simply by stretching two of the existing observations that correlate highly with this principal eigenvector. The observations $[C(T-1), \mathrm{DPI}(T), r(T), \Delta\mathrm{DPI}(T)]$ corresponding to 1969 and 1974 are, for the purposes of this example only, [4527.25, 5135.0, 70.2917, 144.5] and [5520.75, 6028.75, 85.6583, −166.25], respectively; the intercept value of 1 remains unchanged. Also, for use later in the regressions of Section 8.4, the corresponding values of the response variate $C(T)$ are 4569.37 and 5362.29, respectively. These values are generated consistent with the parameter and error-structure estimates in (5.3). It is, of course, impossible to generate these data so as to honor the dynamics inherent in the model (5.2), but that is not important here. It is not a concern whether these observations are real; the purpose of this exercise is to see how well the diagnostic technique works in identifying these observations, which are known to be jointly collinearity influential.

Beginning then, as always, with the **Π** matrix of scaled condition indexes and variance–decomposition proportions given in Exhibit 8.7, we see that there are two very strong near dependencies among these data. In order to save space, only the interesting cases from the one-row-at-a-time diagnostics—the deletion condition indexes $\tilde{\boldsymbol{\eta}}_{(i)}$ and the trace-to-det ratio C_i—are described here. The $\tilde{\boldsymbol{\eta}}_{(i)}$ direct modest attention to 1969, 1970, and 1974. Their values for these observations, along with the base $\tilde{\boldsymbol{\eta}}$, are

$$\begin{aligned}
\tilde{\boldsymbol{\eta}} &= (878,\ 125,\ \ 3,\ 2,\ 1),\\
\tilde{\boldsymbol{\eta}}_{(1969)} &= (967,\ 105,\ \ 9,\ 2,\ 1),\\
\tilde{\boldsymbol{\eta}}_{(1970)} &= (878,\ 173,\ \ 3,\ 2,\ 1),\\
\tilde{\boldsymbol{\eta}}_{(1974)} &= (752,\ \ 97,\ 11,\ 2,\ 1).
\end{aligned}$$

The C_i (not shown) point only mildly to observations 1969 and 1974, showing relative elevations roughly of factors of 2 and 3, respectively, over the

Exhibit 8.7 Scaled Condition Indexes and Variance–Decomposition Proportions: Modified Consumption Function Data

Scaled Condition Index, $\tilde{\eta}$	Proportions of ι var(b_1)	C($T-1$) var(b_2)	DPI(T) var(b_3)	r(T) var(b_4)	ΔDPI(T) var(b_5)
1	.008	.000	.000	.000	.000
2	.047	.000	.000	.000	.012
3	.272	.000	.000	.000	.012
125	.658	.007	.007	.998	.005
878	.015	.993	.993	.002	.971

background level. These one-row-at-a-time diagnostics, then, are not without value in this context but give no striking evidence of major interest.

However, an investigation for high leverage clearly indicates observations 1969, 1970, and 1974. Again to save space, the h_{ii} and u_{ij}^2 only for the high-leverage observations are given in Exhibit 8.8. The u_{ij}^2 show that each of these high-leverage observations is also collinearity influential according to the Walker guidelines. And they clearly show that both 1969 and 1974 are together affecting the largest singular value, indicating the possibility of a pair of masking collinearity-influential observations.

The u_{ij}^2 diagnostic, then, is successful in pointing to the fact that observations 1969 and 1974 are both affecting the largest singular value, as we know by construction they should. Their removal is likely to reduce the condition number of these data. Whether or not this is a desirable thing to do will be discussed in the next section. The scaled condition indexes and variance–decomposition proportions for these data with observations 1969 and 1974 removed are given in Exhibit 8.9. A substantial change results; the condition number is reduced by roughly one full order of magnitude. It is also interesting

Exhibit 8.8 Leverage and Leverage Components u_{ij}^2: Modified Consumption Function Data

Case	Leverage	Leverage Component Associated with $\tilde{\mu}_5$	$\tilde{\mu}_4$	$\tilde{\mu}_3$	$\tilde{\mu}_2$	$\tilde{\mu}_1$
1969	0.976[a]	0.352[b]	0.145	0.460[b]	0.000	0.018
1970	0.504[a]	0.009	0.019	0.010	0.462[b]	0.003
1974	0.993[a]	0.505[b]	0.434[b]	0.051	0.001	0.002

[a]Value greater than cutoff of $2p/n$.
[b]Value greater than cutoff of p/n.

Exhibit 8.9 Scaled Condition Indexes and Variance–Decomposition Proportions: Modified Consumption Function Data. Observations 1969 and 1974 Deleted.

	Proportions of				
Scaled Condition Index, $\tilde{\eta}$	ι var(b_1)	C($T-1$) var(b_2)	**DPI**(T) var(b_3)	**r**(T) var(b_4)	Δ**DPI**(T) var(b_5)
1	.001	.000	.000	.000	.001
6	.077	.000	.000	.000	.167
10	.208	.000	.000	.039	.147
40	.309	.006	.006	.960	.142
360	.405	.994	.994	.001	.543

to note that the scaled condition indexes obtained by deleting 1969 and 1974 together demonstrate far more dramatic drops than those obtained from deleting either observation singly, attesting to the masking nature of this pair. One could also consider the effects of removing observation 1970, but its deletion only increases the second largest condition index.

This, then, provides an example of diagnosing sets of collinearity-influential observations, and in this case, the collinearity-influential observations are what some call "collinearity-creating observations" since their removal improves the data conditioning. Of course, others might consider these observations as providing valuable information about an otherwise unexplored portion of the data space, and we shall have more to say about this in the next section. But for the moment we reiterate the success of the Walker diagnostic based on the u_{ij}^2 of influential observations in determining sets of coexisting collinearity-influential observations.

The International Socioeconomic Data

As a final example, we give brief consideration to the international socioeconomic data set given in Gunst and Mason (1980, data appendix). These data are also used by Walker (1989) as an illustration of diagnosing sets of collinearity-influential points. I include them here in the hope of shedding some new light on their conditioning.

These data consist of observations for 49 countries on the response variate **GNP** (per capita, 1957 dollars) and six socioeconomic explanatory variates: **INFD** (infant deaths per 1000 live births), **PHYS** (number of inhabitants per physician), **DENS** (population per square kilometer), **AGDS** (population per 1000 hectares of agricultural land), **LIT** (percentage literate of population aged 15 and over), and **HIED** (students enrolled in higher education per 100,000 population). Just what theory leads to a model consisting all these variates tossed into a single linear equation leaves me somewhat baffled, but once again, the purpose of the exercise in this context is to see what the diagnostics tell about the data and not to assess the validity or meaningfulness of the model.

Exhibit 8.10 Scaled Condition Indexes and Variance–Decomposition Proportions: International Socioeconomic Data

Scaled Condition Index, $\tilde{\eta}$	Proportions of						
	ι var(b_1)	**INFD** var(b_2)	**PHYS** var(b_3)	**DENS** var(b_4)	**AGDS** var(b_5)	**LIT** var(b_6)	**HIED** var(b_7)
1	.001	.007	.006	.001	.001	.001	.010
2	.000	.004	.000	.010	.012	.001	.008
2	.000	.050	.067	.000	.000	.003	.161
5	.007	.000	.114	.001	.001	.031	.758
5	.002	.702	.363	.001	.001	.005	.037
13	.001	.004	.000	.934	.966	.012	.026
24	.989	.233	.450	.053	.019	.947	.000

Exhibit 8.10 displays the $\mathbf{\Pi}$ matrix of scaled condition indexes and variance–decomposition proportions for the full data set. Unlike the analysis in Walker (1989), these results are applied to the uncentered data and include an intercept term. It is clear that these data are not too badly conditioned. The strongest near dependency involves the intercept, **LIT**, and probably **PHYS** and has a scaled condition number of 24, which is modest at best. The second largest near dependency is quite weak indeed, having a scaled condition index of 13, and involves at least **DENS** and **AGDS**. It is worth noting that, because of Walker's use of centered data, it was this latter weak near dependency that was the strongest (with a centered, scaled condition index of 10) and the focus of his attention. Let us examine these data for the presence of collinearity-influential observations.

An examination of the deletion condition indexes $\tilde{\boldsymbol{\eta}}_{(i)}$ shows little effect except for Hong Kong, Malta, and Singapore. To save room, only a select subset of these diagnostic figures, including a number of unexceptional ones for comparison, are given in Exhibit 8.11. The $\tilde{\boldsymbol{\eta}}_{(i)}$ for all the other countries look similar to those for the first two given here, which in turn are similar to the base condition indexes given in the upper row. The deletion of Malta seems to worsen the conditioning somewhat, while that of either Hong Kong or Singapore makes the conditioning somewhat better. But in all cases no important changes in the conditioning take place due to single-row deletions. Similarly, the trace-to-det-ratio diagnostic (not shown) also demonstrates little effect, India and Malta having slightly elevated values. The one-at-a-time deletion diagnostics, then, give only the mildest indications of single collinearity-influential observations.

There is still, of course, the possibility of sets of masking collinearity-influential observations; Hong Kong and Singapore seem individually to be changing the conditioning in the same way. So, in Exhibit 8.12 we examine the

Exhibit 8.11 Scaled Condition Indexes and Row-Deleted Scaled Condition Indexes: International Socioeconomic Data. Selected Observations.

	Scaled Condition Indexes of **X**					
	$\tilde{\eta}_7$	$\tilde{\eta}_6$	$\tilde{\eta}_5$	$\tilde{\eta}_4$	$\tilde{\eta}_3$	$\tilde{\eta}_2$
	23.8	13.4	5.5	4.7	2.4	1.6
	Scaled Condition Indexes of $\mathbf{X}_{(i)}$					
Case	$\tilde{\eta}_{(i)7}$	$\tilde{\eta}_{(i)6}$	$\tilde{\eta}_{(i)5}$	$\tilde{\eta}_{(i)4}$	$\tilde{\eta}_{(i)3}$	$\tilde{\eta}_{(i)2}$
British Guiana	23.9	13.4	5.5	4.7	2.4	1.6
Bulgaria	23.7	13.3	5.5	4.7	2.4	1.6
Hong Kong	24.1	10.0	5.4	4.7	2.4	1.6
Hungary	23.6	13.4	5.5	4.7	2.4	1.6
India	24.1	13.5	6.5	4.7	2.6	1.6
Japan	23.6	13.3	5.4	4.6	2.4	1.6
Malta	27.2	20.0	5.4	4.7	2.4	1.6
Portugal	24.6	13.3	5.4	4.7	2.4	1.6
Singapore	24.1	11.0	5.4	4.7	2.4	1.6
Switzerland	23.6	13.3	5.5	4.7	2.4	1.6
Taiwan	25.8	13.3	5.4	4.7	2.4	1.6
United States	23.8	13.5	5.6	5.4	2.5	1.6
Yugoslavia	23.6	13.3	5.5	4.7	2.4	1.6

values for leverage and u_{ij}^2. These diagnostic values are given only for those observations demonstrating high leverage.

Malta's main effect here is on the second smallest singular value, so its deletion should only worsen the conditioning. India and the United States both have components in the direction of the same singular value, and this could be a masking pair. Their joint deletion, however, leaves the scaled condition indexes virtually unchanged: $\tilde{\boldsymbol{\eta}} = (24, 14, 7, 5, 3, 2, 1)$. The $\boldsymbol{\Pi}$ matrix for this result is not given since it is quite similar to Exhibit 8.10. Hong Kong and Singapore also seem to be operating as a pair and with considerably larger values for u_{ij}^2. This pair has been noted by Mason and Gunst (1985) and by Walker (1989). Deleting both Hong Kong and Singapore produces the $\boldsymbol{\Pi}$ matrix of scaled condition indexes and variance–decomposition proportions given in Exhibit 8.13.

Comparing Exhibits 8.13 and 8.10, we see that the main effect of deleting these two observations has been to reduce the second largest scaled condition index from 13 to 7 and to reduce somewhat the involvement of **DENS** and **AGDS** in this relation. However, since neither of these scaled condition indexes is very large, this result is really of limited interest. The largest scaled condition index remains virtually unchanged. Here, then, we have an example of multiple influential observations that are not particularly collinearity influential.

Exhibit 8.12 Leverage and Leverage Components u_{ij}^2: Modified Consumption Function Data. High Leverage Observations.

		Leverage Component Associated with						
Case	Leverage	$\tilde{\mu}_7$	$\tilde{\mu}_6$	$\tilde{\mu}_5$	$\tilde{\mu}_4$	$\tilde{\mu}_3$	$\tilde{\mu}_2$	$\tilde{\mu}_1$
Hong Kong	0.511[a]	0.082	0.390[b]	0.000	0.000	0.000	0.014	0.024
India	0.558[a]	0.047	0.010	0.174[b]	0.034	0.283[b]	0.000	0.008
Malta	0.688[a]	0.016	0.008	0.001	0.024	0.002	0.598[b]	0.041
Singapore	0.632[a]	0.081	0.406[b]	0.006	0.006	0.004	0.111	0.018
United States	0.489[a]	0.034	0.018	0.146[b]	0.280[b]	0.008	0.003	0.000

[a]Value greater than cutoff of $2p/n$.
[b]Value greater than cutoff of p/n.

This example does, however, provide an interesting result regarding the behavior of the diagnostics based on the u_{ij}^2. We note from Exhibit 8.12 that Hong Kong and Singapore mainly affect the second largest singular value $\tilde{\mu}_4$. According to Corollary 8.1, then, their deletion should increase the second largest condition index, but we see from Exhibits 8.10 and 8.13 that, either singly or together, their deletion decreases this condition index. The reasons for this are once again that (a) the observations do not lie exactly in the direction of one eigenvector only and (b) these condition indexes are calculated for the $\mathbf{X}_{(i)}$ matrix after it has been rescaled for unit column lengths—the results of Theorem 8.3 and Corollary 8.1 apply to an unscaled $\mathbf{X}_{(i)}$ matrix. The unrescaled condition indexes corresponding to Exhibit 8.11 are given in Exhibit 8.14, and these are indeed much more in accord with the predictions of Corollary 8.1, particularly with respect to Singapore where the second highest condition index is seen definitely to increase.

Exhibit 8.13 Scaled Condition Indexes and Variance-Decomposition Proportions: International Socioeconomic Data. Hong Kong and Singapore Deleted.

Scaled Condition Index, $\tilde{\eta}$	Proportions of ι var(b_1)	INFD var(b_2)	PHYS var(b_3)	DENS var(b_4)	AGDS var(b_5)	LIT var(b_6)	HIED var(b_7)
1	.000	.005	.004	.005	.005	.001	.008
2	.001	.016	.009	.126	.055	.001	.026
3	.001	.041	.067	.000	.000	.003	.167
5	.006	.042	.183	.001	.014	.028	.647
6	.002	.432	.147	.236	.137	.009	.143
7	.002	.218	.148	.615	.785	.005	.009
26	.988	.246	.442	.017	.004	.953	.000

Exhibit 8.14 Scaled Condition Indexes and Row-Deleted Condition Indexes (Unrescaled): International Socioeconomic Data. Selected Observations.

	Scaled Condition Indexes of **X**					
	$\tilde{\eta}_7$	$\tilde{\eta}_6$	$\tilde{\eta}_5$	$\tilde{\eta}_4$	$\tilde{\eta}_3$	$\tilde{\eta}_2$
	23.8	13.4	5.5	4.7	2.4	1.6
	Unrescaled Condition Indexes of $\mathbf{X}_{(i)}$					
Case	$\eta_{(i)7}$	$\eta_{(i)6}$	$\eta_{(i)5}$	$\eta_{(i)4}$	$\eta_{(i)3}$	$\eta_{(i)2}$
British Guiana	23.9	13.3	5.5	4.7	2.4	1.6
Bulgaria	23.8	13.3	5.4	4.7	2.4	1.6
Hong Kong	23.6	13.1	5.3	4.6	2.3	2.1
Hungary	23.7	13.3	5.4	4.8	2.4	1.6
India	23.5	13.0	7.0	4.7	2.6	1.6
Japan	23.6	13.2	5.4	4.7	2.4	1.6
Malta	27.2	20.1	5.4	4.7	2.4	1.6
Portugal	24.6	13.3	5.5	4.7	2.4	1.6
Singapore	23.7	14.6	5.3	4.6	2.4	2.1
Switzerland	23.7	13.3	5.5	4.7	2.4	1.6
Taiwan	25.9	13.3	5.4	4.7	2.4	1.6
United States	23.7	13.3	5.4	4.7	2.4	1.6
Yugoslavia	23.7	13.3	5.4	4.7	2.4	1.6

So an interesting question arises: which diagnostic provides the most relevant information for assessing how the conditioning has actually changed? Has the strength of the relation among the variates involved in this near dependency increased due to the deletion of Hong Kong and/or Singapore, as would be predicted by the unscaled condition indexes of Corollary 8.1, or decreased, as indicated by the (re)scaled condition indexes $\tilde{\boldsymbol{\eta}}_{(i)}$ (or the $\tilde{\boldsymbol{\eta}}_{(I)}$, where I is a subset of deleted observations)? This question can be answered by examining the change in the strength of the auxiliary regression corresponding to this second strongest near dependency. Assuming two near dependencies, we can see from either Exhibit 8.10 or 8.13 that auxiliary regressions of **LIT** and **AGDS** on the remaining variates are appropriate. It is the change in the strength of the second of these, the one corresponding to the weaker near dependency, that is our present interest. The uncentered $\hat{R}^2$s for these auxiliary regressions are .952 when based on the full data set and .819 when Hong Kong and Singapore are deleted (the corresponding centered R^2s are .947 and .627). Thus, the strength of this dependency seems to have weakened, indicating that the (re)scaled condition indexes $\tilde{\boldsymbol{\eta}}_{(i)}$ provide the more relevant information. This is, of course, consistent with all our earlier findings.

The u_{ij}^2, then, seem to be a diagnostic that conveys much useful information in identifying collinearity-influential observations, either singly or in sets, but care must be taken in using them to predict just how the deletions will affect the ultimate conditioning. The u_{ij}^2 can certainly be used to focus attention on the observations of interest, but the full set of scaled deletion condition indexes $\tilde{\boldsymbol{\eta}}_{(i)}$ (or $\tilde{\boldsymbol{\eta}}_{(I)}$) seems to become the best final arbiter.

8.4 AN APPRAISAL

In the preceding sections, the topic of collinearity-influential observations has been treated rather clinically; various diagnostics for collinearity-influential observations have been described and exemplified. The diagnostics seem quite successful in finding observations that have substantive effects on the conditioning of a given problem. Throughout that discussion, however, several underlying questions have lurked that received little attention, such as what value attaches to identifying such observations and what should one do with such observations once identified? To help answer these questions, we turn now to further analyses of collinearity-masking observations, as seen in the cement data example, and collinearity-creating observations, as seen in the modified consumption function example.

Collinearity-Masking Observations

Observation 3 in the cement data of Exhibit 8.2 is seen by all three diagnostics to be what Chatterjee and Hadi (1988) call "collinearity masking." Its deletion increases the scaled condition number of this data set from a quite tolerable 12 to a very substantial 70. Such situations prompt Chatterjee and Hadi to conclude (p. 158): "Therefore, a small condition number does not necessarily mean that **X** is not ill-conditioned." This is an interesting conclusion, for it can, in principle, apply to virtually any data set. Consider, for example, any well-spread (well-conditioned), two-dimensional scatter of data points and remove all points except those that lie near a straight line drawn along a major diagonal of the scatter. The set of points removed is clearly "collinearity influential" since its removal will produce a highly collinear subset. And by the previous statement we must conclude that this is an ill-conditioned data set despite the fact that it is not.

No, I think we must consider a well-conditioned data set to be well conditioned no matter how many ill-conditioned subsets it has. But what Chatterjee and Hadi seem really to be concerned with is a situation in which the ill conditioning is brought on by the deletion of only a very few (in this case one) observations. This concern is stated more clearly when they say (p. 158), "However, like the condition number, the collinearity indices can be substantially influenced by one or a few points in the data set. Thus, for the condition number and collinearity indices to be reliable measures of collinearity, the rows of **X** must have approximately equal influence on κ and κ_j."

Unfortunately, without appropriate justification, the conclusion to this statement is a non sequitur, and the only justification that is offered is the implicit appeal to the reader's intuition made by the statement itself, which is hardly adequate.

I hate to evoke the rather overused image as to whether the glass of water is half empty or half full, but answers do indeed depend upon perspective, and a collinearity-masking observation in one situation can be a collinearity-breaking observation in another. What comprises the difference? Simply whether the indicated observations are good data or bad. Thus, if observation 3 in the cement data were uncorrectably in error, its removal would be necessary, and the data set would be collinear. Notice, however, that the removal of this point did not mask this collinearity in any a priori sense, for there was never a well-conditioned data set to begin with—there was an erroneous data set. The proper data set, once found, was collinear all along; it did not suddenly become so. To be sure, the diagnostics could be considered instrumental in calling attention to this erroneous observation, but note that the influential-data diagnostics (for high leverage) are wholly sufficient to this task; nothing new has been added by the collinearity-influential diagnostics.

If, however, observation 3 were a good data point relative to the model—that is, if, to the best of the user's knowledge, this point is a valid observation generated by the same mechanism that generated all the other data points, and the model is a fair description of this process—, then observation 3 is a collinearity-breaking observation allowing the full data set to be well conditioned. The fact that there is an ill-conditioned subset or that the condition number is low only because of this one observation is irrelevant. And while it would obviously be nicer if one had more observations exploring this and other parts of the data space, the wise practitioner would be delighted to have one such observation rather than none. The attempt, then, to place such "good" leverage points in a pejorative light by classifying them as collinearity masking rather misses the point that such observations almost invariably are providing some of the most significant information regarding certain parameter estimates and, in fact, are breaking up bad conditioning, not masking it.

Let us, therefore, examine the role of observation 3, not in light of its effects on the conditioning of the problem, but in light of its effects on the actual least-squares results, which we recall from the discussion of Section 8.1 is where the ultimate assessment of influence is to be sought. Regressing $\mathbf{y}$ on the $\mathbf{X}_i$ from Exhibit 8.2 without observation 3 gives

$$\begin{aligned} \mathbf{y} = \underset{(0.209)}{0.307\mathbf{X}_1} + \underset{(0.362)}{1.971\mathbf{X}_2} + \underset{(0.252)}{1.251\mathbf{X}_3} + \underset{(0.246)}{0.512\mathbf{X}_4} + \underset{(9.580)}{2.150\mathbf{X}_5}, \\ R^2 = .985, \qquad \text{SER} = 2.774, \qquad \tilde{\kappa}(\mathbf{X}) = 71. \end{aligned} \tag{8.17}$$

These results are mixed. The first coefficient estimate is insignificant, the middle three look reasonable, and the last is noisy indeed. A glance at the $\mathbf{\Pi}$ matrix of scaled condition indexes and variance–decomposition proportions for these data, given in Exhibit 8.15, shows the last four variates to be strongly involved in the strongest near dependency. The first variate is also involved, but more weakly. The insignificance of the last variate, then, is possibly collinearity related, while one must be more cautious in drawing such a conclusion for the first. The middle three estimates seem moderately good despite their involvement in the collinear relation.

Now let us look at these results with observation 3 back in place, namely,

$$\begin{aligned} \mathbf{y} = \underset{(0.171)}{0.326\mathbf{X}_1} + \underset{(0.206)}{2.025\mathbf{X}_2} + \underset{(0.060)}{1.297\mathbf{X}_3} + \underset{(0.050)}{0.558\mathbf{X}_4} + \underset{(1.055)}{0.354\mathbf{X}_5}, \\ R^2 = .986, \qquad \text{SER} = 2.621, \qquad \tilde{\kappa}(\mathbf{X}) = 12. \end{aligned} \tag{8.18}$$

The main effect of introducing this observation has clearly been to produce a substantial decrease in the estimated standard errors. The estimate for the first variate is now marginally significant, those for the middle three are highly significant, and that for the last variate remains very noisy. To assess these results, we must first determine whether observation 3 is good data or bad data, for if this observation is good, it appears to have helped things, and any potential for collinearity in its absence is of little count.

Whether observation 3 is good or bad can only be determined through a thorough investigation of the conditions under which it was observed and the relevance of the model to those conditions. One must look for possible measurement error and model misspecification. For the moment let us assume that the results of such an investigation do not rule out the relevance of observation 3. As a corroborating test, we might next ask whether the results of (8.18) are consistent with those in (8.17). A significant shift in the overall estimated parameter structure would certainly be disquieting. A Chow test on

Exhibit 8.15 Scaled Condition Indexes and Variance–Decomposition Proportions: Cement Data. Without Observation 3.

Scaled Condition Index, $\tilde{\eta}$	Proportions of $\mathbf{X}_1$ var(b_1)	$\mathbf{X}_2$ var(b_2)	$\mathbf{X}_3$ var(b_3)	$\mathbf{X}_4$ var(b_4)	$\mathbf{X}_5$ var(b_5)
1	.003	.002	.000	.000	.000
3	.030	.051	.000	.001	.000
4	.027	.012	.004	.012	.000
10	.722	.322	.025	.011	.002
71	.218	.613	.971	.976	.998

coefficient equality, however, gives a test statistic of 0.036, which, relative to the .05-critical value of 5.32 for an $F(1, 8)$, leaves few concerns on this score.

So we will assume observation 3 is worth keeping and continue the analysis. A glance at the $\mathbf{\Pi}$ matrix for the full cement data (Exhibit 8.3) shows (as we already know) that there is little ill conditioning with observation 3 in place, and in particular, the last variate has been freed from its tight relation with the others. This means that it is unlikely that collinearity is the cause of the insignificance of its estimate. If $H_0: b_5 = 0$ were a permissible hypothesis in this analysis, then the acceptance of this null is far more reasonable with observation 3 in place than without.

Thus, as long as it can be assumed a priori that observation 3 is indeed a good piece of data for this study, the fact that it is collinearity influential is of very little consequence; the analysis is better off for having it present. If, however, observation 3 can be shown to be bad data, then of course, it should be corrected or removed, and the practitioner is fortunate that the influential-data diagnostics helped to highlight it. But note again that the influential-data diagnostics by themselves are sufficient to this purpose; nothing has been added by the collinearity-influential diagnostics.

Collinearity-Creating Observations

We have seen that the modified consumption function data provide an example of what Chatterjee and Hadi call "collinearity-creating" observations (or more appropriately to this case, collinearity-worsening observations—but the idea is the same). Without observations 1969 and 1974, the data are badly conditioned, with a scaled condition number of 360, but with these two observations included, the conditioning becomes a terrible 828 (and the second weakest near dependency increases from 40 to 125). Is this, however, necessarily a bad situation? To be sure, the conditioning has worsened, but if these are good observations, they also provide knowledge about a part of the data space that otherwise goes unexplored, and this increase in information could more than compensate for the worsened conditioning; that is, the increased ill conditioning may not be harmful.

Let us, then, again examine the effects of these observations, not just on the conditioning of the data but also on the estimated results of the regression. The least-squares results with observations 1969 and 1974 deleted are

$$\begin{aligned} \mathbf{C}(T) = \underset{(4.1568)}{5.7602\iota} + \underset{(0.2323)}{0.2300\mathbf{C}(T-1)} + \underset{(0.2017)}{0.7301\mathbf{DPI}(T)} \\ - \underset{(1.8396)}{3.1886\mathbf{r}(T)} + \underset{(0.1975)}{0.0560\,\Delta\mathbf{DPI}(T)}, \end{aligned} \tag{8.19}$$

$$R^2 = .9991, \qquad \text{SER} = 3.406, \qquad \text{DW} = 1.81, \qquad \tilde{\kappa}(\mathbf{X}) = 360,$$

and the results using the full data set are

$$\begin{aligned}\mathbf{C}(T) = \underset{(1.3423)}{6.5018\boldsymbol{\iota}} + \underset{(0.1522)}{0.2257}\mathbf{C}(T-1) + \underset{(0.1377)}{0.7257}\mathbf{DPI}(T)\\ -\underset{(1.6183)}{2.8769}\mathbf{r}(T) + \underset{(0.0969)}{0.1142}\,\Delta\mathbf{DPI}(T),\end{aligned} \tag{8.20}$$

$$R^2 = .9999, \qquad \text{SER} = 3.263, \qquad \text{DW} = 1.89, \qquad \tilde{\kappa}(\mathbf{X}) = 878.$$

Comparing these results makes it clear that, although the use of these two collinearity-influential observations worsens the conditioning, their presence rather substantially reduces the estimated standard errors—provided, of course, they are good data. So the question once again comes down to assessing whether these collinearity-influential observations are good data or bad. A Chow test for coefficient equality here gives a test statistic of 0.096 to be compared with a .05-critical value of 3.49 for an $F(2, 20)$, so both sets of coefficient estimates are consistent with one another. These observations, then, do not significantly alter the parameter estimates but do substantially reduce the estimated standard errors, and it would appear that we are better off using them regardless of the conditioning. The information they add about new regions of the variate space more than compensates for any increase in ill conditioning. So when collinearity-influential observations are good data, the worsened conditioning that accompanies them need not be harmful (in the sense of Chapter 7), and their arbitrary removal simply because they are collinearity influential is ill advised. Of course, if these observations were found to be uncorrectably in error, they should be removed. But doing this would not be removing the ill conditioning, for that was clearly bogus in the first place. We would now simply be reverting to an appropriate data set having whatever conditioning it has.

Summary

In both of the preceding cases we come to the conclusion that the real issue associated with collinearity-influential observations, both collinearity-masking and collinearity-creating ones, is not their effect on conditioning but whether they are good data or bad. If they are deemed bad, we have seen they should be corrected or removed regardless of their influence on collinearity. If they are deemed good, we have seen they should be kept and used regardless of their influence on collinearity (or anything else for that matter). In other words, in the process of highlighting those observations that, because of their influence, deserve to be checked twice as to whether they are bad or good, the collinearity-influential diagnostics seem to add nothing to what is learned by the influential-data diagnostics.

Thus, while the collinearity-influential diagnostics are an interesting idea, there is a serious question as to whether they really play any essential role over and above that already played by the standard influential-data diagnostics.

CHAPTER 9

Collinearity Diagnostics in Models with Logarithms and First Differences

We have seen in Chapter 6 that the collinearity diagnostics are most meaningful as regression diagnostics when applied in the context of the least-squares estimation of the strictly linear, basic model

$$\mathbf{y} = \beta_1 \mathbf{X}_1 + \cdots + \beta_p \mathbf{X}_p + \boldsymbol{\varepsilon}. \tag{9.1}$$

Here, "strictly linear" means the model is linear both in parameters and variates, and "basic" means the variates $\mathbf{X}_i$ are structurally interpretable; that is, the model is parameterized so that its variates are in a form in which numerical changes in their values can be assessed as being unimportant or inconsequential relative to the real-life situation being modeled. When this is so, a large scaled condition number $\tilde{\kappa}(\mathbf{X})$ of the $n \times p$ data matrix $\mathbf{X} \equiv [\mathbf{X}_1 \cdots \mathbf{X}_p]$ is meaningfully interpretable as the potential magnification factor by which small relative changes in the $\mathbf{X}_i$ can be transformed into large relative changes in the least-squares estimates $\mathbf{b} = (\mathbf{X}^T\mathbf{X})^{-1}\mathbf{X}^T\mathbf{y}$, and thus $\tilde{\kappa}(\mathbf{X})$ becomes a meaningful measure of conditioning.

In the event, however, that the $\mathbf{X}_i$ are not themselves structurally interpretable but rather are transformations, either linear or nonlinear—such as $\mathbf{X}_i = \log \mathbf{Z}_i$, or $\mathbf{X}_i = \mathbf{Z}_i^2$, or $\mathbf{X}_i = \Delta \mathbf{Z}_i$—of structurally interpretable $\mathbf{Z}_i$, then the value $\tilde{\kappa}(\mathbf{X})$ could provide little meaningful diagnostic information about the true conditioning of the least-squares estimator. We have already seen this in the context of certain linear transformations in Chapter 6, and we shall encounter the more general problems of nonlinear transformations in Chapter 11. There are, however, two widely employed transformations, logarithms and first differences, that are readily handled by extensions to the collinearity diagnostics as already developed, and we treat them here.

9.1 CONDITIONING IN MODELS WITH LOGARITHMS[1]

In Chapter 11 we shall rather generally define the notion of a conditioning analysis and address the problem of assessing the conditioning of nonlinear models. There, to anticipate somewhat, we learn that

(a) there is no single diagnostic procedure for carrying out a general conditioning analysis (each form of nonlinearity, for example, presents its own concerns) and

(b) unlike estimation, where "nonlinearity" often refers only to nonlinearities in parameters, conditioning analyses are affected by nonlinearities both in the parameters and the variates.

There is, however, one form of nonlinearity in the variates that lends itself to a reasonably straightforward conditioning analysis: the logarithmic transform—a fortunate case since logarithmic transformations are frequently encountered in models used in both the natural and social sciences. Indeed, we find that the collinearity diagnostics of Chapters 3–6 remain relevant once each structurally interpretable $\mathbf{Z}_i$ that enters as a logged term is first *e* scaled, that is, scaled so that the geometric mean g_{z_i} of its elements equals *e*, the base of the natural logarithms. Where a distinction is required between this scaling and the more usually encountered scaling for equal (unit) column length, we shall denote the latter by the term *L* scaling. We first view the various issues involved, then provide a solution, and finally give several illustrations.

A View of the Issues

Assume for the moment that the $\mathbf{X}_i$ in (9.1) are the basic, structurally interpretable variates. Models that differ from (9.1) only in scale (the units in which the $\mathbf{X}_i$ are measured) are clearly structurally equivalent, but as we have noted in Section 6.1, their respective data matrices can have very different condition numbers. Without a canonical scaling, then, the condition number $\kappa(\mathbf{X})$ provides little meaningful diagnostic information since arbitrarily different values can apply to the same "real-life" situation. In the strictly linear context of Chapters 3–6, we were able to resolve this problem through column equilibration—each column of $\mathbf{X}$ is *L* scaled to have the same, usually unit, Euclidean length.

A similar need for a canonical scaling arises when some of the $\mathbf{X}_i$ in (9.1) are logs[2] of structurally interpretable variates, but the resolution is different.

[1]The material in this section is based on Belsley (1988c).

[2]The term *log* is used here generally to mean logarithm. The discussion takes place in terms of natural logs, but the results are readily generalized to apply to logs of any base *r*.

Consider the model

$$\mathbf{y} = \beta_1 \boldsymbol{\iota} + \beta_2 \log \mathbf{Z}_2 + \beta_3 \log \mathbf{Z}_3 + \boldsymbol{\varepsilon}, \tag{9.2}$$

where $\mathbf{Z}_2$ and $\mathbf{Z}_3$ are structurally interpretable. Rescaling the $\mathbf{Z}_i$ as

$$\mathbf{Z}_i^* = c_i \mathbf{Z}_i, \qquad i = 2, 3, \tag{9.3}$$

merely renames the real-life phenomenon represented by the $\mathbf{Z}_i$, and hence (9.2) is structurally equivalent to

$$\mathbf{y} = \tilde{\beta}_1 \boldsymbol{\iota} + \beta_2 \log \mathbf{Z}_2^* + \beta_3 \log \mathbf{Z}_3^* + \boldsymbol{\varepsilon}, \tag{9.4}$$

where $\tilde{\beta}_1 \equiv \beta_1 - \beta_2 \log c_2 - \beta_3 \log c_3$. But the conditioning of the $\log \mathbf{Z}_i^*$ in (9.4) can differ greatly from that of the $\log \mathbf{Z}_i$ in (9.2) even though the correlation between the $\mathbf{Z}_i$ is invariant to such scaling. Hence, a canonical scaling is required for meaningful diagnostics based on the $\log \mathbf{Z}_i$.

We can see the effect of scaling on the conditioning of logged variates from the n-dimensional geometry of Exhibit 9.1. Here the "constant" vector $\boldsymbol{\iota}$ of n ones is drawn vertically with its orthogonal complement represented by the horizontal axis. The variate $\mathbf{Z}_i = (z_{1i}, \ldots, z_{ni})^{\mathrm{T}}$ is assumed to be structurally interpretable, and the vector $\log \mathbf{Z}_i \equiv (\log z_{1i}, \ldots, \log z_{ni})^{\mathrm{T}}$ is derived from it. Since $\log cz = \log c + \log z$, the geometric effect of rescaling $\mathbf{Z}_i$ to $\mathbf{Z}_i^* = c_i \mathbf{Z}_i$ is simply to add to $\log \mathbf{Z}_i$ the vector $\log c_i \boldsymbol{\iota}$; that is, it has the effect of "sliding" $\log \mathbf{Z}_i$ along the affine subspace $\aleph_i$ through $\log \mathbf{Z}_i$ and parallel to $\boldsymbol{\iota}$. Thus, any variate (vector) along $\aleph_i$ is structurally equivalent to $\log \mathbf{Z}_i$ since it results simply from renaming $\mathbf{Z}_i$.

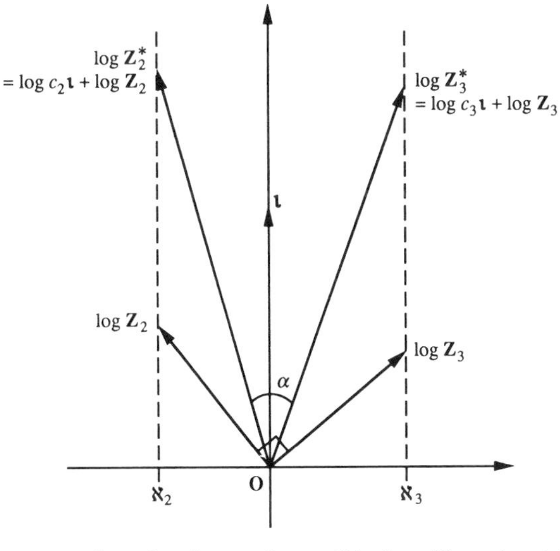

Exhibit 9.1 Effect of scaling on the conditioning of logged variates.

Let us begin, then, as in Exhibit 9.1, with $\log \mathbf{Z}_2$ and $\log \mathbf{Z}_3$ at right angles, so these two variates are perfectly conditioned. Simply by renaming (scaling) the $\mathbf{Z}_i$, we can slide them both upward, reducing the angle α between them and thus arbitrarily worsening their conditioning. This result clearly holds whether or not the three vectors ι, $\log \mathbf{Z}_2$, and $\log \mathbf{Z}_3$ are coplanar as in Exhibit 9.1. The question thus arises as to the scaling for the $\mathbf{Z}_i$ that produces $\log \mathbf{Z}_i^*$ whose conditioning provides meaningful diagnostics.

Measuring Conditioning of Models with Logs

We already know the basis for answering the preceding question from the considerations of Chapter 6. We require a scaling that results in a condition number that is able to signal whether an inconsequentially small relative change in the data can nevertheless produce a large, and presumably consequential, change in the least-squares estimates. In the strictly linear, basic model (9.1), the L-scaled condition number $\tilde{\kappa}(\mathbf{X})$ of the matrix of structurally interpretable $\mathbf{X}_i$ provides just such a measure for the least-squares estimator. That is, a $\tilde{\kappa}(\mathbf{X})$ of 100 indicates that a 1% relative change in the data could produce up to a 100% relative change in the least-squares estimates. If the $\mathbf{X}_i$ are $\log \mathbf{Z}_i$, the same considerations necessarily apply to the $\log \mathbf{Z}_i$; that is, $\tilde{\kappa}(\mathbf{X})$ measures the potential relative sensitivity in $\mathbf{b}$ with respect to small relative changes in the $\log \mathbf{Z}_i$—but not with respect to small relative changes in the $\mathbf{Z}_i$. And if it is the $\mathbf{Z}_i$ that are structurally interpretable, it is this latter measure that is desired.

A solution is suggested by recalling that, for a scalar w and natural logs,

$$\frac{d \log w}{\log w} = \frac{1}{\log w} \frac{dw}{w}. \tag{9.5}$$

Here we see that relative changes in $\log w$ are the same as relative changes in w if and only if w is scaled to equal the base e so that $\log w = 1$. Of course, the $\mathbf{X}_i$ and $\mathbf{Z}_i$ are vectors and not scalars; so taking a hint from the preceding, it seems reasonable to scale the $\mathbf{Z}_i$ so that, at some representative value $\tilde{z}^*$ for the elements of the scaled vector $\mathbf{Z}_i^* \equiv c_i \mathbf{Z}_i$, one has $\log \tilde{z}^* = 1$. This occurs if one chooses

$$c_i \equiv e g_{z_i}^{-1}, \tag{9.6}$$

where g_{z_i} is the geometric mean of the elements of $\mathbf{Z}_i = (z_{1i}, \ldots, z_{ni})^{\mathrm{T}}$, that is

$$g_{z_i} \equiv \left(\prod_j z_{ji} \right)^{1/n}. \tag{9.7}$$

Then the geometric mean $g_{z_i^*}$ of the elements of $\mathbf{Z}_i^* \equiv e g_{z_i}^{-1} \mathbf{Z}_i$ is e, and at the representative value $\tilde{z}^* \equiv g_{z_i^*} = e$ and its corresponding unscaled value $\tilde{z} \equiv e^{-1} g_{z_i} \tilde{z}^* = g_{z_i}$, we have

$$\frac{d \log \tilde{z}^*}{\log \tilde{z}^*} = \frac{d\tilde{z}^*}{\tilde{z}^*} = \frac{d\tilde{z}}{\tilde{z}}. \tag{9.8}$$

The first equality holds since $\log g_{z_i^*} = 1$, and the second holds because scaling does not affect relative values.

The preceding readily generalizes to accommodate logs to any base r. In this case, at the representative value $\tilde{z}^*$, (9.5) becomes

$$\frac{d \log_r \tilde{z}^*}{\log_r \tilde{z}^*} = \frac{\log_r e}{\log_r \tilde{z}^*} \frac{d\tilde{z}^*}{\tilde{z}^*}, \tag{9.5'}$$

and the desired scaling is clearly that which sets $\log_r e = \log_r \tilde{z}^*$, or again, where $\tilde{z}^* \equiv g_{z_i^*} = e$.

For any base, then, the same canonical scaling is indicated as appropriate for assessing the conditioning of the logged variates $\mathbf{Z}_i$: e scaling, or that which makes the geometric mean $g_{z_i^*}$ of the scaled data series $\mathbf{Z}_i^*$ equal to the base e of the natural logarithms.[3] For this scaling, small relative changes in the variates $\log \mathbf{Z}_i^*$ are approximately the same as those in the $\mathbf{Z}_i^*$ and, hence, as those in the structurally interpretable variates $\mathbf{Z}_i$.

A Canonical Scaling for Logs

Assume, then, it is desired to assess the conditioning of a data matrix $\mathbf{X}$, some or all of whose columns are in logged form. The preceding result suggests that, if those structurally interpretable variates entering in logged form are first e scaled (the others left unchanged), resulting in a data matrix that we shall denote $\mathbf{X}^*$, the collinearity diagnostics apply directly to $\mathbf{X}^*$. That is, it suggests using the L-scaled condition number $\tilde{\kappa}(\mathbf{X}^*)$ of $\mathbf{X}^*$ to measure the potential relative change in the least-squares estimator $\mathbf{b}$ that could result from a small relative change in the structurally interpretable $\mathbf{Z}_i$ that enter the model in logged form. Various cases can arise depending upon whether $\mathbf{X}$ contains all logged variates or is a mixture of logged and nonlogged variates.

Case 1. In the data matrix $\mathbf{X} \equiv [\mathbf{X}_1 \cdots \mathbf{X}_p]$, some of the $\mathbf{X}_i$ may be $\log \mathbf{Z}_i$, but it is always the $\mathbf{X}_i$ and not the $\mathbf{Z}_i$ that are structurally interpretable. In this case, no e scaling is necessary; the collinearity diagnostics can be applied directly to the $\mathbf{X}_i$. Of course, all $\mathbf{X}_i$ should first be L scaled as in STEP 2 of Section 5.2, and it

[3]Computational note: The log of the e-scaled vector $\mathbf{Z}^*$ is most easily calculated by noting that $\log \mathbf{Z}^* = \overline{\log \mathbf{Z}} + \iota$, where $\overline{\log \mathbf{Z}} \equiv \log \mathbf{Z} - m_{\log \mathbf{Z}} \iota$ and $m_{\log \mathbf{Z}}$ is the arithmetic mean of the elements of $\log \mathbf{Z}$. That is, $\log \mathbf{Z}$ is first put into deviation-from-mean form $\overline{\log \mathbf{Z}}$, and then 1 is added to each element. In the event that logs to base r are being used, one would instead add $\log_r e$ to each element to give $\overline{\log_r \mathbf{Z}} + \log_r e\, \iota$.

is the L-scaled condition number $\tilde{\kappa}(\mathbf{X})$ and its variance–decomposition proportions that are the desired diagnostic measures for the conditioning analysis.

Case 2. The $\mathbf{X}_i$ are log transforms $\log \mathbf{Z}_i$ of structurally interpretable $\mathbf{Z}_i$. First, e-scale the $\mathbf{Z}_i$ to $\mathbf{Z}_i^*$ having geometric mean $g_{z_i^*} = e$, and then use the collinearity diagnostics to examine the L-scaled condition number $\tilde{\kappa}(\mathbf{X}^*)$ of $\mathbf{X}^* \equiv [\log \mathbf{Z}_1^* \cdots \log \mathbf{Z}_p^*]$ and its variance–decomposition proportions. Thus, the $\mathbf{Z}_i$ are first e scaled to $\mathbf{Z}_i^*$ and then the $\log \mathbf{Z}_i^*$ are L scaled for equal (unit) length. This later scaling does not alter the relative magnitudes of the $d \log \mathbf{Z}_i^* / \log \mathbf{Z}_i^*$ but, as we have seen in Chapter 6, is needed for reasons of column equilibration.

Case 3. (The General Case). Some $\mathbf{X}_i$'s (which may be logged variates) are themselves structurally interpretable while others are $\log \mathbf{Z}_i$ for which it is the $\mathbf{Z}_i$ that are structurally interpretable. Here we have a mixture of the previous two cases. First, e-scale only the structurally interpretable $\mathbf{Z}_i$ variates to $\mathbf{Z}_i^*$ as in case 2; then examine the L-scaled condition number $\tilde{\kappa}(\mathbf{X}^*)$ of the matrix $\mathbf{X}^*$ whose columns are the union of the structurally interpretable $\mathbf{X}_i$ and the $\log \mathbf{Z}_i^*$. It is these scaled condition indexes and associated variance–decomposition proportions that are the desired conditioning diagnostics.

We shall illustrate this procedure shortly, but in order to assess its performance, it is first necessary to examine the magnitude of the condition number that can be anticpated from its use.

The Magnitude of $\tilde{\kappa}(\mathbf{X}^*)$

Consider the two alternate specifications: the model linear in the x's (hereafter known simply as the linear model),

$$y_t = \beta_1 + \beta_2 x_{t2} + \cdots + \beta_p x_{tp} + \varepsilon_t, \tag{9.9}$$

and the model linear in the logged x's (hereafter known simply as the logged model)

$$\log y_t = \tilde{\beta}_1 + \tilde{\beta}_2 \log x_{t2} + \cdots + \tilde{\beta}_p \log x_{tp} + \zeta_t. \tag{9.10}$$

These models are clearly quite similar and, in disciplines like economics, are often considered for alternate specifications for the same real-life phenomenon. Furthermore, since e scaling means that a given relative shift in the x_{ti} in (9.9) will result in roughly the same relative shift in the $\log x_{ti}$ in (9.10), we should be very surprised if the condition number of the e-scaled logged data from (9.10) were not similar to that of the nonlogged data from (9.9). This prospect can be motivated somewhat more formally as follows.

The $\tilde{\beta}_i$ in the logged model (9.10) can be interpreted as the elasticities $\partial \log y_t / \partial \log x_{ti} = (\partial y_t / \partial x_{ti})(x_{ti}/y_t)$ of y_t with respect to the x_{ti}, while the β_i in the linear model (9.9) can be interpreted as the partial derivatives $\partial y_t / \partial x_{ti}$. Let $\tilde{b}_i$ be the least-squares estimates of the elasticities $\tilde{\beta}_i$ and b_i be those of the partials β_i. Then we know that the scaled condition number of the logged data bounds the relative change $\partial\tilde{b}_i/\tilde{b}_i$ in $\tilde{b}_i$ with respect to a given relative change $\partial \log x_{ti}/\log x_{ti}$ in $\log x_{ti}$, while the scaled condition number of the nonlogged data bounds the relative change $\partial b_i/b_i$ in b_i with respect to a given relative shift $\partial x_{ti}/x_{ti}$ in x_{ti}. Of course, with e scaling $\partial \log x_{ti}/\log x_{ti} \approx \partial x_{ti}/x_{ti}$, so it is needed only to compare $\partial\tilde{b}_i/\tilde{b}_i$ and $\partial b_i/b_i$ for a given $\partial x_{ti}/x_{ti}$. This can be done roughly by converting b_i into an elasticity estimate like $\tilde{b}_i$, that is, by considering $\hat{b}_i \equiv b_i(x_{ti}/y_t)$. Now, for a given relative change in x_{ti} (y_t held constant), we would expect the potential sensitivity of the two elasticity estimates to be similar; that is, we would expect $\partial\tilde{b}_i/\tilde{b}_i$ and $\partial\hat{b}_i/\hat{b}_i$ to have a similar bound. But

$$\frac{\partial\hat{b}_i}{\hat{b}_i} = \frac{\partial b_i(x_{ti}/y_t)}{b_i(x_{ti}/y_t)} = \frac{x_{ti}\,\partial b_i + b_i\,\partial x_{ti}}{b_i x_{ti}} = \frac{\partial b_i}{b_i} + \frac{\partial x_{ti}}{x_{ti}}, \tag{9.11}$$

so when the data are ill conditioned and, by definition, $\partial b_i/b_i \gg \partial x_{ti}/x_{ti}$, we must have $\partial\hat{b}_i/\hat{b}_i \approx \partial b_i/b_i$. Under e scaling, then, we expect $\partial\tilde{b}_i/\tilde{b}_i$ and $\partial b_i/b_i$ to have similar bounds, and hence, we anticipate the scaled condition number of the logged e-scaled data to be similar in magnitude to that of the nonlogged data. It remains for the illustrations that follow to show that this indeed occurs in practice.

A Special Case of Interest

The treatment of one special case warrants attention: that of $\log \mathbf{Z}_i$ and $\log \mathbf{Z}_i^2$. With a little thought beforehand, these two variates are not likely to occur together in the same model since $\log \mathbf{Z}_i^2 = 2\log \mathbf{Z}_i$ and they are exactly collinear. The situation could arise, however, if a logged variate happened to be very closely the square of another or, as could all too easily happen, if a given equation containing both $\mathbf{Z}_i$ and $\mathbf{Z}_i^2$ were mechanically "logged" in order to test whether the logged or nonlogged specification seemed better. The issues involved in assessing this situation are most easily made clear geometrically.

Consider first the geometric effect of e scaling. In Exhibit 9.2, the $\boldsymbol{\iota}$ vector is again oriented vertically with its orthogonal complement represented by the horizontal axis. The vector $\log \mathbf{Z} \equiv (\log z_1, \ldots, \log z_n)^{\mathrm{T}}$ is any n-vector of logs.

We know from Chapter 1 that the *arithmetic* mean of *any* vector is determined by its orthogonal projection into the space of $\boldsymbol{\iota}$. The point A, then, is $m_{\log \mathbf{Z}}\,\boldsymbol{\iota}$, where $m_{\log \mathbf{Z}} \equiv n^{-1}\Sigma_i \log z_i$, the arithmetic mean of the elements of $\log \mathbf{Z}$. And from the fact that $m_{\log \mathbf{Z}} = \log(\Pi_i z_i)^{1/n} = \log g_z$, we see that A is also $\log g_z\,\boldsymbol{\iota}$. Thus, for any vector $\log \mathbf{W} \in \aleph_\iota$, the affine subspace through $\boldsymbol{\iota}$ and orthogonal to it, $\log g_w = 1$, and the elements of $\mathbf{W}$ have geometric mean $g_w = e$. Now, recalling that, as we scale $\mathbf{Z}$, we slide $\log \mathbf{Z}$ vertically along $\aleph_{\mathbf{Z}}$, we see that the scaling c that produces a vector $\mathbf{Z}^* \equiv c\mathbf{Z}$ with geometric mean e is that which places $\log \mathbf{Z}^*$ at the intersection of $\aleph_\iota$ and $\aleph_{\mathbf{Z}}$.

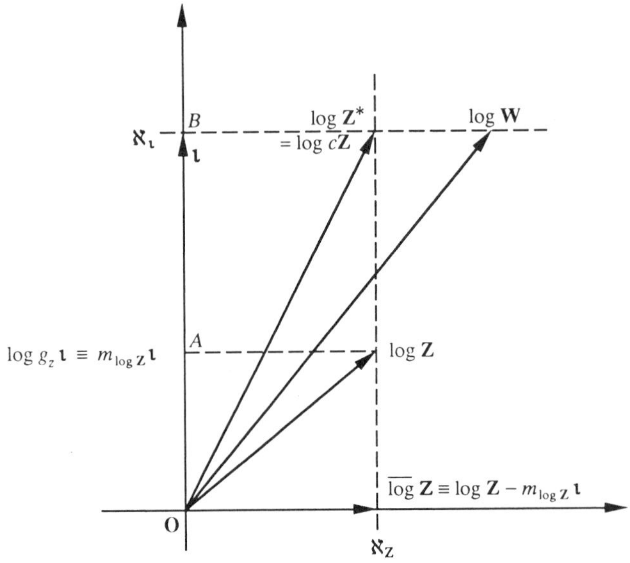

Exhibit 9.2 Effect of *e* scaling.

Consider next, then, applying the preceding to the model

$$\mathbf{y} = \beta_1 \boldsymbol{\iota} + \beta_2 \log \mathbf{Z} + \beta_3 \log \mathbf{Z}^2 + \boldsymbol{\varepsilon}, \tag{9.12}$$

containing both $\log \mathbf{Z}$ and $\log \mathbf{Z}^2 = (\log z_1^2, \ldots, \log z_n^2)^{\mathrm{T}} = 2 \log \mathbf{Z}$. These two vectors are plotted along with $\boldsymbol{\iota}$ in Exhibit 9.3. It is clear that $\log \mathbf{Z}$ and $\log \mathbf{Z}^2$ are perfectly collinear and that such a specification would make little sense. But note that, after $\mathbf{Z}$ is *e* scaled to $\mathbf{Z}^*$, the resulting $\log \mathbf{Z}^*$ and $\log \mathbf{Z}^{2*}$ are no longer collinear. And so it would seem that such scaling would remove our ability correctly to diagnose collinearity in the original data. This is not so, however, since the three vectors $\boldsymbol{\iota}$, $\log \mathbf{Z}^*$, and $\log \mathbf{Z}^{2*}$ remain perfectly collinear (coplanar), and the perfect ill conditioning of $\mathbf{X} \equiv [\boldsymbol{\iota} \ \log \mathbf{Z} \ \log \mathbf{Z}^2]$ also necessarily exists in the scaled matrix $\mathbf{X}^* \equiv [\boldsymbol{\iota} \ \log \mathbf{Z}^* \ \log Z^{2*}]$—but in a different form.

The preceding argument depends on the presence of the constant vector $\boldsymbol{\iota}$ in the model (9.12) (i.e., $\beta_1 \neq 0$) and seems therefore to lack generality, but this is not the case. It is an occasionally unappreciated fact about models containing logged terms that they must always include an intercept. This is because there is no natural unit for measuring any variate—I am always free to "deci" any units you prefer. Thus, if one postulates the homogeneous model

$$\mathbf{y} = \beta_2 \log \mathbf{Z} + \beta_3 \log \mathbf{W} + \boldsymbol{\varepsilon}, \tag{9.13}$$

a simple change of units $\mathbf{Z}^* \equiv c\mathbf{Z}$ ($c \neq 1$) produces the structurally equivalent,

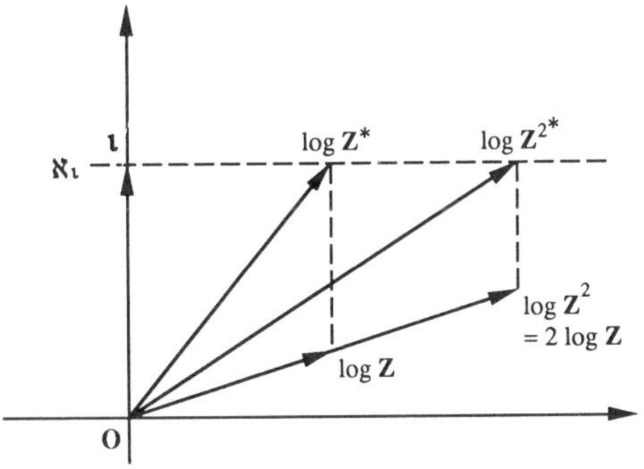

Exhibit 9.3 Effect of e scaling on squared logged variates.

nonhomogeneous model

$$\mathbf{y} = \beta_1 \boldsymbol{\iota} + \beta_2 \log \mathbf{Z}^* + \beta_3 \log \mathbf{W} + \boldsymbol{\varepsilon} \tag{9.14}$$

with $\beta_1 = -\log c \neq 0$. Thus, the homogeneous form (9.13) requires the untenable assumption of unique natural units for the variates. A model with logged variates, then, must always include an intercept, and in assessing the conditioning of the data relevant to such a model, the data matrix being assessed must always include $\boldsymbol{\iota}$, the constant column of ones. Because of this, the method of e scaling proposed above will never fail to detect collinearity between $\log \mathbf{Z}$ and $\log \mathbf{Z}^2$ since it will always materialize as collinearity among $\boldsymbol{\iota}$, $\log \mathbf{Z}^*$, and $\log \mathbf{Z}^{2*}$.

Illustrations

The e scaling introduced above is chosen to provide conditioning diagnostics for models with logs whose magnitudes are able to support an interpretation similar to the one we have found true for strictly linear models, namely, a measure of the relative sensitivity of the resulting least-squares estimator $\mathbf{b}$ to a given relative change in the structurally interpretable variates $\mathbf{Z}_i$ (not the $\log \mathbf{Z}_i$).

The efficacy of e scaling in achieving this goal is illustrated through analyses of three different economic data sets corresponding to a consumption function, a commercial and industrial loans function, and a labor force participation function. The conditioning of these data is assessed and reported first (case I) as if they came from the strictly linear model

$$\mathbf{y} = \beta_1 \boldsymbol{\iota} + \beta_2 \mathbf{X}_2 + \cdots + \beta_p \mathbf{X}_p + \boldsymbol{\varepsilon}, \tag{9.15}$$

and second (case II) as if they came from a model linear in the logs

$$\log \mathbf{y} = \beta_1 \boldsymbol{\iota} + \beta_2 \log \mathbf{X}_2 + \cdots + \beta_p \log \mathbf{X}_p + \boldsymbol{\varepsilon}. \tag{9.16}$$

In this latter instance, the results are reported both with e scaling [case II(a)] and without e scaling [case II(b)] in order to demonstrate that the e scaling is necessary to obtain stable and comparable diagnostic results. Furthermore, assessment is made and reported (case III) assuming the model to possess various mixtures of logged and nonlogged terms. In each instance, it is the $\mathbf{X}_i$ and not the $\log \mathbf{X}_i$ that are assumed structurally interpretable.

The structurally interpretable consumption function data are relevant to the model of the consumption function introduced in Section 5.4. The data are given in Exhibit 5.10, and we reproduce the model (5.2) here in its basic form,

$$\mathbf{C}(T) = \beta_1 \boldsymbol{\iota} + \beta_2 \mathbf{C}(T-1) + \beta_3 \mathbf{DPI}(T) + \beta_4 \mathbf{r}(T) + \beta_5 \, \Delta\mathbf{DPI}(T) + \boldsymbol{\varepsilon}, \tag{9.17}$$

where, we recall,

$$\begin{aligned} \boldsymbol{\iota} &= \text{a constant vector,} \\ \mathbf{C} &= \text{total consumption, 1958 dollars,} \\ \mathbf{DPI} &= \text{disposable personal income, 1958 dollars,} \\ \mathbf{r} &= \text{interest rate (Moody's Aaa).} \end{aligned}$$

The last observation, which has a negative value for $\Delta\mathbf{DPI}$, has been dropped in this application in order to facilitate comparison of this model between its linear and logged forms. Thus, all series are annual, 1948–1973.

The commercial and industrial loans data relate to the model

$$\begin{aligned} \mathbf{MCL}(T) = \beta_1 \boldsymbol{\iota} &+ \beta_2 \mathbf{MCL}(T-1) + \beta_3 \mathbf{YGPPI}(T) \\ &+ \beta_4 [\mathbf{RCL}(T) - \mathbf{RTB}(T)] + \boldsymbol{\varepsilon}, \end{aligned} \tag{9.18}$$

where

$$\begin{aligned} \boldsymbol{\iota} &= \text{a constant vector,} \\ \mathbf{MCL} &= \text{commercial and industrial loans at all commercial banks,} \\ \mathbf{YGPPI} &= \text{gross private product,} \\ \mathbf{RCL} &= \text{average bank rate on short-term commercial and industrial loans,} \\ \mathbf{RTB} &= \text{average yield on 3-month U.S. Treasury bills.} \end{aligned}$$

These data, from Kuh and Schmalensee (1972), are quarterly series from 1959:II to 1969:IV.

The secondary labor force participation data relate to the model

$$\mathbf{LCSLSPOP}(T) = \beta_1 \boldsymbol{\iota} + \beta_2 \mathbf{ER}(T) + \beta_3 \mathbf{LCSLSPOP}(T-1) + \beta_4 \mathbf{TIME}(T) + \boldsymbol{\varepsilon}, \tag{9.19}$$

where

ι = a constant vector,

ER = ratio of civilian employment to total population of working age,

LCSLSPOP = ratio of secondary civilian labor force to secondary population,

TIME = a time dummy.

These data, also from Kuh and Schmalensee (1972), are annual series, 1948–1962.

No effort is made here to explain the meaning of the linear, logged, or mixed-logged regression models behind these data sets. The purpose of this exercise is only to examine the efficacy of the proposed conditioning diagnostics correctly to assess the presence of near dependencies in the data.

The condition indexes for cases I, II (a and b), and III are summarized for each data set in Exhibit 9.4. Most of the comparative information needed for this study can be seen from these results alone. The detailed variance–decomposition proportion matrices are, however, needed for a more complete analysis and are provided in Exhibits 9.5–9.7 of the addendum that follows for those interested in seeing how the different transformations and scalings affect diagnosed variate involvement.

The first important observation comes from comparing columns II(a) and II(b). It is clear that *e* scaling has a major impact on the magnitudes of the condition indexes of the logged variates. It is further seen that *e* scaling can cause these magnitudes either to decrease (as for the first two data sets) or increase (as for the last). That this can happen is readily predictable from Exhibit 9.1.

The second important observation comes from comparing columns I and II(a). Here we see that *e* scaling, consonant with the preceding analysis, always brings the condition indexes of the logged data into line with those of the basic data. Thus *e* scaling successfully achieves the desired end of providing a stable diagnostic measure that meaningfully measures the potential sensitivity of a relative change in the least-squares estimate to a relative shift in the basic, structurally interpretable data as opposed to the logged data. We also see that failure to *e*-scale logged data typically results in grossly distorted diagnostic magnitudes that can be either too high or too low. Although examples of both cases occur here, experience indicates the behavior of the first two data sets to be most representative; that is, failure to *e*-scale logged data most often produces grossly overstated condition indexes.

Comparison of the mixture cases in column III with column I indicates that *e* scaling of logged variates, even when intermixed with nonlogged variates, also tends to produce diagnostic measures whose magnitudes are comparable to those of the basic data.

Exhibit 9.4 Comparative Condition Indexes: Three Illustrative Models

Case I, Basic Nonlogged data	Case II		Case III	
	(a) *e* Scaling	(b) No *e* Scaling	Representative Mixtures	
		Consumption Function Data		
1	1	1	1	1
6	4	9	4	6
10	9	14	9	10
40	40	232	32	46
367	339	1900	113	381
		Commercial and Industrial Loans Data		
1	1	1	1	1
5	5	4	5	5
16	19	43	15	15
146	164	852	101	141
		Secondary Labor Force Participation Data		
1	1	1	1	1
9	12	21	9	12
194	154	156	194	146
333	346	238	331	351

The first mixture in the consumption function data deserves special mention here since these magnitudes seem to be quite different. Examination of the full set of variance–decomposition proportions given in Exhibit 9.5.III(a), however, reveals in this case that, of the three variates **C**, **DPI**, and **ΔDPI** involved in the strongest basic near dependency (the one corresponding to 367 in Exhibit 9.5.I), only the last two have been logged. This differential treatment of the variates involved in a linear dependency will, of course, have a tendency to alter it—in this case weaken it so that its condition index falls from 367 to 113. In the second mixture among the consumption function data, seen in Exhibit 9.5.III(b), only the variate **r** was logged, leaving the strongest near dependency among the other variates untouched, and here we see the *e*-scaled diagnostics are telling a virtually identical story as in the basic case. In the mixture case, then, *e* scaling seems to be working in accord with intuition and expectations.

As to the effect of *e* scaling on the diagnosis of variate involvement, it is readily seen from the variance–decomposition proportion matrices in Exhibits 9.5–9.7 that the *e*-scaled logged data show similar patterns of variance–

decomposition proportions as those for the basic data. Failure to *e* scale can alter these patterns to produce misleading results, as can be seen by comparing the role of the constant term in the consumption function data (Exhibits 9.5.II(a), 9.5.II(b), and 9.5.I), the role of log(**RCL** − **RTB**) in the commercial and industrial loans data (Exhibits 9.6.II(a), 9.6.II(b), and 9.6.I), and the role of **TIME** in the secondary labor force participation data (Exhibits 9.7.II(a), 9.7.II(b), and 9.7.I).

Conclusions

Our goal was to find a canonical scaling for logged data that is able to produce stable conditioning diagnostics having the same interpretative basis and meaning as those of the regular collinearity diagnostics applied to strictly linear, basic models. We find that *e* scaling—scaling structurally interpretable variates that appear in logged terms so that their geometric mean is *e*, the base of the natural logarithms—precisely serves this end. The *e*-scaled logged variates along with the nonlogged variates in the equation provide the proper input to the collinearity diagnostics, the second step of which is to *L*-scale (for unit length) all variates.

It is worth reiterating that *e* scaling should be applied only to structurally interpretable variates $\mathbf{Z}_i$ that enter the equation as $\log \mathbf{Z}_i$. If the variate $\mathbf{X}_i \equiv \log \mathbf{Z}_i$ is itself the structurally interpretable variate, it should be treated normally.

Addendum: The Variance–Decomposition Proportion Matrices

For those interested in examining in greater detail the effect of *e* scaling on determining variate involvement, the full variance–decomposition proportion matrices for the preceding illustrations are presented here in Exhibits 9.5–9.7. The exhibit numbers are keyed by:

9.5 Consumption function data.
9.6 Commercial and industrial loans data.
9.7 Labor force participation data.

- **(I)** Basic nonlogged data.
- **(II)** Logged data:
  - (a) With *e* scaling.
  - (b) Without *e* scaling.
- **(III)** Mixtures, logged and nonlogged.

In the various mixture cases, the variates that have been logged and those that have not are made clear from the column headings.

Consumption Function Data

Exhibit 9.5.I Scaled Condition Indexes and Variance-Decomposition Proportions: Basic Consumption Function Data

Scaled Condition Index, $\tilde{\eta}$	Proportions of ι var(b_1)	**C**($T-1$) var(b_2)	**DPI**(T) var(b_3)	**r**(T) var(b_4)	Δ**DPI**(T) var(b_5)
1	.001	.000	.000	.000	.001
6	.072	.000	.000	.000	.174
10	.206	.000	.000	.033	.110
40	.330	.006	.006	.964	.150
367	.391	.994	.994	.003	.565

Exhibit 9.5.II Scaled Condition Indexes and Variance-Decomposition Proportions: Logged Consumption Function Data

Scaled Condition Index, $\tilde{\eta}$	Proportions of ι var(b_1)	log **C**($T-1$) var(b_2)	log **DPI**(T) var(b_3)	log **r**(T) var(b_4)	log Δ**DPI**(T) var(b_5)
		(a) With e Scaling			
1	.001	.000	.000	.000	.003
4	.023	.000	.000	.000	.282
9	.331	.000	.000	.021	.098
40	.266	.005	.007	.887	.045
339	.379	.995	.993	.092	.572
		(b) Without e Scaling			
1	.000	.000	.000	.000	.001
9	.000	.000	.000	.001	.212
14	.001	.000	.000	.073	.150
232	.944	.005	.007	.835	.064
1900	.055	.995	.993	.091	.573

Exhibit 9.5.III Scaled Condition Indexes and Variance-Decomposition Proportions: Mixed Consumption Function Data

Scaled Condition Index, $\tilde{\eta}$	Proportions of var(b_1)	var(b_2)	var(b_3)	var(b_4)	var(b_5)
	(a) Mixture				
	ι	**C**($T-1$)	log **DPI**(T)	**r**(T)	log Δ**DPI**(T)
1	.002	.000	.000	.000	.007
4	.033	.000	.000	.000	.606
9	.429	.001	.001	.027	.187
32	.361	.015	.074	.645	.133
113	.175	.984	.925	.328	.067
	(b) Mixture				
	ι	**C**($T-1$)	**DPI**(T)	log **r**(T)	Δ**DPI**(T)
1	.001	.000	.000	.000	.002
6	.061	.000	.000	.000	.185
10	.200	.000	.000	.022	.117
46	.243	.006	.008	.902	.181
381	.495	.994	.992	.076	.515

Commercial and Industrial Loans Data

Exhibit 9.6.I Scaled Condition Indexes and Variance–Decomposition Proportions: Basic Commercial and Industrial Loans Data

Scaled Condition Index, $\tilde{\eta}$	Proportions of ι var(b_1)	MCL($T-1$) var(b_2)	YGPPI(T) var(b_3)	RCL − RTB var(b_4)
1	.000	.000	.000	.003
5	.000	.002	.000	.156
16	.057	.009	.000	.539
146	.943	.989	1.000	.302

Exhibit 9.6.II Scaled Condition Indexes and Variance–Decomposition Proportions: Logged Commercial and Industrial Loans Data

Scaled Condition Index, $\tilde{\eta}$	Proportions of ι var(b_1)	log MCL($T-1$) var(b_2)	log YGPPI(T) var(b_3)	log (RCL − RTB) var(b_4)
		(a) With e Scaling		
1	.000	.000	.000	.004
5	.000	.001	.000	.273
19	.186	.007	.000	.599
164	.814	.992	1.000	.124
		(b) Without e Scaling		
1	.000	.000	.000	.010
4	.000	.000	.000	.544
43	.005	.013	.000	.315
852	.995	.987	1.000	.131

Exhibit 9.6.III Scaled Condition Indexes and Variance–Decomposition Proportions: Mixed Commercial and Industrial Loans Data

Scaled Condition Index, $\tilde{\eta}$	Proportions of var(b_1)	var(b_2)	var(b_3)	var(b_4)
		(a) Mixture		
	ι	**MCL**($T-1$)	log **YGPPI**(T)	**RCL** − **RTB**
1	.000	.000	.000	.003
5	.000	.004	.000	.160
15	.060	.033	.001	.506
101	.940	.963	.999	.331
		(b) Mixture		
	ι	**MCL**($T-1$)	**YGPPI**(T)	log(**RCL** − **RTB**)
1	.000	.000	.000	.003
5	.000	.001	.000	.183
15	.059	.009	.000	.557
141	.941	.989	1.000	.257

Secondary Labor Force Participation Data

Exhibit 9.7.I Scaled Condition Indexes and Variance–Decomposition Proportions: Basic Labor Force Participation Data

Scaled Condition Index, $\tilde{\eta}$	Proportions of ι var(b_1)	ER(T) var(b_2)	LCSLSPOP($T-1$) var(b_3)	TIME var(b_4)
1	.000	.000	.000	.001
9	.000	.000	.000	.118
194	.489	.001	.502	.823
333	.511	.999	.498	.058

Exhibit 9.7.II Scaled Condition Indexes and Variance–Decomposition Proportions: Logged Labor Force Participation Data

Scaled Condition Index, $\tilde{\eta}$	Proportions of ι var(b_1)	log ER(T) var(b_2)	log LCSLSPOP($T-1$) var(b_3)	log TIME var(b_4)
		(a) With e Scaling		
1	.000	.000	.000	.000
12	.001	.000	.000	.181
154	.399	.003	.319	.654
346	.600	.997	.681	.165
		(b) Without e Scaling		
1	.000	.000	.000	.000
21	.000	.002	.003	.081
156	.542	.431	.005	.201
238	.458	.567	.992	.718

Exhibit 9.7.III Scaled Condition Indexes and Variance-Decomposition Proportions: Mixed Labor Force Partitipation Data

Scaled Condition Index, $\tilde{\eta}$	Proportions of			
	var(b_1)	var(b_2)	var(b_3)	var(b_4)
	(a) Mixture			
	ι	log **ER**(T)	**LSCLSPOP**($T-1$)	**TIME**
1	.000	.000	.000	.001
9	.000	.000	.000	.118
194	.487	.001	.511	.825
331	.513	.999	.489	.056
	(b) Mixture			
	ι	**ER**(T)	**LSCLSPOP**($T-1$)	log **TIME**
1	.000	.000	.000	.001
12	.001	.000	.000	.189
146	.331	.001	.323	.661
351	.668	.999	.677	.149

9.2 FIRST DIFFERENCING[4]

We turn now to an analysis of another commonly employed transformation of time-series data: first differences. This transformation provides some statistical advantages when there is high serial correlation in the error structure $\boldsymbol{\varepsilon}$ of the standard linear model $\mathbf{y} = \mathbf{X}\boldsymbol{\beta} + \boldsymbol{\varepsilon}$, as often is the case in models containing highly trended variates.[5] It is also thought that this transformation, by looking only at the changes, can help reduce the collinearity that typically accompanies such variates. Clearly, whether this is the case or not must turn on whether it is the levels variates or their first differences that are structurally interpretable. We begin with an examination of the issues involved and obtain an approximate solution, which shows that the better conditioning that often attends first-differenced data can be highly misleading. An example is then given using the equation for the price deflator for consumption of autos and parts from the Michigan Quarterly Econometric Model (MQEM).

Differencing and Conditioning

Consider first the general first-differenced model

$$\Delta\mathbf{y} = \beta_1\boldsymbol{\iota} + \beta_2\Delta\mathbf{X}_2 + \cdots + \beta_p\Delta\mathbf{X}_p + \boldsymbol{\varepsilon}, \tag{9.20}$$

where $\Delta\mathbf{y} \equiv [y_1 - y_0, \ldots, y_n - y_{n-1}]^{\mathrm{T}}$, $\Delta\mathbf{X}_i \equiv [X_{i,1} - X_{i,0}, \ldots, X_{i,n} - X_{i,n-1}]^{\mathrm{T}}$ and β_1 is an intercept that may or may not be present, depending upon whether there is a trend term in the corresponding levels, or integral, model. Denote by $\Delta\mathbf{X}$ the first-differenced data matrix $[\boldsymbol{\iota}\ \Delta\mathbf{X}_2 \cdots \Delta\mathbf{X}_p]$. We know from previous considerations that the scaled condition number $\tilde{\kappa}(\Delta\mathbf{X})$ assesses the relative sensitivity of the least-squares estimates $\mathbf{b}$ of (9.20) to small relative changes in the differenced data $\Delta\mathbf{X}_i$. If the differenced data are themselves structurally interpretable, then this may well be the desired diagnostic information. But if, as is often the case, it is the levels data $\mathbf{X}_i$ that are structurally interpretable, then the differenced data may not be, and $\tilde{\kappa}(\Delta\mathbf{X})$ provides misleading diagnostic information. That the $\Delta\mathbf{X}_i$'s often lack structural interpretability is readily seen by noting that in many instances one must return to the $\mathbf{X}_i$'s to determine whether a 1% shift in $\Delta\mathbf{X}_i$ is consequential or inconsequential within the underlying real situation being modeled.

It is to be stressed that our concern here is not with levels models that may also include some structurally interpretable first-differenced terms, as might be the case, for example, for a model involving an investment term that is defined as the time rate of change of the capital stock. If such a term is indeed structurally interpretable and is the appropriate object of a diagnostic analysis, then it can be treated just like any other structurally interpretable variate in applying the

[4]The material in this section is based on Belsley (1986a).
[5]See, e.g., Pindyck and Rubinfeld (1981).

collinearity diagnostics. Rather, our concern is with models that are structurally interpretable when formulated originally in terms of the levels variates but that are then first differenced "wholesale" to achieve some other statistical objective.

It is also interesting to note that, in terms of signal-to-noise from Chapter 7, the first-differencing process itself clearly has a tendency to transform highly trended variates with a large constant component into short data. Under these circumstances, it is quite possible for $\tilde{\kappa}(\mathbf{\Delta X})$ to be pleasantly small even though there may be other causes of weak data. Thus, even should the first-differenced data be structurally interpretable, care must be taken in using $\tilde{\kappa}(\mathbf{\Delta X})$ to assess the overall presence of data weaknesses.

Let us therefore suppose that it is the levels data, the $\mathbf{X}_i$'s, that are structurally interpretable and that the first differences are not, so that $\tilde{\kappa}(\mathbf{\Delta X})$ gives an incorrect measure of the conditioning of (9.20) with respect to the $\mathbf{X}_i$. What provides the correct measure? Here we show that a lower bound for this conditioning is given by $\tilde{\kappa}(\mathbf{X})$, the scaled condition number of the data matrix of the corresponding levels, or integral, model, that is, the scaled condition number of $\mathbf{X} \equiv [\boldsymbol{\iota}\ \mathbf{t}\ \mathbf{X}_2 \cdots \mathbf{X}_p]$, where $\mathbf{t}$ is the trend variate $[1, 2, \ldots, n]^T$.

Diagnosing the Conditioning of First Differences with Respect to the Basic Data

Consider the levels, or integral, equation corresponding to (9.20), namely,

$$\begin{aligned} \mathbf{y} &= \mathbf{X}\boldsymbol{\beta} + \boldsymbol{\zeta} \\ &= \beta_0\boldsymbol{\iota} + \beta_1\mathbf{t} + \beta_2\mathbf{X}_2 + \cdots + \beta_p\mathbf{X}_p + \boldsymbol{\zeta}, \end{aligned} \tag{9.21}$$

where β_0 is the constant of integration, and the variate $\mathbf{t} \equiv [1, 2, \ldots, n]^T$ is the partial sum (integral) of the constant in (9.20) and will be present only if an intercept term is included there. The $n \times (p+1)$ matrix $\mathbf{X} = [\boldsymbol{\iota}\ \mathbf{t}\ \mathbf{X}_2 \cdots \mathbf{X}_p]$ is the matrix of levels, or integral, data.

Assume that the $\mathbf{X}$ and $\mathbf{y}$ data are such that $\mathbf{X}$ has full rank and a regression of $\mathbf{y}$ on $\mathbf{X}$ produces a perfect fit, that is,

$$\mathbf{Xb} = \mathbf{y} \quad \text{where } \mathbf{b} = \mathbf{X}^+\mathbf{y}, \tag{9.22}$$

and we will examine the sensitivity of $\mathbf{b}$ to changes in $\mathbf{X}$. Of course, we know from (3.19) that relaxing the assumption of a perfect fit can only increase the potential sensitivity, so for this and reasons to follow, the measure we obtain can only provide a lower bound for the true conditioning.

Now, from among all the possible perturbations in $\mathbf{X}$, consider the subset $\mathscr{S}$ of rank-preserving perturbations $\boldsymbol{\delta}\mathbf{X}$ such that $\mathbf{y} \in C(\mathbf{X} + \boldsymbol{\delta}\mathbf{X})$, the column space of $\mathbf{X} + \boldsymbol{\delta}\mathbf{X}$. The assumption of rank preservation does not further restrict the directions of possible perturbations, only their magnitudes, for we know from Section 3.1 that it is sufficient for $\text{rank}(\mathbf{X}) = \text{rank}(\mathbf{X} + \boldsymbol{\delta}\mathbf{X})$ that $\|\boldsymbol{\delta}\mathbf{X}\|/\|\mathbf{X}\|$

$< \kappa^{-1}(\mathbf{X})$. For this subset of perturbations, then, we must also have, for some $\mathbf{b} + \delta\mathbf{b} \neq \mathbf{0}$,

$$(\mathbf{X} + \delta\mathbf{X})(\mathbf{b} + \delta\mathbf{b}) = \mathbf{y}. \tag{9.23}$$

Of course, (9.22) and (9.23) must also hold when premultiplied by any reweighting matrix $\mathbf{A}$ of full rank, that is,

$$\begin{aligned} \mathbf{AXb} &= \mathbf{Ay}, \\ \mathbf{A}(\mathbf{X} + \delta\mathbf{X})(\mathbf{b} + \delta\mathbf{b}) &= \mathbf{Ay}. \end{aligned} \tag{9.24}$$

Thus, any response $\delta\mathbf{b}$ of $\mathbf{b}$ to a change $\delta\mathbf{X} \in \mathscr{S}$ that satisfies (9.22) and (9.23) must also satisfy (9.24), and hence the sensitivity of the solution to (9.24) over $\mathscr{S}$ must be at least that of (9.23). The relevance of this to assessing the conditioning of first differences with respect to changes in $\mathbf{X}$ is seen directly by noting that the first-difference equation (9.20) is derived from the levels equation (9.21) with the $(n-1) \times n$ transformation matrix of full rank,

$$\mathbf{A} = \begin{bmatrix} -1 & 1 & 0 & \cdots & 0 & 0 \\ 0 & -1 & 1 & \cdots & 0 & 0 \\ \vdots & \vdots & \vdots & & \vdots & \vdots \\ 0 & 0 & 0 & \cdots & -1 & 1 \end{bmatrix} \tag{9.25}$$

The sensitivity of the solutions, then, of both the first differences—(9.24) with $\mathbf{A}$ from (9.25)—and the levels (9.22) to at least some subset of permutations $\mathscr{S}$ is derivable from (9.22) and (9.23) as

$$\delta\mathbf{b} = -\mathbf{X}^{+}\delta\mathbf{X}(\mathbf{b} + \delta\mathbf{b}). \tag{9.26}$$

And using the properties of matrix norms given in Section 3.1, we get

$$\frac{\|\delta\mathbf{b}\|}{\|\mathbf{b} + \delta\mathbf{b}\|} \leqslant \kappa(\mathbf{X}) \frac{\|\delta\mathbf{X}\|}{\|\mathbf{X}\|}. \tag{9.27}$$

Among the subset of perturbations $\mathscr{S}$, then, the sensitivity of either the first-differenced estimates or the levels estimates with respect to relative changes in the basic data $\mathbf{X}$ (not $\Delta\mathbf{X}$) is bounded by $\kappa(\mathbf{X})$, and thus this diagnostic provides at least a lower bound for the desired measure of conditioning. As always, the bound will be lowest and most meaningful when $\mathbf{X}$ has been column equilibrated, so that it becomes the scaled condition number $\tilde{\kappa}(\mathbf{X})$ that is used.

Given the preceding, a question naturally arises as to how the incorrect measure $\tilde{\kappa}(\Delta\mathbf{X})$ of the conditioning of first-differenced data relates to $\tilde{\kappa}(\mathbf{X})$ and, thus, the way its diagnostic information can be misleading. Unlike the case of mean centering, where it can be shown unambiguously that $\tilde{\kappa}(\tilde{\mathbf{X}}) \leqslant \tilde{\kappa}(\mathbf{X})$ (see Section 6.9), we find here that $\tilde{\kappa}(\mathbf{\Delta X})$ can bear any relation to $\tilde{\kappa}(\mathbf{X})$. The example

$$\mathbf{X} = \begin{bmatrix} 1 & 1 & 0 & 0 \\ 1 & 0 & 1 & 0 \\ 1 & 0 & 0 & 1 \\ 1 & 0 & 0 & 0 \\ 1 & 0 & 0 & 0 \end{bmatrix}, \qquad \mathbf{\Delta X} = \begin{bmatrix} -1 & 1 & 0 \\ 0 & -1 & 1 \\ 0 & 0 & -1 \\ 0 & 0 & 0 \end{bmatrix},$$

provides a case where $\tilde{\kappa}(\mathbf{\Delta X}) = 3.73 > \tilde{\kappa}(\mathbf{X}) = 2.81$. And for highly trended time series, a case where first differencing is often employed, we typically find that $\tilde{\kappa}(\mathbf{\Delta X}) \ll \tilde{\kappa}(\mathbf{X})$, making the conditioning appear very much better than it actually is. We see this to be the case in the following example, which illustrates both the ineffectiveness of $\tilde{\kappa}(\mathbf{\Delta X})$ and the usefulness of $\tilde{\kappa}(\mathbf{X})$ for diagnosing the conditioning of first-differenced data.

An Illustration with a Price Deflator Equation

Consider the following equation for the price deflator for consumption of autos and parts:

$$\mathbf{\Delta PCDA} = \beta_1 \iota + \beta_2 \mathbf{\Delta PPNF} + \beta_3 \mathbf{\Delta PAUTO} + \varepsilon, \tag{9.28}$$

where

$$\begin{aligned} \mathbf{PCDA} &= \text{deflator for consumption of autos and parts, 1972} = 100, \\ \mathbf{PPNF} &= \text{deflator for business, nonfarm GNP, 1972} = 100, \\ \mathbf{PAUTO} &= \text{consumer price index for new autos, 1967} = 100. \end{aligned}$$

All data are quarterly 1954:I–1979:IV and are taken from the MQEM. These levels data are structurally interpretable and thus constitute the basic data for this example.

We examine first the estimation and conditioning of the differenced model (9.28). Least squares gives

$$\begin{aligned} \mathbf{\Delta PCDA} = \underset{(0.131)}{0.208} + \underset{(0.111)}{0.175}\,\mathbf{\Delta PPNF} + \underset{(0.131)}{0.708}\,\mathbf{\Delta PAUTO}, \\ R^2 = .652, \qquad \tilde{\kappa}(\mathbf{\Delta X}) = 3. \end{aligned} \tag{9.29}$$

We immediately note that the scaled condition number of the differenced data matrix $\mathbf{\Delta X} = [\iota \;\; \mathbf{\Delta PPNF} \;\; \mathbf{\Delta PAUTO}]$ is excellent, indicating a lack of sensitivity of the least-squares estimates to small relative changes in the first-differenced data. That the same insensitivity does not, however, occur with respect to small relative changes in the structurally interpretable levels data **PPNF** and **PAUTO** is seen simply by perturbing these data by small relative amounts. Thus, construct

$$\begin{aligned} \mathbf{PPNF}^{\circ} &= \mathbf{PPNF} + \boldsymbol{\xi}, \\ \mathbf{PAUTO}^{\circ} &= \mathbf{PAUTO} + \mathbf{v}, \end{aligned} \tag{9.30}$$

where $\boldsymbol{\xi}$ and $\mathbf{v}$ are chosen randomly but scaled so that their effects are of small consequence, $\|\boldsymbol{\xi}\| = 0.01\|\mathbf{PPNF}\|$ and $\|\mathbf{v}\| = 0.01\|\mathbf{PAUTO}\|$.

The least-squares estimates of the perturbed, differenced model are

$$\underset{}{\Delta\mathbf{PCDA}} = \underset{(0.140)}{0.423} + \underset{(0.071)}{0.151}\,\Delta\mathbf{PPNF}^{\circ} + \underset{(0.057)}{0.434}\,\Delta\mathbf{PAUTO}^{\circ}. \tag{9.31}$$

Letting $\mathbf{b}$ and $\mathbf{b}^{\circ}$ denote the unperturbed and perturbed estimates, respectively, we calculate that $\|\mathbf{b} - \mathbf{b}^{\circ}\|/\|\mathbf{b}\| = .4603$. Thus, the 1% relative shift in the basic **X**'s has produced over a 46% shift in the estimated parameters, a degree of sensitivity that is certainly not indicated by the excellent condition number of 3 belonging to the differenced data $\Delta\mathbf{X}$. This occurs, of course, as it did in Section 6.7 for the centered data, because small proportionate changes in the $\mathbf{X}_i$'s can produce gigantic proportionate changes in the $\Delta\mathbf{X}_i$'s. For these data, for example, a 1% change in **PPNF** produces a 96% relative change in Δ**PPNF**. The corresponding figure for Δ**PAUTO** is 82%. Clearly, then, we see that $\tilde{\kappa}(\Delta\mathbf{X})$ is not the desired diagnostic measure.

According to the diagnostic suggested above, however, let us consider the conditioning of the data matrix corresponding to the levels, or integral, model

$$\mathbf{PCDA} = \beta_0\boldsymbol{\iota} + \beta_1\mathbf{t} + \beta_2\mathbf{PPNF} + \beta_3\mathbf{PAUTO} + \boldsymbol{\varepsilon}. \tag{9.32}$$

We find that the condition number $\tilde{\kappa}(\mathbf{X})$ of this levels data matrix $\mathbf{X} = [\boldsymbol{\iota}\ \mathbf{t}\ \mathbf{PPNF}\ \mathbf{PAUTO}]$ is 77, a magnitude that is clearly a far better indicator of the degree of sensitivity of the least-squares estimates of (9.28) to small relative changes in the basic data observed above.

Conclusions

When the levels data **X** are structurally interpretable and the first-differenced data $\Delta\mathbf{X}$ are not, the scaled condition number of the first-differenced data $\tilde{\kappa}(\Delta\mathbf{X})$ can be quite misleading as to the true conditioning of the first-differenced model with respect to the basic, structurally interpretable data. This diagnostic can be either too high or too low, but with highly trended time series, $\tilde{\kappa}(\Delta\mathbf{X})$ tends to understate the desired conditioning, thus giving the illusion that first differencing has improved the conditioning. The scaled condition number of the levels, or integral, data **X**, however, appears to provide an appropriate lower bound on this conditioning.

Rather more generally, we have seen that many data and model transformations that are legitimately undertaken for purposes other than conditioning can nevertheless cause the collinearity diagnostics to be less than meaningful when applied to the transformed data. It is important, then, no matter how the model may be transformed for, say, estimation purposes to adopt an appropriate form for assessing the conditioning of that model—*even for assessing the conditioning of that model in its transformed state.* Thus, in Chapter 6 we saw that the conditioning of the mean-centered model is best assessed through diagnosis of the uncentered, basic data (even if these data are not used for estimation). And at the beginning of this chapter, we saw that the conditioning of models using log transforms is best assessed through diagnosis of the *e*-scaled logged data (even if these data are not used for estimation). Similarly, we have just seen that the conditioning of models written in first-differenced form of structurally interpretable variates is best assessed through diagnosis of the corresponding levels, or integral, data (even if these data are not used for estimation). In each case, the guiding principle for determining the proper form in which to assess conditioning has been to choose that form in which the data are structurally interpretable.

CHAPTER 10

Corrective Action and Case Studies

The ability we have now acquired to diagnose the presence, composition, and severity of collinear relations is also the crucial first step in determining and applying appropriate corrective action. We begin this chapter with a discussion of means for correcting for collinearity once it has been discovered. Then we examine several substantive case studies in which the collinearity diagnostics are used as an integral part of the statistical analysis, both for diagnosing collinearity and for improving estimation or forecasting quality.

10.1 CORRECTIVE MEASURES

Historically, collinearity is among the first-encountered and separately named problems of econometric estimation. However, collinearity is really very closely related to the more general problem of identification. Quite generally, an equation is unidentifiable, and therefore inestimable, when it is not possible to determine its parameters uniquely from knowledge of the conditional distribution of $\mathbf{y}$ given $\mathbf{X} \in \Theta$, where Θ is the sample space of relevant outcomes on $\mathbf{X}$.[1] But even though an equation may be identifiable, there can be specific $\mathbf{X}$, usually contained in a subset of Θ of measure zero, on which estimation is impossible. For example, in the event that there is exact collinearity, so that some variates in $\mathbf{X}$ are indistinguishable from linear combinations of the others, the otherwise identifiable least-squares estimator is not uniquely determined at that $\mathbf{X}$, and the equation is inestimable. And further, in the event that there is near collinearity, so that $\mathbf{X}$ is of full rank but poorly conditioned, the equation is estimable, but weakly so. Thus, there is a relation between the strength of the data and the degree of estimability, or what we might call the "degree of effective identification." We will pursue this relation in greater detail in the next chapter, but for the moment we need only realize that corrective measures for ill conditioning must be like corrective measures for identification; namely, some identifying

[1]See Malinvaud (1970).

prior or auxiliary information must be introduced to compensate for the lack of information inherent in the data.

We begin, then, by discussing the two major sources that exist for such condition-identifying information: additional, better-conditioned data and Bayes-like methods for introducing prior information. We can also use this perspective to show that a frequently suggested solution for collinearity, dropping some of the collinear variates, is typically inappropriate—if correcting the collinearity problem requires the addition of identifying information, one can readily imagine what value attaches to a "solution" that throws information away.

Introduction of New Data

The most direct and obvious means for improving data conditioning is through the collection and use of additional data. Of course, not all new data are equally helpful for this purpose. Intuition certainly suggests that new data possessing near dependencies similar to those in the original data will be of less value than new data that provide information on novel or underrepresented portions of Θ, the sample space for $\mathbf{X}$. We can see from Theorem 8.1 of Section 8.2, for example, that it is observations that lie in the directions of the eigenvectors corresponding to the smallest singular values of $\mathbf{X}$ (eigenvalues of $\mathbf{X}^T\mathbf{X}$) that are the most useful for reducing existing ill conditioning.[2] Such observations are the ones that will help to "fill out" the degenerate dimensions in situations like those depicted in Exhibits 3.3 and 3.4. But of course, no new observations are worthless, for all added data can help to reduce the variance of the least-squares estimates (loosely, to increase the overall length of $\mathbf{X}$ relative to σ^2 and so to increase the signal-to-noise).

Unfortunately, the remedy of obtaining new data is rarely useful to the many users of least squares who have only a few precious observations and who can obtain new data only at substantial cost, either in terms of the time needed to wait for new observations to materialize or in terms of the cost needed to sample additional, less-collinear observations. Furthermore, even if new data were obtainable, there is often no guarantee that they will be consistent with the original data or that they will indeed provide strongly independent information. Applied statisticians are often unable to control their experiments, whereas nature often closely replicates hers.

The introduction of new data, therefore, is not a fix available in many applications of least squares. However, when it is possible to provide new data, the corrective action is straightforward, obvious, and simple.

Bayesian-Type Techniques

Short of new data, the introduction of appropriate prior information is the best available solution to the collinearity problem. At least three Bayes-like pro-

[2]See also Silvey (1969) and Oldford (1982, 1983).

cedures exist for introducing the experimenter's subjective prior information: (1) a pure Bayesian technique, (2) a mixed-estimation technique, and (3) the technique of ridge regression. These procedures are listed in order of increasing ease of use and of decreasing generality and suitability. We use the mixed-estimation technique in dealing with several of the examples given shortly.

Pure Bayes

The use of Bayesian estimation for introducing the identifying information that is needed to improve data conditioning is explained in the work of Zellner (1971) and Leamer (1973, 1978). These excellent studies show that the seemingly insoluble problem of collinearity can in fact be dealt with if the investigator possesses—and is willing to use—subjective prior information on the parameters of the model. The drawbacks of this method are severalfold. First, it relies on subjective information that many researchers simply distrust or feel they do not possess. Second, its use may require a rather complete statement of the prior distribution, an apparent degree of detail that many find too exacting to be realistic. Third, it draws on a statistical theory not as widely understood or appreciated as classical techniques, and fourth, it requires computer software and estimation methods that are not so widely available or so easily (mechanically?) employed as standard techniques. In practice, the first two drawbacks are far more psychological than real, and efforts such as Kadane et al. (1980) have produced techniques that allow the user efficiently and straightforwardly to answer questions that adequately reveal his subjective priors. Indeed, such research will moderate all four drawbacks to strict Bayesian estimation, and one can hope to see these techniques gain wider acceptance in the future.

Mixed Estimation

Theil and Goldberger (1961) and Theil (1963, 1971) have suggested a Bayes-like technique called mixed estimation. With this procedure, prior,[3] or auxiliary, information is added directly to the data matrix. Mixed estimation is simple to employ and need not require a full specification of the prior distribution.

Beginning with the linear model

$$\mathbf{y} = \mathbf{X}\boldsymbol{\beta} + \boldsymbol{\varepsilon}, \tag{10.1}$$

with $E\boldsymbol{\varepsilon} = \mathbf{0}$ and $\mathbf{V}(\boldsymbol{\varepsilon}) \equiv \boldsymbol{\Sigma}_1$, it is assumed the investigator can construct r linear prior restrictions on the elements of $\boldsymbol{\beta}$ in the form

$$\mathbf{c} = \mathbf{R}\boldsymbol{\beta} + \boldsymbol{\xi}, \tag{10.2}$$

[3]The use of the term *prior* here follows Theil (1971). For an approximation that allows for a Bayesian interpretation of mixed estimation, see Theil (1971, pp. 670–672). Also, Tiao and Zellner (1964) have shown this procedure to provide an approximation to the posterior mean of a pure Bayesian estimator whose prior is a posterior from a previous study (with known variance) and is otherwise diffuse.

with $E\boldsymbol{\xi} = \mathbf{0}$ and $\mathbf{V}(\boldsymbol{\xi}) \equiv \boldsymbol{\Sigma}_2$. Here $\mathbf{R}$ is a matrix of rank r of known constants, $\mathbf{c}$ is an r-vector of specifiable values, and $\boldsymbol{\xi}$ is a random vector, independent of $\boldsymbol{\varepsilon}$, with mean zero and variance–covariance matrix $\boldsymbol{\Sigma}_2$, also assumed to be stipulated by the investigator.

In the method of mixed estimation as suggested by Theil and Goldberger,[4] estimation of (10.1) subject to (10.2) proceeds by augmenting $\mathbf{y}$ and $\mathbf{X}$ to give

$$\begin{bmatrix} \mathbf{y} \\ \mathbf{c} \end{bmatrix} = \begin{bmatrix} \mathbf{X} \\ \mathbf{R} \end{bmatrix} \boldsymbol{\beta} + \begin{bmatrix} \boldsymbol{\varepsilon} \\ \boldsymbol{\xi} \end{bmatrix}, \tag{10.3}$$

where

$$\mathbf{V}\begin{pmatrix} \boldsymbol{\varepsilon} \\ \boldsymbol{\xi} \end{pmatrix} = \begin{bmatrix} \boldsymbol{\Sigma}_1 & \mathbf{0} \\ \mathbf{0} & \boldsymbol{\Sigma}_2 \end{bmatrix} \equiv \boldsymbol{\Sigma}.$$

If $\boldsymbol{\Sigma}_1$ and $\boldsymbol{\Sigma}_2$ are known, generalized least squares applied to (10.3) results in the unbiased mixed-estimation estimator

$$\mathbf{b}_{\mathrm{ME}} = (\mathbf{X}^{\mathrm{T}}\boldsymbol{\Sigma}_1^{-1}\mathbf{X} + \mathbf{R}^{\mathrm{T}}\boldsymbol{\Sigma}_2^{-1}\mathbf{R})^{-1}(\mathbf{X}^{\mathrm{T}}\boldsymbol{\Sigma}_1^{-1}\mathbf{y} + \mathbf{R}^{\mathrm{T}}\boldsymbol{\Sigma}_2^{-1}\mathbf{c}). \tag{10.4}$$

The Aitken-type augmented data matrix equivalent to this procedure is $\begin{bmatrix} \boldsymbol{\Sigma}_1^{-1/2}\mathbf{X} \\ \boldsymbol{\Sigma}_2^{-1/2}\mathbf{R} \end{bmatrix}$, and it is the improved conditioning of this matrix that should reflect the usefulness of the prior information in undoing any ill effects of collinearity. In practice, an estimate $\mathbf{S}_1$ is substituted for $\boldsymbol{\Sigma}_1$ (usually in the form $s^2\mathbf{I}$), and $\boldsymbol{\Sigma}_2$ is specified by the investigator as part of the prior information. In the event that $\boldsymbol{\Sigma}_1$ is estimated by $\mathbf{S}_1 = s^2\mathbf{I}$, one can equivalently use $\begin{bmatrix} \mathbf{X} \\ \mathbf{DR} \end{bmatrix}$, where $\mathbf{D} = s\boldsymbol{\Sigma}_2^{-1/2}$.

Ridge Regression

Ridge regression is a biased, minimum mean-squared error estimation technique with roots extending back to Stein (1956) and developed fully in Hoerl and Kennard (1970) and Dempster et al. (1977) and surveyed in Vinod (1978). The ridge-regression estimator, with the single ridge parameter k, is simply

$$\mathbf{B}_{\mathrm{R}} = (\mathbf{X}^{\mathrm{T}}\mathbf{X} + k\mathbf{I})^{-1}\mathbf{X}^{\mathrm{T}}\mathbf{y}. \tag{10.5}$$

While differing in its philosophical basis, the ridge estimator is seen to be computationally equivalent to the mixed-estimation estimator with $\mathbf{R} = \mathbf{A}$ (where $\mathbf{A}$ is any column-orthogonal matrix—$\mathbf{A}^{\mathrm{T}}\mathbf{A} = \mathbf{I}$), $\boldsymbol{\Sigma}_1 = \sigma^2\mathbf{I}$, $\boldsymbol{\Sigma}_2 = \lambda^2\mathbf{I}$, and $\mathbf{c} = \mathbf{0}$. In this event, $k = \sigma^2/\lambda^2$. Of course, in mixed estimation, $\mathbf{c}$ is taken to be stochastic with $E\mathbf{c} = \mathbf{R}\boldsymbol{\beta}$, whereas here $\mathbf{c}$ is taken as a set of constants (zeros), which results in a biased estimator. Computationally, therefore, mixed estimation provides a compromise between the rigors of full Bayesian estimation and the rather inflexible ridge estimator, whose single parameter k arbitrarily

[4]See Theil (1971, pp. 347–352).

forces the weight attached to all prior constraints to be equal. This restriction is relaxed somewhat with the generalized ridge estimator

$$\mathbf{b}_{GR} = (\mathbf{X}^T\mathbf{X} + \boldsymbol{\Delta})^{-1}\mathbf{X}^T\mathbf{y}, \tag{10.6}$$

where $\boldsymbol{\Delta}$ is a positive-definite matrix. But both ridge and generalized ridge, in assuming $\mathbf{c} = \mathbf{0}$, effectively impose a prior mean of zero on $\boldsymbol{\beta}$, a value whose relevance will typically be a happy accident.[5] In practice, then, mixed estimation allows prior information to be introduced in a more natural, meaningful, and flexible fashion than ridge estimation and with little additional computational effort.

Shouldn't Priors be Prior?

A legitimate question can be advanced against considering the use of prior information as a remedy for ill conditioning, for if one has this information, why not use it from the outset? Or viewed slightly differently, is it not odd, having encountered a problem with ill conditioning, suddenly to realize that one may have additional information to help fill the gap?[6] While these questions are indeed legitimate, the statistical practice that lies behind them is pure, beautiful, and wondrous to behold and, like all things pure, beautiful, and wondrous to behold, very rarely seen in practice.

It is far more usual for an investigator to conceive of some relevant variates, collect the indicated data, and run a regression. If the results look good, he gives thanks and uses them. If they look bad, he then begins looking for reasons why. Now whether this is good or bad statistical practice cannot wholly be the issue, for it would be very difficult to show that this is not often the way of life. And in fairness, there are good reasons for such a course of action. First, from an economic perspective (as opposed to a pristine statistical one), there may be large costs in time and effort needed to develop prior information, costs that are well avoided unless they are seen to be necessary. Second, many practitioners hesitate to publish results that depend on prior information, fearing that their readers (or superiors) will disagree with the priors and discount the results. And this is indeed a legitimate fear. Thus, practitioners often wish to ignore prior information at the outset, taking the time and trouble to develop it only when there is no other acceptable way to proceed. And of course, there are less legitimate reasons for such behavior, namely, that the user may not even be concerned with prior information, even in principle, until forced to be.

Thus, whether for legitimate reasons or not, the practitioner often finds himself in a problemful situation and needs to know how best to proceed. In this case the questions arise (a) what is the source of the trouble and (b) what prior

[5]See Holland (1973).

[6]Of course, the same question could be raised against considering the use of new, better-conditioned data as a remedy for ill conditioning—if you could have gotten the data, why didn't you?

information is most effective in helping out? The remedial nature of the collinearity diagnostics helps to answer both of these questions. If, in answer to (a), the diagnostics show collinearity to be the culprit (which is indeed something that must be shown), then the information gotten from the diagnostics in answer to (b) can be compared with the available prior information, and one can proceed in an orderly and meaningful fashion to introduce this prior information into the estimation process. In this light, there is no question that the techniques given here are indeed remedies for ill conditioning.

A Nonsolution: Occam's Hatchet

Before examining various case studies that exemplify the use of the collinearity diagnostics for remedial purposes, it is important to discuss one popular "solution" to the collinearity problem that rarely has real legitimacy: the removal of some of the collinear variates. A survey of statistics and econometrics texts dealing with the least-squares procedure will show this to be a commonly advocated procedure. However, it has never been particularly clear to me what logic lies behind this suggestion. If an investigator has reason for including a variate in the regression model in the first place, there is just that much reason for not excluding it capriciously. And if, otherwise, the investigator has no reason for including the variate, I suppose that situation speaks for itself. The following study should help make this clear.

This example[7] deals with the care and use of Occam's razor in dealing with collinearity. It is a concocted example that nevertheless closely parallels a data series that arose in a real-life study (Belsley, 1969b). Occam's razor, of course, is the tool that, over the centuries, has been used to trim away unneeded model fat. In paraphrase it says: when two models are equally good at explaining the same thing, use the simpler one—or somewhat more closely to the original, do not add complications without reason.[8]

But what is the "thing" being explained? The data? or something broader? This example shows that it is something broader: in addition to the data, the model must account for those elements of the data-generating mechanism that are maintained to be true a priori, *even if they are not observed in the given data.* Here, the data-generating mechanism (dgm) refers to the real-life process that is presumed to generate the given data set. This idea is certainly not inconsistent with Occam's razor, but it flies directly in the face of those who would model specifically to achieve a "parsimonious" set of variates that produces a good fit or those who would mechanically cast out collinear variates.

Consider, then, the data given in Exhibit 10.1. We assume that they have been given to a statistical investigator for the purpose of forecasting the effect of a change in $\mathbf{w}$ on $\mathbf{y}$. This investigator is not an economist (with whatever other sins

[7]Adapted from Belsley (1988b).

[8]William of Occam's fourteenth-century original, *entia non sunt multiplicanda praeter necessitatem*, renders as "objects ought not to be multiplied except of necessity."

Exhibit 10.1 Data for Example of Occam's Hatchet

Observation	y	w	r
1	19.223	100	101
2	19.316	103	99
3	19.085	105	98
4	21.258	111	97
5	18.315	112	94
6	25.870	114	90
7	25.178	118	88
8	24.472	125	87
9	26.151	127	82
10	28.448	128	81
11	26.511	129	79
12	30.175	135	76
13	30.419	135	75
14	31.223	138	74
15	32.715	143	71
16	33.146	148	66
17	34.759	150	64
18	39.384	150	62
19	37.858	155	61
20	40.024	160	56

we imbue him, this is not one of them), and he bothers only to know that they are economic data explaining some sort of output measure **y** in terms of a related index of the real price of capital **r** and an appropriate index of the real wage rate **w**. We, however, are in a more fortunate position and know these data in fact to have been generated by a perfectly competitive production process according to

$$\mathbf{y} = 100 - 0.15\mathbf{w} - 0.65\mathbf{r} + \boldsymbol{\varepsilon}, \tag{10.7}$$

where $\boldsymbol{\varepsilon}$ is iid Gaussian with mean zero and variance 2.0. Further, over the 20 years of observation, economic forces have been at work that have caused real wages steadily to rise and the real cost of capital steadily to fall.

So when the investigator, as part of his routine data analysis, examines a scatter of **r** against **w**, the situation is as in Exhibit 10.2, well evidencing strong collinearity. No doubt, then, the analyst anticipates the trouble that occurs when a least-squares regression of **y** on **w** and **r** is run to obtain

$$\begin{aligned} \mathbf{y} = \underset{(40.717)}{107.175} - \underset{(0.174)}{0.174}\mathbf{w} - \underset{(0.228)}{0.705}\mathbf{r}, \\ R^2 = .960, \qquad \text{SER} = 1.447. \end{aligned} \tag{10.8}$$

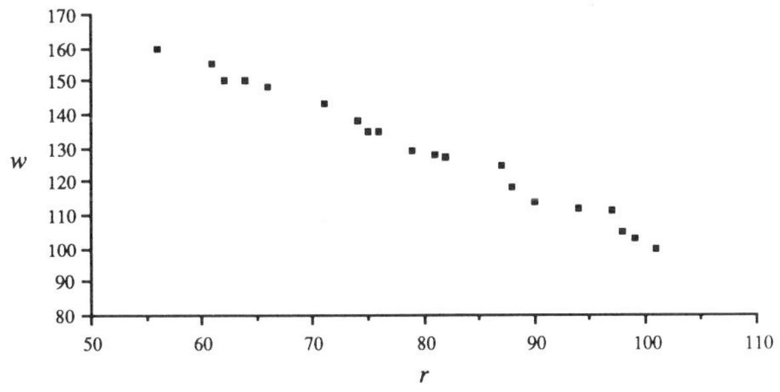

Exhibit 10.2 A scatter of **w** vs. **r**.

Comparing these results to (10.7), we see these estimates are not really too bad. But the investigator does not know this, and he is upset. He has been asked to forecast the effect of a change in **w** on **y**, and the estimated coefficient of **w** is insignificant. Armed, however, with knowledge of the collinear relation between **r** and **w** and with Occam's broadsword, the analyst considers **r** "superfluous" and tosses it out. The result is

$$\mathbf{y} = -18.822 + 0.363\mathbf{w}, \quad (10.9)$$
$$\quad (2.892) \quad (0.022)$$

$$R^2 = .937, \qquad \text{SER} = 1.759, \qquad \text{DW} = 2.198.$$

This indeed appears to be a parsimonious parameterization. Here R^2 still looks good, there is no evidence of serial correlation, and the two coefficients, including that of the all-important **w** term, are highly significant. And so the analyst uses these estimates as the basis of his forecast and draws the conclusion that a unit increase in wages **w** would increase output **y** by 36.3 tons.

Needless to say, an analyst with a greater awareness of the basic economic process behind these data would be led instinctively to smile (perhaps rather broadly) at such a conclusion: the idea that an increase in wages would result in an increase in output would be considered absurd on theoretical grounds. And those same theoretical grounds would insist upon the relevance and retention of both **r** and **w** despite **w**'s insignificance. It is true that the removal of **r** has given the appearance of a more significant set of estimates in (10.9), but in fact it has only created a situation in which an omitted variate has resulted in a badly biased estimate of the coefficient of **w**. And from the vantage of knowing the truth, we can see that, though noisy, the estimates in (10.8) are really a far better set of figures. They make economic sense, and although there is collinearity and **w**'s estimate is insignificant, the analyst would do better not only in using these results for forecasting but also in learning from them more fully the true nature of the economic process they describe.

This example has two important morals. First, care must be exercised in applying Occam's razor: its purpose is to give a shave, not a face lift. Strong prior belief in the relevance of a variate in a model is ample reason for its continued inclusion in a model even if its presence is not well evidenced in the data; that is, even if its coefficient is not statistically significant. Throwing out such variates invariably risks biased estimates possessing no advantages either for forecasting or for understanding the structural elements of the model being estimated.

Second, tossing out collinear variates is quite generally neither a good nor a recommended solution to the collinearity problem. If collinearity can be shown to be adversely affecting the estimate of the coefficient of a variate of interest or rendering some a priori important effect insignificant, the appropriate conclusion is that the data lack the information needed to accomplish the statistical task at hand with precision, not that the model must be molded into a form that looks good relative to the data. Under these circumstances, as previously noted, one must either get better data or introduce appropriate prior information.

10.2 REMEDIAL ACTION IN THE CONSUMPTION FUNCTION

As a first case study in applying remedial action to blunt the ill effects of collinearity, let us return once again to the consumption function data introduced in Exhibit 5.10 of Section 5.4. We recall that these data are very poorly conditioned, having a scaled condition number $\tilde{\kappa}$ of 376. They possess two strong near dependencies, shown in Exhibits 5.12 and 5.13, and result in a least-squares regression (5.3) displaying only one statistically significant coefficient, that of **DPI**(T). Since economists have strong prior beliefs about the consumption function, this situation is ripe for remedial action that introduces prior information. In this study, we employ mixed estimation for this purpose. In addition to attempting to improve the quality of the estimates, we seek also to determine how well the diagnostics reflect the improved conditioning that results from the introduction of the prior information.

In the exercise that follows, three levels of prior information are specified and examined: a very loose (high-variance) set of priors, a middle set, and a very tight (relatively low-variance) set. Initially these priors are specified as ranges or intervals in which the specific parameters, or linear combinations of them, are assumed to lie with probability .95. These ranges are then translated into a form usable for mixed estimation by assuming that the error terms $\boldsymbol{\xi}$ in (10.2) are independent zero-mean Gaussians and that $\mathbf{c}$ and $\mathbf{R}$ are set so that the expected values of the c_i are equal to the midpoints of the specified prior intervals and the (implied) variances of the c_i cause their 95% concentration intervals and these prior intervals to coincide.

Before proceeding, it is worth noting that it is not the purpose of this exercise to present a highly refined and sophisticated set of priors. We are primarily interested in seeing how the increased information embodied in successively stronger priors, which should manifest itself in better-conditioned augmented data matrices, is in fact reflected in the diagnostics and the regression results.

However, an attempt has been made to make these priors economically meaningful and to have different levels that are interesting and reasonable contrasts, and it is felt that there is much of economic value to be learned from this study.

The consumption function given in (5.2) was introduced in Section 5.4 without comment. We reproduce it here:

$$\mathbf{C}(T) = \beta_1 \mathbf{\iota} + \beta_2 \mathbf{C}(T-1) + \beta_3 \mathbf{DPI}(T) + \beta_4 \mathbf{r}(T) + \beta_5 \Delta\mathbf{DPI}(T) + \boldsymbol{\varepsilon}(T). \quad (10.10)$$

This consumption function, which allows for Friedmanesque (1957) dynamics in its inclusion of $\mathbf{C}(T-1)$ and $\Delta\mathbf{DPI}(T)$, is somewhat unorthodox in its inclusion of interest rates $\mathbf{r}(T)$. We will have more to say on these matters as we proceed.

Prior Restrictions

Among the five parameters, we shall impose three prior restrictions. As noted, the error terms attached to these restrictions are assumed to be independent and normal.

1. *The Short-Run Marginal Propensity to Consume, β_3, Has an Expected Value of* 0.7. The short-run marginal propensity to consume, the rate of change of current consumption with respect to current disposable income, is simply β_3 in this model, and it is a concept about which economists know a good deal. Three intensifying levels for the 95% prior intervals and their implied variances are given in Exhibit 10.3.

Exhibit 10.3 95% Ranges and Implied Variances: Prior, $\beta_3 = 0.7 + \xi_1$

Prior	95% Range	Implied Variance
Loose	0.50–0.90	0.0104
Medium	0.55–0.85	0.0058
Tight	0.60–0.80	0.0026

The loose prior here is loose indeed, allowing a short-term marginal propensity to consume between 0.5 and 0.9 with high odds. The tight extreme is not excessive.

2. *The Long-Run Marginal Propensity to Consume, $\beta_3/(1-\beta_2)$, Has Expected Value* 0.9. The long-run marginal propensity to consume is simply the rate of change of consumption with respect to disposable income in the steady state. In the steady state, time-dimensioned variates remain constant, so

ignoring the error term, (10.10) becomes

$$\mathbf{C}^* = \frac{\beta_1}{1-\beta_2} + \frac{\beta_3}{1-\beta_2}\,\mathbf{DPI}^* + \frac{\beta_4}{1-\beta_2}\,\mathbf{r}^*, \tag{10.11}$$

where the asterisks indicate steady-state values. Thus, the steady-state marginal propensity to consume is simply $\beta_3/(1-\beta_2)$.

One must make an approximation here to mold this prior restriction into the linear framework of (10.2). If $\beta_3/(1-\beta_2) = 0.9 + u$, then $\beta_3 + 0.9\beta_2 = 0.9 + \xi_2$, resulting in an error structure, $\xi_2 \equiv (1-\beta_2)u$, dependent on β_2. To capture the flavor of this result in the necessary linear approximation, we treat β_2 in the error structure as a known constant (but in the error structure only) and set it equal to its least-squares estimate from (5.3), that is, 0.2454. This procedure is, of course, in violation of a strict application of mixed estimation, which requires that the prior information be independent of the original data. Thus, we approximate $\mathrm{var}((1-\beta_2)u)$ by $(1-\beta_2)^2\sigma_u^2 \approx 0.6\sigma_u^2$. Exhibit 10.4 presents the 95% prior intervals and the corresponding implied variances for u and ξ_2.

Exhibit 10.4 95% Ranges and Implied Variances: Prior, $\boldsymbol{\beta_3/(1-\beta_2) = 0.9 + u}$; Incorporated as $\boldsymbol{\beta_3 + 0.9\beta_2 = 0.9 + \xi_2}$.

Prior	95% Range	Implied Variance, σ_u^2	$\sigma_{\xi_2}^2 \approx 0.6\sigma_u^2$
Loose	0.80–1.00	0.002603	0.0015618
Medium	0.85–0.95	0.000651	0.0003904
Tight	0.89–0.91	0.000026	0.0000156

The prior variances are specified on u and reflect strong prior information even in the loose case, depicting a strong prior belief in the 0.9 figure. In the tightest case, little variation is allowed.

3. *Twenty-Five Cents of Every "Windfall" Dollar Is Spent; That Is, β_5 Has Expected Value* 0.25. Prior information is not too strong on this parameter. If one believed that all windfall income were saved and only permanent income determines consumption behavior, then $\beta_5 = 0$ would be an appropriate value. To capture this position, the loose case is allowed to encompass this possibility with high probability as well as the possibility that $\beta_5 < 0$. This latter case would occur if instead of consuming part of windfall income, consumers added to windfalls with increased savings in preparation for even larger purchases at a later date. Exhibit 10.5 presents the three prior levels and corresponding implied variances for this parameter. The middle case allows marginally for the $\beta_5 = 0$ hypothesis, and the tight case depicts prior beliefs that people do indeed spend some of their windfall income.

Exhibit 10.5 95% Ranges and Implied Variances: Prior, $\boldsymbol{\beta}_5 = 0.25 + \xi_3$

Prior	95% Range	Implied Variance
Loose	−0.25–0.75	0.0650
Medium	0.00–0.50	0.0162
Tight	0.15–0.35	0.0026

Ignored Information

No prior information is included for either the constant term β_1 or the effect of interest rates β_4. The former merely reflects doubts about the validity of this linear model at the origin, a data point that is quite far removed from the center of the $\mathbf{X}$ data. The data series are, however, measured relative to origins that allow them to be structurally interpretable and are suitable for the collinearity diagnostics.

The lack of prior information about the effect of interest rates, however, is of significant interest. Interest rates are often ignored in simple versions of consumption function studies; this is a characteristic feature, for example, of most elementary Keynesian models, where $\mathbf{r}$ is determined in the investment equation. And indeed, many statistical studies fail to observe the significance of $\mathbf{r}$ in a consumption function. There are, of course, good theoretical reasons for its inclusion. The consumption data used here are total-consumption figures, total goods (durables plus nondurables) and services, and the relevance of $\mathbf{r}$ to decisions on consumer durables has long been recognized. Further, any intertemporal theory of consumption would suggest that, ceteris paribus, the greater is $\mathbf{r}$, the more expensive is current consumption relative to future consumption. Both of these reasons would argue that $\mathbf{r}$ should enter the consumption function with a negative coefficient. Indeed, β_4 is estimated as negative in (5.3), but the coefficient is insignificant. Although no prior has been placed on β_4 here, a more sophisticated analysis would surely provide at least a one-sided diffuse prior on β_4. It has also been decided to ignore any prior specification on the possible off-diagonal elements of $\boldsymbol{\Sigma}_2$ in this study.

Summary of Prior Data

The prior restrictions described above result in the following specification for (10.2) relative to mixed estimation:

$$\mathbf{R} = \begin{bmatrix} 1 & 0 & 1 & 0 & 0 \\ 0 & 0.9 & 1 & 0 & 0 \\ 0 & 0 & 0 & 0 & 1 \end{bmatrix}, \qquad \mathbf{c} = \begin{bmatrix} 0.7 \\ 0.9 \\ 0.25 \end{bmatrix},$$

Σ_1 is estimated as $s^2\mathbf{I}$, where s^2 is the estimated variance from (5.3), that is, 12.6534, and the Σ_2, for the three levels of prior, are:

Σ_2, loose prior, $$\begin{bmatrix} 0.0104 & 0 & 0 \\ 0 & 0.0015618 & 0 \\ 0 & 0 & 0.0650 \end{bmatrix},$$

Σ_2, medium prior, $$\begin{bmatrix} 0.0058 & 0 & 0 \\ 0 & 0.0003904 & 0 \\ 0 & 0 & 0.0162 \end{bmatrix},$$

Σ_2, tight prior, $$\begin{bmatrix} 0.0026 & 0 & 0 \\ 0 & 0.0000156 & 0 \\ 0 & 0 & 0.0026 \end{bmatrix}.$$

Regression Results and Variance–Decomposition Proportions for the Mixed-Estimation Consumption Function Data

Exhibits 10.6–10.8 present, respectively, the regression output, the variance–decomposition proportions, and appropriate auxiliary regressions for the original data and these data augmented by the three degrees of stringency of prior information. These latter variance–decomposition proportions are based

Exhibit 10.6 Least-Squares and Mixed-Estimation Results[a]: Consumption Function Data and Various Mixed-Estimation Augmentations

	ι b_1	C($T-1$) b_2	DPI(T) b_3	r(T) b_4	ΔDPI(T) b_5
Original data	6.72 [1.76]	0.245 [1.03]	0.698 [3.36]	−2.210 [−1.20]	0.161 [0.88]
Data augmented by					
Loose prior	7.21 [2.51]	0.243 [2.35]	0.693 [7.88]	−1.754 [−1.18]	0.178 [1.86]
Medium prior	7.60 [3.10]	0.241 [3.14]	0.688 [10.39]	−1.354 [−1.26]	0.198 [2.69]
Tight prior	7.77 [3.89]	0.240 [4.82]	0.684 [15.38]	−1.079 [−2.32]	0.229 [5.35]

[a]Since interest here centers on tests of significance, t-statistics are given in the square brackets.

on a column-equilibrated transformation of the augmented data matrix $\begin{bmatrix} \mathbf{X} \\ \mathbf{DR} \end{bmatrix}$, where $\mathbf{D} \equiv s\Sigma_2^{-1/2}$.

Conclusions

The following conclusions are indicated:

1. The introduction of prior information greatly sharpens the least-squares

Exhibit 10.7 Scaled Condition Indexes and Variance–Decomposition Proportions: Consumption Function Data and Various Mixed-Estimation Augmentations

Scaled Condition Index, $\tilde{\eta}$	Proportions of ι var(b_1)	**C**($T-1$) var(b_2)	**DPI**(T) var(b_3)	**r**(T) var(b_4)	Δ**DPI**(T) var(b_5)
			Original Consumption Function Data		
1	.001	.000	.000	.000	.001
4	.004	.000	.000	.002	.136
8	.310	.000	.000	.013	.001
39	.264	.005	.005	.984	.048
376	.421	.995	.995	.001	.814
			Mixed-Estimation Consumption Function Data		
			Loose Prior		
1	.002	.000	.000	.000	.006
4	.007	.000	.000	.002	.466
8	.550	.000	.000	.019	.002
31	.270	.015	.018	.967	.092
160	.171	.985	.982	.012	.434
			Medium Prior		
1	.003	.000	.000	.001	.009
4	.008	.000	.000	.003	.650
8	.751	.000	.000	.033	.002
22	.134	.013	.015	.958	.043
120	.104	.987	.985	.005	.296
			Tight Prior		
1	.006	.000	.000	.004	.016
3	.002	.000	.000	.003	.846
5	.285	.002	.002	.065	.040
8	.688	.000	.000	.896	.000
83	.019	.998	.998	.032	.098

Exhibit 10.8 Auxiliary Regressions: Consumption Function Data and Various Mixed-Estimation Augmentations

Dependent Variate	Coefficient of ι	$C(T-1)$	$\Delta DPI(T)$	$\hat{R}^2$	$\tilde{\eta}$
	Original Consumption Function Data				
DPI(T)	−11.547 [−4.9]	1.138 [164.9]	0.804 [11.9]	.9999	376
r(T)	−1.024 [−3.9]	0.017 [22.3]	−0.015 [−1.9]	.9945	39
	Mixed-Estimation Consumption Function Data				
	Loose Prior				
DPI(T)	−11.152 [−2.2]	1.139 [75.5]	0.752 [5.2]	.9996	160
r(T)	−0.899 [−2.9]	0.017 [19.0]	−0.012 [−1.4]	.9917	31
	Medium Prior				
DPI(T)	−10.027 [−1.5]	1.141 [59.3]	0.630 [3.6]	.9994	120
r(T)	−0.553 [−1.3]	0.016 [13.2]	−0.007 [−0.68]	.9839	22
	Tight Prior				
DPI(T)	0.215 [0.03]	1.124 [69.1]	0.322 [1.8]	.9987	83
r(T)	2.864 [4.5]	0.005 [3.2]	0.009 [0.5]	.9112	8

estimates, as seen in Exhibit 10.6. On the basis of the original, poorly conditioned data, only the estimate of β_3 is statistically significant. Even the information from the loose prior, however, brings b_1 and b_2 into significance, and the tight prior provides sufficient information to make all estimates significant, even that for b_4, the coefficient of the interest rate!—about which, more follows.

2. The improved conditioning of the data through the introduction of increasingly strong prior information is clearly and dramatically reflected in the scaled condition indexes of Exhibit 10.7. The loose prior reduces the scaled condition index of the dominant relation by more than one-half. The medium prior adds slightly more conditioning information, mainly to the second near dependency, and the tight prior has reduced the scaled condition index of the

dominant relation to a more nearly manageable 83 and has all but undone the second near dependency involving the interest rate variate, whose scaled condition index is now 8. There is, therefore, a strong parallel between the diagnostic indications of improved conditioning and the sharpened least-squares estimates that result as increasingly stringent prior information is introduced.

3. Examination of the patterns of variance–decomposition proportions in Exhibit 10.7 reveals two important changes. The first is the ever-decreasing proportion of var(b_1) determined in association with $\tilde{\eta}_5$, the largest scaled condition index, and the second is the ever-increasing proportion of var(b_5) determined in association with $\tilde{\eta}_2$, the second from the smallest scaled condition index, and away from $\tilde{\eta}_5$. On the basis of these changes, we would expect both of these variates to play a decreasing role in the dominant near dependency associated with $\tilde{\eta}_5$, and this is verified by examining the auxiliary regressions in Exhibit 10.8.

4. We also see in the auxiliary regressions how the decreasing scaled condition indexes are reflected in decreasing $\hat{R}^2$s, quite in line with the progression we have come to expect.

5. Finally, it is very interesting to note that the most stringent prior information has provided sufficient conditioning information so that even the coefficient of the interest rate, $\mathbf{r}(T)$, about which no prior information has been introduced, has become significant. In essence, the introduction of prior information on the other coefficients that are involved along with $\mathbf{r}(T)$ in the second near dependency appears to have "freed up" what independent information there is in the original data on $\mathbf{r}(T)$ to allow its coefficient to be more precisely estimated.

This last phenomenon is also faithfully reflected in the corresponding diagnostic results—as one would hope it would be. We have already seen in Section 5.4 that $\mathbf{r}(T)$ is involved only in the second of the two near dependencies present in the original data. This is clearly seen in the auxiliary regressions for the original data in Exhibit 10.8, where it is also seen that this second near dependency becomes weaker and weaker as the increasingly stronger prior information is introduced, finally reduced to having a scaled condition index of only 8. Thus, the prior information has broken the collinear relation involving $\mathbf{r}(T)$, making it possible to test effectively its significance in the estimated consumption function.

10.3 COLLINEARITY AND FORECASTING

In this next case study, the collinearity diagnostics are used to help direct the introduction of prior information to increase the accuracy of a forecasting model.[9]

[9]This study is based on Belsley (1984c). (Copyright © 1984, John Wiley & Sons, Ltd., used by permission.)

It is well known that collinearity need not harm forecasts, even if it has harmed structural estimation, *as long as the pattern of collinearity continues into the forecast period.* If, however, it is determined that collinearity exists that is unlikely to continue into the forecast period and has harmed structural estimates over the estimation period (e.g., produced unreasonable coefficients or incorrect signs), then some means to improve structural estimates in line with prior information will likely result in more meaningful forecasts.

Model builders have often attempted to deal with this issue somewhat informally by dropping variates thought to be collinear or by replacing subsets of the collinear variates by specific linear combinations, such as principal components. As we have seen, however, such reductions often violate maintained hypotheses (e.g., that the dropped variates really do belong in the model), thereby leading to specification errors with their consequent estimation biases. These estimation biases become forecasting biases when projecting into situations where the collinearity no longer prevails. In this study, then, we examine a more appropriate procedure for dealing with collinearity under these circumstances.

Rather generally, the collinearity diagnostics along with the test for signal-to-noise given in Chapter 7 can be used to determine the presence and the structure of harmful collinearity over the estimation period. This information helps to assess whether the collinearity is likely to continue into the forecast period and, if not, where prior information can most meaningfully be employed to produce structural estimates from which more useful forecasts can be made.

This process—both diagnosing the collinearity and correcting for it—is illustrated here by an example provided by the world events of the mid-1970s that caused energy prices to rise sharply relative to other prices, altering the nature of the strong collinearity characteristic of price movements before that time. We use the period up to the mid-1970s as the estimation period and show that collinearity, including that among the relevant price series, is indeed involved in producing poor structural estimates of an equation for manufacturing production. We can then see that these estimates, in turn, result in systematic errors for forecasts projected into the late 1970s and early 1980s where the same collinear structure for the data no longer prevails. The model is then reestimated with mixed estimation, using prior information that would indeed have been available to a skilled forecaster in the mid-1970s, and the resulting forecasts are seen to be markedly improved.

The Model

It is shown in Belsley (1969a) that distinct differences characterize the production and inventory-investment behavior of industries that produce to order and those that produce to stock. A model relevant to production to stock takes the form

$$\tilde{\mathbf{P}}(T) = \alpha_0 + \alpha_1 \tilde{\mathbf{P}}(T-1) + \alpha_2 \tilde{\mathbf{H}}(T-1) + \alpha_3 \tilde{\mathbf{S}}^e(T) + \tilde{\varepsilon}(T), \qquad (10.12)$$

where

$\tilde{\mathbf{P}}$ = real output,
$\tilde{\mathbf{H}}$ = real stock of finished-goods inventories on hand at the end of the period,
$\tilde{\mathbf{S}}^e$ = expected real sales over the next period.

For various reasons, some now repudiated by the author, factor prices were not originally introduced into this model; but the sharp rise in relative energy prices in the mid-1970s focuses attention on their role in this context, particularly for forecasting under current conditions. Following a neoclassical tradition, then, relative factor prices for labor and energy are appended to (10.12) to give, as a linear approximation,

$$\tilde{\mathbf{P}}(T) = \alpha_0 + \alpha_1 \tilde{\mathbf{P}}(T-1) + \alpha_2 \tilde{\mathbf{H}}(T-1) + \alpha_3 \tilde{\mathbf{S}}^e(T) + \alpha_4 \frac{\mathbf{W}(T-1)}{\mathbf{\Pi}(T-1)} + \alpha_5 \frac{\mathbf{E}(T-1)}{\mathbf{\Pi}(T-1)} + \tilde{\boldsymbol{\varepsilon}}(T), \tag{10.13}$$

where

$\mathbf{W}$ = wage rate,

$\mathbf{\Pi}$ = price of output,

$\mathbf{E}$ = a price index for energy costs.

Since the shipments and inventories series are published in dollar terms, we avoid unnecessarily introducing a deflation error by estimating the essentially equivalent homogeneous model

$$\mathbf{P}(T) = \alpha_1 \mathbf{P}(T-1) + \alpha_2 \mathbf{H}(T-1) + \alpha_3 \mathbf{S}^e(T) + \alpha_0 \mathbf{\Pi}(T-1) + \alpha_4 \mathbf{W}(T-1) + \alpha_5 \mathbf{E}(T-1) + \boldsymbol{\varepsilon}(T), \tag{10.14}$$

where $\mathbf{P}$, $\mathbf{H}$, and $\mathbf{S}^e$ denote current-dollar measures.

The Data

We estimate (10.14) with data for total, nondurable goods manufacturing. Although this aggregate is not composed wholly of industries that produce solely to stock—three of its nine two-digit SIC component industries, textiles (22), paper (26), and printing (27), have some production to order—, it is considerably more homogeneous in this regard than any other readily available set of data. Following Belsley (1969a), actual sales are used as a proxy for expected sales $\mathbf{S}^e(T)$, and the following series are employed:

S = manufacturers' shipments, nondurable goods industries, Manufacturer's Shipments and Inventories Series, Department of Census;

H = manufacturers' finished-goods inventories, nondurable goods industries, Manufacturer's Shipments and Inventories Series, Department of Census;

$\mathbf{P} \equiv \mathbf{S} + \Delta\mathbf{H}$,

$\mathbf{\Pi}$ = producer price index, total nondurable goods, Producer Price Indexes, Bureau of Labor Statistics;

W = index of average hourly earnings, nondurable goods manufacturing, Employment and Earnings, Bureau of Labor Statistics;

E = industrial price index, fuels and related products and power, Producer Price Indexes, Bureau of Labor Statistics.

All series are monthly, beginning March 1953. These series are available in the Citibank economic database as MNS (**S**), INMN (**H**), PWMND ($\mathbf{\Pi}$), LE6HMN (**W**), and PWFUEL (**E**). The shipments and inventories series are seasonally adjusted. The three price series show little, if any, seasonal variation and are left unadjusted.

An Overview of the Analysis

For the purposes of this illustration, we assume that we are in the twelfth month of 1974. This is one year after the beginning of the Arab oil embargo, and it has become clear that energy prices are rising swiftly relative to other prices and that most observers expect this rise to continue. Forecasts of the effects of these sharply rising energy prices are therefore of great interest. But as we shall see, the data from the estimation period 1953:3 to 1974:12 do not allow good structural estimation of (10.14). The collinearity diagnostics show these least-squares estimates to be severely harmed by several strong near dependencies. Mixed estimation is therefore employed to provide structural estimates that are in line with a priori information, and the ex post forecasts of the least-squares estimates and the mixed-estimation estimates are then compared.

Analyzing the Initial Estimates

Estimating (10.14) over the period 1953:3 to 1974:12 with least squares gives

$$\begin{aligned}\mathbf{P}(T) = {} & \underset{(0.019)}{0.057}\mathbf{P}(T-1) + \underset{(0.012)}{0.003}\mathbf{H}(T-1) + \underset{(0.019)}{0.972}\mathbf{S}^e(T) \\ & - \underset{(1.281)}{1.178}\mathbf{\Pi}(T-1) - \underset{(0.816)}{2.912}\mathbf{W}(T-1) + \underset{(0.919)}{2.637}\mathbf{E}(T-1), \end{aligned} \tag{10.15}$$

$$\hat{R}^2 = .999, \qquad \text{SER} = 84.9, \qquad \tilde{\kappa}(\mathbf{X}) = 266.$$

Looking first at the coefficient estimates, several surprises immediately confront us, the most important of which are the perverse signs of the coefficients of two of the price terms, **Π**$(T-1)$ and **E**$(T-1)$. As noted above, the relevance of these variates is not in question; theoretical considerations preclude their being dropped. Their unreasonable estimates, therefore, could indicate severe collinearity problems or weak data, a fact yet to be examined. In any event, these incorrect signs do not augur well for forecasting into a period where energy prices are rising rapidly relative to other prices. Also, the coefficient of the inventory term **H**$(T-1)$ is insignificant and incorrectly signed. The experience in Belsley (1969a) with individual two-digit industry data would strongly indicate a value of approximately -0.06 for this coefficient, since the estimates for this parameter for the six two-digit production-to-stock components that comprise the total nondurable aggregate being used here all cluster about that value. The coefficients of **P**$(T-1)$ and **S**(T) are of the a priori correct sign and are significant according to standard t-tests. Again, the experience of Belsley (1969a) would suggest that the coefficient of shipments **S**(T) is somewhat high (0.9 would be more reasonable) and that of lagged production **P**$(T-1)$ is too low (0.1 would be better).

Collinearity Diagnostics

The perverse signs of two of the price terms might be due to collinearity, and a glance at the scaled condition number $\tilde{\kappa}(\mathbf{X}) = 266$ indeed confirms the presence of severe collinearity problems. Thus, it is not a question of whether there is collinearity, but of how many near dependencies there are and whether any of the troubled estimates noted above are associated with them. This information is given in Exhibit 10.9, which lists the scaled condition indexes and variance–decomposition proportions for the data over the estimation period.

Here we see the presence of three near dependencies among the six variates. The first is very strong, having a scaled condition index $\tilde{\eta} = 266$, and involves

Exhibit 10.9 Scaled Condition Indexes and Variance–Decomposition Proportions: Nondurable Manufacturing Production Data

Scaled Condition Index, $\tilde{\eta}$	Proportions of **P**$(T-1)$ var(a_1)	**H**$(T-1)$ var(a_2)	**S**(T) var(a_3)	**Π**$(T-1)$ var(a_0)	**W**$(T-1)$ var(a_4)	**E**$(T-1)$ var(a_5)
1	.000	.000	.000	.000	.000	.000
11	.000	.002	.000	.010	.000	.018
25	.003	.021	.003	.012	.019	.064
56	.003	.243	.008	.190	.007	.340
122	.015	.526	.002	.757	.812	.577
266	.979	.208	.987	.031	.162	.001

P($T-1$), **S**(T), and probably **H**($T-1$). The second is quite strong, having a scaled condition index $\tilde{\eta} = 122$, and involves **H**($T-1$) and the three price terms **Π**($T-1$), **W**($T-1$), and **E**($T-1$). It could also contain **P**($T-1$) and **S**(T), their potential involvement being masked by their presence in the first and dominating near dependency. The third is of moderate-to-strong intensity, having a scaled condition index $\tilde{\eta} = 56$, and involves at least the two price terms, **Π**($T-1$) and **E**($T-1$).

The full structure of these relations is displayed in Exhibit 10.10, which shows the auxiliary regressions formed by selecting **P**($T-1$), **H**($T-1$), and **Π**($T-1$) to regress on the remaining variates. It will be noted that these are not the variates that would be picked using the selection procedure described in Section 5.3. In this instance, that method would choose **S**(T), **W**($T-1$), and **E**($T-1$), thereby preventing any one of the auxiliary regressions from displaying the interrelation among all of the three price terms together. Therefore, the selections are made as follows: **P**($T-1$) is picked because it has the greatest combined proportions across the two stronger near dependencies; **Π**($T-1$) is picked because it has the greatest combined proportions across the two weaker near dependencies; and **H**($T-1$) is picked because it is clearly involved in all three relations and its choice leaves the two remaining price terms among the three auxiliary regressors.

Here we confirm that **P**($T-1$) is involved in a very strong near dependency with **S**(T), which is seen also to include **E**($T-1$) more weakly. Both **S**(T) and **E**($T-1$) enter into the second near dependency along with **H**($T-1$) and **W**($T-1$). Finally, all of the price terms are strongly involved in the third near dependency. Collinearity is therefore present, and its structure is readily displayed.

Testing for Harmful Collinearity

From the point of view of forecasting, these results might be acceptable if each of these collinear relations could be counted upon to continue into the forecast

Exhibit 10.10 Auxiliary Regressions: Nondurable Manufacturing Production Data

Dependent Variate	Coefficient of S(T)	Coefficient of W($T-1$)	Coefficient of E($T-1$)	$\hat{R}^2$	$\tilde{\eta}$
P($T-1$)	0.992	−1.039	3.566	.9999	266
	[117.82]	[−1.26]	[3.60]		
H($T-1$)	0.182	39.609	−13.017	.999	122
	[11.85]	[26.31]	[−7.19]		
Π($T-1$)	−0.003	0.319	0.735	.998	56
	[−19.32]	[24.02]	[46.12]		

period beyond 1974:12. At that time, however, few observers would have considered this likely. It is therefore particularly distressing that the coefficient of energy prices **E**($T-1$) in (10.15) is so poorly estimated with incorrect sign. It is likely that collinearity has seriously harmed this structural estimate and that forecasts based on this estimated equation would be unsuitable for assessing the role to be played by energy prices.

Thus, let us turn to a systematic analysis of the harm collinearity has caused the estimation of (10.14) through the use of the test for harmful collinearity developed in Chapter 7. To this end, we will adopt a value of 0.05 for α, the size for the test for signal-to-noise, and a value of 0.999 for γ, the level of adequacy. It is useful to begin with a joint test on the overall adequacy of the signal-to-noise. From the collinearity diagnostics, we are already well aware that there are three strong collinear relations among the six explanatory variates. Hence, if collinearity is to be deemed generally harmful, it should produce inadequate signal-to-noise relative to the joint estimation of some set of three variates known to be strongly involved in one or more of these relations. The three variates **P**($T-1$), **H**($T-1$), and **Π**($T-1$) chosen for the regressands in the auxiliary regressions of Exhibit 10.10 fit this criterion and provide an obvious first choice for this test. The ϕ^2 from (7.12) relevant to this group is 4.97, which is less than the appropriate $F_{0.05}$ critical value of approximately 11.8 taken from Exhibit 7.9*e* with $n-p=256$ and $p_2=3$ degrees of freedom. Hence, we reject the presence of adequate signal-to-noise and accept, rather generally, the presence of harmful collinearity. Only if no such selection of three "involved" variates results in a failed test for adequate signal-to-noise can it be concluded that the three collinear relations, though present, are not jointly harmful. One would then turn to two and, finally, only to one harmful collinear relation. In this example, however, the joint harmfulness of the three collinear relations is shown straightaway.

A second joint test of interest is that for the coefficients of the three price terms **Π**($T-1$), **W**($T-1$), and **E**($T-1$). The ϕ^2 for this test is 18.06, which exceeds the 11.8 critical value ($p_2=3$), and hence, collinearity has not removed all the information contained in the price series but seems certainly to have impaired our ability to obtain good structural estimates for the separate price terms. Let us now turn to these tests on individual coefficients.

Exhibit 10.11 lists the ϕ^2 test values for each of the individual coefficients in (10.15). These are to be compared to the $F_{0.05}$ critical value of 24.9 ($n-p=256$ and $p_2=1$). Since all of the estimates have already been determined to have been degraded by the presence of collinearity, a value of ϕ^2 less than this critical value also accepts the presence of harmful collinearity relative to the estimation of this particular coefficient. Here, five of the parameters, those of **P**($T-1$), **H**($T-1$), **Π**($T-1$), **W**($T-1$), and **E**($T-1$), give evidence of being harmed by collinearity, and it is to be strongly supposed that the introduction of prior information to alleviate the serious estimation problems that result will greatly aid the forecasting ability of this model, recalling that the model is assumed to be otherwise properly specified.

Exhibit 10.11 Tests for Adequate Signal-to-Noise: Individual Coefficients[a]

Coefficient of	ϕ^2	Harmful collinearity? ($\phi^2 \leqslant F_{0.05}$)
$\mathbf{P}(T-1)$	9.00	yes
$\mathbf{H}(T-1)$	0.06	yes
$\mathbf{S}(T)$	2617.13	no
$\mathbf{\Pi}(T-1)$	0.84	yes
$\mathbf{W}(T-1)$	12.74	yes
$\mathbf{E}(T-1)$	8.23	yes

[a] $F_{0.5}$ critical value ≈ 24.9.

We recall that the ϕ^2 values given in Exhibit 10.11 for these tests of individual parameters are simply the squares of the usual t-statistics. We see here several coefficients, like those for $\mathbf{W}(T-1)$ and $\mathbf{P}(T-1)$, for which it is possible for the estimates to be significantly different from zero by conventional test standards (their t's being, respectively, -3.57 and 3.00) but still to give evidence of inadequate signal-to-noise and hence to indicate the presence of data weaknesses—in this case collinearity. In this application, of course, since all variates are maintained to be relevant to the model a priori, the standard tests of significance are irrelevant.

Reestimation with Prior Information

The two conditions highlighted at the outset of this study are clearly met: collinearity has harmed structural estimation over the estimation period, and the collinear relations are not expected to continue into the forecast period. Good forecasts, therefore, require more meaningful structural estimates, and it seems reasonable to introduce prior information to this end. A glance at Exhibit 10.11 reminds us that prior information dealing with any of the five coefficients of $\mathbf{P}(T-1)$, $\mathbf{H}(T-1)$, $\mathbf{\Pi}(T-1)$, $\mathbf{W}(T-1)$, and $\mathbf{E}(T-1)$ will be the most helpful, and Exhibit 10.9 reminds us that the variances associated with all of these are likely being determined in more than one near dependency, so that prior information on any of them will help simultaneously to counteract several sources of weakness in the data. Clear first candidates are priors for the coefficients of $\mathbf{H}(T-1)$, $\mathbf{\Pi}(T-1)$, and $\mathbf{E}(T-1)$, which we know, at the very least, must be negative, positive, and negative, respectively. We again shall use mixed estimation as a means for introducing the prior information. This technique is both convenient and flexible, but any preferred means could be used at this point.

Equation (10.14) is therefore reestimated by mixed estimation using three priors on the coefficients of $\mathbf{H}(T-1)$, $\boldsymbol{\Pi}(T-1)$, and $\mathbf{E}(T-1)$. The priors on the two price terms $\boldsymbol{\Pi}(T-1)$ and $\mathbf{E}(T-1)$ are formed by assuming that these variates have relatively small elasticities (evaluated at the point of means) of 0.25 and -0.25, respectively, with variances that give each prior a substantial weight relative to the data in the final outcome. Details on the formation of these priors are given in the addendum to this section. The third mixed-estimation constraint places the coefficient of the inventory term $\mathbf{H}(T-1)$ narrowly about an expected value of -0.06. As already noted, this value arises from strong prior beliefs derived from Belsley (1969a) about the strength of the role played by finished-goods inventories in determining production to stock. The covariances between and among these three mixed-estimation relations are assumed to be zero. This results in

$$\begin{aligned}\mathbf{P}(T) = {} & \underset{(0.018)}{0.121}\mathbf{P}(T-1) - \underset{(0.004)}{0.047}\mathbf{H}(T-1) + \underset{(0.018)}{0.935}\mathbf{S}^{e}(T) \\ & + \underset{(1.030)}{5.312}\boldsymbol{\Pi}(T-1) - \underset{(0.459)}{2.764}\mathbf{W}(T-1) - \underset{(0.805)}{3.357}\mathbf{E}(T-1), \qquad (10.16) \\ & \hat{R}^2 = .9997.\end{aligned}$$

These results are quite in line with prior expectations. The price terms and inventory term are now of the correct sign and are reasonable in magnitude. In addition, the magnitudes of $\mathbf{P}(T-1)$ and $\mathbf{S}(T)$ have come more into line with the prior beliefs expressed earlier, this despite the fact that no prior information has directly been applied to them. Equation (10.16) is therefore used for generating the mixed-estimation forecasts to be compared with those from least squares given in (10.15).

Comparing Forecasts

It remains only to compare the forecasts that result from the least-squares estimates (10.15) and the mixed-estimation estimates (10.16). For this purpose, it is quite reasonable to use ex post forecasts over some period beyond the estimation period for which the actual data exist for the predetermined variates. For this period, we use 1975:1–1980:11. The least-squares and mixed-estimation forecast errors are plotted in Exhibit 10.12. It is quite clear that the mixed-estimation forecasts are vastly superior to the least-squares forecasts, which systematically diverge from actual output over time. This is to be expected, since the least-squares estimated model cannot deal correctly with the curtailment of output that must occur, ceteris paribus, as a result of the increase in energy costs that occurs over this projection period. The mixed-estimation forecasts, by contrast, do capture this price effect whose presence, as of 1974:12, could be known from a priori considerations but not from what the data alone could divulge.

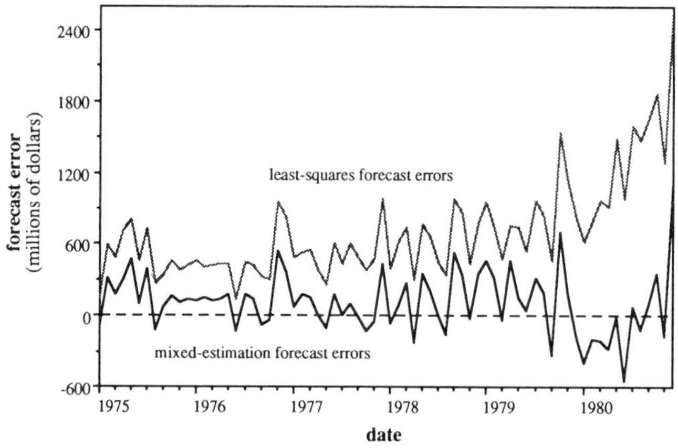

Exhibit 10.12 Forecast errors: least squares vs. collinearity-corrected mixed estimation.

Looking at this from another point of view, we see from Exhibit 10.13 that the collinearity-corrected, mixed-estimation (ME) forecasts reduce the least-squares (LS) RMS forecast error (RMSE) by a factor of 3 over the entire period and by a factor of 3.6 over the last 23 months. The bias reduction, as anticipated, is even greater, particularly in the latter period.

Conclusions

As common sense would have it, the introduction of prior information can substantially improve the quality of estimates and forecasts when collinearity has harmed structural estimation over the estimation period and the collinearity is not expected to continue into the forecast period. It is to be emphasized, however, that this process is not just an exercise in using prior information. The collinearity diagnostics and the test for harmful signal-to-noise enter this process in important and integral ways. First, they establish the presence of the collinear relations and identify their structure. This is essential to the process of assessing, from previous knowledge of the situation, whether the collinear relations are expected to continue into the forecast period. Second, they

Exhibit 10.13 RMSE and Bias: Least-Squares and Mixed-Estimation Forecast Errors

	RMSE			Bias		
Period	LS	ME	LS/ME	LS	ME	LS/ME
1975:1–1980:11	774.7	253.8	3.05	679.9	98.0	6.94
1975:1–1978:12	554.6	225.2	2.46	514.3	127.9	4.02
1979:1–1980:11	1100.2	304.9	3.61	1025.6	35.6	28.81

determine whether the collinear relations can be held responsible for the imprecise or untenable estimates—recalling that it is assumed that the model is correctly specified and that all parameters of interest are maintained a priori to be nonzero. Third, they help to direct attention to those places where prior information will be most useful in counteracting the deleterious effects of collinearity.

Addendum: Forming the Priors

Choosing specific priors and the way this information is to be incorporated into the estimation process is rarely a straightforward and mechanical procedure and is, in fact, one of the places where considerable judgment and art must be applied. Although the mixed-estimation procedure employed here is not always ideal, particularly, for example, in incorporating the one-sided diffuse priors such as might have been stipulated for the price terms, it is nevertheless a relatively simple and flexible means for approximating many interesting situations. Here we discuss in greater detail the formation and incorporation of the priors used in this example.

As a preliminary, let us note that there are two interpretations that can be given to the general process of mixed estimation that can help to set prior values for various of the required parameters: the specific-distribution interpretation inherent in the original Theil–Goldberger formulation and a weighted-regression interpretation.

Recall from (10.1), (10.2), and (10.4) that the mixed-estimation estimator $\mathbf{b}_{\mathrm{ME}} = (\mathbf{X}^{\mathrm{T}}\boldsymbol{\Sigma}_1^{-1}\mathbf{X} + \mathbf{R}^{\mathrm{T}}\boldsymbol{\Sigma}_2^{-1}\mathbf{R})^{-1}(\mathbf{X}^{\mathrm{T}}\boldsymbol{\Sigma}_1^{-1}\mathbf{y} + \mathbf{R}^{\mathrm{T}}\boldsymbol{\Sigma}_2^{-1}\mathbf{c})$ blends the information from the user-specified linear restrictions $\mathbf{c} = \mathbf{R}\boldsymbol{\beta} + \boldsymbol{\xi}$ with the sample data assumed to arise from $\mathbf{y} = \mathbf{X}\boldsymbol{\beta} + \boldsymbol{\varepsilon}$. The weighting factors are the inverses of $\boldsymbol{\Sigma}_1$ and $\boldsymbol{\Sigma}_2$, the respective variance–covariance matrices of the error terms $\boldsymbol{\varepsilon}$ and $\boldsymbol{\xi}$. In this work, we assume that $\boldsymbol{\Sigma}_1 = \sigma^2\mathbf{I}$, and we estimate σ^2 with s^2 from a previous least-squares regression of $\mathbf{y}$ on $\mathbf{X}$. It remains, therefore, to specify $\mathbf{c}$, $\mathbf{R}$, and $\boldsymbol{\Sigma}_2$.

In the specific-distribution interpretation of mixed estimation, it is assumed that the researcher has adequate probabilistic information to stipulate $E\mathbf{c} = \mathbf{R}\boldsymbol{\beta}$ and $\mathbf{V}(\mathbf{c}) = \boldsymbol{\Sigma}_2$. Often, however, this will not be the case—particularly for the information needed to determine the various variances and covariances in $\boldsymbol{\Sigma}_2$. In this instance, it is useful to view (10.4) as a weighted regression in which the variate $\binom{\mathbf{y}}{\mathbf{c}}$ is regressed on the augmented data matrix $\begin{bmatrix}\mathbf{X}\\ \mathbf{R}\end{bmatrix}$ with weights

$$\begin{bmatrix}\sigma^{-1}\mathbf{I} & \mathbf{0}\\ \mathbf{0} & \boldsymbol{\Sigma}_2^{-1/2}\end{bmatrix},$$

or, since scaling does not affect the outcome, with weights

$$\begin{bmatrix}\mathbf{I} & \mathbf{0}\\ \mathbf{0} & \sigma\boldsymbol{\Sigma}_2^{-1/2}\end{bmatrix}.$$

If we assume the priors to be independent pieces of information, $\mathbf{\Sigma}_2$ becomes diagonal, and the smaller a particular diagonal element of $\mathbf{\Sigma}_2$ is made relative to σ, the greater will be the weight given in the final outcome to its corresponding prior constraint in (10.2), becoming, in the limit, a nonstochastic equality constraint—actually, a linear constraint with zero variance.

Suppose now that only one prior constraint is introduced, so that $\mathbf{c}$ is a scalar and $\mathbf{R}$ a row vector. In this case, we see that mixed estimation has the effect of adding a data point (observation) to the original data matrices $\mathbf{y}$ and $\mathbf{X}$. If the weight given to this new data point is similar to that attached to each of the original rows of $\mathbf{y}$ and $\mathbf{X}$, the prior will be but one of $n + 1$ data points and need not have much to say in the final result. Thus, if we wish the prior information to have weight comparable to $k > 0$ times the observed data combined, we wish to give it weight w^2, so that $\sigma^2/w^2 \approx k^2 n$. This rule of thumb is admittedly coarse, and research into more suitable methods is desirable, but it seems to provide a useful order of magnitude for w and is easily put into practice by estimating σ^2 by s^2 from a previous least-squares regression of $\mathbf{y}$ on $\mathbf{X}$.

In this example, each of the two prior constraints on the price terms is given roughly twice the weight of the observed data, giving $w \approx 85/(2\sqrt{262}) = 2.6$, or as an order of magnitude, $w^2 = 10$. The weight attached to the third prior, that on the inventory term, is determined in a fashion similar to that used with the preceding consumption function data; namely, the weight ($w^2 \approx 2 \times 10^{-5}$) is found that gives the coefficient a .99 probability of lying in the interval -0.06 ± 0.01, assuming it to be normally distributed. In this latter case, sufficient prior information is available to employ the specific-distribution interpretation of mixed estimation. It is assumed throughout that the off-diagonal elements of $\mathbf{\Sigma}_2$ are zero.

In picking the values for $\mathbf{c}$, it has been assumed that the expected price elasticities would be small, 0.25 for $\mathbf{\Pi}(T-1)$ and -0.25 for $\mathbf{E}(T-1)$ in (10.14), such prices rarely showing substantial elasticities. When evaluated at the point of means, this implies values for $\mathbf{c}$ of 48 for $E\alpha_0$ from (10.14) and -47 for $E\alpha_5$. The prior expectation for α_3, as already noted, is -0.06. Thus, the mixed-estimation results in (10.16) are calculated from (10.4) with $\mathbf{\Sigma}_1 = s^2\mathbf{I}$, $s = 85$, and

$$\mathbf{c} = \begin{bmatrix} 48 \\ -47 \\ -0.06 \end{bmatrix},$$

$$\mathbf{R} = \begin{bmatrix} 0 & 0 & 0 & 1 & 0 & 0 \\ 0 & 0 & 0 & 0 & 0 & 1 \\ 0 & 1 & 0 & 0 & 0 & 0 \end{bmatrix},$$

$$\mathbf{\Sigma}_2 = \begin{bmatrix} 10 & 0 & 0 \\ 0 & 10 & 0 \\ 0 & 0 & 2 \times 10^{-5} \end{bmatrix}.$$

As a matter of procedure, it is interesting to note that the sequence of operations given here and in the preceding example resulted directly in these parameter values and in the estimates of (10.16). Since these coefficients were considered reasonable, no further action was indicated. If, however, the price terms had not been estimated with reasonable magnitudes and correct signs, the weight attached to the relevant priors would certainly have been increased.

10.4 AN ANALYSIS OF ENERGY CONSUMPTION

This final case study makes use of an extremely interesting data set introduced in an article entitled "Combining Robust and Traditional Least Squares Methods: A Critical Evaluation" (Janson, 1988). As the title indicates, the paper focuses on a number of issues extending far beyond the scope of this current work, but it also provides an excellent context in which to demonstrate the value of the collinearity diagnostics as an integral and coordinated tool for critical model evaluation and reformulation, and that is the focus taken here. It will first be necessary to summarize briefly the relevant parts of Janson's study to set the scene for the subsequent discussion.

Background

The Janson study examines robust versus traditional least-squares estimation techniques in the context of explaining annual energy consumption by buildings sampled by a midwestern energy agency. An initial additive model, which we shall call model 1, is adapted from the literature in the form

$$\mathbf{TotCon} = \beta_1 \iota + \beta_2 \mathbf{WLRFFN} + \beta_3 \mathbf{WLRFINS} + \beta_4 \mathbf{AC} + \beta_5 \mathbf{Fuel} + \beta_6 \mathbf{DD}, \tag{10.17}$$

where

TotCon = total annual energy consumption in 1000 Btu/ft^2;

WLRFFN = (wall area + roof area + fenestrated area)/heated area;

WLRFINS = (wall area × dummy 1 + roof area × dummy 2)/heated area, where dummy 1 is 1 when wall is insulated and 0 otherwise, and dummy 2 is 1 when roof is insulated and 0 otherwise;

AC = percentage of building air conditioned × hours of operation;

Fuel = fuel dummy, which is 1 for steam and 0 for all other fuels;

DD = yearly heating degree days, based on a 65°F base.

This model is estimated with least squares using the "estimation data set" given in Exhibit 10.14. This is a randomly selected subset of the total sample

Exhibit 10.14 Energy Consumption Data: Estimation Data Set

	Energy[a]		Dimension (ft^2)						Building Occupancy			
			Wall		Roof		Glass					
Building Number	Total Consumption	Electrical Consumption	Area	Ins?	Area	Ins?	Area	Ins?[b]	Hours	Work Days	Full Days[c]	Part Days[c]
1019	112.0	45.0	13,897	Yes	12,159	Yes	2,059	D	13	365	7	45
1020	156.1	47.2	6,125	Yes	17,487	Yes	2,331	D	14	338	33	75
1060	135.1	48.2	4,444	Yes	10,932	Yes	410	D	13	313	7	50
1143	102.2	2.9	3,950	Yes	1,700	Yes	550	D	3	208	2	0
1611	116.7	15.1	3,655	No	2,700	No	736	D	8	300	6	20
1700	168.8	29.5	7,965	No	7,796	No	1,038	S	9	300	12	300
1827	68.0	19.6	9,501	Yes	14,658	Yes	1,922	S	10	312	12	0
1876	73.3	37.0	3,468	Yes	9,204	Yes	1,368	D	11	303	9	4
2046	120.9	10.2	4,183	No	3,150	Yes	965	S	12	302	14	0
2123	153.2	38.6	13,225	Yes	6,020	Yes	1,406	D	12	312	10	15
2149	144.0	95.2	6,050	Yes	14,788	Yes	2,502	D	11	312	6	30
2150	167.1	56.3	5,856	Yes	14,478	Yes	1,326	D	11	312	6	20
2152	176.0	22.0	2,880	No	1,981	No	576	D	5	312	2	50
2164	194.2	31.3	2,910	No	3,820	No	790	S	11	313	3	3
2185	232.9	67.4	10,135	Yes	13,212	Yes	1,345	D	11	312	6	20

2189	106.5	20.2	4,380	Yes	5,086	Yes	324	S	7	312	4	8
2190	109.6	47.0	10,500	Yes	15,000	Yes	2,000	D	11	312	6	20
2195	90.7	15.2	7,788	No	4,963	Yes	824	D	5	286	3	0
2231	125.6	9.4	1,227	No	1,394	No	488	D	4	260	1	1
2288	74.3	6.3	2,536	Yes	3,300	Yes	324	S	8	312	1	2
2299	125.9	125.9	1,667	Yes	2,576	Yes	383	D	4	261	2	2
2336	149.6	20.8	1,072	Yes	1,650	Yes	320	D	5	260	1	1
2362	186.6	67.4	6,818	No	12,326	Yes	1,510	D	11	312	6	20
2363	123.4	123.4	6,990	Yes	15,340	Yes	905	D	11	312	6	10
2516	179.0	3.2	1,686	No	1,941	No	381	S	5	200	1	15
2523	55.4	9.1	2,379	No	4,131	Yes	1,053	D	9	312	4	4
2832	88.0	26.7	3,102	No	1,887	Yes	418	S	6	301	1	5
2897	155.1	3.6	8,230	No	5,302	No	1,240	D	8	251	20	30
3091	169.8	6.9	15,520	No	15,000	No	6,040	S	6	308	1	1
3107	98.3	23.6	10,650	No	3,050	No	1,449	S	13	304	0	0
3240	85.0	24.2	4,895	Yes	11,459	Yes	1,632	D	9	253	6	51
3268	115.0	19.3	2,292	No	4,184	No	328	D	9	253	4	13
3333	139.5	25.0	3,445	No	7,725	Yes	1,722	D	11	280	10	77
3348	109.8	4.6	1,696	No	1,173	No	80	S	6	260	0	1
3365	159.9	9.1	750	No	910	No	58	D	5	300	1	12
3428	167.0	20.9	3,950	No	4,690	Yes	420	D	6	300	2	15
9099	140.3	4.8	3,061	Yes	1,620	Yes	503	D	6	261	2	1

Exhibit 10.14 (Continued)

	Heating					Original Air Conditioning			Corrected Air Conditioning		
Building Number	Heated Area (ft^2)	Steam Heat?	Degree Days	Temperature Occupied (°F)	Temperature Unoccupied (°F)	Air Condi-tioning	Air-Conditioned Area (ft^2)	Air-Conditioned Hours	Air Condi-tioning	Air-Conditioned Area (ft^2)	Air-Conditioned Hours
1019	21,726	No	8511	68	63	Yes	21,726	13	Yes	21,726	13
1020	18,578	No	8511	68	63	Yes	17,102	15	Yes	17,102	15
1060	10,932	No	8555	68	68	Yes	10,932	24	Yes	10,932	24
1143	3,400	No	9988	68	64	No	0	0	No	0	0
1611	5,760	No	9690	69	65	No	0	0	No	0	0
1700	11,984	No	8511	68	60	Yes	8,754	10	Yes	8,754	10
1827	28,288	Yes	8430	65	65	Yes	26,894	5	Yes	26,894	5
1876	17,000	Yes	9028	70	70	Yes	17,000	10	Yes	17,000	10
2046	6,300	No	9461	68	65	Yes	3,150	24	Yes	3,150	24
2123	17,150	No	8511	68	68	Yes	17,150	24	Yes	17,150	24
2149	13,923	No	8511	70	70	Yes	13,923	24	Yes	13,923	24
2150	14,478	No	8511	68	60	Yes	13,858	11	Yes	13,858	11
2152	2,400	No	8010	68	68	Yes	1,600	6	Yes	1,600	6
2164	4,142	No	8549	68	60	No	0	0	Yes	4,142	12
2185	13,455	No	8511	68	65	Yes	12,783	24	Yes	12,783	24
2189	5,086	No	9333	72	72	Yes	5,086	24	Yes	5,086	24
2190	16,552	No	8511	68	60	Yes	14,288	24	Yes	14,288	24

2195	6,400	No	9461	70	62	No	0	0	No	0	0
2231	2,818	No	8886	68	65	No	1,394	4	No	1,394	4
2288	6,050	No	9640	68	68	No	0	0	No	0	0
2299	2,576	No	8606	68	68	Yes	2,576	4	Yes	2,576	4
2336	1,650	No	8368	70	55	No	0	0	Yes	1,650	4
2362	11,713	No	8511	70	70	No	0	0	Yes	11,713	24
2363	15,066	No	8511	70	70	Yes	15,066	24	Yes	15,066	24
2516	1,941	No	8400	72	62	No	750	4	No	750	4
2523	7,644	No	8511	68	65	No	0	0	No	0	0
2832	3,774	No	8926	70	na	Yes	1,887	6	Yes	1,887	6
2897	10,902	No	9973	70	65	No	0	0	No	0	0
3091	15,000	No	8500	65	65	Yes	15,000	0	Yes	15,000	0
3107	11,236	Yes	8493	68	68	Yes	11,236	14	Yes	11,236	14
3240	11,760	No	8511	68	68	Yes	11,760	10	Yes	11,760	10
3268	8,367	No	8511	68	68	Yes	4,184	10	Yes	4,184	10
3333	14,451	No	8511	68	68	Yes	14,451	12	Yes	14,451	12
3348	1,817	No	9200	67	60	Yes	1,817	4	Yes	1,817	4
3365	910	No	8227	60	60	Yes	910	0	Yes	910	0
3428	4,800	No	8932	65	55	No	0	0	Yes	4,800	6
9099	3,240	No	8932	65	60	Yes	1,620	6	Yes	1,620	6

Source: Janson (1988).
[a]Annual consumption in 1000 Btu/ft^2.
[b]S = single, D = double.
[c]In 1000 people per square foot.

Exhibit 10.15 Energy Consumption Data: Validation (Hold-out) Data Set

	Energy[a]		Dimension (ft^2)						Building Occupancy			
			Wall		Roof		Glass					
Building Number	Total Consumption	Electrical Consumption	Area	Ins?	Area	Ins?	Area	Ins?[b]	Hours	Work Days	Full Days[c]	Part Days[c]
1002	136.9	21.4	8,623	No	8,235	No	1,539	S	12	287	11	400
1018	190.5	39.0	2,384	No	4,807	Yes	1,044	S	12	338	2	32
1030	99.6	24.9	5,406	Yes	8,124	Yes	473	D	11	300	4	4
1120	89.2	24.5	4,111	Yes	2,845	Yes	425	S	13	365	4	6
1212	116.6	9.8	2,424	Yes	3,025	Yes	216	D	0	260	1	0
1375	105.8	2.3	6,180	No	4,750	No	980	D	8	261	2	0
1411	83.0	34.0	5,725	No	12,350	Yes	385	D	10	300	14	185
1573	145.7	6.2	7,408	No	5,370	Yes	1,208	S	12	299	17	70
1634	132.9	9.2	920	No	1,740	No	50	S	4	260	0	1
1723	175.7	21.9	2,870	No	1,956	No	574	D	6	312	2	50
1803	141.4	40.8	4,361	No	5,466	Yes	229	D	11	302	6	9
1804	187.7	30.7	19,640	No	24,550	No	4,910	D	8	355	41	9
1816	147.3	17.1	3,873	No	4,740	Yes	1,257	D	6	260	2	25
1960	138.4	47.1	4,743	Yes	10,492	Yes	631	D	12	312	5	20

1979	139.2	14.6	4,177	No	1,970	No	268	D	7	251	3	30
2079	92.4	6.6	3,388	No	2,184	No	532	D	7	312	2	35
2148	158.3	56.3	9,009	Yes	16,486	Yes	1,395	S	11	312	5	20
2153	163.1	69.1	7,267	Yes	14,697	Yes	3,145	D	11	312	6	30
2154	167.8	60.2	3,859	Yes	8,364	Yes	1,690	D	11	312	6	20
2186	122.9	37.4	6,224	Yes	12,544	Yes	100	S	11	312	6	20
2187	174.6	61.6	8,517	Yes	14,471	Yes	1,003	D	11	312	6	20
2199	64.0	37.2	3,760	Yes	4,800	Yes	160	D	6	365	2	20
2207	50.4	10.0	7,174	No	14,232	No	2,752	D	12	288	10	150
2285	123.4	7.3	3,005	Yes	2,015	Yes	451	D	8	312	1	11
2341	177.0	28.0	4,558	No	1,756	No	611	S	8	288	5	0
2353	83.8	24.2	10,860	Yes	13,824	Yes	1,752	D	11	303	26	4
2518	167.6	59.4	12,016	Yes	11,664	Yes	965	D	10	312	9	4
2520	83.7	13.2	2,121	No	4,480	Yes	843	S	9	312	6	5
2521	100.1	9.7	5,301	No	6,440	Yes	433	S	9	312	4	3
2616	154.8	28.4	3,557	No	3,190	Yes	221	D	8	261	2	5
2740	50.4	10.0	7,174	No	14,232	No	2,752	D	12	312	10	150
3101	83.5	24.0	10,860	Yes	13,824	Yes	1,752	D	12	303	26	4
3306	92.6	25.3	2,719	Yes	5,671	Yes	465	D	11	289	1	45
3424	166.0	26.1	1,339	Yes	1,440	Yes	69	D	6	261	1	30

Exhibit 10.15 (Continued)

	Heating					Original Air Conditioning			Corrected Air Conditioning		
Building Number	Heated Area (ft^2)	Steam Heat?	Degree Days	Temperature Occupied (°F)	Temperature Unoccupied (°F)	Air Condi-tioning	Air-Conditioned Area (ft^2)	Air-Conditioned Hours	Air Condi-tioning	Air-Conditioned Area (ft^2)	Air-Conditioned Hours
1002	13,658	No	8,535	70	70	Yes	13,658	24	No	0	0
1018	4,382	No	8,511	68	63	Yes	4,383	11	Yes	4,383	11
1030	13,766	No	8,511	68	68	Yes	13,766	11	Yes	13,766	11
1120	4,256	No	8,527	68	64	Yes	2,837	10	Yes	2,837	10
1212	2,722	No	8,741	70	70	Yes	2,722	5	Yes	2,722	5
1375	9,600	No	10,388	72	68	No	0	0	No	0	0
1411	20,300	No	8,947	68	60	Yes	18,600	6	Yes	18,600	6
1573	16,036	No	8,527	68	65	No	0	0	No	0	0
1634	1,800	No	8,276	72	66	No	0	0	No	0	0
1723	2,400	No	8,010	68	68	Yes	1,600	6	Yes	1,600	6
1803	6,609	No	8,511	72	72	Yes	6,609	11	Yes	6,609	11
1804	24,550	No	8,511	70	70	Yes	24,550	8	Yes	24,550	8
1816	4,740	No	8,535	70	65	Yes	4,740	24	Yes	4,740	24
1960	10,907	No	10,406	68	68	Yes	10,907	24	Yes	10,907	24
1979	3,988	No	8,548	68	60	Yes	1,994	7	Yes	1,994	7

2079	4,284	Yes	10,260	68	65	Yes	4,284	8	Yes	4,284	8
2148	13,904	No	8,511	70	62	Yes	13,904	11	Yes	13,904	11
2153	14,647	No	8,511	70	70	Yes	14,647	24	Yes	14,647	24
2154	8,386	No	8,511	70	70	Yes	8,074	11	Yes	8,074	11
2186	11,952	No	8,511	72	65	Yes	11,952	11	Yes	11,952	11
2187	14,471	No	8,511	68	68	Yes	13,859	24	Yes	13,859	24
2199	9,600	No	8,609	68	68	Yes	9,600	24	Yes	9,600	24
2207	29,213	Yes	10,388	65	65	No	0	0	No	0	0
2285	4,030	No	8,555	68	62	No	0	0	No	0	0
2341	3,930	No	9,447	65	65	Yes	3,930	8	Yes	3,930	8
2353	26,112	No	9,700	68	na	Yes	26,112	11	Yes	26,112	11
2518	15,990	No	8,511	68	65	Yes	15,990	10	Yes	15,990	10
2520	6,630	No	8,511	68	65	No	0	0	No	0	0
2521	6,440	No	8,511	68	65	No	0	0	No	0	0
2616	3,000	No	8,511	68	62	Yes	3,000	8	Yes	3,000	8
2740	29,213	Yes	10,388	65	65	No	0	0	No	0	0
3101	26,112	No	9,700	68	68	Yes	26,112	12	Yes	26,112	12
3306	5,505	No	10,450	65	50	Yes	5,505	9	Yes	5,505	9
3424	1,436	No	8,700	65	55	Yes	1,436	6	Yes	1,436	6

Source: Janson (1988).
[a]Annual consumption in 1000 Btu/ft^2.
[b]S = single, D = double.
[c]In 1000 people per square foot.

comprising roughly one-half the total observations, the other half being used for a hold-out, or validation, data set, which is given in Exhibit 10.15. The initial least-squares results are

$$\begin{aligned}\textbf{TotCon} = {} & \underset{(110.19)}{273.75\iota} + \underset{(14.72)}{40.45}\textbf{WLRFFN} - \underset{(9.72)}{13.56}\textbf{WLRFINS} \\ & + \underset{(0.7600)}{0.7736}\textbf{AC} - \underset{(20.95)}{42.94}\textbf{Fuel} - \underset{(0.0115)}{0.0222}\textbf{DD}, \\ & R^2 = .44.\end{aligned} \tag{10.18}$$

The **AC** variate shows insignificance here, and the sign of the coefficient of **DD** violates prior expectations.

In the next step of the analysis, deletion techniques are applied to these results, and various buildings show up as possessing large influential-data diagnostics. Some aspects of the influential-data diagnostics have been examined in Chapter 8, but it is beyond our current purpose to explain them in detail. The interested reader is directed to Belsley et al. (1980) and Cook and Weisberg (1982). However, with their use, buildings 2185, 2523, and 2897 are singled out as possessing some combination of large studentized residuals (a normalized least-squares residual), Cook's distance (a measure of parameter estimate sensitivity), or leverage (a measure of prediction sensitivity).

As a result, these observations, as well as various variates, are removed from the analysis in several steps. First, the most troublesome building, 2185, is removed because it has the largest studentized residual, and following it, the insignificant variate **AC** is removed because of its insignificance. Then the other two suspect data points, 2523 and 2897, are deleted because they are suspect, and once these are gone, **WLRFINS** becomes insignificant, and so it too is removed. The final version of this additive model, which we shall call model 1a, is of the form

$$\textbf{TotCon} = \beta_1\iota + \beta_2\textbf{WLRFFN} + \beta_3\textbf{Fuel} + \beta_4\textbf{DD}. \tag{10.19}$$

The least-squares estimation of this model, which is unnecessary to display, is then used to predict the energy consumption of the buildings in the hold-out data set of Exhibit 10.15, and a summary figure of mean-square prediction error (MSPE) is used as a basis for comparing model performance. The MSPE for this additive model 1a is 1144.

In the next stage of the study, concerns are raised, quite legitimately, about the adequacy of the specification of the original additive model 1 in (10.17). The insignificance of **AC** in (10.18) is disturbing, as is the perverse sign for **DD**. Using an engineering model, it is recognized that **DD** should enter the model multiplicatively rather than additively. And it is also discovered that the air-conditioning data seem in error, and so those data are corrected, but for some reason, the corrected data are not used. The result is a second model, the multiplicative model, which we shall call model 2,

$$\textbf{TotCon} = \beta_1 \iota + \beta_2 \textbf{WLRFFNDD} + \beta_3 \textbf{WLRFINSDD} + \beta_4 \textbf{AC} + \beta_5 \textbf{Fuel}, \tag{10.20}$$

where we add to the variates defined previously

$$\textbf{WLRFFNDD} = (\text{wall area} + \text{roof area} + \text{fenestrated area}) \times \text{degree days}/(\text{heated area} \times 10{,}000),$$

$$\textbf{WLRFINSDD} = (\text{wall area} \times \text{dummy } 1 + \text{roof area} \times \text{dummy } 2) \times \text{degree days}/(\text{heated area} \times 10{,}000).$$

Subjecting this model to the same kind of influential-data analysis as the additive model 1, the preferred least-squares estimates are those that result using the estimation data set with observations 2185, 2189, 2195, and 2523 removed as being unduly influential. The estimates are

$$\begin{aligned} \textbf{TotCon} = \underset{(22.02)}{72.80}\iota + \underset{(14.80)}{50.56}\textbf{WLRFFNDD} - \underset{(8.41)}{26.07}\textbf{WLRFINSDD} \\ + \underset{(0.59)}{1.63}\textbf{AC} - \underset{(16.36)}{41.44}\textbf{Fuel}, \end{aligned} \tag{10.21}$$

$$R^2 = .55.$$

These results are claimed to be preferable to those from model 1a because the model "includes air-conditioning, wall and roof insulation, degree-day effects on energy consumption, and type of fuel used for space heating. The parameter estimates carry the correct signs, positive for the air-conditioning and degree-day effects and negative for wall and roof insulation and type of fuel." [Janson (1988, p. 425)]. Further, when this model is used to predict over the hold-out set, the MSPE is reduced quite substantially to 799.

Now, of course, the original Janson study has many other and different points to make regarding the value of robust estimation relative to least-squares, and the preceding summary cannot possibly do justice to all the points made there. But our need here is simply to make clear those of its aspects that are relevant to a reanalysis that we shall now undertake incorporating the collinearity diagnostics. This analysis is adapted from Belsley and Welsch (1988).

Model 1 Revisited

It is quite true that model 1 leaves much to be desired, but we shall see that it is not quite so bad as it first appears. When treated properly, model 1 works quite well—not as well as some others that we shall derive—but actually better than the preferred model 2.

A Conditioning Analysis

The Janson study does not conduct a conditioning analysis of the data for the various models, but we can see from the scaled condition indexes and variance–decomposition proportions given in Exhibit 10.16 that the data for model 1 are not well conditioned. There is one moderate-to-strong near dependency with a scaled condition index of 59. And from the variance–decomposition proportions, we see the strong involvement in it of degree days **DD** and the intercept term ι as well as the weaker involvement of both **WLRFFN** and **AC**. This relation can be displayed in greater detail through the auxiliary regression

$$\begin{aligned}\mathbf{DD} = {} & \underset{[25.3]}{9273.0\iota} - \underset{[-1.3]}{279.3\mathbf{WLRFFN}} + \underset{[0.9]}{138.6\mathbf{WLRFINS}} \\ & - \underset{[-1.8]}{20.3\mathbf{AC}} - \underset{[-0.7]}{213.7\mathbf{Fuel}}, \\ & \hat{R}^2 = .997.\end{aligned} \tag{10.22}$$

Here we see the very tight relation between the intercept term ι and **DD** and the lesser involvement of **AC**. It is not surprising, then, that these are the two variates that cause problems for estimation in model 1.

Several points about collinearity analyses are made very clear by this interesting example and ably illustrate a number of the concerns raised in Chapter 6. First, the involvement of **DD** in a strong collinear relation with the intercept would be completely overlooked by those who argue that collinearity should be assessed using mean-centered data (or correlation matrices), data with their so-called "nonessential" collinearity removed. The scaled condition number of the mean-centered data here is 2. A similar blindness affects variance inflation factors. The largest VIF for the centered data is 1.4, whereas that appropriately based on the uncentered $\hat{R}^2$ given in (10.22) is 378.4.

Exhibit 10.16 Scaled Condition Indexes and Variance–Decomposition Proportions: Estimation Data Set, Model 1

	Proportions of					
Scaled Condition Index, $\tilde{\eta}$	ι var(b_1)	**WLRFFN** var(b_2)	**WLRFINS** var(b_3)	**AC** var(b_4)	**Fuel** var(b_5)	**DD** var(b_6)
1	.000	.002	.011	.012	.004	.000
2	.000	.001	.008	.000	.787	.000
3	.000	.012	.069	.358	.001	.000
5	.000	.002	.883	.495	.049	.000
11	.008	.872	.004	.023	.131	.015
59	.992	.111	.025	.112	.028	.985

Second, we see that it is equally incorrect to argue that uncentered data are relevant to a collinearity analysis only when the constant term is of interest. The constant term is not of inherent interest to this study, but the variates collinear with it are! And their estimates, as well as that of the constant term, are being adversely affected by the collinear relation.

Third, those who argue that collinearity with the constant term should be obvious from the data without diagnostics had better sit on their hands. Apparently it was not.

What do we learn from the collinearity diagnostics in this case? Immediately, we see that the solution adopted for the **DD** problem, namely, to make it multiplicative, in effect has cut **DD** from the model. In multiplying **WLRFFN** by the relatively constant **DD** in constructing model 2, **WLRFFN** has effectively been rescaled while **DD** has been otherwise dropped from the model. We also learn that collinearity could be the reason **DD** is not behaving according to expectation. We have already noted that many textbooks advocate dropping such collinear variates, but we have also seen that when the variates have a priori relevance to the analysis, as does **DD** in this situation, such advice is usually misplaced. Rather, we must realize that **DD** is simply weak data. Its presence is neither helpful nor harmful, and its incorrect sign cannot be used as an indication of model misspecification. We have two choices. If, on the one hand, we feel the nature of the data weakness for **DD** to be true for the full data set (including the validation set), we can use **DD** just as it is. If, on the other hand, we believe **DD** has problems only for the estimation set, then we can compensate for its weakness through prior information that describes how we feel **DD** properly should behave. Since the estimation and validation sets are picked randomly, we have no a priori reason to believe that data relations characterizing the estimation set should not also occur in the validation set. In other words, **DD** could (and should) very well be allowed to stand unaltered.

Deletion Diagnostics

It is also notable that a decision was made to delete three buildings from the estimation data set in determining the final estimates of model 1a. As mentioned, it is beyond the current study to examine these influential-data diagnostic tools in detail. But we recall from Chapter 8 that these diagnostic measures allow one to determine when certain observations (rows of the **X** matrix) have a particularly large (influential) effect on some estimated result, such as the least-squares residuals or the least-squares coefficients. And the decision to delete these observations appears to have been made simply because these buildings have certain combinations of "high" influential-data diagnostic values. It should be obvious, however, that an observation is not "bad" merely because it has high influential-data diagnostics. Such diagnostic values are merely suggestive that an observation is suitable for further investigation in order to determine its appropriateness to the analysis. Clearly, an observation that is both influential and demonstrably inappropriate is one that can have important adverse effects on the results, and appropriate action should be taken. But so far as we know—

or are led to believe—, there is nothing inappropriate with the data for these three buildings beyond their large diagnostic values. There is, therefore, no reason to throw them out, and in so doing, one could be ridding onself of important pieces of information. These three observations should stay.

A decision to keep these three observations has other important ramifications. Recall that it is their deletion that makes the estimate of the coefficient of **WLRFINS** insignificant and leads to the removal of this variate. This, then, is an interesting situation, for it appears that the estimate of the insulation parameter is being determined only by a few observations. This could be problematic if we had meaningful doubts about either the insulation variate's relevance or the observations determining it. But in this case, we have neither. This variate should stay.

Correct Air-Conditioning Data

Janson's analysis of model 1 led him to question the validity of the air-conditioning data, and indeed several observations were found to be in error. Inexplicably, however, the corrected air-conditioning data were not used in the estimation of either model 1 or model 2. Instead, the insignificance of air conditioning in model 1 is noted and the variate is consequentially dropped. We have seen that this practice is inadvisable. The relevance of air conditioning for such a model has unquestionable a priori support. It should need far more than mere insignificance to warrant removal of such a variate. Combining this variate's erroneous observations with its involvement, albeit weak, in a collinear relation, we should certainly hang on to it and give it a proper chance.

Model 1 Reestimated

So let us take the three preceding considerations into account and reestimate model 1. Specifically, we examine the additive model 1—but we include **DD**, use the correct air-conditioning data, and delete no observations from the estimation data set. We will call this model 1*. The least-squares estimates are

$$\begin{aligned}\mathbf{TotCon} = \underset{(105.6)}{217.384\iota} &+ \underset{(13.61)}{41.146\mathbf{WLRFFN}} - \underset{(9.14)}{19.689\mathbf{WLRFINS}} \\ &+ \underset{(0.72)}{1.661\mathbf{CorAC}} - \underset{(19.68)}{42.822\mathbf{Fuel}} - \underset{(0.011)}{0.016\mathbf{DD}}, \qquad (10.23)\\ &R^2 = .51, \qquad \text{SER} = 29.95,\end{aligned}$$

where

$$\mathbf{CorAC} = \text{air-conditioning variate based on corrected data.}$$

Exhibit 10.17 shows the mean-squared prediction errors (over the validation data set) of the models considered here. The first two rows repeat Janson's best model 1a and model 2 results, namely 1144 and 799, respectively. The next row shows the MSPE that results from using the estimates of model 1* from (10.23),

Exhibit 10.17 Mean-Squared Prediction Errors: All Models

Model	Description	MSPE
1a	Janson's model 1a (10.19)	1144
2	Janson's model 2 (10.20)	799
1*	Equation (10.23)	652
3	Equation (10.30)	585
4	Equation (10.31)	582

namely 652. Clearly, model 1* is doing a better job by this criterion than even Janson's preferred model 2.[10] But we can do better yet.

Model 3

Janson uses engineering information to introduce **DD** multiplicatively into model 2. This is an excellent step, but it can be taken further. These considerations in fact argue for a completely multiplicative model. Heat is stored in the mass of an object and radiated at the surface. Thus, in simplest terms, we might have

$$\mathbf{E} \propto \mathbf{A}\mathbf{V}^{-1}, \tag{10.24}$$

where **E** is energy dissipation, **A** is the surface area, and **V** is the volume. Of course, we are dealing with a complex shape here, so we might enrich the model to

$$\mathbf{E} = \mathbf{C}\mathbf{A}^{\alpha}\mathbf{V}^{\beta}, \tag{10.25}$$

where **C** is a scale and correction factor to be explained further.

In this case, **V** is measured by **HeatedArea**, presuming a standard ceiling height. The area **A**, however, is a complex of heterogeneous substances: walls, roofs, and glass—with the possible presence of insulation that can alter the heat transference properties of all of these. Since areas are necessarily additive, one reasonable way of modeling this is to construct an effective area

$$\begin{aligned}\mathbf{EffArea} \equiv \mathbf{WallArea} &+ \gamma_1\mathbf{RoofArea} + \gamma_2\mathbf{GlassArea} \\ &+ \delta_1\mathbf{WallIns} + \delta_2\mathbf{RoofIns} + \delta_3\mathbf{GlassIns},\end{aligned} \tag{10.26}$$

where the "**Ins**" variates are insulation dummies having the value zero if there is no insulation and the corresponding "**Area**" otherwise. The coefficient of

[10] For completeness, it should be noted that the inclusion of the four deleted observations in the estimation of model 2 does not improve its predictive performance. The MSPE of model 2 based on all observations is 841 as compared to 799 when the four are deleted.

WallArea is normalized at 1. Of course, introducing (10.26) into (10.25) presents a nonlinear estimation problem, but that should not be allowed to be a hindrance.

The **C** term catches all of the heat transmission rate correction factors, appropriately multiplicative; thus

$$\mathbf{C} = \nu(\mathbf{DD})^{\lambda}\mathbf{SK}, \tag{10.27}$$

where ν is a scale factor, **DD** is degree days, **S** is a dummy for steam heat, and **K** is a dummy for air conditioning. We model **S** simply as exp(ω**Fuel**) and **K** = exp(υ**CorAC**). The elements of these variates have value 1 in the absence of the steam heat or air conditioning, respectively, and are otherwise appropriately scaled positive values with limits determined by ω and υ, which are to be estimated. Once (10.26) and (10.27) are introduced into (10.25), we get the following nonlinear model, which we shall denote model 3, transformed logarithmically as

$$\begin{aligned}\log \mathbf{TctCon} &= \log \nu\, \iota + \alpha \log \mathbf{EffArea} + \beta \log \mathbf{HeatedArea}\\ &\quad + \omega\mathbf{Fuel} + \upsilon\mathbf{CorAC} + \lambda \log \mathbf{DD}.\end{aligned} \tag{10.28}$$

There is not much hope that this model can provide good estimates for the separate parameters in **EffArea** given in (10.26). A collinearity analysis of these area data with their dummies, displayed in Exhibit 10.18, shows a modest condition index of 21 (but corresponding to an $\hat{R}^2$ of .97 in the auxiliary regression) with complex involvement of the roof- and glass-related variates. We are therefore asking a lot from these data in this nonlinear context. Indeed, a full nonlinear estimation of (10.28) turns out to be insignificantly different [$F(4, 26) = 1.11$] from a simpler, constrained model, namely, one using

$$\mathbf{EffArea} \equiv \mathbf{Area} + \delta\mathbf{Insulation}, \tag{10.29}$$

where **Area** ≡ (**WallArea** + **RoofArea** + **GlassArea**) and **Insulation** ≡ (**WallIns** + **RoofIns** + **GlassIns**). Using this definition of **EffArea** and the full-estimation data set, the mildly nonlinear model 3 is estimated as

$$\begin{aligned}\log \mathbf{TotCon} = {} & \underset{(7.29)}{15.842\iota} + \underset{(0.157)}{0.427} \log (\mathbf{Area} - \underset{(0.165)}{0.427}\mathbf{Insulation})\\ & + \underset{(0.006)}{0.011}\mathbf{CorAC} - \underset{(0.144)}{0.406} \log \mathbf{HeatedArea}\\ & - \underset{(0.163)}{0.359}\mathbf{Fuel} - \underset{(0.798)}{1.244} \log \mathbf{DD},\\ & R^2 = .534, \qquad \text{SER} = 0.242.\end{aligned} \tag{10.30}$$

Exhibit 10.18 Scaled Condition Indexes and Variance–Decomposition Proportions: Effective Area Data

	Proportions of						
Scaled Condition Index, $\tilde{\eta}$	ι var(b_1)	**Wall Area** var(b_2)	**Roof Area** var(b_3)	**Glass Area** var(b_4)	**Wall Ins** var(b_5)	**Roof Ins** var(b_6)	**Glass Ins** var(b_7)
1	.006	.002	.001	.001	.004	.001	.005
3	.054	.014	.000	.041	.075	.010	.027
4	.087	.042	.000	.005	.201	.004	.191
5	.492	.003	.004	.096	.038	.002	.020
6	.017	.017	.026	.004	.029	.050	.586
11	.320	.819	.005	.451	.511	.012	.033
21	.024	.103	.964	.402	.142	.921	.138

These estimates all accord with expectations, except, of course, that of the coefficient of **DD**, which is insignificant. The scaled condition number of the data for this model (with the logged data appropriately e scaled as indicated in Chapter 9) is 65, and the variance–decomposition proportions are effectively the same as those for the basic data given in Exhibit 10.16. Again, the single strong near dependency is principally between log **DD** and the constant ι. It is interesting to note that, in accordance with the prior expectations of (10.24), the coefficient of **HeatedArea** is, even without constraint, effectively the negative of that of **EffArea**. Exhibit 10.17 shows the MSPE of 585 for this model 3 over the validation sample. By this criterion, model 3 is superior to all other models so far considered, and its specification is merely the full extension of the principles that lead Janson to his model 2.

Model 4

Actually, once the simplification (10.29) is adopted in place of (10.26), it is possible to consider another model reformulation that results in a linear estimation. It is reasonable to question whether the insulation dummy should enter **A** in (10.25), as we currently have it, or in **C** as a factor that modifies the rate of heat transfer from the surface. This would produce a model like model 3 except with **A** ≡ **Area** and $\mathbf{C} = \nu(\mathbf{DD})^{\lambda}\mathbf{SKI}$, where $\mathbf{I} = \exp(\phi\mathbf{NormIns})$. Here, **NormIns** ≡ **Insulation/Area**, so that the heat transmission rate adjustment factor in **I** equals 1 (no adjustment) for a building with no insulation and e^{ϕ} for a completely insulated building, where ϕ is to be estimated. The corresponding model, linear in logs, which will be called model 4, is estimated as

$$
\begin{aligned}
\log \textbf{TotCon} = {} & \underset{(7.205)}{15.640\iota} + \underset{(0.173)}{0.408} \log \textbf{Area} - \underset{(0.109)}{0.258}\textbf{NormIns} \\
& + \underset{(0.006)}{0.012}\textbf{CorAC} - \underset{(0.159)}{0.388} \log \textbf{HeatedArea} \\
& - \underset{(0.164)}{0.368}\textbf{Fuel} - \underset{(0.790)}{1.219} \log \textbf{DD}, \qquad (10.31) \\
& R^2 = .540, \qquad \text{SER} = 0.240.
\end{aligned}
$$

These estimates are wholly consonant with those in (10.30), and as is seen in Exhibit 10.17, this model 4 provides a MSPE of 582 over the validation sample that is very marginally better than that for model 3.

Internal Cross Validation

We indeed seem to have been able to use various diagnostic techniques, including the collinearity diagnostics, to achieve a superior model specification and better quality least-squares estimates. We need not accept this conclusion, however, without further test. For in this case, where we have both an estimation data set and a validation data set, we can put our success to the trial by going the other way round, making use of the validation data set for estimation and cross validating it on the original estimation data set. Our interest is in comparing model 1* (10.23), model 2 (10.20), and model 4 (10.31). Models 1 (10.17) and 1a (10.19), with their incorrect data and incorrectly deleted variates, need not be considered here.

A preliminary analysis with the influential-data diagnostics for model 1* estimated with the validation data set shows up buildings 1573 and 2199 as being worthy of further attention, both showing substantial influence on the least-squares estimates. Model 2 highlights the same observations. Model 4 indicates buildings 1573, 2079, 2199, and 3424. Thus, building 2199 is clearly indicated by all models, and indeed this observation is found by Janson to be in error. It is of particular interest that model 4 most strongly highlights 2199 while model 2 is least effective here. None of the other indicated observations, however, is found to be corrupted.

Ideally we would like to proceed by getting the correct data for this erroneous observation 2199, but unfortunately, they are not available. Here, then, we have an example of a proper reason for deleting an observation: the observation is not only influential but also uncorrectably in error. After deleting it, we estimate the three models and then use them to forecast in the training sample (the original estimation data set). It is not necessary to report the regression results since they are more or less close to the estimates obtained with the original estimation sample. This closeness is very great for model 4, a gratifying result

Exhibit 10.19 Mean-squared Prediction Errors: Internal Cross-Validation Results[a]

Model	Description	MSPE
2	Janson's model 2 (10.20)	940
1*	Equation (10.23)	775
4	Equation (10.31)	741

[a]Observation 2199 deleted.

indeed, reasonably close for model 1*, and least close for model 2. Finally, we calculate the mean-squared prediction errors for these models over the original estimation sample, which are given in Exhibit 10.19. Here we see that model 4 does best (MSPE = 741), followed closely by model 1* (MSPE = 775) and only distantly chased by Janson's preferred model 2 (MSPE = 940).

Conclusions

This case study has served several purposes. First, it has provided an excellent illustration of several of the concerns voiced elsewhere in this book. The use of mean-centered data, for example, would have prevented our finding the most damaging source of weak data for this study, that with the constant. Likewise, a propensity to throw out insignificant (but a priori valid) variates without determining if their estimation weakness could be due to data weakness can lead to unwarranted model respecifications. Second, we have seen how we can use the collinearity diagnostics in coordination with other diagnostic techniques to come to a better understanding of the nature of the data problems and, as a result, to adopt a more valid model specification and obtain better regression estimates and forecasts.

CHAPTER 11

General Conditioning and Extensions to Nonlinearities and Simultaneity

In this chapter we forge new directions, defining notions of conditioning that substantially generalize the concept as it has been used so far and introducing applications that extend these notions into the realms of nonlinear and simultaneous-equations models. Many of the ideas here are relatively new and have not had the same opportunity for being tested so thoroughly in the field as the basic collinearity diagnostics of the earlier chapters. This chapter, therefore, also provides numerous areas and suggestions for research. We begin with a discussion of the general problem of ill conditioning in statistical analyses. We then turn to a theoretical basis for extending the basic collinearity diagnostics for assessing the conditioning of an arbitrary statistical model, linear or nonlinear, single-equation or simultaneous system. Finally, two extensions relating to the estimation of simultaneous systems are made: a measure based on signal-to-noise for the "degree of effective identification" and a use of the condition number as a means for choosing between the two- and three-stage least-squares estimators.

11.1 THE GENERAL PROBLEM OF ILL CONDITIONING

The term *conditioning* has so far been used virtually synonymously with the notion of collinearity. In Chapter 1, however, it is noted that collinearity deals with the existence of nearly linear relationship among a set of variates, whereas conditioning is concerned with the sensitivity (or insensitivity) of a given relation to shifts (perturbations) in the underlying data. So, while the two concepts are obviously very closely related, it is clear that they are not the same. And indeed, we shall now see that the collinearity diagnostics are but one type of conditioning analysis useful in statistical applications.

Thus, this section—which is based on Belsley and Oldford (1986) and owes much to my association with R. Wayne Oldford—starts with a general definition of a conditioning analysis and then proceeds to distinguish among three different kinds of conditioning analyses that are of interest in statistics: data, estimator, and criterion conditioning. We shall see that these three concepts coincide in the oft-encountered case of the least-squares estimation of a linear model—the LS/linear case—, and this is the reason the two concepts have been able effectively to be identified with one another until now. But we shall also see that the two concepts can and do diverge otherwise—with estimators that are not least squares or with models that are not linear. This section concludes with an example that shows how a computer-intensive means for perturbation analyses would seem to allow the practical conditioning of virtually any situation to be assessed.

A General Notion of Ill Conditioning

Suppose rather generally we have any system of continuous equations

$$\boldsymbol{\lambda} = \mathbf{f}(\boldsymbol{\omega}), \tag{11.1}$$

where $\boldsymbol{\lambda}$, $\boldsymbol{\omega}$, and $\mathbf{f}$ are vectors and/or matrices. In this formulation, the elements of $\boldsymbol{\omega}$ could be any combination of data, parameters, or random variables. Thus, (11.1) might describe an estimator, such as $\mathbf{b} = (\mathbf{X}^T\mathbf{X})^{-1}\mathbf{X}^T\mathbf{y}$, or a nonlinear stochastic model, such as $\mathbf{y} = \beta_1\boldsymbol{\iota} + \beta_2\mathbf{X}_1^{\alpha}\mathbf{X}_2^{\gamma} + \boldsymbol{\varepsilon}$, or a system of data dependencies, such as $\mathbf{d} = \mathbf{Xc}$, or in general, any system of interest in which the elements $\boldsymbol{\lambda}$ are assumed to depend continuously upon the elements $\boldsymbol{\omega}$. In some applications this relation might be defined implicitly as $\mathbf{f}(\boldsymbol{\lambda}, \boldsymbol{\omega}) = \mathbf{0}$.

Frequently, it is of interest to know something about the sensitivity of $\boldsymbol{\lambda}$ to particular changes in $\boldsymbol{\omega}$: changes that belong to some specified set Ω. In earlier chapters, for example, we were interested in knowing the kinds of changes that could occur in the least-squares estimator $\mathbf{b}$ when the elements of $\mathbf{X}$ are changed by a small amount, or the elements of $\mathbf{y}$. And typically, concerns arise when disproportionately large or small changes can occur in $\boldsymbol{\lambda}$ as a result of a given change in $\boldsymbol{\omega}$. When this happens, we shall say that $\boldsymbol{\lambda}$ is ill conditioned with respect to Ω, a notion that receives formal definition shortly. But for the moment it is worth highlighting that "disproportionately large or small" can only be meaningfully defined relative to some assessment of what is important in some appropriate underlying context, and hence a meaningful analysis of conditioning must necessarily involve not only an underlying context but also elements of one's interpretative understanding of that context.

Here, then, the notion of ill conditioning has been generalized considerably over its use so far. In the case of the least-squares estimator $\mathbf{b} = (\mathbf{X}^T\mathbf{X})^{-1}\mathbf{X}^T\mathbf{y}$ mentioned above, for example, $\boldsymbol{\omega}$ could represent the data $\mathbf{X}$, and the set Ω might consist of changes in $\mathbf{X}$ corresponding to the measurement accuracy of the data. Here, $\boldsymbol{\lambda} = \mathbf{b}$ is an estimator based on these data, and so $\boldsymbol{\lambda}$'s being adjudged ill conditioned with respect to Ω would indicate that the given estimate (or estimator) is not well determined on account of the inadequate quality of the observed data. Notice that it is the estimator $\mathbf{b}$ that is being called ill conditioned in this illustration, not the data. In the previous chapters, we have implicitly switched between these two notions, talking at one time about collinearity in $\mathbf{X}$ and at another about the ill conditioning of the least-squares estimator $\mathbf{b}$. Now we are in a position to make the relation between these two concepts clear as well as to examine other kinds of statistically interesting forms of ill conditioning.

More formally, the situation can be represented as follows: Suppose the quantities $\boldsymbol{\omega}$ and $\boldsymbol{\lambda}$ are related according to (11.1) so that an additive perturbation $\delta\boldsymbol{\omega}$ in $\boldsymbol{\omega}$ results in a perturbation in $\boldsymbol{\lambda}$ equal to $\delta\boldsymbol{\lambda} \equiv \mathbf{f}(\boldsymbol{\omega} + \delta\boldsymbol{\omega}) - \mathbf{f}(\boldsymbol{\omega})$. For fixed $\boldsymbol{\omega}$, a function $\mathbf{g}(\delta\boldsymbol{\omega}) \equiv \mathbf{f}(\boldsymbol{\omega} + \delta\boldsymbol{\omega}) - \mathbf{f}(\boldsymbol{\omega})$ may be defined that maps the elements of $\delta\boldsymbol{\omega}$ of a given domain Ω to elements $\delta\boldsymbol{\lambda}$ in a corresponding range set Λ. That is,

$$\mathbf{g}\colon \delta\boldsymbol{\omega} \mapsto \delta\boldsymbol{\lambda},$$

or

$$\mathbf{g}\colon \Omega \to \Lambda.$$

Now, conceptually, there exists a third set Λ^* that consists of all those perturbations $\delta\boldsymbol{\lambda}$ that are considered a priori to be reasonable given the set Ω. And concern arises when, corresponding to some $\delta\boldsymbol{\omega}$ in Ω, there exist $\delta\boldsymbol{\lambda}$ in Λ that are not in Λ^*; that is, when perturbations $\delta\boldsymbol{\omega} \in \Omega$ can produce unreasonable responses $\delta\boldsymbol{\lambda} \notin \Lambda^*$. A sensitivity analysis becomes interesting, for example, when the set Ω consisting of all "reasonably small" perturbations $\delta\boldsymbol{\omega}$ of $\boldsymbol{\omega}$ contains some $\delta\boldsymbol{\omega}$ that can nevertheless produce perturbations $\delta\boldsymbol{\lambda}$ of $\boldsymbol{\lambda}$ that are not "reasonably small."

These considerations suggest the following general definitions:

A Conditioning Analysis. The specification of the *conditioning triple* $K \equiv \{\mathbf{f}, \Omega, \Lambda^*\}$ followed by a determination of whether $\boldsymbol{\lambda}$ is ill conditioned.

Ill Conditioning. Given K and its implied Λ, $\boldsymbol{\lambda}$ is said to be ill conditioned with respect to $\boldsymbol{\omega}$ (or Ω) if $\Lambda \not\subset \Lambda^*$. Equivalently, one can call the system $\mathbf{f}$ ill conditioned.

The relevance of a conditioning analysis depends critically on the determination of the sets Ω and Λ^* in K. And as we shall see, these sets *must* be determined within the context of the problem at hand. To determine them otherwise is to render a conditioning analysis meaningless.

For many practical applications, relative perturbations and relative responses are appropriate, and hence the following specifications of Ω and Λ^* are often useful:

$$\Omega = \{\delta\boldsymbol{\omega} : \|\delta\boldsymbol{\omega}\|/\|\boldsymbol{\omega}\| \leqslant m_1\} \tag{11.2}$$

and

$$\Lambda^* = \{\delta\boldsymbol{\lambda} \;:\; \|\delta\boldsymbol{\lambda}\|/\|\boldsymbol{\lambda}\| \leqslant m_2\}, \tag{11.3}$$

where $m_1, m_2 \geqslant 0$ are real constants and $\|\cdot\|$ denotes some norm, such as the Euclidean or spectral norms, both of which will be used here. Cases can also be considered in which either or both of the inequalities in (11.2) and (11.3) are replaced by $\geqslant$ or $=$ depending on the kind of conditioning being investigated.

The specification of Λ^* typically follows naturally once Ω has been defined, so long as the perturbations $\delta\boldsymbol{\omega}$ are meaningfully interpretable within the context of the problem. The notion of *structural interpretability* has already been introduced in Chapter 6 to describe data for which this is so, but let us review this concept in this context.

Relevance of the Conditioning Analysis

We see from the preceding definition that the conditioning triple $K \equiv \{\mathbf{f}, \Omega, \Lambda^*\}$ completely specifies a conditioning analysis, so changing any of its three elements produces a completely different conditioning analysis. In specifying a conditioning analysis K, then, it is essential to make each of its three parts clear and explicit, anything less necessarily rendering the analysis irrelevant. This may require more consideration than first appears to be the case, for both Ω and Λ^* clearly involve subjective components. Typically Ω is a set of "reasonable" perturbations and Λ^* is a set of "reasonable" outcomes given Ω. Such notions of reasonable can only be obtained through the analyst's assessment of the underlying context being analyzed. This section highlights two critical but frequently overlooked aspects in specifying K.

The first is the necessity of specifying clearly what sensitivity is being addressed by the analysis, since a conditioning analysis intended for one purpose can be quite misleading for (and indeed often confused with) another. We have seen in Chapter 6, for example, how the condition number of mean-centered data can inadvertently mask an intended analysis of the sensitivity of least-squares estimates to perturbations in the basic, or uncentered, data.

The second is the necessity of having the elements of K be interpretable if the conditioning analysis itself is to be so. Each element of K must be defined so as to be meaningfully interpretable within the larger context in which the conditioning analysis is being conducted. Only against such a backdrop (which is typically some real-life subject matter) can one hope to argue the meaningfulness of the results of a particular conditioning analysis.

To amplify on this second point, recall that, in specifying Ω, this set is intended to represent additive changes $\delta\boldsymbol{\omega}$ in $\boldsymbol{\omega}$ that are considered to have some special meaning or interpretation within the associated context. For example, one's knowledge of the underlying context may suggest a set of perturbations $\delta\boldsymbol{\omega} \in \Omega$ that can be considered inconsequential to the analysis, or those that may be considered especially large. So care must be taken in the selection both of $\boldsymbol{\omega}$

and Ω. The changes $\delta\boldsymbol{\omega}$ are, in turn, defined relative to $\boldsymbol{\omega}$, and so they can be argued to be interpretable only if the $\boldsymbol{\omega}$ are. Suppose, for example, that Ω is defined as in (11.2). The value given to m_1 will depend upon both $\boldsymbol{\omega}$ and $\delta\boldsymbol{\omega}$. In the absence of any supporting context to provide a basis for interpretation, one might simply adopt a psychologically small number as representing an inconsequential proportionate change in $\boldsymbol{\omega}$, say, a 1% change, or $m_1 = 0.01$. For the $\boldsymbol{\omega}$ under consideration, however, knowledge of the underlying context may in fact suggest that values as large as 4.0 or even 40.0 should be regarded as inconsequential. What constitutes the set of inconsequential changes Ω, then, depends on $\boldsymbol{\omega}$ and its interpretation within the overall context, not just on some conventional psychological notion of small.

For exactly the same reasons, $\boldsymbol{\lambda} = \mathbf{f}(\boldsymbol{\omega})$ must be interpretable within the underlying context before any defensible choice for Λ^* can be made. Here, Λ^* is defined to be a set of perturbations that are deemed reasonable a priori, a phrase that we have seen can only be given meaning relative to the broader context in which the conditioning analysis is being conducted.

The reader will please be patient if this discussion seems unnecessarily laborious; it is being stressed because experience suggests that overlooking its message is the source of much needless confusion regarding conditioning analyses. We have seen in Chapter 6, for example, that a failure to see the need for data to be structurally interpretable has led some to advocate assessing collinearity on the basis of mean-centered data. Thus, it is precisely to facilitate reference to this important concept that we denote those elements in K that can be meaningfully interpreted in terms of an underlying or associated context by the terms *contextually* or *structurally interpretable.* The term contextually interpretable is to be viewed broadly, indicating quite generally any meaning derived from the broader context in which the conditioning analysis is being conducted, even if that context exists only conceptually. The slightly narrower term, structurally interpretable, refers specifically to those cases in which the context is some underlying real-life situation, as is typical of applied statistics. Thus, the relevance of a conditioning analysis must be argued on the basis of a contextually interpretable K, and it is incumbent upon the investigator to convince the reader of the relevance of this interpretation.

We turn now to three kinds of conditioning analyses that are often of interest in statistical applications: data conditioning, estimator conditioning, and criterion conditioning. They are first treated quite generally and are then discussed again for the particular case of least-squares linear regression in the following section.

Data, Estimator, and Criterion Conditioning: General Case

To fix ideas, we now examine three different kinds of conditioning analyses. The first, which we call data conditioning, is directly associated with collinearity. The others, estimator and criterion conditioning, forge new directions.

Data Conditioning (Collinearity)

Given $\mathbf{X}_1, \ldots, \mathbf{X}_p$, observed n-vectors on p variates (endogenous, exogenous, or both), we wish to know if there exists an exact, or nearly exact, linear relation among them. If so, the variates $\mathbf{X}_1, \ldots, \mathbf{X}_p$, or the columns of the matrix $\mathbf{X} = [\mathbf{X}_1 \cdots \mathbf{X}_p]$, are said to be collinear. For this purpose, it is useful to adopt Gunst's (1983) definition of collinearity that is given in (3.30) of Section 3.3 and is shown there to be equivalent to a definition based on the scaled condition number $\tilde{\kappa}(\mathbf{X})$. Recall that, given $\|\mathbf{X}_i\| = 1$ for all i, collinearity exists among the $\mathbf{X}_i$'s if, for a suitably small predetermined $\gamma > 0$, there exist constants $\mathbf{c}^{\mathrm{T}} = [c_1, \ldots, c_p]$, not all zero, such that

$$\mathbf{Xc} = \mathbf{a} \tag{11.4}$$

and

$$\|\mathbf{a}\| < \gamma\|\mathbf{c}\|. \tag{11.5}$$

This definition of collinearity is equivalent to the following conditioning analysis. Let $\boldsymbol{\lambda}$ and $\boldsymbol{\omega}$ be defined so that

$$\boldsymbol{\lambda} = \mathbf{f}(\boldsymbol{\omega}) \equiv \mathbf{X}\boldsymbol{\omega}. \tag{11.6}$$

And consider the perturbations

$$\delta\boldsymbol{\lambda} = \mathbf{g}(\delta\boldsymbol{\omega}) \equiv \mathbf{X}\,\delta\boldsymbol{\omega} \tag{11.7}$$

having domain set chosen as

$$\Omega = \{\delta\boldsymbol{\omega} : \|\delta\boldsymbol{\omega}\|/\|\boldsymbol{\omega}\| = m_1\} \tag{11.8}$$

and acceptable response set chosen as

$$\Lambda^* = \{\delta\boldsymbol{\lambda} : \|\delta\boldsymbol{\lambda}\|/\|\boldsymbol{\lambda}\| \geqslant m_2\}. \tag{11.9}$$

Under this conditioning analysis, perturbations $\delta\boldsymbol{\omega}$ of fixed relative size m_1 are required to result in perturbations $\delta\boldsymbol{\lambda}$ whose length relative to $\boldsymbol{\lambda}$ is not less than m_2, some small number. Otherwise, the data $\mathbf{X}$ are ill conditioned with respect to Ω of (11.8) with a degree of ill conditioning that depends upon the selection of m_1 and m_2; the larger is m_1 relative to m_2, the worse the conditioning. Heuristically, these choices of Ω and Λ^* may be motivated as follows: If $\mathbf{X}$ were perfectly ill conditioned, that is, of less than full rank, then we know there exists some $\mathbf{c} \neq \mathbf{0}$ such that $\mathbf{X}(\alpha\mathbf{c}) = \mathbf{0}$ for all α. Now consider any $\boldsymbol{\omega} \neq \mathbf{0}$ that produces a $\boldsymbol{\lambda} \equiv \mathbf{X}\boldsymbol{\omega} \neq \mathbf{0}$, and choose α so that $\|\alpha\mathbf{c}\| = m_1\|\boldsymbol{\omega}\|$. Then the perturbation $\delta\boldsymbol{\omega} \equiv \alpha\mathbf{c}$ lies in Ω since $\|\delta\boldsymbol{\omega}\|/\|\boldsymbol{\omega}\| = m_1$ but results in $\delta\boldsymbol{\lambda} = \mathbf{X}\,\delta\boldsymbol{\omega} = \mathbf{X}(\alpha\mathbf{c}) = \mathbf{0}$, and hence, $\|\delta\boldsymbol{\lambda}\|/\|\boldsymbol{\lambda}\| = 0 < m_2$. Thus, if $\mathbf{X}$ is perfectly ill conditioned, it is always possible to find perturbations in $\delta\boldsymbol{\omega} \in \Omega$ that result in responses $\delta\boldsymbol{\lambda} \in \Lambda$ that are not contained in a set Λ^* as defined by (11.9), and hence $\Lambda \not\subset \Lambda^*$ becomes an appropriate criterion for defining ill conditioning.

More formally, the equivalence of this notion of data conditioning to that of Gunst is shown as follows: Suppose first the data $\mathbf{X}$ are ill conditioned as above; that is, for some $\|\delta\boldsymbol{\omega}\|/\|\boldsymbol{\omega}\| = m_1$ we observe $\|\delta\boldsymbol{\lambda}\|/\|\boldsymbol{\lambda}\| < m_2$ for m_2 chosen small. This implies $\|\delta\boldsymbol{\lambda}\|/\|\delta\boldsymbol{\omega}\| < (m_2/m_1)\|\boldsymbol{\lambda}\|/\|\boldsymbol{\omega}\|$, and thus, taking $\gamma \equiv (m_2/m_1)\|\boldsymbol{\lambda}\|/\|\boldsymbol{\omega}\|$, we find the data to be collinear according to Gunst's definition with (11.4) replaced by (11.7). The larger is m_1 relative to m_2, the smaller will be γ and the greater the degree of collinearity. Conversely, suppose we observe $\|\delta\boldsymbol{\lambda}\| < \gamma\|\delta\boldsymbol{\omega}\|$ for some "suitably small" $\gamma > 0$ and $\delta\boldsymbol{\omega} \neq \mathbf{0}$. Choose $\|\delta\boldsymbol{\omega}\|$ so that $\|\delta\boldsymbol{\omega}\| = m_1\|\boldsymbol{\omega}\|$ for some $\|\boldsymbol{\omega}\|$ and m_1, so we observe $\|\delta\boldsymbol{\lambda}\| < \gamma m_1\|\boldsymbol{\omega}\|$, or letting $m_2 = \gamma m_1\|\boldsymbol{\omega}\|/\|\boldsymbol{\lambda}\|$, $\|\delta\boldsymbol{\lambda}\|/\|\boldsymbol{\lambda}\| < m_2$.

Estimator Conditioning

Consider an estimator

$$\hat{\boldsymbol{\theta}} = \mathbf{f}(\mathbf{X}, \mathbf{Y}) \tag{11.10}$$

of parameters $\boldsymbol{\theta}$, where $\mathbf{X}$ is a matrix of observed exogenous variates and $\mathbf{Y}$ is a matrix of observed endogenous variates. System (11.10) pairs up with (11.1) in the obvious fashion with $\boldsymbol{\lambda} \equiv \hat{\boldsymbol{\theta}}$ and $\boldsymbol{\omega} \equiv [\mathbf{X}, \mathbf{Y}]$. Our interest here is to determine the potential sensitivity of $\hat{\boldsymbol{\theta}}$ with respect to perturbations in $\mathbf{X}$ or $\mathbf{Y}$ or both, that is, the conditioning of $\hat{\boldsymbol{\theta}}$ with respect to $\boldsymbol{\omega} \equiv [\mathbf{X}, \mathbf{Y}]$.

In the case where numerical problems are an issue, interest often attaches to the sets

$$\Omega = \{\delta\boldsymbol{\omega} \equiv [\delta\mathbf{X}, \delta\mathbf{Y}] \;:\; \|\delta\mathbf{X}\|/\|\mathbf{X}\| \leqslant m_1,\ \|\delta\mathbf{Y}\| = 0\} \tag{11.11}$$

and

$$\Lambda^* = \{\delta\boldsymbol{\lambda} \equiv \delta\hat{\boldsymbol{\theta}} \;:\; \|\delta\hat{\boldsymbol{\theta}}\|/\|\hat{\boldsymbol{\theta}}\| \leqslant m_2\}, \tag{11.12}$$

where m_1 and m_2 are usually chosen to indicate proportionate changes of marginal consequence to the analysis at hand. To determine whether $\hat{\boldsymbol{\theta}}$ is ill conditioned with respect to Ω, we consider $m \equiv \sup_{\delta\boldsymbol{\omega} \in \Omega} \|\delta\hat{\boldsymbol{\theta}}\|/\|\hat{\boldsymbol{\theta}}\|$. If $m > m_2$, then $\hat{\boldsymbol{\theta}}$ is ill conditioned with respect to Ω.

With regard to shifts in $\mathbf{Y}$, Ω might be chosen as

$$\Omega = \{\delta\boldsymbol{\omega} \equiv [\delta\mathbf{X}, \delta\mathbf{Y}] : \|\delta\mathbf{Y}\|/\|\mathbf{Y}\| \leqslant m_1,\ \|\delta\mathbf{X}\| = 0\}, \tag{11.13}$$

where m_1 determines a region relative to the stochastically generated $\delta\mathbf{Y}$ so that $\text{Prob}(\Omega) = .95$, or some other such probabilistically defined region. This choice of Ω suggests a way of viewing the intimate relation known to exist between ill conditioning and high-variance estimates.

Criterion Conditioning

Parameters $\boldsymbol{\theta}$ are often estimated by minimizing some criterion function of the data and parameters. Let this function be $Q(\mathbf{X}, \mathbf{Y}, \boldsymbol{\theta})$, where, as above, $\mathbf{X}$ is exogenous, $\mathbf{Y}$ endogenous, $\boldsymbol{\theta} \in \Theta$ is the vector of parameters to be estimated, and $\boldsymbol{\Theta}$ is its domain. Suppose $\hat{\boldsymbol{\theta}}$ is selected to satisfy

$$Q(\mathbf{X}, \mathbf{Y}, \hat{\boldsymbol{\theta}}) = \inf_{\boldsymbol{\theta} \in \Theta} Q(\mathbf{X}, \mathbf{Y}, \boldsymbol{\theta}). \tag{11.14}$$

Frequently employed criteria $Q(\mathbf{X}, \mathbf{Y}, \boldsymbol{\theta})$ include

(i) $[\mathbf{y} - \mathbf{f}(\mathbf{X}, \mathbf{y}, \boldsymbol{\theta})]^T[\mathbf{y} - \mathbf{f}(\mathbf{X}, \mathbf{y}, \boldsymbol{\theta})]$ for least-squares estimates,

(ii) $-\log L(\boldsymbol{\theta})$, where $L(\boldsymbol{\theta})$ is the likelihood function of $\boldsymbol{\theta}$ for maximum-likelihood estimates,

(iii) $\rho(\mathbf{X}, \mathbf{y}, \boldsymbol{\theta})$, for M-estimation, where ρ is some function chosen for its robustness properties.

In each case, it is desirable that parameter values $\boldsymbol{\theta}$ that represent large deviations $\delta\hat{\boldsymbol{\theta}} \equiv \boldsymbol{\theta} - \hat{\boldsymbol{\theta}}$ from $\hat{\boldsymbol{\theta}}$ should be detectable by the selected criterion function, for a failure to do so would indicate a poor determination of the parameter estimate. The following perturbation sets Ω and Λ^* are therefore of interest:

$$\Omega = \{\delta\boldsymbol{\theta} : \|\delta\hat{\boldsymbol{\theta}}\| = m_1 > 0\} \tag{11.15}$$

and

$$\Lambda^* = \{\delta Q : \|\delta Q\| / \inf_{\delta\hat{\boldsymbol{\theta}} \in \Omega} \|\delta Q\| \leqslant m_2\}. \tag{11.16}$$

If $\|\hat{\boldsymbol{\theta}}\| \neq 0$, then $\|\delta\hat{\boldsymbol{\theta}}\|$ in (11.15) might be more meaningfully replaced by $\|\delta\hat{\boldsymbol{\theta}}\|/\|\hat{\boldsymbol{\theta}}\|$. Note that the denominator in the definition of Λ^* should not be $\|Q(\mathbf{X}, \mathbf{Y}, \hat{\boldsymbol{\theta}})\|$, since $Q(\mathbf{X}, \mathbf{Y}, \boldsymbol{\theta})$ itself could be replaced by $Q(\mathbf{X}, \mathbf{Y}, \boldsymbol{\theta}) + c$, where $c > 0$ is an arbitrary constant, without changing $\hat{\boldsymbol{\theta}}$ or δQ. In this case, the constant m_2 would be difficult to assign meaningfully, since its value would depend on c. Of course, the denominator could be set equal to 1 if absolute perturbations in the criterion were deemed important.

The appeal of Λ^* as defined in (11.16) lies in its assessing the effect of any perturbation in relation to the worst possible effect—the worst case becomes a standard of acceptability. If, however, perturbations $\delta\hat{\boldsymbol{\theta}} \in \Omega$ produce δQ's that are not in Λ^*, then there must exist $\boldsymbol{\theta}^*$ that differ substantially from $\hat{\boldsymbol{\theta}}$ but are relatively indistinguishable from $\hat{\boldsymbol{\theta}}$ by criterion $Q(\cdot)$. If $\inf_{\delta\hat{\boldsymbol{\theta}} \in \Omega} \|\delta Q\| = 0$, then $\boldsymbol{\theta}$ could be said to be *inestimable* with respect to this criterion and these data for those $\delta\hat{\boldsymbol{\theta}}$ for which $\|\delta Q\| = 0$.

As an example, consider

$$Q(\mathbf{X}, \mathbf{Y}, \boldsymbol{\theta}) = -\log L(\boldsymbol{\theta}). \tag{11.17}$$

Here $\delta Q = -\log L(\hat{\boldsymbol{\theta}} + \delta\hat{\boldsymbol{\theta}}) + \log L(\hat{\boldsymbol{\theta}})$, so that δQ represents the drop in log-likelihood due to setting $\boldsymbol{\theta} = \hat{\boldsymbol{\theta}} + \delta\hat{\boldsymbol{\theta}}$ relative to $\boldsymbol{\theta} = \hat{\boldsymbol{\theta}}$, or the log of the likelihood-ratio statistic of $\boldsymbol{\theta} = \hat{\boldsymbol{\theta}}$ versus $\boldsymbol{\theta} = \hat{\boldsymbol{\theta}} + \delta\hat{\boldsymbol{\theta}}$. When ill conditioning of the maximum-likelihood estimate occurs according to Ω and Λ^* defined as above, there must be some perturbation $\delta\hat{\boldsymbol{\theta}}_1 \in \Omega$ for which the log-likelihood ratio can distinguish the difference between $\hat{\boldsymbol{\theta}}$ and $\hat{\boldsymbol{\theta}} + \delta\hat{\boldsymbol{\theta}}_1$ better than it can distinguish the difference between $\hat{\boldsymbol{\theta}}$ and $\hat{\boldsymbol{\theta}} + \delta\hat{\boldsymbol{\theta}}_2$ for some other perturbation $\delta\hat{\boldsymbol{\theta}}_2 \in \Omega$.

More generally, to determine the criterion conditioning for the Λ^* defined as before, one must evaluate

$$\sup_{\delta\hat{\boldsymbol{\theta}} \in \Omega} \|\delta Q\| \Big/ \inf_{\delta\hat{\boldsymbol{\theta}} \in \Omega} \|\delta Q\|. \tag{11.18}$$

Since this is rarely easily done, we approximate this quantity to provide a rough guide to the possibility of ill conditioning. The first few terms of a Taylor expansion of $Q(\mathbf{X}, \mathbf{Y}, \hat{\boldsymbol{\theta}} + \delta\hat{\boldsymbol{\theta}})$ about $\hat{\boldsymbol{\theta}}$ (assuming that $\|\delta\hat{\boldsymbol{\theta}}\|$ or m_1 is suitably small) yields

$$\begin{aligned} \delta Q &= Q(\mathbf{X}, \mathbf{Y}, \hat{\boldsymbol{\theta}} + \delta\hat{\boldsymbol{\theta}}) - Q(\mathbf{X}, \mathbf{Y}, \hat{\boldsymbol{\theta}}) \\ &\approx \delta\hat{\boldsymbol{\theta}}^{\mathrm{T}} \left(\frac{\partial Q(\mathbf{X}, \mathbf{Y}, \boldsymbol{\theta})}{\partial \boldsymbol{\theta}} \right)\Bigg|_{\boldsymbol{\theta} = \hat{\boldsymbol{\theta}}} + \delta\hat{\boldsymbol{\theta}}^{\mathrm{T}} \left(\frac{\partial^2 Q(\mathbf{X}, \mathbf{Y}, \boldsymbol{\theta})}{\partial \boldsymbol{\theta}\, \partial \boldsymbol{\theta}^{\mathrm{T}}} \right)\Bigg|_{\boldsymbol{\theta} = \hat{\boldsymbol{\theta}}} \delta\hat{\boldsymbol{\theta}}. \end{aligned} \tag{11.19}$$

For those many criterion functions for which the first-order term is zero at $\boldsymbol{\theta} = \hat{\boldsymbol{\theta}}$, (11.19) becomes

$$\partial Q \approx \delta\hat{\boldsymbol{\theta}}^{\mathrm{T}} \mathbf{A}\, \delta\hat{\boldsymbol{\theta}}, \tag{11.20}$$

where $\mathbf{A} \equiv (\partial^2 Q(\mathbf{X}, \mathbf{Y}, \boldsymbol{\theta})/\partial \boldsymbol{\theta}\, \partial \boldsymbol{\theta}^{\mathrm{T}})|_{\boldsymbol{\theta} = \hat{\boldsymbol{\theta}}}$ is the Hessian matrix of Q with respect to $\boldsymbol{\theta}$ at $\boldsymbol{\theta} = \hat{\boldsymbol{\theta}}$.[1] Using (11.20) in (11.18) and recalling the well-known extremum properties of eigenvalues discussed in Section 3.1, we see that (11.18) is equivalent to $\kappa(\mathbf{A})$, the condition number of the matrix $\mathbf{A}$. In practice, then, $\kappa(\mathbf{A})$ may be used as a rough guide to assess the criterion conditioning.

Data, Estimator, and Criterion Conditioning: Linear Case

The three kinds of conditioning whose general cases have been considered so far are easily specialized for the standard linear model

$$\mathbf{y} = \mathbf{X}\boldsymbol{\beta} + \boldsymbol{\varepsilon} \tag{11.21}$$

[1]This is, of course, the sample information matrix if $Q = -\log L(\hat{\boldsymbol{\theta}})$.

when $\boldsymbol{\beta}$ is estimated by least squares, the LS/linear case. By now, we should not be too surprised to discover that a single number, the condition number $\kappa(\mathbf{X})$, is found to be central to assessing all three forms of conditioning in this LS/linear case—a coincidence that does not typically occur otherwise, as we shall also demonstrate.

In this analysis, we draw repeatedly upon the important inequality (3.19) introduced in Section 3.1 in our discussion of the sensitivity of inexact linear systems to perturbations in the data. It is repeated here for convenience:

$$\|\delta\mathbf{b}\|/\|\mathbf{b}\| \leqslant \kappa(\mathbf{X})\hat{R}^{-1}[2 + (1 - \hat{R}^2)^{1/2}\kappa(\mathbf{X})]\nu + O(\nu^2), \qquad (11.22)$$

where, we recall, $\nu = \max(\|\delta\mathbf{y}\|/\|\mathbf{y}\|, \ \|\delta\mathbf{X}\|/\|\mathbf{X}\|)$ and $\hat{R}$ is the uncentered multiple-correlation coefficient (2.4a). We now examine in the LS/linear context the three forms of conditioning.

Data Conditioning

In (11.21), the data conditioning of interest is that of the data matrix $\mathbf{X}$, which we assume here to be of full rank. Let $\mathbf{X}^+ \equiv (\mathbf{X}^T\mathbf{X})^{-1}\mathbf{X}^T$ denote the Moore–Penrose inverse of $\mathbf{X}$ as defined in Section 3.1. Take Ω and Λ^* as in (11.8) and (11.9), and for any $\boldsymbol{\omega}$, define $\boldsymbol{\lambda}$ by $\boldsymbol{\lambda} \equiv \mathbf{X}\boldsymbol{\lambda}$, so that $\boldsymbol{\omega}$ and $\boldsymbol{\lambda}$ also obey $\boldsymbol{\omega} = \mathbf{X}^+\boldsymbol{\lambda}$. This latter expression is, of course, the least-squares solution of a regression of $\boldsymbol{\lambda}$ on $\mathbf{X}$, and hence, applying inequality (11.22) with only $\boldsymbol{\lambda}$ perturbed yields $\|\delta\boldsymbol{\lambda}\|/\|\boldsymbol{\lambda}\| \geqslant \frac{1}{2}\kappa^{-1}(\mathbf{X})\|\delta\boldsymbol{\omega}\|/\|\boldsymbol{\omega}\|$, where we note in this case that $\hat{R}$, the uncentered multiple correlation of $\boldsymbol{\lambda}$ regressed on $\mathbf{X}$, necessarily equals 1 since the "fit" $\boldsymbol{\lambda} \equiv \mathbf{X}\boldsymbol{\omega}$ is perfect by construction. From this inequality, which may be an equality for certain $\boldsymbol{\lambda}$, it is readily seen that the larger is the condition number $\kappa(\mathbf{X})$, the smaller must m_2 in (11.9) be defined relative to m_1 in (11.8), and therefore the greater the ill conditioning of $\mathbf{X}$ with respect to Ω.

Estimator Conditioning

The estimator to be examined in this case is the standard least-squares estimator $\mathbf{b} = \mathbf{X}^+\mathbf{y}$. Two $\boldsymbol{\omega}$'s are of immediate interest to perturb, namely $\boldsymbol{\omega} = \mathbf{X}$ and $\boldsymbol{\omega} = \mathbf{y}$. A third, $\boldsymbol{\omega} = E(\mathbf{y}) = \mathbf{X}\boldsymbol{\beta}$, will also be considered.

$\boldsymbol{\omega} = \mathbf{X}$

As is usual in a conditioning analysis, three elements must be specified: the sets Ω and Λ^* and the relation $\boldsymbol{\lambda} = \mathbf{f}(\boldsymbol{\omega})$ (from which we get $\delta\boldsymbol{\lambda} \equiv \mathbf{g}(\delta\boldsymbol{\omega})$). When $\mathbf{X}$ is perturbed, these quantities are taken to be

$$\Omega = \{\delta\mathbf{X} : \|\delta\mathbf{X}\|/\|\mathbf{X}\| \leqslant m_1\} \qquad (11.23)$$

$$\Lambda^* = \{\delta\mathbf{b} : \|\delta\mathbf{b}\|/\|\mathbf{b}\| \leqslant m_2\} \qquad (11.24)$$

and

$$\delta\mathbf{b} = \mathbf{g}(\delta\mathbf{X}) = (\mathbf{X} + \delta\mathbf{X})^+\mathbf{y} - \mathbf{X}^+\mathbf{y}, \qquad (11.25)$$

where m_1 and m_2 are picked to be relevant to the underlying context according to the previous discussion. If the range Λ of $\delta\mathbf{b}$ given by (11.25) based on the Ω of (11.23) contains any element not in Λ^*, the least-squares estimate is ill conditioned with respect to Ω. Thus, if reasonable relative changes in the $\mathbf{X}$ matrix can produce unreasonable relative changes in the estimate, the estimate is said to be ill conditioned. This is, of course, precisely the notion of ill conditioning as it affects the least-squares estimate that has been used somewhat informally throughout the previous chapters. To determine whether the least-squares estimate is ill conditioned in any particular instance, we must calculate

$$\sup_{\delta\mathbf{X}\in\Omega} \|\delta\mathbf{b}\|/\|\mathbf{b}\|. \tag{11.26}$$

Should this quantity be larger than m_2, $\Lambda \not\subset \Lambda^*$, and $\mathbf{b}$ is ill conditioned. In practice (11.26) is not easily evaluated, but by (11.22) with only $\mathbf{X}$ perturbed, it is known to be bounded from above by

$$m_1\kappa(\mathbf{X})\hat{R}^{-1}[2 + (1 - \hat{R}^2)^{1/2}\kappa(\mathbf{X})] \tag{11.27}$$

and may in fact be equal to (11.27) for some $\mathbf{X}$, $\delta\mathbf{X}$, and $\mathbf{y}$. Thus, as a rough guide to the conditioning of $\mathbf{b}$ with respect to Ω, the quantity $2m_1\kappa(\mathbf{X})\hat{R}^{-1}$ could be compared to m_2. If it is much larger than m_2, then $\mathbf{b}$ will be said to be ill conditioned. This guide is particularly good—and becomes essentially $2m_1\kappa(\mathbf{X})$—when the fit is good, that is, when $\hat{R}$ is near unity. But it could understate the extent of ill conditioning when the fit is poor and the $\kappa^2(\mathbf{X})$ term dominates. In any event, we note that the condition number $\kappa(\mathbf{X})$ is an important multiplicative factor, it being possible, for example, that a 1% ($m_1 = 0.01$) relative change in $\mathbf{X}$ could produce a $\kappa(\mathbf{X})$ percent change in $\|\delta\mathbf{b}\|/\|\mathbf{b}\|$. And this is precisely the reason the condition number has been so useful in the previous chapters in this regard.

$\boldsymbol{\omega} = \mathbf{y}$

The second quantity to be perturbed is $\mathbf{y}$, where Ω, Λ^*, and $\mathbf{g}(\delta\mathbf{y})$ are as follows:

$$\Omega = \{\delta\mathbf{y} : \|\delta\mathbf{y}\|/\|\mathbf{y}\| \leqslant m_1\}, \tag{11.28}$$

$$\Lambda^* = \{\delta\mathbf{b} : \|\delta\mathbf{b}\|/\|\mathbf{b}\| \leqslant m_2\}, \tag{11.29}$$

$$\delta\mathbf{b} = \mathbf{g}(\delta\mathbf{y}) = \mathbf{X}^{+}\delta\mathbf{y}. \tag{11.30}$$

By proceeding in a manner entirely analogous to the preceding, the relevant bound again becomes $2m_1\kappa(\mathbf{X})\hat{R}^{-1}$, in which the condition number $\kappa(\mathbf{X})$ remains an important factor.

$\boldsymbol{\omega} = E(\mathbf{y})$

Consider now a third perturbation for this estimator, again involving $\mathbf{y}$. This time, however, we do not perturb $\mathbf{y}$ about observed values but rather about its expected value $E(\mathbf{y}) = \mathbf{X\beta}$. Since the perturbations are taken about $\mathbf{X\beta}$, it is reasonable, when constructing the set Ω, to take into account elements of the stochastic mechanism that generates $\mathbf{y}$. That is, $\delta\mathbf{y}$ could be taken to be equal to a possible $\boldsymbol{\varepsilon}$ of (11.21). In this case

$$\Omega = \{\delta\mathbf{y} : \|\delta\mathbf{y}\|/\|\mathbf{X\beta}\| \equiv \|\boldsymbol{\varepsilon}\|/\|\mathbf{X\beta}\| \leqslant m_1\}. \tag{11.31}$$

Since, in this instance $\|\delta\mathbf{y}\|/\|\mathbf{X\beta}\| \equiv \|\boldsymbol{\varepsilon}\|/\|\mathbf{X\beta}\|$ has the dimensions of the inverse of signal-to-noise, m_1^{-1} could be chosen to be the minimum "signal-to-noise" ratio expected to be encountered in model (11.21). The set Λ^* is now taken to be

$$\Lambda^* = \{\delta\mathbf{b} : \|\delta\mathbf{b}\|/\|\boldsymbol{\beta}\| \leqslant m_2\}. \tag{11.32}$$

To distinguish the basis for this conditioning analysis from the others, it proves useful to call it a *stochastically based* conditioning analysis.

Since $\mathbf{y}$ has been taken to be $\mathbf{X\beta}$ and $\mathbf{y} + \delta\mathbf{y} = \mathbf{X\beta} + \boldsymbol{\varepsilon}$, we have

$$\delta\mathbf{b} = \mathbf{X}^+(\mathbf{y} + \delta\mathbf{y}) - \mathbf{X}^+\mathbf{y} = \mathbf{X}^+\delta\mathbf{y} = \hat{\boldsymbol{\beta}} - \boldsymbol{\beta}, \tag{11.33}$$

where $\hat{\boldsymbol{\beta}}$ is the least-squares estimate based on the realization $\mathbf{X\beta} + \boldsymbol{\varepsilon}$. Therefore, $\|\delta\mathbf{b}\|^2 = (\hat{\boldsymbol{\beta}} - \boldsymbol{\beta})^T(\hat{\boldsymbol{\beta}} - \boldsymbol{\beta})$ is the squared error of the least-squares estimator, and hence determining whether $\mathbf{b}$ is ill conditioned in this situation is equivalent to determining whether, for those possible or probable realizations of $\boldsymbol{\varepsilon}$, the maximal squared error of the resulting least-squares estimator can exceed some amount $m_2\|\boldsymbol{\beta}\|$. A guide to the possibility of such a high squared error again results from applying (11.22) to (11.33), yielding

$$\|\delta\mathbf{b}\|/\|\boldsymbol{\beta}\| = \|\hat{\boldsymbol{\beta}} - \boldsymbol{\beta}\|/\|\boldsymbol{\beta}\| \leqslant 2\kappa(\mathbf{X})\|\delta\mathbf{y}\|/\|\mathbf{X\beta}\| = 2\kappa(\mathbf{X})\|\boldsymbol{\varepsilon}\|/\|\mathbf{X\beta}\|, \tag{11.34}$$

where again $\hat{R} = 1$ in this application.

For $\delta\mathbf{y} \in \Omega$, the last quantity in (11.34) is less than $2m_1\kappa(\mathbf{X})$, and again we see the dominant role played by the condition number $\kappa(\mathbf{X})$ in determining estimator conditioning; the larger is $\kappa(\mathbf{X})$, the greater the possible size of the squared error of the least-squares estimates.

Criterion Conditioning

In least squares, the criterion to be minimized is

$$Q(\mathbf{X}, \mathbf{Y}, \boldsymbol{\beta}) = (\mathbf{y} - \mathbf{X\beta})^T(\mathbf{y} - \mathbf{X\beta}). \tag{11.35}$$

In this case, the Taylor series expansion of (11.19) is exact, and (11.18) can be easily evaluated. Taking Ω and Λ^* as in (11.15) and (11.16), where now $\boldsymbol{\theta} = \boldsymbol{\beta}$, $\hat{\boldsymbol{\theta}} = \hat{\boldsymbol{\beta}} = \mathbf{X}^+\mathbf{y}$, and $\delta Q = \mathbf{g}(\delta\hat{\boldsymbol{\beta}}) = 2\,\delta\hat{\boldsymbol{\beta}}^T(\mathbf{X}^T\mathbf{X})\delta\hat{\boldsymbol{\beta}}$, (11.18) simply becomes $\kappa^2(\mathbf{X})$, and so the criterion (11.35) is ill conditioned with respect to Ω if $\kappa^2(\mathbf{X})$ is greater than m_2. Once again, it is $\kappa(\mathbf{X})$ that provides information on the ill conditioning in the LS/linear case.

Nonlinearities

We now have a general notion of conditioning that can be used to show the relevance of at least three forms of conditioning: data, estimator, and criterion conditioning. And at least in the special LS/linear case, it is seen that the condition number coincidentally provides important information for assessing all three types of conditioning. Further, the results of Section 9.1 can be directly interpreted to show that at least one form of nonlinearity, logarithms, can be normalized so that its conditioning admits to a similar analysis. However, it is readily seen that this is not true for all forms of nonlinearities and that, rather generally, the simple expedient of a condition number is not always available for assessing the conditioning of a given problem. The several simple examples that follow illustrate the types of problems that can arise and the divergences among the different types of conditioning that can occur when nonlinearities (in variables and/or parameters) are allowed.

Consider first the two orthogonal vectors $\mathbf{X}_1 = (1, 1, 1, 1)^{\mathrm{T}}$ and $\mathbf{X}_2 = (-1, 1, -1, 1)^{\mathrm{T}}$, which for the analysis at hand are considered to be contextually interpretable, and therefore basic, data. Being orthogonal, these basic data would be very suitable for estimating the linear (in both parameters and variables) model

$$\mathbf{y} = \beta_1 \mathbf{X}_1 + \beta_2 \mathbf{X}_2 + \boldsymbol{\varepsilon} \tag{11.36}$$

but would be useless if the model were

$$\mathbf{y} = \beta_1 \mathbf{X}_1 + \beta_2 \mathbf{X}_2^2 + \boldsymbol{\varepsilon}. \tag{11.37}$$

They would again become suitable for estimating

$$\mathbf{y} = \beta_1 \mathbf{X}_1^2 + \beta_2 \mathbf{X}_2^3 + \boldsymbol{\varepsilon} \tag{11.38}$$

but not for

$$\mathbf{y} = \beta_1 \mathbf{X}_2 + \beta_2 \mathbf{X}_1 \mathbf{X}_2 + \boldsymbol{\varepsilon}. \tag{11.39}$$

From this we see that, in the assessment of estimator conditioning when there are nonlinearities, both the data and the nature of the model must be considered. The divergence between the perfect data conditioning and the very imperfect estimator conditioning in the context of estimating models (11.37) and (11.39) is clear.

Further problems can arise when there are nonlinearities in the parameters.

Consider

$$\mathbf{y} = \beta_1 \mathbf{X}_1 + \beta_2 \mathbf{X}_2^{\alpha} + \boldsymbol{\varepsilon}. \tag{11.40}$$

For the basic data $\mathbf{X}_1$ and $\mathbf{X}_2$ as before, it is clear that these data would be suitable for estimation if $\alpha = 1$ but could be problemful if $\alpha = 2$. Unfortunately, α must be estimated, and hence any measure of the suitability of the $\mathbf{X}$ data must depend on an estimate $\hat{\alpha}$ of α. This is further complicated by the fact that even if $\alpha = 1$, there is no guarantee that $\hat{\alpha}$ will be near 1. Indeed, in most cases there will still be a nonzero probability that $\hat{\alpha}$ will be arbitrarily close to 2, leaving one with uncertain hopes for correctly assessing the conditioning, say, by using the condition number of $\mathbf{X} \equiv [\mathbf{X}_1 \;\; \mathbf{X}_2^{\hat{\alpha}}]$. In such cases, assessment of conditioning may require the introduction of prior information on α. Thus, if one's prior information puts α strongly in the vicinity of 2, these data would be ill conditioned, whereas they are well conditioned relative to a prior that places α in a neighborhood of 1.

An Example of a Computational Alternative

We have seen that it is possible to provide a mathematical solution to the conditioning problem in the LS/linear case—$\kappa(\mathbf{X})$, for example, figures prominently in a theoretically derived bound on the sensitivity of $\|\delta\mathbf{b}\|/\|\mathbf{b}\|$ to changes in $\mathbf{X}$ or $\mathbf{y}$. However, such convenient solutions are not generally available: either the mathematics becomes too cumbersome or the breadth of perturbations deemed relevant invalidates simple approximations through a few terms of a Taylor expansion. In this section we suggest and exemplify a more generally applicable computational alternative that, in principle, can be applied in any situation.

A Computational Alternative

We recall that any conditioning analysis consists in determining the triple $K = \{\mathbf{f}, \Omega, \Lambda^*\}$, where Ω is the set of a priori "reasonable" perturbations and Λ^* is the set of a priori reasonable responses. The term *reasonable* usually denotes "inconsequential to the analysis at hand," so that the objective becomes one of determining whether inconsequential perturbations $\delta\boldsymbol{\omega} \in \Omega$ can result in consequential responses $\delta\boldsymbol{\lambda} \notin \Lambda^*$. A straightforward, albeit computer-intensive, way of conducting such an analysis is to select elements $\delta\boldsymbol{\omega}$ randomly from Ω, calculate their corresponding $\delta\boldsymbol{\lambda} = \mathbf{g}(\delta\boldsymbol{\omega})$, and check to see if any fall outside Λ^*.

This procedure has many advantages. First, it is universally applicable. So long as the conditioning triple K can be defined, the method can be employed. Second, for highly ill-conditioned problems, a few random picks for $\delta\boldsymbol{\omega}$ from Ω should suffice to find a response $\delta\boldsymbol{\lambda}$ lying outside Λ^*. Experience with the use of this procedure overwhelmingly supports this. Third, it allows for complete generality in the way perturbations are defined. This point is especially relevant in practice and deserves some added discussion. In the mathematical solution to the conditioning of the LS/linear case given before, it is necessary to assume that all perturbations are in terms of relative shifts $\|\delta\boldsymbol{\omega}\|/\|\boldsymbol{\omega}\|$ and that the appropriate measure is a Euclidean (or some) vector norm. Such restrictions will often severely strain the structural interpretability of the resulting perturbations. Quite possibly, the appropriate structurally interpretable perturbations will differ for different elements of $\boldsymbol{\omega}$, as we shall see in the example to follow. Fourth, in analyzing nonlinear models **f**, this computational alternative does not rely on linear approximations. Thus, it can properly accommodate large perturbations $\delta\boldsymbol{\omega}$ that are reasonable for the problem at hand ($\delta\boldsymbol{\omega} \in \Omega$) but too large for a Taylor approximation.

The method, of course, has drawbacks. Like any computationally intensive procedure, it can be expensive. Of greater import is the fact that, whereas the method seems readily to show the presence of ill conditioning, it cannot (without a complete examination of Ω) demonstrate the absence of ill conditioning. At best, after many draws from Ω, a reasonable presumption may be allowed to the statement that ill conditioning is absent.

An Example

As an illustration of the suggested computational method, consider again a conditioning analysis of the U.S. consumption function. In this case, however, we shall specify the model **f** nonlinearly as

$$C_t = \beta_1 C_{t-1}^{\beta_2} \mathrm{DPI}_t^{\beta_3} r_t^{\beta_4} \left(\frac{\mathrm{DPI}_t}{\mathrm{DPI}_{t-1}}\right)^{\beta_5} + \varepsilon_t, \qquad t = 1, \ldots, n, \tag{11.41}$$

where, with a slight change in the time-scripting notation that proves useful in what follows, the variates are as defined in Section 5.4.

Here, the consumption function used in Section 5.4 is given a Cobb–Douglas form with an additive error. This function differs from the linear-in-the-logs formulation often employed in econometric analysis only in the specification of an additive rather than a multiplicative error. This alteration, however, makes (11.41) an essentially nonlinear function: one incapable of simple transformation into a form amenable to linear estimation and a standard conditioning analysis. Had the error been multiplicative, the methods of Section 7.1 for models with logged variates would be directly applicable to a logged transform of (11.41). Forms like (11.41) have tended to be ignored in econometric studies less on the grounds of economic plausibility than the added complications necessitated by nonlinear estimation.

To complete the specification of a conditioning analysis, we must specify Ω and Λ^*. This is readily done here, since each of these economic time series is in a structurally interpretable form. Their magnitudes, and therefore the changes in their magnitudes, can be meaningfully assessed as being large or small through prior knowledge of the underlying economic phenomena they measure.

Thus, perturbations δC_t and $\delta \mathrm{DPI}_t$ that are within $\pm 0.1\%$ (± 0.001) of C_t and DPI_t, respectively, can be considered small. Not only are such magnitudes of little macroeconomic consequence, but also they would be perceived by most economists as lying within the bounds of measurement error. For these two variates, relative perturbations make sense. By contrast, we assume that perturbations δr_t that lie in an interval of one basis point (± 0.05 of a percentage point) are reasonably considered small in measuring interest rates. For this variate we are assuming that additive perturbations make sense. Thus, Ω is chosen as

$$\Omega = \{(\delta C_t,\ \delta \mathrm{DPI}_t,\ \delta r_t)\ \forall t:\ \delta C_t \in \pm 0.1\% \text{ of } C_t,\ \delta \mathrm{DPI}_t \in \pm 0.1\% \text{ of } \mathrm{DPI}_t,\ \delta r_t \in \pm 0.05\}. \tag{11.42}$$

To pick Λ^*, we merely state that a relative response to such perturbations by any coefficient estimate in excess of 10% is too large (e.g., would yield a substantively different policy analysis.)

The sensitivity analysis is now straightforward. First, estimate (11.41) with nonlinear least squares (NLS) using the basic data $\boldsymbol{\omega} = [\mathbf{C}, \mathbf{DPI}, \mathbf{r}]$ to obtain base estimates $\mathbf{b}$. Then repeatedly reestimate (11.41) with perturbed data $\boldsymbol{\omega} + \delta\boldsymbol{\omega}$ determined by random draws $\delta\boldsymbol{\omega}$ from Ω given in (11.42). This can be accomplished through uniform selections ($t = 1, \ldots, n$) from

$$\begin{aligned} \delta C_t &\sim U(0.999C_t,\ 1.001C_t), \\ \delta \mathrm{DPI}_t &\sim U(0.999\mathrm{DPI}_t,\ 1.001\mathrm{DPI}_t), \\ \delta r_t &\sim U(-0.05,\ 0.05). \end{aligned} \tag{11.43}$$

Each reestimation produces a new estimate $\mathbf{b}_i^*$ and a resulting $\delta\mathbf{b}_i \equiv \mathbf{b}_i^* - \mathbf{b}$, and our interest centers on whether any of the $\delta\mathbf{b}_i/\mathbf{b}$ fall outside the 10% level chosen for Λ^*. Exhibit 11.1 shows the extreme results over 30 replications for each of the parameters in (11.41).

The base estimates $\mathbf{b}$ are shown in column 1. These estimates are completely compatible with the estimates of the analogous linear model (5.3) analyzed in Section 5.4. Column 2 shows the range of the perturbed estimates over the 30

Exhibit 11.1 Extreme Responses to Random Perturbations in Ω: Nonlinear Consumption Function (11.41)

Coefficient and Base Estimate, $\mathbf{b}$	Range of $\delta\mathbf{b}$	Maximal Percentage Increase	Maximal Percentage Decrease
$b_1 = 0.975$	0.064	3.8	2.8
$b_2 = 0.130$	0.081	24.6	37.8
$b_3 = 0.867$	0.080	5.2	4.0
$b_4 = -0.022$	0.012	32.1	22.9
$b_5 = 0.097$	0.081	44.4	39.3

replications, and columns 3 and 4 show, respectively, the largest percentage increase and the largest percentage decrease for the particular coefficient. It is clear that the 10% target level for Λ^* has been exceeded in both directions for b_2, b_4, and b_5, while both b_3 and the constant b_1 (or scale factor) seem more stably determined. These results are wholly consonant with the conditioning analysis for the analogous linear model of the consumption function given in Section 5.4. Indeed, the same patterns of instability are exhibited.

We can also plot scatter diagrams showing how the instability in the estimate of one coefficient relates to that of another over the different perturbations. Such scatter plots are given in Exhibit 11.2. The tight dependency pairs (b_2, b_3), (b_1, b_4), (b_2, b_5), and (b_3, b_5) draw immediate attention, each showing that instability in the estimate of one of the pair tends to be accompanied by covariant instability in the other. It should be noted, however, that the proportionate variation in b_1 and b_3 is considerably less than for the other estimates.

These scatter plots provide useful auxiliary information to a conditioning analysis and are, in this day and age, quickly acquired. They give similar information for two-dimensional relations that one gets from the variance–decomposition proportion matrices and auxiliary regressions of the standard collinearity diagnostics, as will be clear from examining Exhibit 5.12, where all of these covariant pairs can be seen. Unlike the variance–decomposition proportion matrices, however, these two-dimensional scatter plots can (but need not) overlook joint dependencies involving three or more parameter estimates. But at the same time, the scatter plots can provide visual indications of nonlinear dependencies (as possibly between b_4 and b_5) and bifurcated dependencies (as between b_2 and b_5 or between b_3 and b_5) that could never be seen in a table of variance–decomposition proportions.

Conclusion

A conditioning analysis, then, is a sensitivity analysis carefully constructed to guarantee that its results relate meaningfully to the problem at hand through the appropriate selection of Ω and Λ^*. Such a conditioning analysis can be directed at many interesting elements of a given statistical analysis, as exemplified by data, estimator, and criterion conditioning, all of which provide the analyst with useful and important information in assessing different aspects of the validity and reliability of the statistical results.

In some circumstances, conditioning can be assessed mathematically, through the derivation of some measure that bounds potential sensitivity. This is seen to be the case for the LS/linear problem, for which the condition number $\kappa(\mathbf{X})$ applied to structurally interpretable data conveniently provides the needed measure for all three: data, estimator, and criterion conditioning.

In more general (e.g., nonlinear) contexts, however, such mathematically derived bounds need not be forthcoming. But in these cases it should always be possible to investigate any form of conditioning empirically using the computationally intensive method described.

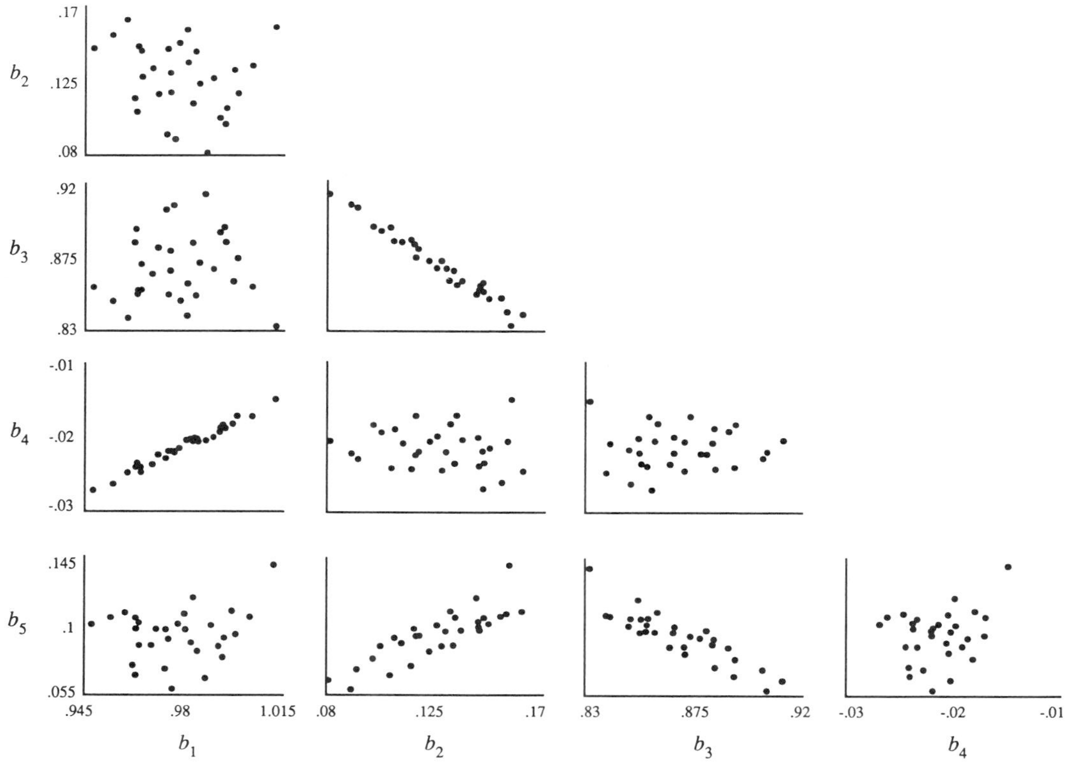

Exhibit 11.2 Scatterplot matrix of coefficient estimates.

11.2 A MORE GENERAL CONDITIONING DIAGNOSTIC

We turn now to another suggestion for a conditioning diagnostic that can be applied quite generally—to nonlinear estimators as well as to simultaneous-equations estimators. The detailed behavior of this diagnostic has yet to be fully investigated and constitutes a good area for research. It is based on an obvious generalization of the standard collinearity diagnostics of the previous chapters.

Consider, then, the general model

$$\mathbf{g}(\mathbf{Z}, \boldsymbol{\theta}, \mathbf{u}) = \mathbf{0}, \tag{11.44}$$

where $\mathbf{g}$ can be a vector-valued nonlinear stochastic function in the p parameters $\boldsymbol{\theta}$, $\mathbf{u}$ is an n-vector error term, and $\mathbf{Z}$ is an $n \times k$ matrix of k variates, which may be endogenous or exogenous. This formulation encompasses nonlinear and simultaneous-equations models. In this latter case, it is assumed that the system has been "stacked" into its single-equation equivalent with identifying restrictions.[2] In the event that interest focuses directly on $\boldsymbol{\theta}$, we can rewrite (11.44) more simply as $\mathbf{g}(\boldsymbol{\theta})$.

Let

$$\hat{\boldsymbol{\theta}} = \boldsymbol{\phi}(\mathbf{Z}) \tag{11.45}$$

be any consistent estimator of $\boldsymbol{\theta}$ with variance–covariance matrix $\mathbf{V}(\hat{\boldsymbol{\theta}})$, and let $\mathbf{S}(\hat{\boldsymbol{\theta}})$ be any consistent estimator of $\mathbf{V}(\hat{\boldsymbol{\theta}})$.

In the LS/linear case, (11.44) takes the familiar form

$$\mathbf{y} = \mathbf{X}\boldsymbol{\beta} + \boldsymbol{\varepsilon}, \qquad \mathbf{V}(\hat{\boldsymbol{\theta}}) = \sigma^2(\mathbf{X}^T\mathbf{X})^{-1}, \qquad \mathbf{S}(\hat{\boldsymbol{\theta}}) = s^2(\mathbf{X}^T\mathbf{X})^{-1}, \tag{11.46}$$

where $s^2 = \mathbf{e}^T\mathbf{e}/(n-p)$. And we know from above and from the previous chapters that, in this case, conditioning can be assessed by the $\tilde{\eta}_i(\mathbf{X})$, the scaled condition indexes of $\mathbf{X}$, and that variate involvement can be discerned through the variance–decomposition proportions of $\mathbf{X}$. Equivalently, as we have seen in Chapter 3, these diagnostics can be determined from $\mathbf{X}^T\mathbf{X}$. The scaled condition indexes are simply the ratios $(\lambda_{\max}/\lambda_i)^{-1/2}$—positive square roots of ratios of eigenvalues of $\mathbf{X}^T\mathbf{X}$—, and the variance–decomposition proportions are derived by applying (3.23) and (3.24) directly to the matrix $\mathbf{V}$ of the eigenvectors of $\mathbf{X}^T\mathbf{X}$ and $\mu_i = \lambda_i^{-1/2}$. Thus, in the LS/linear case, the conditioning diagnostics are obtainable from $\mathbf{S}^{-1}(\mathbf{b}) = s^{-2}\mathbf{X}^T\mathbf{X}$ after $\mathbf{X}$ has been column equilibrated, so that the diagonal elements of $\mathbf{X}^T\mathbf{X}$ are all equal. The scale factor of s^{-2} is irrelevant since it cancels from all the ratios determining the condition indexes and the variance–decomposition proportions.

[2]See Theil (1971, Chapter 10).

A Suggested Diagnostic

This, then, suggests using $\mathbf{S}^{-1}(\hat{\boldsymbol{\theta}})$ rather generally as a basis for a conditioning diagnostic. Specifically, relative to (11.44) and (11.45), consider the following steps:

1. Obtain $\hat{\boldsymbol{\theta}}$ and its estimated variance–covariance matrix $\mathbf{S}(\hat{\boldsymbol{\theta}})$.
2. Scale $\mathbf{S}^{-1}(\hat{\boldsymbol{\theta}})$ symmetrically so that it has unit diagonal elements; that is, determine $\mathbf{W} \equiv \mathbf{D}^T\mathbf{S}^{-1}(\hat{\boldsymbol{\theta}})\mathbf{D}$, where $\mathbf{D} \equiv \text{diag}(d_1, \ldots, d_p)$, and d_i is the inverse of the positive square root of the ith diagonal element of $\mathbf{S}^{-1}(\hat{\boldsymbol{\theta}})$.
3. Obtain the eigenvalues λ_i and eigenvectors $\mathbf{V}$ of $\mathbf{W}$.
4. Form the condition indexes $\tilde{\eta}_i = (\lambda_{\max}/\lambda_i)^{-1/2}$, $i = 1, \ldots, p$, and the variance–decomposition proportions as in (3.23) and (3.24) with $\mu_i = \lambda_i^{-1/2}$.

This procedure is clearly a proper generalization of the diagnostics of the preceding chapters, reducing to them in the LS/linear case. The results of step 4 can be brought together into a $\boldsymbol{\Pi}$ matrix identical in form to those used earlier. Care, however, must be taken in interpreting just what these figures mean, and we will be better able to see this after examining a special case.

Gauss–Newton with a Linearized Model

In the case that (11.44) takes the often-specified form

$$\mathbf{g}(\mathbf{Z}, \boldsymbol{\theta}) = \mathbf{u}, \tag{11.47}$$

$\boldsymbol{\theta}$ can be estimated through a minimum-distance (least-squares) approach that finds the $\hat{\boldsymbol{\theta}}$ that, given $\mathbf{Z}$, minimizes $\mathbf{u}^T\mathbf{u} = \mathbf{g}^T(\boldsymbol{\theta})\mathbf{g}(\boldsymbol{\theta})$. One method for obtaining this solution is that of Gauss–Newton, using the first two terms of a Taylor expansion of (11.47) to form a linear approximation to the model about $\boldsymbol{\theta}^\circ$ as

$$\mathbf{z} = \mathbf{J}(\boldsymbol{\theta}^\circ)\boldsymbol{\theta} - \mathbf{u}, \tag{11.48}$$

where $\mathbf{z} \equiv \mathbf{J}(\boldsymbol{\theta}^\circ)\boldsymbol{\theta}^\circ - \mathbf{g}(\boldsymbol{\theta}^\circ)$ and $\mathbf{J}(\boldsymbol{\theta}^\circ) \equiv (\partial\mathbf{g}/\partial\boldsymbol{\theta}^\circ)$, the $n \times p$ Jacobian matrix of $\mathbf{g}$ with respect to $\boldsymbol{\theta}$ evaluated at $\boldsymbol{\theta}^\circ$.[3] It is well known that iteratively reestimating (11.48) with least squares as $\boldsymbol{\theta}_r = (\mathbf{J}^T(\boldsymbol{\theta}_{r-1})\mathbf{J}(\boldsymbol{\theta}_{r-1}))^{-1}\mathbf{J}^T(\boldsymbol{\theta}_{r-1})\mathbf{z}_r$ results in $\hat{\boldsymbol{\theta}}$.[4] Further, the estimated variance–covariance matrix at this solution is simply $s^2(\mathbf{J}^T(\hat{\boldsymbol{\theta}})\mathbf{J}(\hat{\boldsymbol{\theta}}))^{-1}$, which, as long as the derivatives in $\mathbf{J}$ are predetermined, can be shown to be a consistent estimator of $\mathbf{V}(\hat{\boldsymbol{\theta}})$. Thus, asymptotically, any $\mathbf{S}(\hat{\boldsymbol{\theta}})$ used

[3] See Bard (1974) or Judge et al. (1982, Chapter 24).
[4] See Judge et al. (1982, Chapter 24).

in the diagnostic suggested above is approximated by $s^2(\mathbf{J}^T(\hat{\boldsymbol{\theta}})\mathbf{J}(\hat{\boldsymbol{\theta}}))^{-1}$, and the conditions of estimation are approximated linearly by

$$\mathbf{z} = \mathbf{J}(\hat{\boldsymbol{\theta}})\boldsymbol{\theta} - \mathbf{u}. \tag{11.49}$$

Interpreting the Diagnostics

The preceding means that, in the context of (11.47), the diagnostics obtained by the procedure suggested above are approximations to the standard collinearity diagnostics applied to (11.49). That is, they tell us something about the conditioning of the pseudodata that comprise the Jacobian matrix $\mathbf{J}(\hat{\boldsymbol{\theta}})$, specifically, the number and strengths of the near dependencies that exist in $\mathbf{J}(\hat{\boldsymbol{\theta}})$ and the involvement of the various columns of $\mathbf{J}(\hat{\boldsymbol{\theta}})$ in those near dependencies. These pseudodata may or may not be structurally interpretable; they are, after all, derivatives of various nonlinear relations involving the $\mathbf{Z}$ variates that are themselves likely to be the structurally interpretable forms, and we have seen that such transformations often destroy structural interpretability. Thus, rather generally, more care must be taken in interpreting this information. However, the information obtained is by no means useless. For we have seen that estimation of (11.47) is akin to estimating (11.49) by least squares, and hence data weaknesses, however they arise, that can degrade this least-squares estimation can affect all estimators and are worth knowing about.

Here, data weakness does not refer, for example, to collinearity or short data among the basic $\mathbf{Z}$ variates but rather to collinearity or short data among the columns of $\mathbf{J}(\hat{\boldsymbol{\theta}})$, the result of the particular nonlinear transformations that $\mathbf{Z}$ undergoes in the formulation (11.47). Thus, the individual variance–decomposition proportions in the $\boldsymbol{\Pi}$ matrix cannot be associated with a specific $\mathbf{Z}$ variate, but they still can be associated with a specific estimated parameter variance—in the terms of the preceding section, even if these diagnostics are not directly suitable for assessing data conditioning, they remain suitable for assessing estimator conditioning. Even so, high variance–decomposition proportions linked up with high condition indexes are still symptomatic that the information on the $\mathbf{Z}$ data, after undergoing the transformations imposed by the given model, are unable to be used effectively for estimating certain of its parameters. Furthermore, if it is possible to associate specific $\mathbf{Z}$'s with particular columns of $\mathbf{J}(\hat{\boldsymbol{\theta}})$ that are involved in near dependencies, then it becomes possible to see more clearly why the existing $\mathbf{Z}$ data are inadequate and what corrective steps are most likely to prove useful.

The magnitudes of the scaled condition indexes that result from the suggested diagnostic can be more meaningfully interpreted as diagnostics for data conditioning if information can be obtained regarding structurally interpretable changes in the columns of $\mathbf{J}(\hat{\boldsymbol{\theta}})$. Let us suppose that the $\mathbf{Z}$'s are structurally interpretable, so that we can determine a set $\Omega_{\mathbf{z}}$ of reasonable perturbations. Further, let us write $\mathbf{J}(\hat{\boldsymbol{\theta}})$ as $\mathbf{J}(\hat{\boldsymbol{\theta}}, \mathbf{Z})$ to highlight its dependence on the basic $\mathbf{Z}$

data. Then perturbations $\delta\mathbf{Z} \in \Omega_z$ map into a set $\Omega_\mathbf{J}$ of induced reasonable perturbations in $\mathbf{J}(\hat{\boldsymbol{\theta}}, \mathbf{Z})$, namely, $\Omega_\mathbf{J} = \{\delta\mathbf{J} \equiv \mathbf{J}(\hat{\boldsymbol{\theta}}, \mathbf{Z} + \delta\mathbf{Z}) - \mathbf{J}(\hat{\boldsymbol{\theta}}, \mathbf{Z}) : \delta\mathbf{Z} \in \Omega_\mathbf{Z}\}$. It is the implied relative changes in the set $\Omega_\mathbf{J}$ that are best used in interpreting the magnitude of the scaled condition indexes along the lines described in Section 6.4.

It should be noted that the interpretative considerations derived from the Gauss–Newton approximation technically hold only if the Jacobian matrix $\mathbf{J}(\hat{\boldsymbol{\theta}})$ is predetermined, that is, only if the derivatives in $\mathbf{J}(\hat{\boldsymbol{\theta}})$ are not functions of currently endogenous variates. The diagnostic procedure suggested above, however, can in principle be applied to any appropriate simultaneous-equations estimator.

11.3 DIAGNOSING THE DEGREE OF EFFECTIVE IDENTIFICATION

In Chapter 10 it was noted that collinearity was akin to the identification problem. We exploit this relation here to derive a measure of the degree of effective identification for simultaneous systems of equations, making use of the measure for signal-to-noise introduced in Chapter 7.

Motivation

Of course, an equation from a linear system of equations is either identifiable or not, due to zero restrictions, depending on whether the rank condition holds. But the rank condition reflects a property of the actual conditional distribution of the endogenous variates given the exogenous variates, and as such, it is equivalent to having an infinite sample. As such, it cannot inform us of any practical problems that one might encounter in estimating a given equation with a specific, finite data set. This point can be seen readily from the following simple example. Suppose the equation

$$\mathbf{y}_1 = \gamma_1 + \gamma_2\mathbf{y}_2 + \beta_1\mathbf{x}_1 + \boldsymbol{\varepsilon} \tag{11.50}$$

is part of a simultaneous system of equations, where the $\mathbf{y}_i$'s are endogenous and $\mathbf{x}_1$ is exogenous. Further, assume this equation is identifiable. Now suppose that $\mathbf{x}_2$ is an exogenous variate excluded from this equation and that it is the only other exogenous variate in the system. All of the identifying information, then, is embodied in $\mathbf{x}_2$. Now, if $\mathbf{x}_1$ and $\mathbf{x}_2$ just happen to be highly collinear in the particular data set available for estimating (11.50), it is clear that although $\mathbf{x}_2$ is excluded in theory, most of its information is already contained in $\mathbf{x}_1$—and indeed, in the limiting case of perfect correlation between $\mathbf{x}_1$ and $\mathbf{x}_2$, all of its information is so contained—, and so it is not very "excluded" in fact. Thus, despite its being identifiable in theory, the given equation is poorly identified in

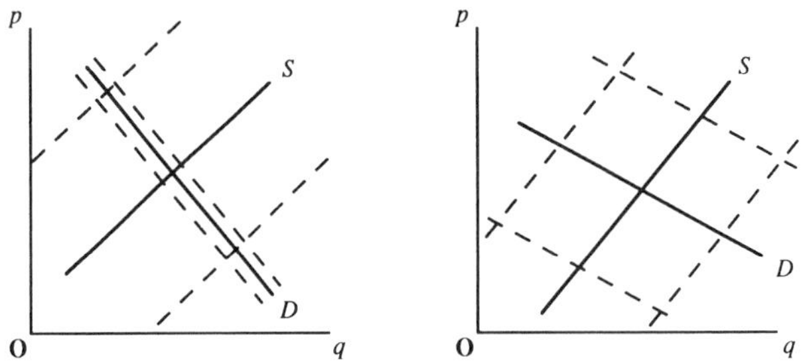

Exhibit 11.3 (*a*) Strong effective identification of the demand function *D*. (*b*) Weak effective identification of the demand function **D**.

practice, and the greater the degree of collinearity between $\mathbf{x}_1$ and $\mathbf{x}_2$, the less the degree of effective identification.

This could occur, for example, in a supply-and-demand analysis in which (11.50) represents the demand equation with endogenous variates price and quantity and an exogenous variate $\mathbf{x}_1$ representing tastes. We suppose the accompanying supply equation contains the same two endogenous variates along with an exogenous variate $\mathbf{x}_2$ representing weather. If now, for whatever reasons, the particular data set were characterized by a high degree of collinearity between weather and tastes—perhaps consumer's optimism is affected by the amount of rain—, then these two effects cannot be well separated, and there is no real identifying information to allow estimation to proceed with precision.[5]

Correlation between $\mathbf{x}_1$ and $\mathbf{x}_2$ is not the only reason why there may be ineffective identifying information contained in the excluded exogenous variates. It may be the case, for example, that the noise inherent in the error term of the given equation is so great that the signal available from the identifying excluded exogenous variates is overshadowed. Again considering the familiar textbook case of supply and demand, one recalls that the demand equation becomes identifiable when its excluded exogenous variates are able to shift the supply equation strongly in relation to it, as in Exhibit 11.3*a*.

The dotted lines here indicate the bounds of probable variation, and hence values for **p** and **q** will tend to lie in the intersection of these two dotted regions. In the situation of Exhibit 11.3*a*, the data on **p** and **q** will be clustered in a narrow band around the demand equation, allowing it to be well estimated. If, however, the additive noise in the demand equation *D* is great, then even the presence of additional exogenous variation in the supply equation *S* need not permit effective identification of *D* for small samples, as seen in Exhibit 11.3*b*. Here, the data on **p** and **q** will be dispersed about the large area bounded by the dotted

[5]Sims (1980, p. 6) provides a closely related concept when he writes, "When the strictly exogenous variables have low explanatory power, estimates of the endogenous-on-exogenous regressions are likely to be subject to great sampling error, and the identification may be said to be weak."

lines, displaying little of the shape of the demand curve, even though it is technically identifiable.

The degree of effective identification, then, has to do with whether or not the excluded (identifying) exogenous variates provide adequate signal-to-noise, either by themselves or over and above that already embodied in the included exogenous variates, to allow estimation to proceed with precision, that is, to allow identification to take place as a practical matter. This suggests the relevance of the measure for signal-to-noise given in Chapter 7 as a means for assessing the degree of effective identification.

The Diagnostic Measure

We begin with a very brief review of the relevant econometric background. Readers unfamiliar with the standard results regarding the identification and estimation of simultaneous systems are referred to such texts as Theil (1971) or Johnston (1984).

Consider, then, the simultaneous system of linear equations

$$\mathbf{Y\Gamma} + \mathbf{XB} + \mathbf{U} = \mathbf{0}, \tag{11.51}$$

where $\mathbf{Y}$ is an $n \times g$ matrix of g endogenous variates, $\mathbf{X}$ is an $n \times k$ matrix of k predetermined variates, $\mathbf{U}$ is an $n \times g$ matrix of error terms having independent rows with mean zero and nonsingular variance–covariance matrix $\mathbf{\Sigma}$, and $\mathbf{\Gamma}$ and $\mathbf{B}$ are, respectively, $g \times g$ and $g \times k$ matrices of parameters to be estimated.

We shall be concerned with one of these equations, which we assume to be identifiable. When normalized and written with all zero restrictions, this equation is

$$\mathbf{y} = \mathbf{Y}_1\boldsymbol{\gamma}_1 + \mathbf{X}_1\boldsymbol{\beta}_1 + \mathbf{u}, \tag{11.52}$$

where $\mathbf{y}$ is the endogenous variate with respect to which the equation is normalized, $\mathbf{Y}_1$ is the set of $g_1 - 1$ included endogenous variates, $\mathbf{X}_1$ is the set of k_1 included exogenous variates, $\mathbf{u}$ is the appropriate column of $\mathbf{U}$, and $\boldsymbol{\gamma}_1$ and $\boldsymbol{\beta}_1$ are the appropriate nonzero—more properly, not known a priori to be zero—elements from $\mathbf{\Gamma}$ and $\mathbf{B}$ after normalization. In addition, there are the matrices of g_2 excluded endogenous variates $\mathbf{Y}_2$ and k_2 excluded exogenous variates $\mathbf{X}_2$ such that $\mathbf{X} \equiv [\mathbf{X}_1\ \mathbf{X}_2]$ and $\mathbf{Y} \equiv [\mathbf{y}\ \mathbf{Y}_1\ \mathbf{Y}_2]$, $g = g_1 + g_2$, and $k = k_1 + k_2$.

Postmultiplying (11.51) by $\mathbf{\Gamma}^{-1}$, which is assumed to exist, produces the reduced form

$$\mathbf{Y} = \mathbf{X\Pi} + \mathbf{V}, \tag{11.53}$$

where $\mathbf{\Pi} \equiv -\mathbf{B\Gamma}^{-1}$ and $\mathbf{V} \equiv -\mathbf{U\Gamma}^{-1}$. When partitioned commensurately with the partitions of **Y** and **X** given above, this becomes

$$[\mathbf{y}\ \mathbf{Y}_1\ \mathbf{Y}_2] = [\mathbf{X}_1\ \mathbf{X}_2]\begin{bmatrix}\boldsymbol{\Pi}_{10} & \boldsymbol{\Pi}_{11} & \boldsymbol{\Pi}_{12}\\ \boldsymbol{\Pi}_{20} & \boldsymbol{\Pi}_{21} & \boldsymbol{\Pi}_{22}\end{bmatrix} + [\mathbf{v}\ \mathbf{V}_1\ \mathbf{V}_2]. \tag{11.54}$$

Thus, we can write

$$\mathbf{Y}_1 = \mathbf{X}_1\boldsymbol{\Pi}_{11} + \mathbf{X}_2\boldsymbol{\Pi}_{21} + \mathbf{V}_1, \tag{11.55}$$

where $\boldsymbol{\Pi}_{11}$ is $k_1 \times (g_1 - 1)$ and $\boldsymbol{\Pi}_{21}$ is $k_2 \times (g_1 - 1)$. The assumption that (11.52) is identifiable is equivalent to the condition, known as the rank condition, that this latter matrix $\boldsymbol{\Pi}_{21}$ has full rank $g_1 - 1 \leqslant k_2$. The relevance of this condition is simply seen by noting that, for the columns of $\boldsymbol{\Pi} \equiv -\mathbf{B}\boldsymbol{\Gamma}^{-1}$ corresponding to the given equation, one has $\boldsymbol{\Pi}\boldsymbol{\gamma} = -\boldsymbol{\beta}$, or partitioned with the appropriate zero restrictions for the omitted variates $\mathbf{Y}_2$ and $\mathbf{X}_2$,

$$\begin{bmatrix}\boldsymbol{\Pi}_{10} & \boldsymbol{\Pi}_{11} & \boldsymbol{\Pi}_{12}\\ \boldsymbol{\Pi}_{20} & \boldsymbol{\Pi}_{21} & \boldsymbol{\Pi}_{22}\end{bmatrix}\begin{pmatrix}-1\\ \boldsymbol{\gamma}_1\\ \mathbf{0}\end{pmatrix} = -\begin{pmatrix}\boldsymbol{\beta}_1\\ \mathbf{0}\end{pmatrix}. \tag{11.56}$$

This gives the two equations

$$\begin{aligned}-\boldsymbol{\Pi}_{10} + \boldsymbol{\Pi}_{11}\boldsymbol{\gamma}_1 &= -\boldsymbol{\beta}_1,\\ -\boldsymbol{\Pi}_{20} + \boldsymbol{\Pi}_{21}\boldsymbol{\gamma}_1 &= \mathbf{0}.\end{aligned} \tag{11.57}$$

Clearly, $\boldsymbol{\gamma}_1$ is uniquely derivable from the second equation, which is the essence of identifiability, if and only if $\boldsymbol{\Pi}_{21}$ has full rank. Then $\boldsymbol{\beta}_1$ is derivable from the first equation.

So, substituting (11.55) into (11.52) and gathering terms gives

$$\begin{aligned}\mathbf{y} &= \mathbf{X}_1(\boldsymbol{\Pi}_{11}\boldsymbol{\gamma}_1 + \boldsymbol{\beta}_1) + \mathbf{X}_2\boldsymbol{\Pi}_{21}\boldsymbol{\gamma}_1 + \boldsymbol{\xi}\\ &= \mathbf{X}_1\boldsymbol{\delta}_1 + \tilde{\mathbf{X}}_2\boldsymbol{\gamma}_1 + \boldsymbol{\xi},\end{aligned} \tag{11.58}$$

where $\boldsymbol{\xi} = \mathbf{u} + \mathbf{V}_1\boldsymbol{\gamma}_1 \equiv \mathbf{v}$, $\boldsymbol{\delta}_1 = \boldsymbol{\Pi}_{11}\boldsymbol{\gamma}_1 + \boldsymbol{\beta}_1$, and $\tilde{\mathbf{X}}_2 \equiv \mathbf{X}_2\boldsymbol{\Pi}_{21}$. In the form of the last equality, we see that all of the identifying information—which is contained in the excluded exogenous variates $\mathbf{X}_2$—is introduced through $\tilde{\mathbf{X}}_2 \equiv \mathbf{X}_2\boldsymbol{\Pi}_{21}$, which is directly associated with $\boldsymbol{\gamma}_1$. And it is here that we can see the relation between true identifiability and the degree of effective identification due to data weaknesses. Clearly, estimation of $\boldsymbol{\gamma}_1$ in (11.58) cannot take place if $\tilde{\mathbf{X}}_2$ is of less than full rank (or more generally, if $\tilde{\mathbf{X}}_2$ is ill conditioned). And $\tilde{\mathbf{X}}_2$ will be of less than full rank if either $\boldsymbol{\Pi}_{21}$ is rank deficient or $\mathbf{X}_2$ is. The rank deficiency of $\boldsymbol{\Pi}_{21}$ is directly associated with identifiability, for (11.52) is identifiable if and only if $\boldsymbol{\Pi}_{21}$ is of full rank, a condition that has nothing to do with the particular data set

[Y X] available for estimation. However, $\tilde{\mathbf{X}}_2 \equiv \mathbf{X}_2\mathbf{\Pi}_{21}$ will also be of less than full rank (or ill conditioned) if $\mathbf{X}_2$ is, a condition that is clearly directly related only to the particular data available for estimation and having nothing directly to do with the rank of $\mathbf{\Pi}_{21}$. Thus, both unidentifiability and the ill conditioning of $\mathbf{X}_2$ have the same effect in rendering (11.52) inestimable.

That $\boldsymbol{\gamma}_1$ should be the focus of the identifying information is quite understandable, since identification really has only to do with the ability to determine the coefficients of the included endogenous variates. In the discussion surrounding (11.57), we saw that identifiability hinges first on the ability to solve for $\boldsymbol{\gamma}_1$ uniquely from the second equation. The solution for $\boldsymbol{\beta}_1$ then follows directly from the first equation. From the point of view of estimation this can be seen by noting that, once estimates $\hat{\boldsymbol{\gamma}}_1$ are obtained for $\boldsymbol{\gamma}_1$, estimation of $\boldsymbol{\beta}_1$ can proceed by least squares of (11.52) in the form of $\mathbf{y} - \mathbf{Y}_1\hat{\boldsymbol{\gamma}}_1 = \mathbf{X}_1\boldsymbol{\beta}_1 + \mathbf{u}$.[6]

Given that the equation is identifiable, then, it is clear as a practical matter that identification will be effective only as long as the $\tilde{\mathbf{X}}_2$ are not weak data, that is, as long as the columns of $\tilde{\mathbf{X}}_2$ are not short data or involved in strong collinear relations, either among themselves or with the columns of $\mathbf{X}_1$. But as we have seen in Chapter 7, this is precisely the condition that the estimator of $\boldsymbol{\gamma}_1$ possess adequate signal-to-noise. Thus, if we knew $\mathbf{\Pi}_{21}$ (and hence $\tilde{\mathbf{X}}_2$), Equation (11.58) would be amenable to estimation by least squares, and we could apply the test for adequate signal-to-noise given in Section 7.4 directly to the least-squares estimate $\hat{\boldsymbol{\gamma}}_1$ of $\boldsymbol{\gamma}_1$ in order to provide a test for effective identification. Remember, the issue here is not identifiability; Equation (11.52) is assumed to be identifiable. The issue is whether there is adequate novel information in the excluded exogenous variates to be able effectively to exploit this identifiability.

Unfortunately, we do not typically know $\mathbf{\Pi}_{21}$, but fortunately this suggested test for effective identification can take place without knowing $\mathbf{\Pi}_{21}$. Rewrite (11.58) as

$$\mathbf{y} = \mathbf{X}_1\boldsymbol{\delta}_1 + \mathbf{X}_2\boldsymbol{\delta}_2 + \boldsymbol{\xi}, \tag{11.59}$$

where $\boldsymbol{\delta}_2 \equiv \mathbf{\Pi}_{21}\boldsymbol{\gamma}_1$. This equation is seen to be that from the reduced form (11.54) corresponding to the variate $\mathbf{y}$. And under the identifying restrictions, we see that the least-squares estimator $\hat{\boldsymbol{\delta}}_2$ of $\boldsymbol{\delta}_2$ must be degenerately distributed by a degree exactly equal to the degree of overidentification $k_2 - g_1 + 1$, since $E\hat{\boldsymbol{\delta}}_2 = \boldsymbol{\delta}_2 \equiv \mathbf{\Pi}_{21}\boldsymbol{\gamma}_1$. And therefore,

$$\mathbf{V}(\hat{\boldsymbol{\delta}}_2) = \mathbf{\Pi}_{21}\mathbf{V}(\hat{\boldsymbol{\gamma}}_1)\mathbf{\Pi}_{21}^{\mathrm{T}}, \tag{11.60}$$

where $\mathbf{V}(\hat{\boldsymbol{\gamma}}_1)$ is the variance–covariance matrix of the least-squares estimator $\hat{\boldsymbol{\gamma}}_1$ of $\boldsymbol{\gamma}_1$ in (11.58) and where $\mathbf{V}(\hat{\boldsymbol{\delta}}_2)$ is the $k_2 \times k_2$ variance–covariance matrix of $\hat{\boldsymbol{\delta}}_2$,

[6]Compare this with the least generalized variance ratio estimator (LGVRE) as derived in Goldberger (1964, p. 338).

whose rank is $g_1 - 1 \leqslant k_2$ (since, we recall, that the identifiability of (11.52) assumes $\mathbf{\Pi}_{21}$ to have full rank equal to $g_1 - 1$).

Because of (11.60), we see that the signal-to-noise τ^2 for $\hat{\boldsymbol{\gamma}}_1$, for whose adequacy we wish to test, is exactly that for $\hat{\boldsymbol{\delta}}_2$, that is,

$$\begin{aligned}\tau^2 &\equiv \boldsymbol{\gamma}_1^{\mathrm{T}}\mathbf{V}^{-1}(\hat{\boldsymbol{\gamma}}_1)\boldsymbol{\gamma}_1 \\ &= \boldsymbol{\gamma}_1^{\mathrm{T}}\mathbf{\Pi}_{21}^{\mathrm{T}}\mathbf{V}^{+}(\hat{\boldsymbol{\delta}}_2)\mathbf{\Pi}_{21}\boldsymbol{\gamma}_1 \\ &= \boldsymbol{\delta}_2^{\mathrm{T}}\mathbf{V}^{+}(\hat{\boldsymbol{\delta}}_2)\boldsymbol{\delta}_2,\end{aligned} \tag{11.61}$$

where $\mathbf{V}^{+}(\hat{\boldsymbol{\delta}}_2)$ is the generalized inverse of $\mathbf{V}(\hat{\boldsymbol{\delta}}_2)$ from (11.60). The first equality in (11.61) is obtained as follows:

Lemma 11.1. Let $\mathbf{P}$ be an $m \times n$ matrix of full rank $n \leqslant m$. Then $\mathbf{P}^{+}\mathbf{P} = \mathbf{I}$.

Proof. From the singular-value decomposition, $\mathbf{P} = \mathbf{UDV}^{\mathrm{T}}$ and $\mathbf{P}^{+} = \mathbf{VD}^{+}\mathbf{U}^{\mathrm{T}} = \mathbf{VD}^{-1}\mathbf{U}^{\mathrm{T}}$. Hence $\mathbf{P}^{+}\mathbf{P} = \mathbf{VD}^{-1}\mathbf{U}^{\mathrm{T}}\mathbf{UDV}^{\mathrm{T}} = \mathbf{I}$. □

Lemma 11.2. Let $\mathbf{A} = \mathbf{PBP}^{\mathrm{T}}$, where $\mathbf{P}$ is as in Lemma 11.1 and $\mathbf{B}$ is nonsingular. Then $\mathbf{A}^{+} = \mathbf{P}^{+\mathrm{T}}\mathbf{B}^{-1}\mathbf{P}^{+}$ and $\mathbf{B}^{-1} = \mathbf{P}^{\mathrm{T}}\mathbf{A}^{+}\mathbf{P}$.

Proof. We must verify that $\mathbf{P}^{+\mathrm{T}}\mathbf{B}^{-1}\mathbf{P}^{+}$ satisfies the four conditions given in Section 3.1 that define the generalized inverse:

(i) $(\mathbf{PBP}^{\mathrm{T}})(\mathbf{P}^{+\mathrm{T}}\mathbf{B}^{-1}\mathbf{P}^{+})(\mathbf{PBP}^{\mathrm{T}}) = \mathbf{PBB}^{-1}\mathbf{BP}^{\mathrm{T}} = \mathbf{PBP}^{\mathrm{T}}$, since by virtue of Lemma 11.1, $\mathbf{P}^{\mathrm{T}}\mathbf{P}^{+\mathrm{T}} = (\mathbf{P}^{+}\mathbf{P})^{\mathrm{T}} = \mathbf{I}$.

(ii) $(\mathbf{P}^{+\mathrm{T}}\mathbf{B}^{-1}\mathbf{P}^{+})(\mathbf{PBP}^{\mathrm{T}})(\mathbf{P}^{+\mathrm{T}}\mathbf{B}^{-1}\mathbf{P}^{+}) = \mathbf{P}^{+\mathrm{T}}\mathbf{B}^{-1}\mathbf{BB}^{-1}\mathbf{P}^{+} = \mathbf{P}^{+\mathrm{T}}\mathbf{B}^{-1}\mathbf{P}^{+}$.

(iii) $(\mathbf{PBP}^{\mathrm{T}})(\mathbf{P}^{+\mathrm{T}}\mathbf{B}^{-1}\mathbf{P}^{+}) = \mathbf{PP}^{+}$ is symmetric by virtue of $\mathbf{P}^{+}$'s being a generalized inverse.

(iv) $(\mathbf{P}^{+\mathrm{T}}\mathbf{B}^{-1}\mathbf{P}^{+})(\mathbf{PBP}^{\mathrm{T}}) = \mathbf{P}^{+\mathrm{T}}\mathbf{P}^{\mathrm{T}} = (\mathbf{PP}^{+})^{\mathrm{T}}$ is symmetric for the same reason. □

Thus, the desired measure of signal-to-noise that is unattainable in the form of the first equality in (11.61)—because this required knowing $\tilde{\mathbf{X}}_2$ and, hence, $\mathbf{\Pi}_{21}$ to conduct the least-squares estimation of (11.58)—is attainable in the form of the final equality, which needs only to estimate (11.59) by least squares. So we may test for the adequacy of the signal-to-noise for $\boldsymbol{\gamma}_1$, and hence the degree of effective identification, by testing for adequacy of the signal-to-noise of $\boldsymbol{\delta}_2$ from the least-squares estimation of an appropriately partitioned reduced-form equation (11.59).

There will be one minor modification in the direct application of the techniques of Section 7.4 in this context. Because the rank of $\mathbf{V}^{+}(\hat{\boldsymbol{\delta}}_2)$ is

$g_1 - 1 \leqslant k_2$, the k_2-vector $\boldsymbol{\delta}_2$ will generally be degenerately distributed, and hence we have

$$(\hat{\boldsymbol{\delta}}_2 - \boldsymbol{\delta}_2)^{\mathrm{T}}\mathbf{V}^{+}(\hat{\boldsymbol{\delta}}_2)(\hat{\boldsymbol{\delta}}_2 - \boldsymbol{\delta}) \overset{A}{\sim} \chi^2_{g_1 - 1}, \tag{11.62}$$

a chi-squared distribution with $g_1 - 1$ degrees of freedom. Hence the appropriate threshold of adequacy used in (7.20) will be

$$\tau^2_\gamma \equiv {}_\gamma\chi^2_{g_1 - 1}, \tag{11.63}$$

rather than ${}_\gamma\chi^2_{k_2}$, the difference being the degree of overidentification. This means that the appropriate critical values for this test for the degree of effective identification are determined from Exhibits 7.7–7.10 with $p_2 = g_1 - 1$ and $n - p = n - k$.

Thus, the test for adequate signal-to-noise provides the basis for a measure of the degree of effective identification. The higher the γ for which it is possible to accept adequate signal-to-noise (the closer γ is to 1), the higher the degree of effective identification. Furthermore, tests for adequate signal-to-noise for subsets of $\boldsymbol{\gamma}_1$ (subsets of $\boldsymbol{\delta}_1$) can be interpreted as tests for those coefficients for which there is effective identification versus those for which there is not. This information is useful in determining where additional identifying information, say, through the introduction of prior information, would be most effective in correcting weak identification through zero restrictions. In a related vein, those excluded exogenous variates corresponding to elements of $\boldsymbol{\delta}_2$ that display the greatest signal-to-noise are clearly those that are the best instruments to employ in instrumental-variable estimation, say in the case of an undersized sample, when two-stage least squares cannot proceed.

11.4 DIAGNOSTICS TO CHOOSE BETWEEN TWO AND THREE STAGES OF LEAST SQUARES

In this section we make use of the condition number, along with two other measures of multiple linear dependencies, as a diagnostic for helping to choose between two-stage least squares (2SLS) and three-stage least squares (3SLS) when estimating a simultaneous system of equations.[7] This is a straightforward application of the conditioning diagnostics. Again, readers unfamiliar with estimation of simultaneous systems, and the 2SLS and 3SLS estimators in particular, are referred to any standard econometrics text, such as Theil (1971) or Johnston (1984).

In estimating a system of simultaneous equations, the question often arises whether to use two or three stages of least squares. Two-stage least squares is computationally cheaper, and whereas 3SLS is known asymptotically to be

[7]The material in this section draws freely upon Belsley (1988a). (Reprinted by permission of Kluwer Academic Publishers.

more efficient, this need not be so for small samples. Thus, 3SLS becomes the estimator of choice only when (1) the researcher considers a gain in efficiency to be important relative to computational cost and (2) the potential for such a gain is high. In this section, three measures for this potential are suggested: the condition number, the minimum singular value, and the determinant of the disturbance correlation matrix. These measures are at once both simple and more general than the previously suggested diagnostic: high pairwise correlations.

That 3SLS need not possess greater efficiency than 2SLS for small samples is readily motivated as follows. It is well known that 2SLS and 3SLS are equivalent when there is no cross-equation covariation.[8] The asymptotic efficiency of 3SLS arises, then, from exploiting nonzero cross-equation covariation. In practice, of course, samples are finite, and this cross-equation covariation must be estimated. Thus, when the true but unknown cross-equation covariation is small, it can be more efficient to impose the restriction that it is zero, which is what 2SLS does, than to use an estimate of it, as does 3SLS. As a result, one would expect 2SLS to be more efficient when the cross-equation covariation is small and for 3SLS to become more worthwhile as this covariation becomes larger. Indeed, for a two-equation simultaneous system, Mikhail (1975) demonstrates with Monte Carlo studies that 2SLS has the smaller mean-square error when the between-equation correlation is $\rho = .18$, but 3SLS becomes the winner when $\rho = .76$.

It would seem, then, that estimates of the cross-equation correlation coefficients should indicate when 3SLS is likely to be worthwhile. And indeed, high cross-equation correlations are sufficient indicators of this condition, but they are not necessary. For, as we shall see, in larger systems of equations, it is quite possible for all cross-equation correlations to be small even though the equations' error terms are tightly linked through more general multiple correlations. Of course, 2SLS continues to ignore these more general relations, while 3SLS does not. Hence, 3SLS can have a greater small-sample efficiency than 2SLS even when pairwise correlations are small. A more general measure than pairwise correlations is therefore needed if one is to be apprised of when this is so, and we suggest several such generalizations here.

First, we show the inadequacy of pairwise correlations for indicating the presence of more general correlations among the error terms. Then we discuss three possible measures that are more appropriate indicators of the presence and strength of these multiple relations. Finally, Monte Carlo results are described that demonstrate the efficacy of these suggested diagnostics.

Inadequacy of Pairwise Correlations

Consider again the general linear model (11.51) of g simultaneous equations. The rows $\mathbf{u}^{\mathrm{T}}(t)$, $t = 1, \ldots, n$, of the $n \times g$ error matrix $\mathbf{U}$ are g-vectors that are assumed to be independently distributed with mean $\mathbf{0}$ and variance–covariance

[8] See Theil (1971).

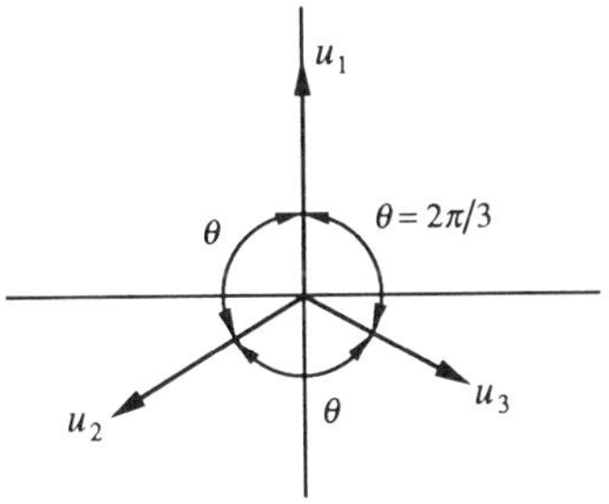

Exhibit 11.4 Pairwise correlations among three perfectly linearly related random variables.

matrix $\boldsymbol{\Sigma}$. The off-diagonal elements of $\boldsymbol{\Sigma}$ are the cross-equation covariances. Let us denote by $\mathbf{P}$ the correlation matrix corresponding to $\boldsymbol{\Sigma}$, that is, $\mathbf{P} \equiv \mathbf{D}^T\boldsymbol{\Sigma}\mathbf{D}$, where $\mathbf{D}^{-1}$ is the diagonal matrix whose diagonal elements are those of $\boldsymbol{\Sigma}^{1/2}$.

Our immediate goal is simply to show that it is possible for the off-diagonal elements of $\mathbf{P}$ (the pairwise correlations between the elements of $\mathbf{u}$) to be small even when there is mcre generally a strong multiple correlation among the elements of $\mathbf{u}$ (some element of $\mathbf{u}$ is highly correlated with some linear combination of several other elements of $\mathbf{u}$). This demonstration is important, of course, to show that the off-diagonal elements of $\mathbf{P}$ are a sufficient but not necessary indicator of the potential for 3SLS.

Assume, then, that a perfect linear relation exists among the elements of $\mathbf{u}$; that is, there is some $\mathbf{c} \neq \mathbf{0}$ such that $\text{var}(\mathbf{c}^T\mathbf{u}) = 0$. A little reflection will convince the reader that, under these conditions, we can minimize the largest absolute pairwise correlation among the elements of $\mathbf{u}$ by making them all equal. This is seen for $g = 3$ in Exhibit 11.4, where the elements of $\mathbf{u} \equiv (u_1, u_2, u_3)$ are plotted. The perfect linear relation among the u's is depicted by their lying in a two-dimensional space, and the pairwise correlations between the elements of $\mathbf{u}$ are depicted by the angles between them. It is clear that, starting from the equal-angle situation depicted, no one correlation may be made smaller without making another one larger. Thus, the largest absolute pairwise correlation here may be as small as $.5 = |\cos 2\pi/3|$.

Indeed, more generally, we find for the g-vector $\mathbf{u}$ that it is possible for there to be a perfect linear relation among its elements while the largest absolute pairwise correlation can be as small as $1/(g-1)$. Thus, if there are 20 equations in the system, there could be perfect cross-equation covariation even though no two equations' error terms are absolutely correlated by more than $\frac{1}{19} \approx .05$.

This result is readily seen by noting that the equi-correlation matrix

$$\mathbf{P}^*(\rho) \equiv \begin{bmatrix} 1 & \rho & \rho & \cdots & \rho \\ \rho & 1 & \rho & \cdots & \rho \\ \rho & \rho & 1 & \cdots & \rho \\ \vdots & \vdots & \vdots & \ddots & \vdots \\ \rho & \rho & \rho & \cdots & 1 \end{bmatrix} \tag{11.64}$$

is (a) singular for $\rho = 1$ and $\rho = -1/(g-1)$, and (b) positive definite for $-1/(g-1) < \rho < 1$. Both of these properties follow directly from the fact that the g eigenvalues of $\mathbf{P}^*(\rho)$ are $1 - \rho$ (with multiplicity $g - 1$) and $1 + (g-1)\rho$, all of which are positive for $-1/(g-1) < \rho < 1$ and some of which are zero for $\rho = 1$ and $\rho = -1/(g-1)$.[9]

Three Indicators for 3SLS

From the preceding, we know that small absolute pairwise correlations between the error terms in a simultaneous system of equations need not indicate the adequacy of 2SLS; more general relations may still exist among the elements of $\mathbf{u}$ that could give 3SLS the edge. To detect these more general dependencies, a measure more general than pairwise correlations is needed. Three such measures can be suggested, all based on the error correlation matrix $\mathbf{P}$: the determinant $\det(\mathbf{P})$, the smallest eigenvalue $\lambda_{\min}$ of $\mathbf{P}$, and the condition number $\kappa(\mathbf{P})$.

det(P)

Clearly, $\det(\mathbf{P}) = 0$ if and only if there is some $\mathbf{c} \neq \mathbf{0}$ such that $\text{var}(\mathbf{c}^T\mathbf{u}) = 0$. This situation defines quite generally what we have loosely been calling a *perfect linear relation* among the elements of $\mathbf{u}$. Also, we note that $\det(\mathbf{P}) = 1$ when there is no correlation of any sort among the elements of $\mathbf{u}$, for here $\mathbf{P} = \mathbf{I}$. In general, $0 \leqslant \det(\mathbf{P}) \leqslant 1$. The first inequality follows simply from $\mathbf{P}$'s being nonnegative definite. The second follows since $\det(\mathbf{P}) = \Pi_i\lambda_i$ and $\text{tr}(\mathbf{P}) = \Sigma_i\lambda_i = g$, where the λ_i are the eigenvalues of $\mathbf{P}$.[10] It is straightforward to show that the maximum of a product of nonnegative magnitudes whose sum is constant occurs when their values are equal, and in this instance, this means each $\lambda_i = 1$. Thus $\det(\mathbf{P})$ is an index of general multiple correlation, a fact that we employed in Section 7.6 in discussing the four causes of low signal-to-noise. The closer is $\det(\mathbf{P})$ to zero, the tighter the existence of some general correlation among the elements of $\mathbf{u}$. The closer is $\det(\mathbf{P})$ to unity, the smaller any such general level of correlation.

λ_{min}

A second indicator arises from the recognition that a tight correlation among the elements of $\mathbf{u}$ occurs if there exists a unit vector $\mathbf{c}$ such that $\text{var}(\mathbf{c}^T\mathbf{u})$ is small. Without normalization, however, this notion has a major drawback: for a given $\mathbf{c}$, without changing any of the angles (pairwise correlations) between the elements of $\mathbf{u}$, we can make $\text{var}(\mathbf{c}^T\mathbf{u})$ as small or as large as we wish merely by rescaling the u_i's. To normalize the problem, we consider $\text{var}(\mathbf{c}^T\boldsymbol{\xi})$

[9] See Rao (1973, p. 53).

[10] These relations are readily derived from the singular-value decomposition of $\mathbf{P}$. Because of the symmetry of $\mathbf{P}$, we have $\mathbf{P} = \mathbf{VDV}^T$. Hence, $\det(\mathbf{P}) = \det(\mathbf{VDV}^T) = \det(\mathbf{D}) = \Pi_i\lambda_i$ due to the orthogonality of $\mathbf{V}$ and the diagonality of $\mathbf{D}$. And $g = \text{tr}(\mathbf{P}) = \text{tr}(\mathbf{VDV}^T) = \text{tr}(\mathbf{DV}^T\mathbf{V}) = \text{tr}(\mathbf{D}) = \Sigma_i\lambda_i$. The fact that $\text{tr}(\mathbf{P}) = g$ follows since $\mathbf{P}$ is a correlation matrix with ones along the diagonal.

of the standardized variables $\boldsymbol{\xi} = (\xi_1, \xi_2, \xi_3) \equiv \{u_1/\sqrt{\text{var}(u_1)},\ u_2/\sqrt{\text{var}(u_2)},\ u_3/\sqrt{\text{var}(u_3)}\}$. Our interest is in examining when the minimum of

$$\text{var}(\mathbf{c}^{\mathrm{T}}\boldsymbol{\xi}) = \mathbf{c}^{\mathrm{T}}\mathbf{P}\mathbf{c} \tag{11.65}$$

is small subject to $\mathbf{c}^{\mathrm{T}}\mathbf{c} = 1$. The solution to this problem, we recall from Section 3.1, is the familiar result for eigenvalues, namely that the minimum of (11.65) subject to $\mathbf{c}^{\mathrm{T}}\mathbf{c} = 1$ is $\lambda_{\min}$, the minimum eigenvalue of $\mathbf{P}$. Thus, a tight correlation exists among the elements of $\mathbf{u}$ when $\lambda_{\min}$ is close to zero. Furthermore, it is readily shown that $\lambda_{\min} \leqslant 1$ and that the upper bound of 1 is assumed if and only if $\mathbf{P} = \mathbf{I}$, that is, if and only if there is an absence of any cross-equation correlation. This first result follows from the facts that $\text{tr}(\mathbf{P}) = \Sigma_i \lambda_i = g$ and that $\lambda_i > 0$ for all i; the second reflects the necessary existence of an orthogonal matrix $\mathbf{C}$ giving $\mathbf{C}^{\mathrm{T}}\mathbf{P}\mathbf{C} = \boldsymbol{\Lambda} \equiv \text{diag}(\lambda_1, \ldots, \lambda_g)$. Thus, $\lambda_{\min}$ also behaves like a measure of general correlation, ranging between zero and unity, with the extremes reflecting perfect correlation and noncorrelation, respectively.

κ(P)

A third measure of general multiple relations is the now familiar condition number of $\mathbf{P}$,

$$\kappa(\mathbf{P}) \equiv \frac{\lambda_{\max}}{\lambda_{\min}} \geqslant 1. \tag{11.66}$$

The motivation for $\kappa(\mathbf{P})$ in this context is similar to that just given for $\lambda_{\min}$ except here we accept the presence of a general multiple correlation among the elements of $\mathbf{u}$ (equivalently $\boldsymbol{\xi}$) if there are two unit vectors $\mathbf{c}_1$ and $\mathbf{c}_2$ such that $\text{var}(\mathbf{c}_1^{\mathrm{T}}\boldsymbol{\xi}) \ll \text{var}(\mathbf{c}_2^{\mathrm{T}}\boldsymbol{\xi})$. We know from Section 3.1 that (11.65) is minimized at $\lambda_{\min}$ and maximized at $\lambda_{\max}$. Hence, a strong multiple correlation among the elements of $\mathbf{u}$ occurs when $\kappa(\mathbf{P})$ is large. Further, it is clear that $\kappa(\mathbf{P})$ assumes its minimum value 1 when $\mathbf{P} = \mathbf{I}$.[11]

The parallel between the use of the condition number here and that in the collinearity diagnostics of Section 3.1 is immediately clear. The standardization of the u_i's to the ξ_i's used above is analogous to the column equilibration used there. And indeed, the argument of Section 3.5 can be used directly to show that, among all the possible scalings of the u_i's, that which produces the correlation matrix $\mathbf{P}$ is the one that is guaranteed to produce a condition number that is nearly minimum and therefore most meaningful for our current needs. Rather

[11]It is worth noting that no simple transformation of $\kappa(\mathbf{P})$ provides a meaningful correlation-like behavior. One might try to base such a transform on a single-parameter matrix like $\mathbf{P}^*(\rho)$ defined in (11.64). Recalling that the roots of $\mathbf{P}^*(\rho)$ are $1 - \rho$ with a multiplicity $g - 1$ and $1 + (g - 1)\rho$, for $\rho > 0$ we have $\kappa(\mathbf{P}^*(\rho)) = (1 + (g - 1)\rho)/(1 - \rho)$, while for $\rho < 0$, $\kappa(\mathbf{P}^*(\rho)) = (1 - \rho)/(1 + (g - 1)\rho)$. Only for $g = 2$ is $\mathbf{P}^*(\rho)$ always the actual correlation matrix $\mathbf{P}$ and does a symmetric relation exist between $\kappa(\mathbf{P})$ and ρ, namely, $\kappa(\mathbf{P}) = (1 + |\rho|)/(1 - |\rho|)$ or $\rho = (\kappa(\mathbf{P}) - 1)/(\kappa(\mathbf{P}) + 1)$.

more generally, the dualism that exists between a variable space like that used here for the u_i's and an observation space like that used in Chapter 3 for the $\mathbf{X}_i$'s is examined in Dempster (1969).

The detailed behavior of these three indicators in differing situations is a matter of empirical research yet to be done. Some specifics are, however, clear. For example, $\lambda_{\min}$ necessarily provides the least information since it is based only on one of **P**'s eigenvalues. By contrast, det(**P**) is the product of all **P**'s eigenvalues, but this measure can become very small through the joint presence of several modest eigenvalues, no one of which need be very small. And finally, $\kappa(\mathbf{P})$ must necessarily tell different stories about situations that would be treated the same by $\lambda_{\min}$. Using the same information that showed $\lambda_{\min} \leqslant 1$, we can show that $\lambda_{\max} \leqslant g$, For a given $\lambda_{\min}$, then, since $\Sigma_i \lambda_i = g$, $\kappa(\mathbf{P})$ must be larger the greater the number of other small λ's, that is, the greater the number of different multiple correlations that coexist among the elements of **u**.

In any event, it is reasonable to suppose that these several measures, based as they are on the correlation matrix **P** of the error structure of the system (11.51), are more general indicators of the potential for gains in efficiency through using 3SLS. Presumably, the closer is any one of them to indicating a strong general correlation among the u_i's, the greater is this potential regardless of the absolute magnitudes of the pairwise correlations.

A Monte Carlo Experiment

Here we conduct a Monte Carlo experiment to demonstrate the previously described phenomena, namely, (1) that 3SLS can have greater small-sample efficiency than 2SLS despite low cross-equation correlations so long as a strong, more general correlation exists and (2) that this situation can be effectively assessed by the indicators suggested above. No attempt at completeness is made here. It is the purpose of this study only to demonstrate these phenomena and to suggest tentative conclusions. A final section for speculative considerations, however, suggests those areas where further research will be most fruitful.

The Model

The following six-equation model is employed in the Monte Carlo experiment:

$$\begin{aligned}
y_1 &= 5 + 0.3y_2 + 0.5y_6 - 0.6x_1 + 1.0x_2 + u_1,\\
y_2 &= 10 + 0.5y_1 - 0.3x_3 + 0.7x_4 + u_2,\\
y_3 &= -6 - 0.4y_1 + 0.2y_4 + 1.0x_5 + 0.5x_6 + u_3,\\
y_4 &= 40 + 0.5y_3 - 2.0x_2 + 0.6x_3 + u_4,\\
y_5 &= -9 + 0.4y_3 - 0.3y_6 + 0.6x_4 + 0.3x_5 + u_5,\\
y_6 &= 15 + 0.2y_5 + 0.7x_1 + 0.2x_6 + u_6,
\end{aligned} \tag{11.67}$$

with $\mathbf{u} \sim N_6(\mathbf{0}, \boldsymbol{\Sigma})$.

This model was chosen to be large enough ($g = 6$) to be interesting, to insure each equation to be overidentified, and to be close to a model that could occur in common econometric practice.

The Experiment

The equi-correlation matrix introduced in (11.64),

$$\mathbf{P}^*(\rho) = (1 - \rho)\mathbf{I} + \rho\iota\iota^T, \tag{11.68}$$

affords an excellent test environment for this study, for we can move smoothly from a situation where the u_i's are completely independent when $\rho = 0$ to one where they possess a single perfect multiple correlation as ρ goes to $-1/(g - 1) = -.2$, recalling here that $g = 6$. Thus, the largest absolute pairwise correlation never exceeds .2 while all the other indicators attain their extreme values: $\lambda_{\min} \to 0$, $\det(\mathbf{P}^*(\rho)) \to 0$, and $\kappa(\mathbf{P}^*(\rho)) \to \infty$. In the following experiments, then, nine values of ρ are chosen to span this range: 0, $-.02$, $-.05$, $-.07$, $-.10$, $-.125$, $-.15$, $-.19$, and $-.199$.

For each value of ρ, a run of 100 replications is made. Each replication consists of 40 observations whose **y**'s are generated subject to (11.67) for a fixed set of **x**'s and whose **u**'s are generated according to $N_6(\mathbf{0}, \mathbf{\Sigma})$. Here, $\mathbf{\Sigma}$ is determined to have a corresponding correlation matrix (11.68) and to have variances that produce a signal-to-noise of 20 for the structural equations (11.67). This value is chosen to mimic good-quality economic data providing R^2s in the neighborhood of .9 for the estimated structural equations. Following Mikhail (1975), the **x**'s are chosen randomly from uniform distributions of differing ranges. In particular, the x_i's are chosen, respectively, from uniforms 5–10, 0.5–5.0, 15–30, 2–8, 8–22, and 10–14.

The data for each replication are used to estimate (11.67) by both 2SLS and 3SLS, providing for each ρ 100 estimates of each of the 27 parameters by each estimator.

The Results

Exhibits 11.5–11.7 summarize the results. For each ρ, a root-mean-square error (RMSE) about the true parameter value is calculated for each parameter and each estimator. A relative RMSE is then calculated as RRMSE $\equiv$ $\text{RMSE}_{2SLS}/\text{RMSE}_{3SLS}$. Clearly, a RRMSE > 1 favors 3SLS, while RRMSE $<$ 1 favors 2SLS.

Exhibit 11.5 shows for each ρ the number of coefficients out of the 27 for which 3SLS proved superior (RRMSE > 1). A dashed line between 13 and 14 is drawn to denote the halfway point. Exhibit 11.6 shows for each ρ the mean relative RMSE (across the 27 parameters) as well as the largest and smallest relative RMSE. A line is drawn at 1.0; mean relative RMSEs above this line denote when, on the average, 3SLS outperforms 2SLS and inversely.

Both of these summary figures show that 3SLS breaks even with 2SLS for ρ somewhere between $-.1$ and $-.125$ and is wholly in the lead by the time ρ

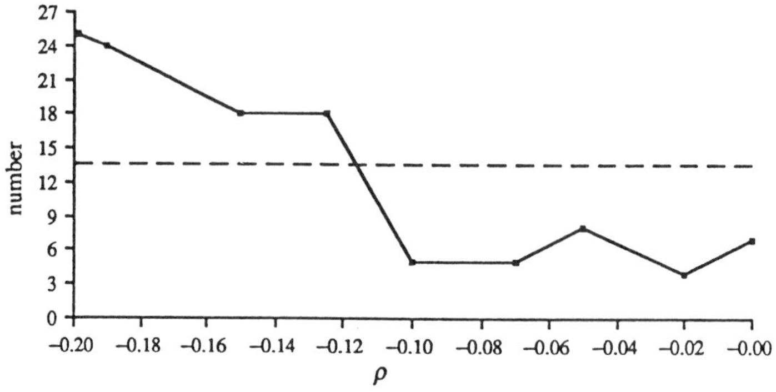

Exhibit 11.5 Number of relative RMSEs > 1.

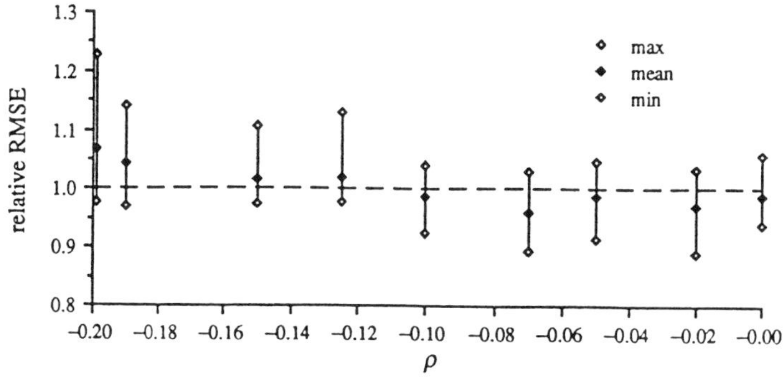

Exhibit 11.6 Mean relative RMSE and relative RMSE spread.

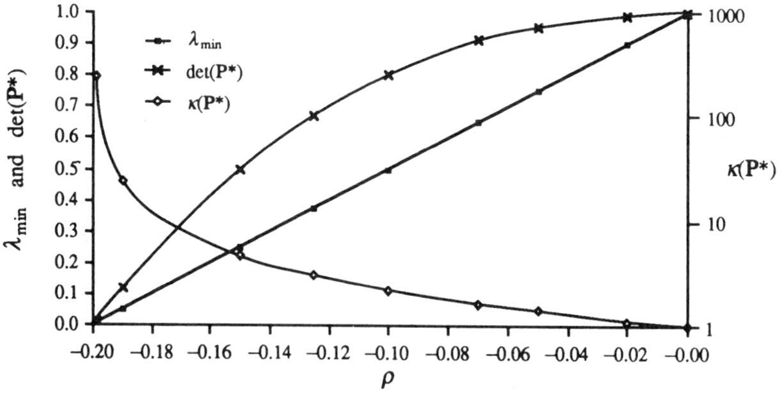

Exhibit 11.7 Indicators of multiple correlations.

approaches $-.2$. It is clear that this situation could not be assessed solely from examination of the low magnitude of ρ. Examination of Exhibit 11.7, however, shows that all three of the suggested general indicators point to this outcome. Taking very conservative values, for example, we note that 3SLS is certainly dominant by the time $\lambda_{\min}$ has dropped to 0.05, det($\mathbf{P}^*$) has dropped to 0.12, and $\kappa(\mathbf{P}^*)$ has exceeded 23. These values are all extremes in their respective ranges, while they correspond to an absolute ρ of only .19, a value most would consider small indeed.

Conclusions and Research Issues

Consonant with expectations, an indicator more general than pairwise correlations among the errors of a system of equations is necessary to determine when 3SLS is likely to possess greater small-sample efficiency than 2SLS and, thus, when the added computational cost of 3SLS is likely to be worthwhile. For larger systems of equations, the presence of strong multiple correlations among these error terms can give the edge to 3SLS even when the largest absolute pairwise correlation is small.

Any of the three suggested general indicators $\lambda_{\min}$, det($\mathbf{P}$), or $\kappa(\mathbf{P})$, would seem to provide this more appropriate information. While it is beyond the present study to provide a detailed comparison of these three measures or interpretation of their values, it would seem safe to say that 3SLS would possess good small-sample relative efficiency for values of $\lambda_{\min}$ and det($\mathbf{P}$) in the neighborhood of 0.1 and for values of $\kappa(\mathbf{P})$ above 20–30. This latter magnitude is wholly consistent with values found in the experiments of Chapter 4 to signal the presence of strong collinear dependencies among columns of a data matrix in linear regression.

It should be emphasized that the experimental situation employed in this study is narrow by design, serving only to demonstrate that pairwise correlations cannot adequately indicate when the added computation of 3SLS becomes worthwhile whereas the three suggested general indicators can. In this it succeeds, but it is not intended, and does not pretend, to provide definitive information about comparing and interpreting these three indicators in more general contexts. Such a study would necessarily have to be a substantial research project addressing the following issues:

1. The general rules for how small $\lambda_{\min}$ and det($\mathbf{P}$) must be and how large $\kappa(\mathbf{P})$ must be before 3SLS has greater small-sample efficiency than 2SLS.
2. How these rules depend on:
   - **a.** The number of equations, g.
   - **b.** The number of coexisting multiple correlations within $\mathbf{u}$.
   - **c.** Sample size.
   - **d.** Degree of overidentification.
3. Whether information from one of these indicators dominates that of another or whether there are differing situations favoring specific indicators.

4. Whether there are wide classes of econometric situations in which simple rules of interpretation can be effective.

Some speculation on these issues is in order. Issues 2c and 2d arise since it is well known that the performance of 3SLS bests that of 2SLS as the sample size increases indefinitely but the two are equivalent as the degree of overidentification goes to zero. One could conjecture, then, that a stronger degree of intercorrelation among the errors would increase the attractiveness of 3SLS the smaller the sample size and the greater the degree of overidentification.

Issues 2a and 2b are of central interest. We can see from the eigenvalues of $\mathbf{P}^*(\rho)$ that all three indicators depend on g. Different rules or thresholds may be needed, therefore, for different system sizes. It would obviously be nicer if single thresholds would apply (at least over a wide range of practical values for g).

In the example presented above, there is only one multiple correlation among the elements of $\mathbf{u}$ as $\rho \to -.2$ (since only one of the g eigenvalues of $\mathbf{P}^*(\rho)$ goes to zero). In typical econometric situations there could be coexisting multiple correlations. It is to be supposed that the more such relations, the greater the relative advantage for 3SLS. The three indicators, however, differ in their ability to assess such distinctions. Thus, $\lambda_{\min}$ can easily remain unchanged as the number of such multiple correlations increases while both $\det(\mathbf{P})$ and $\kappa(\mathbf{P})$ will respond to such alterations. For this reason, it is to be expected that these latter two indicators will be more generally useful. Indeed, one can further bolster the value of $\kappa(\mathbf{P})$ in this regard by considering a full set of condition indexes for $\mathbf{P}$ according to Section 3.2 rather than the condition number $\kappa(\mathbf{P})$ alone.

Bibliography

Anderson, T. W. (1958), *An Introduction to Multivariate Statistical Analysis*, Wiley: New York.

Andrews, D. W. K. (1989), "Power in Econometric Applications," *Econometrica* **57**, 1059–1090.

Askin, R. G. (1982), "Multicollinearity in Regression: Review and Examples," *Journal of Forecasting* **1**, 281–292.

Bard, Y. (1974), *Nonlinear Parameter Estimation*, Academic: New York.

Bargmann, R. E. and S. P. Ghosh (1964), "Noncentral Statistical Distribution Programs for a Computer Language," Report no. RC-1231, IBM Watson Research Center, Yorktown Heights, NY.

Bauer, F. L. (1963), "Optimally Scaled Matrices," *Numerische Mathematik* **5**, 73–87.

Bauer, F. L. (1971), "Elimination with Weighted Row Combinations for Solving Linear Equations and Least Squares Problems," in *Handbook for Automatic Computation*, Vol. 2, *Linear Algebra*, J. H. Wilkinson and C. Reisch, Eds., Springer-Verlag: New York, 119–133.

Beaton, A. E., D. B. Rubin, and J. L. Barone (1976), "The Acceptability of Regression Solutions: Another Look at Computational Accuracy," *Journal of the American Statistical Association* **71**, 158–168.

Becker, R., N. Kaden, and V. Klema (1974), "The Singular Value Analysis in Matrix Computation," Working paper 46, Computer Research Center, National Bureau of Economic Research, Cambridge, MA.

Belsley, D. A. (1969a), *Industry Production Behavior: the Order–Stock Distinction*, North-Holland: Amsterdam.

Belsley, D. A. (1969b), "An Econometric Study of the Silver Market," Various mimeographed reports, Charles River Associates, Cambridge, MA.

Belsley, D. A. (1974), "Estimation of Systems of Simultaneous Equations, and Computational Specifications of GREMLIN," *Annals of Economic and Social Measurement* **3**, 551–614.

Belsley, D. A. (1976), "Multicollinearity: Diagnosing Its Presence and Assessing the Potential Damage It Causes Least-Squares Estimation," Working paper no. 154, Computer Research Center, National Bureau of Economic Research, Cambridge, MA.

Belsley, D. A. (1982), "Assessing the Presence of Harmful Collinearity and Other Forms of Weak Data through a Test for Signal-to-Noise," *Journal of Econometrics* **20**, 211–253.

Belsley, D. A. (1984a), "Eigenvector Weaknesses and Other Topics for Assessing Conditioning Diagnostics," Letters, *Technometrics* **26**, 297–299.

Belsley, D. A. (1984b), "Demeaning Conditioning Diagnostics through Centering," with accompanying comments and author's reply, *American Statistician* **38**, 73–93.

Belsley, D. A. (1984c), "Collinearity and Forecasting," *Journal of Forecasting* **3**, 183–196.

Belsley, D. A. (1986a), "Centering, the Constant, First-Differencing, and Assessing Conditioning," in *Model Reliability*, E. Kuh and D. A. Belsley, Eds., MIT Press: Cambridge, MA.

Belsley, D. A. (1986b), "Model Selection in Regression Analysis, Regression Diagnostics and Prior Knowledge," *International Journal of Forecasting* **2**, 41–46.

Belsley, D. A. (1987), "Comment: Well-conditioned Collinearity Indices?" *Statistical Science* **2**, 86–91.

Belsley, D. A. (1988a), "Two- or Three-Stages of Least Squares?" *Computer Science in Economics and Management* **1**, 21–30.

Belsley, D. A. (1988b), "Modelling and Forecasting Reliability," *International Journal of Forecasting* **4**, 427–447.

Belsley, D. A. (1988c), "Conditioning in Models with Logs," *Journal of Econometrics* **38**, 127–143.

Belsley, D. A. and V. C. Klema (1974), "Detecting and Assessing the Problems Caused by Multicollinearity: A Use of the Singular-Value Decomposition," Working paper no. 66, Computer Research Center, National Bureau of Economic Research, Cambridge, MA.

Belsley, D. A. and W. R. Oldford (1986), "The General Problem of Ill Conditioning and Its Role in Statistical Analysis," *Computational Statistics and Data Analysis* **4**, 103–120.

Belsley, D. A. and R. E. Welsch (1988), "Comment: Modelling Energy Consumption: Using and Abusing Regression Diagnostics," *Journal of Business and Economic Statistics* **6**, 442–447.

Belsley, D. A., E. Kuh, and R. E. Welsch (1980), *Regression Diagnostics: Identifying Influential Data and Sources of Collinearity*, Wiley: New York.

Berk, K. N. (1977), "Tolerance and Condition in Regression Computations," *Journal of the American Statistical Association* **72**, 863–866.

Businger, P. and G. H. Golub (1965), "Linear Least Squares Solutions by Householder Transformations," *Numerische Mathematik* **7**, 269–276.

Chatterjee, S. and A. S. Hadi (1988), *Sensitivity Analysis in Linear Regression*, Wiley: New York.

Chatterjee, S. and B. Price (1977), *Regression by Example*, Wiley: New York.

Cook, R. D. and S. Weisberg (1982), *Residuals and Influence in Regression*, Chapman & Hall: London.

Cramér, N. (1946), *Mathematical Methods of Statistics*, Princeton University Press: Princeton, NJ.

Critchley, F. (1985), "Influence in Principal Component Analysis," *Biometrika* **72**, 627–636.

Crombrugghe, Denis de (1983), "The Correlation Matrix of Estimated Coefficients," CORE discussion paper no. 8307, Université Catholique de Louvain, Belgium.

Dempster, A. P. (1969), *Elements of Continuous Multivariate Analysis*, Addison-Wesley: Reading, MA.

Dempster, A. P., M. Schatzoff, and N. Wermuth (1977), "A Simulation Study of Alternatives to Ordinary Least Squares," *Journal of the American Statistical Association* **72**, 77–91.

Dongarra, J. J., J. R. Bunch, C. B. Moler, and G. W. Stewart, (1979), *Linpack Users Guide*, Society for Industrial and Applied Mathematics: Philadelphia, PA.

Dorsett, D. (1982), "Resistant *M*-Estimators in the Presence of Influential Points," Ph.D. dissertation, Department of Statistics, Southern Methodist University, Dallas, TX.

Eckart, G. and G. Young (1936), "The Approximation of One Matrix by Another of Lower Rank," *Psychometrika* **1**, 211–218.

Faddeeva, V. N. (1959), *Computational Methods of Linear Algebra* (translated by C. D. Berster), Dover: New York.

Farrar, D. E. and R. R. Glauber (1967), "Multicollinearity in Regression Analysis: The Problem Revisited," *Review of Economics and Statistics* **49**, 92–107.

Forsythe, G. E. and C. B. Moler (1967), *Computer Solutions of Linear Algebraic Systems*, Prentice-Hall: Englewood Cliffs, NJ.

Forsythe, G. E. and E. G. Straus (1955), "On Best Conditioned Matrices," *Proceedings of the American Mathematical Association* **6**, 340–345.

Friedman, B. (1977), "Financial Flow Variables and the Short Run Determination of Long Term Interest Rates," *Journal of Political Economy* **85**, 661–689.

Friedman, M. (1957), *A Theory of the Consumption Function*, Princeton University Press: Princeton, NJ.

Frisch, R. (1934), *Statistical Confluence Analysis by Means of Complete Regression Systems*, Publication no. 5, University Institute of Economics, Oslo.

Garbow, B. S. et al., Eds., (1977), *Matrix Eigensystem Routines–EISPACK Guide Extension*, Vol. 51, Springer-Verlag: New York.

Gardner, J. R. and S. H. Hymans (1978), "An Econometric Model of the U.S. Monetary Sector," RSQE research report, University of Michigan, Ann Arbor, MI.

Goldberger, A. S. (1964), *Econometric Theory*, Wiley: New York.

Golub, G. H. (1969), "Matrix Decompositions and Statistical Calculations," in *Statistical Computation*, R. C. Milton and J. A. Nelder, Eds., Academic: New York, 365–397.

Golub, G. H. and C. Reinsch (1970), "Singular Value Decomposition and Least-Squares Solutions," *Numerische Mathematik* **14**, 403–420.

Golub, G. H. and C. F. Van Loan (1983), *Matrix Computations*, Johns Hopkins University Press: Baltimore.

Golub, G. H., V. Klema, and S. Peters (1980), "Rules and Software for Detecting Rank Degeneracy," *Journal of Econometrics* **12**, 41–48.

Golub, G. H., V. Klema, and G. W. Stewart (1976), "Rank Degeneracy and Least-Squares Problems," Computer Science Technical Report Series, #TR-456, University of Maryland, College Park, MD.

Greene, T. (1985), "Simplicity and Structure in Underlying Systems of Linear Relations," Ph.D. dissertation, Cornell University, Ithaca, NY.

Greene, T. (1986a), "Simple Structure of Matroids," Technical report no. 245, University of Kentucky, Lexington, KY.

Greene, T. (1986b), "The Depiction of Multivariate Structure by Matroids, I and II," Technical report nos. 250 and 251, University of Kentucky, Lexington, KY.

Greene, T. (1987), "Descriptively Sufficient Subcollections of Flats in Matroids," Technical report no. 262, University of Kentucky, Lexington, KY.

Greene, T. (1988), "The Depiction of Linear Association by Matroids," Technical report no. 270, University of Kentucky, Lexington, KY.

Gunst, R. F. (1983), "Regression Analysis with Multicollinear Predictor Variables," *Communications in Statistics* **A12**, 2217–2260.

Gunst, R. F. and R. L. Mason (1980), *Regression Analysis and Its Application*, Marcel Dekker: New York.

Haavelmo, T. (1944), "The Probability Approach in Econometrics," *Econometrica* **12** (Suppl.).

Hadi, A. S. (1988), "Diagnosing Collinearity-Influential Observations," *Computational Statistics and Data Analysis* **7**, 143–159.

Hadi, A. S. and P. F. Velleman (1987), "Diagnosing Near Collinearities in Least Squares Regression," *Statistical Science* **2**, 93–98.

Hadi, A. S. and M. T. Wells (1990), "Assessing the Effects of Multiple Rows on the Condition Index of a Matrix," *Journal of the American Statistical Association* **85**, 786–792.

Haitovsky, Y. (1969), "Multicollinearity in Regression Analysis: Comment," *Review of Economics and Statistics* **50**, 486–489.

Hanson, R. J. and C. L. Lawson (1969), "Extensions and Applications of the Householder Algorithm for Solving Linear Least Squares Problems," *Mathematics of Computation* **23**, 787–812.

Hawkins, D. M. (1973), "On the Investigation of Alternative Regressions by Principal Component Analysis," *Applied Statistics* **22**, 275–286.

Hoaglin, D. C. and Welsch, R. E. (1978), "The Hat Matrix in Regression and ANOVA," *American Statistician* **32**, 17–22.

Hocking, R. R. (1983), "Response to 'Developments in Linear Regression Methodology: 1959–1982,'" *Technometrics* **25**, 248–249.

Hoerl, A. E. and R. W. Kennard (1970), "Ridge Regression: Biased Estimation for Nonorthogonal Problems," *Technometrics* **12**, 55–68.

Holland, P. W. (1973), "Weighted Ridge Regression: Combining Ridge and Robust Regression Models," Working paper no. 11, Computer Research Center, National Bureau of Economic Research, Cambridge, MA.

Hotelling, H. (1931), "The Generalization of Student's Ratio," *Annals of Mathematical Statistics* **2**, 360–378.

Huber, P. J. (1981), *Robust Statistics*, Wiley: New York.

Huber, P. J. (1985), "Projection Pursuit," *Annals of Statistics* **13**, 435–525.

Janson, M. A. (1988), "Combining Robust and Traditional Least Squares Methods: A Critical Evaluation," *Journal of Business and Economic Statistics* **6**, 415–427.

Johnston, J. (1984), *Econometric Methods*, 3rd ed., McGraw-Hill: New York.

Judge, G. G., R. C. Hill, W. E. Griffiths, H. Lütkepohl, and T.-C. Lee (1982), *Introduction to the Theory and Practice of Econometrics*, Wiley: New York.

Kadane, J. B., J. M. Dickey, R. L. Winkler, W. S. Smith, and S. C. Peters (1980), "Interactive Elicitation of Opinion for a Normal Linear Model," *Journal of the American Statistical Association* **75**, 845–854.

Kaiser, H. F. (1958), "The Varimax Criterion for Analytic Rotation in Factor Analysis," *Psychometrika* **23**, 187–200.

Kempthorne, P. J. (1985), "Assessing the Influence of Single Cases on the Condition Number of a Design Matrix," Memorandum NS-509, Department of Statistics, Harvard University, Cambridge, MA.

Kempthorne, P. J. (1989), "Identifying Rank-Influential Groups of Observations in Linear Regression Modeling," Sloan working paper no. 3018-89-MS, Sloan School, Massachusetts Institute of Technology, Cambridge, MA.

Kendall, M. G. (1957), *A Course in Multivariate Analysis*, Griffen: London.

Kloeck, T. and L. B. M. Mennes (1960), "Simultaneous Equations Estimation Based on Principal Components of Predetermined Variables," *Econometrica* **28**, 45–61.

Kuh, E. and R. Schmalensee (1972), *An Introduction to Applied Macroeconomics*, North-Holland: Amsterdam.

Kumar, T. K. (1975a), "The Problem of Multicollinearity: A Survey," Unpublished mimeo, Abt Associates, Cambridge, MA.

Kumar, T. K. (1975b), "Multicollinearity in Regression Analysis," *Review of Economics and Statistics* **57**, 365–366.

Laub, A. and V. Klema (1980), "The Singular Value Decomposition: Its Computation and Some Applications," *IEEE Transactions on Automatic Control* **15**, 164–176.

Lawson, C. R. and R. J. Hanson (1974), *Solving Least-Squares Problems*, Prentice-Hall: Englewood Cliffs, NJ.

Leamer, E. E. (1973), "Multicollinearity: A Bayesian Interpretation," *Review of Economics and Statistics* **55**, 371–380.

Leamer, E. E. (1978), *Specification Searches*, Wiley: New York.

Lehmer, E. (1944), "Inverse Tables of Probabilities of Errors of the Second Kind," *Annals of Mathematical Statistics* **15**, 388–398.

Lesage, J. P. and S. D. Simon (1985), "Numerical Accuracy of Statistical Algorithms for Microcomputers," *Computational Statistics and Data Analysis* **3**, 47–57.

Longley, J. W. (1967), "An Appraisal of Least-Squares Programs for the Electronic Computer from the Point of View of the User," *Journal of the American Statistical Association* **62**, 819–831.

Malinvaud, E. (1970), *Statistical Methods of Econometrics*, 2nd rev. ed., North-Holland: Amsterdam.

Marquardt, D. W. (1980), "You Should Standardize the Predictor Variables in Your Regression Models," *Journal of the American Statistical Association* **75**, 74–103.

Mason, R. L. and R. F. Gunst (1985), "Outlier-induced Collinearities," *Technometrics* **27**, 401–407.

Menger, K. (1959), "Mensuration and Other Mathematical Connections of Observable Material," in *Measurement: Definitions and Theory*, C. W. Churchman and P. Ratoosh, Eds., Wiley: New York.

Mikhail, W. M. (1975), "A Comparative Monte Carlo Study of the Properties of Econometric Estimators," *Journal of the American Statistical Association* **70**, 94–104.

Mirsky, L. (1960), "Symmetric Gauge Functions and Unitarily Invariant Norms," *Quarterly Journal of Mathematics* **11**, 50–59.

Montgomery, D. C. and E. A. Peck (1982), *Introduction to Linear Regression Analysis*, Academic: New York.

Nyquist, H. (1989), "On Diagnosing Collinearity-Influential Points in Linear Regression," Unpublished manuscript, Department of Statistics, University of Umeå, Sweden.

O'Brien, R. J. (1975), "The Sensitivity of OLS and Other Econometric Estimators," Unpublished manuscript, University of Southampton, Southampton.

O'Hagan, J. and B. McCabe (1975), "Tests for the Severity of Multicollinearity in Regression Analysis: A Comment," *Review of Economics and Statistics* **57**, 369–370.

Oldford, R. W. (1982), "Effective Dimension: A Theory for the Non-stochastic Examination of Linear Regression," Ph.D. dissertation, University of Toronto.

Oldford, R. W. (1983), "Effective Dimensionality in Linear Regression," Technical report no. 38, Center for Computational Research in Economics and Management Science, MIT, Cambridge, MA.

Oldford, R. W. and S. Peters (1984), "Building a Statistical Knowledge Based System with Mini-Mycin," Technical report no. 42, Center for Computational Research in Economics and Management Science, MIT, Cambridge, MA.

Oldford, R. W. and S. Peters (1985), "DINDE: Towards More Statistically Sophisticated Software," Technical report no. 55, Center for Computational Research in Economics and Management Science, MIT, Cambridge, MA.

Pindyck, R. S. and D. L. Rubinfeld (1981), *Econometric Models and Economic Forecasts*, 2nd ed., McGraw-Hill: New York.

Ploberger, W. and W. Krämer (1987), "Mean Adjustment and the CUSUM Test for Structural Change," *Economics Letters* **25**, 255–258.

Rao, C. R. (1973), *Linear Statistical Inference and Its Applications*, 2nd ed., Wiley: New York.

Schall, R. and T. T. Dunne (1988), "Variance Inflation and Collinearity in Regression," Unpublished manuscript, Institute for Biostatistics, SA Medical Research Council, Tygerberg, RSA.

Silvey, S. D. (1969), "Multicollinearity and Imprecise Estimation," *Journal of the Royal Statistical Society, Series B* **31**, 539–552.

Simon, S. D. and J. P. Lesage (1988a), "Benchmarking Numerical Accuracy of Statistical Algorithms," *Computational Statistics and Data Analysis* **7**, 197–209.

Simon, S. D. and J. P. Lesage (1988b), "The Impact of Collinearity Involving the Intercept Term on the Numerical Accuracy of Regression," *Computer Science in Economics and Management* **1**, 137–152.

Sims, C. A. (1980), "Macroeconomics and Reality," *Econometrica* **48**, 1–48.

Smith, B. T., et al., Eds. (1976), *Matrix Eigenvalue Routines—Eispack Guide*, Springer-Verlag: New York.

Smith, G. and F. Campbell (1980), "A Critique of Some Ridge Regression Methods," *Journal of the American Statistical Association* **75**, 74–103.

Stein, C. M. (1956), "Inadmissibility of the Usual Estimator for the Mean of a Multivariate Normal Distribution," *Proceedings of the Third Berkeley Symposium on Mathematical Statistics and Probability* **1**, 197–206.

Stewart, G. W. (1973), *Introduction to Matrix Computations*, Academic: New York.

Stewart, G. W. (1987), "Collinearity and Least Squares Regression," *Statistical Science* **1**, 68–100.

Strang, G. (1980), *Linear Algebra and Its Applications*, 2nd ed., Academic: New York.

Theil, H. (1963), "On the Use of Incomplete Prior Information in Regression Analysis," *Journal of the American Statistical Association* **58**, 401–414.

Theil, H. (1971), *Principles of Econometrics*, Wiley: New York.

Theil, H. and A. S. Goldberger (1961), "On Pure and Mixed Statistical Estimation in Economics," *International Economic Review* **2**, 65–78.

Tiao, G. C. and A. Zellner (1964), "Bayes Theorem and the Use of Prior Information in Regression Analysis," *Biometrika* **51**, 219–230.

Toro-Vizcarrendo, C. and T. D. Wallace (1968), "A Test of the Mean Square Error Criterion for Restrictions in Linear Regression," *Journal of the American Statistical Association* **63**, 558–572.

van der Sluis, A. (1969), "Condition Numbers and Equilibration of Matrices," *Numerische Mathematik* **14**, 14–23.

van der Sluis, A. (1970), Conditioning, Equilibration and Pivoting in Linear Algebraic Systems," *Numerische Mathematik* **15**, 74–88.

Vetterling, W. T. and W. H. Press (1988), *Numerical Recipes Example Book (C)*, Cambridge University Press: Cambridge.

Vinod, H. D. (1978), "A Survey of Ridge Regression and Related Techniques for Improvements over Ordinary Least Squares," *Review of Economics and Statistics* **60**, 121–131.

Walker, E. (1989), "Detection of Collinearity-Influential Observations," *Communications in Statistics, Theory and Methods* **18**(5), 1675–1690.

Wang, S.-G. and H. Nyquist (1989), "Effects on the Eigenstructure of a Data Matrix When Deleting an Observation," Unpublished manuscript, Department of Statistics, University of Umeå, Sweden.

Webster, J. T., R. F. Gunst, and R. L. Mason (1974), "Latent Root Regression Analysis," *Technometrics* **16**, 513–522.

Webster, J. T., R. F. Gunst, and R. L. Mason (1976), "A Comparison of Least-Squares and Latent Root Regression Estimators," *Technometrics* **18**, 75–83.

Weisberg, S. (1980), *Applied Linear Regression*, Wiley: New York.

Wilkinson, J. H. (1965), *The Algebraic Eigenvalue Problem*, Oxford University Press: Oxford.

Wilks, S. S. (1962), *Mathematical Statistics*, Wiley: New York.

Wonnacott, R. J. and T. H. Wonnacott (1979), *Econometrics*, 2nd ed., Wiley: New York.

Woods, H., H. H. Steinour, and H. R. Starke (1932), "Effect of Composition of Portland Cement on Heat Evolved during Hardening," *Industrial Engineering and Chemistry* **24**, 1207–1214.

Zellner, A. (1971), *An Introduction to Bayesian Inference in Econometrics*, Wiley: New York.

Author Index

Subject Index

Applied Probability and Statistics (Continued)

JOHNSON and KOTZ • Distributions in Statistics
Discrete Distributions
Continuous Univariate Distributions—1
Continuous Univariate Distributions—2
Continuous Multivariate Distributions
JUDGE, GRIFFITHS, HILL, LÜTKEPOHL, and LEE • The Theory and Practice of Econometrics, *Second Edition*
JUDGE, HILL, GRIFFITHS, LÜTKEPOHL, and LEE • Introduction to the Theory and Practice of Econometrics, *Second Edition*
KALBFLEISCH and PRENTICE • The Statistical Analysis of Failure Time Data
KASPRZYK, DUNCAN, KALTON, and SINGH • Panel Surveys
KAUFMAN and ROUSSEEUW • Finding Groups in Data: An Introduction to Cluster Analysis
KEENEY and RAIFFA • Decisions with Multiple Objectives
KISH • Statistical Design for Research
KISH • Survey Sampling
KUH, NEESE, and HOLLINGER • Structural Sensitivity in Econometric Models
LAWLESS • Statistical Models and Methods for Lifetime Data
LEAMER • Specification Searches: Ad Hoc Inference with Nonexperimental Data
LEBART, MORINEAU, and WARWICK • Multivariate Descriptive Statistical Analysis: Correspondence Analysis and Related Techniques for Large Matrices
LEVY and LEMESHOW • Sampling of Populations: Methods and Applications
LINHART and ZUCCHINI • Model Selection
LITTLE and RUBIN • Statistical Analysis with Missing Data
McNEIL • Interactive Data Analysis
MAGNUS and NEUDECKER • Matrix Differential Calculus with Applications in Statistics and Econometrics
MAINDONALD • Statistical Computation
MALLOWS • Design, Data, and Analysis by Some Friends of Cuthbert Daniel
MANN, SCHAFER, and SINGPURWALLA • Methods for Statistical Analysis of Reliability and Life Data
MARTZ and WALLER • Bayesian Reliability Analysis
MASON, GUNST, and HESS • Statistical Design and Analysis of Experiments with Applications to Engineering and Science
MILLER • Beyond ANOVA, Basics of Applied Statistics
MILLER • Survival Analysis
MILLER, EFRON, BROWN, and MOSES • Biostatistics Casebook
MONTGOMERY and PECK • Introduction to Linear Regression Analysis
NELSON • Accelerated Testing, Statistical Models, Test Plans, and Data Analyses
NELSON • Applied Life Data Analysis
OCHI • Applied Probability and Stochastic Processes in Engineering and Physical Sciences
OSBORNE • Finite Algorithms in Optimization and Data Analysis
OTNES and ENOCHSON • Digital Time Series Analysis
PANKRATZ • Forecasting with Univariate Box-Jenkins Models: Concepts and Cases
POLLOCK • The Algebra of Econometrics
RAO and MITRA • Generalized Inverse of Matrices and Its Applications
RÉNYI • A Diary on Information Theory
RIPLEY • Spatial Statistics
RIPLEY • Stochastic Simulation
ROSS • Introduction to Probability and Statistics for Engineers and Scientists
ROUSSEEUW and LEROY • Robust Regression and Outlier Detection
RUBIN • Multiple Imputation for Nonresponse in Surveys
RUBINSTEIN • Monte Carlo Optimization, Simulation, and Sensitivity of Queueing Networks
RYAN • Statistical Methods for Quality Improvement
SCHUSS • Theory and Applications of Stochastic Differential Equations
SEARLE • Linear Models
SEARLE • Linear Models for Unbalanced Data
SEARLE • Matrix Algebra Useful for Statistics
SKINNER, HOLT, and SMITH • Analysis of Complex Surveys
SPRINGER • The Algebra of Random Variables